Encapsulation Technologies for Active Food Ingredients and Food Processing

Nicolaas Jan Zuidam • Viktor A. Nedović
Editors

Encapsulation Technologies for Active Food Ingredients and Food Processing

Springer

Editors
Nicolaas Jan Zuidam
Unilever Research and Development
Vlaardingen
The Netherlands
klaas-jan.zuidam@unilever.com

Viktor A. Nedović
Department of Food Technology
and Biochemistry
Faculty of Agriculture
University of Belgrade
Serbia
vnedovic@agrif.bg.ac.rs

ISBN 978-1-4419-1007-3 e-ISBN 978-1-4419-1008-0
DOI 10.1007/978-1-4419-1008-0
Springer New York Dordrecht Heidelberg London

Library of Congress Control Number: 2009929702

Printed on acid-free paper

Springer is part of Springer Science+Business Media (www.springer.com)

Contents

Contributors

Artur Bartkowiak
Chair of Food Packaging and Biopolymers, Faculty of Food Sciences and Fisheries, The Westpomeranian University of Technology, Kazimierza Krolewicza 4, 71-550 Szczecin, Poland, artur.bartkowiak@tz.ar.szczecin.pl

Christiaan M. Beindorff
Unilever R&D Vlaardingen, Olivier van Noortlaan 120, 3133 AT Vlaardingen, The Netherlands, christiaan.beindorff@unilever.com

Véronique Breguet
Baxter BioScience Manufacturing Sarl, Neuchatel, Switzerland
veronique_breguet@baxter.com

Karlis Briviba
Institute of Nutritional Physiology, Max Rubner Institute, Federal Research Institute of Nutrition and Food, Haid-und-Neu-Str. 9, 76131 Karlsruhe, Germany
karlis.briviba@mri.bund.de

Branko Bugarski
Department of Chemical Engineering, Faculty of Technology and Metallurgy, University of Belgrade, Karnegijeva 4, 11000 Belgrade, Serbia, branko@tmf.bg.ac.yu

Claude P. Champagne
Agriculture and Agri-Food Canada, 3600 Casavant, St-Hyacinthe, QC, Canada J2S 8E3, claude.champagne@agr.gc.ca

Freddy R. Delvaux
Centre for Malting and Brewing Science, KU Leuven, Kasteelpark Arenberg 22, 3001 Heverlee, Belgium, freddy.delvaux@biw.kuleuven.be

Robert Engel
BASF SE, GCT/P – L549, 67056 Ludwigshafen, Germany
robert.engel@basf.com

Stephen E. Harding
National Centre for Macromolecular Hydrodynamics,
University of Nottingham, School of Biosciences, Sutton Bonington, UK
steve.harding@nottingham.ac.uk

Emmanuel Heinrich
Unilever R&D Vlaardingen, Olivier van Noortlaan 120, 3133 AT Vlaardingen,
The Netherlands, emmanuel.heinrich@unilever.com

Kasipathy Kailasapathy
Probiotics and Encapsulated Functional Foods Unit, Centre for Plant and Food
Science, School of Natural Sciences, University of Western Sydney, Hawkesbury
Campus, Locked Bag 1797, Penrith South DC, NSW 1797, Australia
k.kailasapathy@uws.edu.au

Yiannis Kourkoutas
Department of Molecular Biology & Genetics, Democritus University
of Thrace, Alexandroupolis GR-68100, Greece, ikourkou@mbg.duth.gr

Daniel Law
Department of Chemical Engineering, University of Birmingham,
Edgbaston, Birmingham B15 2TT, UK

Byong H. Lee
Agriculture and Agri-Food Canada, 3600 Casavant, St-Hyacinthe,
QC, Canada J2S 8E3, byong.lee@agr.gc.ca

Ida Laskošek-Čukalović
Department of Food Technology and Biochemistry, Faculty of Agriculture,
University of Belgrade, Nemanjina 6, PO Box 127, 11081 Belgrade-Zemun,
Serbia

Guoping Lian
Unilever Discover, Colworth Science Park, Sharnbrook,
Bedford MK44 1LQ, UK, guoping.lian@unilever.com

Verica Manojlović
Department of Chemical Engineering, Faculty of Technology and Metallurgy,
University of Belgrade, Karnegijeva 4, 11000 Belgrade, Serbia
vmanojlovic@tmf.bg.ac.yu

Ian W. Marison
School of Biotechnology, Dublin City University, Glasnevin, Dublin 9, Ireland
ian.marison@dcu.ie

Gabrie M.H. Meesters
DSM Food Specialties, PO Box 1, 2600 MA Delft, The Netherlands
Nano Structured Materials Group, Faculty Applied Sciences, Delft University
of Technology, Julianalaan 136, 2628 BL Delft, The Netherlands
gabrie.meesters@dsm.com

Viktor A. Nedović
Department of Food Technology and Biochemistry, Faculty of Agriculture, University of Belgrade, Nemanjina 6, PO Box 127, 11081 Belgrade-Zemun, Serbia
vnedovic@agrif.bg.ac.rs

Henelyta S. Ribeiro
Unilever Discover, Colworth Science Park, Sharnbrook, Bedford MK44 1LQ, UK
henelyta.ribeiro@unilever.com

Linda Saucier
Department of Animal Science, Paul Comtois Building, 2425 Agriculture Street, Université Laval, Quebec City, QC, Canada G1V 0A6
linda.saucier@fsaa.ulaval.ca

Heike P. Schuchmann
Institute of Process Engineering in Life Sciences, Food Process Engineering, KIT/University of Karlsruhe, Kaiserstr. 12, D-76131 Karlsruhe, Germany
heike.schuchmann@kit.edu

Eyal Shimoni
Laboratory of Functional Foods, Nutraceuticals, and Food Nanoscience, Faculty of Biotechnology and Food Engineering, Technion – Israel Institute of Technology, Haifa 32000, Israel, eshimoni@tx.technion.ac.il

Pieter J. Verbelen
Centre for Malting and Brewing Science, KU Leuven, Kasteelpark Arenberg 22, 3001 Heverlee, Belgium, pieter.verbelen@biw.kuleuven.be

Vojislav Vojinovic
Nestle Research Centre, Nestec SA, Vers-Chez-les-Blanc Case postale 44, 1000 Lausanne 26, Vaud, Switzerland, vojislav.vojinovic@rdls.nestle.com

Elke Walz
Department of Food and Bioprocess Engineering, Max Rubner-Institute (MRI), Federal Research Institute of Nutrition and Food, Karlsruhe, Germany
elke.walz@mri.bund.de

Christine Wandrey
Ecole Polytechnique Fédérale de Lausanne, Laboratoire de Médecine Régénérative et de Pharmacobiologie, CH-1015 Lausanne, Switzerland
christine.wandrey@epfl.ch

Ronnie Willaert
Structural Biology Brussels, Vrije Universiteit Brussel, Flanders Institute for Biotechnology, Pleinlaan 2, B-1050 Brussels, Belgium
ronnie.willaert@vib-vub.be

Erich J. Windhab
Laboratory of Food Process Engineering, Institute of Food Science and Nutrition, Swiss Federal Institute of Technology (ETH), Zürich, Switzerland
erich.windhab@ilw.agrl.ethz.ch

Zhibing Zhang
Department of Chemical Engineering, University of Birmingham, Edgbaston, Birmingham B15 2TT, UK, z.zhang@bham.ac.uk

Michael B. Zimmermann
Laboratory for Human Nutrition, Institute for Food Science and Nutrition, Schmelzbergstrasse 7, LFV E19, Swiss Federal Institute of Technology (ETH), Zürich, Switzerland, michael.zimmermann@ilw.agrl.ethz.ch

Nicolaas Jan Zuidam
Unilever R&D Vlaardingen, Olivier van Noortlaan 120, 3133 AT Vlaardingen, The Netherlands, klaas-jan.zuidam@unilever.com

Chapter 1
Introduction

Nicolaas Jan Zuidam and Viktor A. Nedović

Consumers prefer food products that are tasty, healthy and convenient. Encapsulation, a process to entrap active agents into particles, is an important way to meet these demands by delivering food ingredients at the right time and place. For example, this technology may allow taste and aroma differentiation, mask bad tasting or bad smelling components, stabilize food ingredients and/or increase their bioavailability. Encapsulation may also be used to immobilize cells or enzymes in the production of food materials or products, as in fermentation or metabolite production.

This book provides a detailed overview of the technologies used in the preparation and characterization of encapsulates for food active ingredients to be used in food products, processing, or production. This book aims to inform people, with both a limited and an advanced knowledge of the field, who work in the academia or R&D of companies on the delivery of food actives via encapsulation and on food processing using immobilized cells or enzymes.

The first part of the book reviews the general encapsulation technologies – food-grade materials and characterization methods for encapsulates.

Chapter 2 by Zuidam and Shimoni introduces the readers to the most common encapsulation technologies and the general criteria to select a proper encapsulation technology for a certain application.

Chapter 3 by Wandrey, Bartkowiak, and Harding discusses food-grade materials to be used for encapsulation.

Chapter 4 by Zhang, Law, and Lian describes the principle behind the methods used to characterize properties of encapsulates, including their applications. Furthermore, the release mechanism of actives from encapsulates are also described.

The second part of the book discusses encapsulates of active ingredients, i.e., aroma, fish oil, minerals, carotenoids, enzymes, peptides, and probiotics, for specific food applications. The group of actives is chosen so that they represent different classes of actives. This part of the book is intended to serve as a guide to a food

N.J. Zuidam (✉)
Unilever R&D Vlaardingen, Olivier van Noortlaan 120, 3133 AT, Vlaardingen, The Netherlands
e-mail: klaas-jan.zuidam@unilever.com

V.A. Nedović
Department of Food Technology and Biochemistry, Faculty of Agriculture,
University of Belgrade, Nemanjina 6, PO Box 127, 11081 Belgrade-Zemun, Serbia
e-mail: vnedovic@agrif.bg.ac.rs

N.J. Zuidam and V.A. Nedović (eds.), *Encapsulation Technologies for Active Food Ingredients and Food Processing*,
DOI 10.1007/978-1-4419-1008-0_1,

scientist or developer looking for a specific solution to fulfill his or her needs. As encapsulation technologies may change rapidly, emphasis is laid on strategy, assuming that strategy does not change as fast. Most chapters include application possibilities of the encapsulation technologies in specific food products.

Chapter 5, written by Zuidam and Heinrich, highlights aroma or flavor encapsulates. Most of the food encapsulates used are aroma encapsulates. This chapter also discusses how encapsulates could be used to retain aroma during production, storage, and cooking of food products and to release aroma during eating.

Chapter 6, written by Beindorff and Zuidam, discusses fish oil microencapsulates. In this chapter, the authors provide an overview of possible encapsulation technologies used for fish oil and the criteria to select them for different food applications.

Chapter 7 on iron encapsulation is written by Zimmermann and Windhab, and covers the use of encapsulates containing iron and other micronutrients for food fortification of, e.g., salt and staple cereals.

Chapter 8, authored by Ribeiro, Schuchmann, Engel, Walz, and Briviba highlights the use of encapsulates in delivering carotenoids. Carotenoids are instable, natural pigments that are insoluble in water and hardly soluble in oil. Formulation of carotenoids in emulsions or encapsulates influences these characteristics, and improves their bioavailability.

Chapter 9 reviews the encapsulation of enzymes and peptides, including the different types of drying and agglomeration processes. The author, Meesters, also provides examples of their use in industry.

Chapter 10 discusses the encapsulation of probiotics by Manojlović, Nedović, Kailasapathy, and Zuidam. These living and large actives need to survive the food process, storage, and food intake before they can be useful. Examples of the use of encapsulated probiotics in food products are provided.

The last part of the book describes immobilization technologies of cells or enzymes for use in food processing and production.

Chapter 11, authored by Verbelen, Nedović, Manojlović, Delvaux, Laskošek-Čukalović, Bugarski, and Willaert, shows how immobilization of yeast cells can be used in the fermentation of beer.

Chapter 12 reviews how encapsulation of microbial cells can be used for alcoholic and malolactic fermentation of wine and cider. It is written by Kourkoutas, Manojlović, and Nedović.

Chapter 13 by Champagne, Lee, and Saucier describes how immobilization of cells and enzymes can be utilized in dairy and meat fermentation processes.

Chapter 14 presents the view of Breguet, Vojinovic, and Marison on the use of encapsulates for food bioconversions and metabolite production.

The editors are grateful to all the authors for their willingness, time, and effort in contributing to this book! Without their contributions, the book would not have been of such an outstanding quality. We would also like to thank Prof. Denis Poncelet from ENITIAA (France), who as President & Coordinator of Bioencapsulation Research Group and the EU-sponsored action COST 865, supported the idea of writing this book. Many authors who contributed are active in these networks. Finally, many thanks to the editorial staff at Springer for their valuable help throughout this project.

We hope you will enjoy reading this book and that it may help you in choosing the right encapsulation solution to fulfill your need!

Chapter 2
Overview of Microencapsulates for Use in Food Products or Processes and Methods to Make Them

Nicolaas Jan Zuidam and Eyal Shimoni

2.1 Definitions and Benefits of Microencapsulates in Food Products

Encapsulation may be defined as a process to entrap one substance within another substance, thereby producing particles with diameters of a few nm to a few mm. The substance that is encapsulated may be called the core material, the active agent, fill, internal phase, or payload phase. The substance that is encapsulating may be called the coating, membrane, shell, carrier material, wall material, external phase, or matrix. The carrier material of encapsulates used in food products or processes should be food grade and able to form a barrier for the active agent and its surroundings. Please see Chap. 3 for more information on this.

Two main types of encapsulates might be distinguished, i.e., the reservoir type and the matrix type (see Fig. 2.1). The reservoir type has a shell around the active agent. This type is also called capsule, single-core, mono-core or core-shell type. Application of pressure can lead to breakage of the reservoir type of encapsulates and thus to the release of its contents. Poly- or multiple-core type of encapsulates with several reservoir chambers in one particle also exist. The active agent in the matrix type is much more dispersed over the carrier material; it can be in the form of relatively small droplets or more homogenously distributed over the encapsulate. Active agents in the matrix type of encapsulates are in general also present at the surface (unless they have an additional coating, see Fig. 2.1), in contrast to those in the reservoir type. For simplification, Fig. 2.1 shows only spherical shaped encapsulates, but they can also be cylindrical, oval or irregular shaped.

N.J. Zuidam (✉)
Unilever R&D Vlaardingen, Olivier van Noortlaan 120, 3133 AT, Vlaardingen, The Netherlands
e-mail: klaas-jan.zuidam@unilever.com

E. Shimoni
Laboratory of Functional Foods, Nutraceuticals and Food Nanoscience, Faculty of Biotechnology and Food Engineering, Technion - Israel Institute of Technology, Haifa 32000, Israel

N.J. Zuidam and V.A. Nedović (eds.), *Encapsulation Technologies for Active Food Ingredients and Food Processing*,
DOI 10.1007/978-1-4419-1008-0_2,

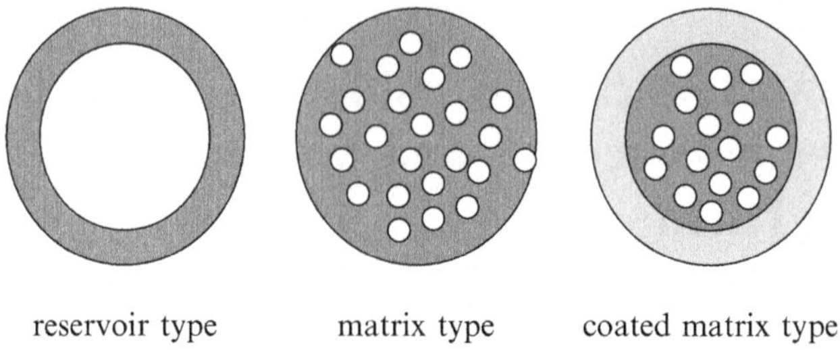

Fig. 2.1 Reservoir type (*left*), matrix type (*middle*), and coated matrix type (*right*) encapsulates. The latter is a combination of the first two. Only spherical shaped encapsulates are shown but other forms are also possible. Here the active is indicated in white, and the carrier material in gray. The active in the matrix type of encapsulates might be in the form of tiny droplets or is dispersed at the molecular level throughout the particle

Encapsulates might also be defined by their particle size, e.g., nanoparticles, microcapsules, microreservoir, etc.

The possible benefits of microencapsulated ingredients in the food industry could be:

- Superior handling of the active agent (e.g., conversion of liquid active agent into a powder, which might be dust free, free flowing, and might have a more neutral smell)
- Immobility of active agent in food processing systems
- Improved stability in final product and during processing (i.e., less evaporation of volatile active agent and/or no degradation or reaction with other components in the food product such as oxygen or water)
- Improved safety (e.g., reduced flammability of volatiles like aroma, no concentrated volatile oil handling)
- Creation of visible and textural effects (visual cues)
- Adjustable properties of active components (particle size, structure, oil- or water-soluble, color)
- Off-taste masking
- Controlled release (differentiation, release by the right stimulus)

Such benefits should overcome the following possible negatives:

- Additional costs
- Increased complexity of production process and/or supply chain
- Undesirable consumer notice (visual or touch) of the encapsulates in food products
- Stability challenges of encapsulates during processing and storage of the food product.

Because of these possible negatives, encapsulates should generally not be seen as a first option when designing food formulations. Only when other, simple options fail one may consider encapsulation. Nevertheless, because encapsulates facilitate formulations of food products that are healthier, tastier and more convenient, the demand for encapsulation has been growing since the last few decades (Frost and Sullivan 2005).

2.2 Encapsulation Processes

This section aims to provide a short overview of commonly used processes to encapsulate food active agent. It is certainly not a complete list. More details about these processes can be found in the references, and their use for specific applications can be found in the other chapters of this book.

Many encapsulation processes are based on making first droplets of the active (in gas, liquid or powder form) and these droplets are subsequently surrounded by carrier material in a gas or liquid phase via different physico-chemical processes (see Table 2.1 and below). The preparation of melt extrudates, liposomes, inclusion complexation technologies, and the use of natural encapsulates like yeast cells (see Chap. 5) might be the exceptions.

Table 2.1 Overview of common microencapsulation processes

Technology	Process steps	Morphology	Load (%)	Particle size (μm)
Spray-drying	1. Disperse or dissolve active in aqueous coating solution 2. Atomize 3. Dehydrate	Matrix	5–50	10–400
Fluid bed coating	1. Fluidize active powder 2. Spray coating 3. Dehydrate or cool	Reservoir	5–50	5–5,000
Spray-chilling/cooling	1. Disperse or dissolve active in heated lipid solution 2. Atomize 3. Cool	Matrix	10–20	20–200
Melt injection	1. Melt the coating 2. Disperse or dissolve active in the coating 3. Extrude through filter 4. Cooling and dehydrating	Matrix	5–20	200–2,000
Melt extrusion	1. Melt the coating 2. Disperse or dissolve active in the coating 3. Extrude with twin-screw extruder 4. Cool	Matrix	5–40	300–5,000
Emulsification	1. Dissolve active and emulfiers in water or oil phase 2. Mix oil and water phases under shear	Matrix	1–100	0.2–5,000
Preparation of emulsions with multilayers	1. Prepare o/w emulsions with lipophilic active in oil phase and ionic emulsifiers 2. Mix with aqueous solution containing oppositely charged polyelectrolytes 3. Remove excess of free polyelectrolytes (option) 4. Repeat steps 2 and 3	Reservoir	1–90	0.2–5,000

Table 2.1 (continued)

Technology	Process steps	Morphology	Load (%)	Particle size (μm)
Coacervation	1. Prepare o/w emulsions with lipophilic active in oil phase 2. Mix under turbulent conditions 3. Induce three immiscible phases 4. Cool 5. Crosslink (optionally)	Reservoir	40–90	10–800
Preparation of microspheres via extrusion or dropping	1. Dissolve or disperse active in alginate solution 2. Drop into gelling bath	Matrix	20–50	200–5,000
Preparation of microspheres via emulsification	1. Emulsify water with biopolymer in oil phase 2. Add gelling agent under shear	Matrix	20–50	10–1,000
Co-extrusion	1. Dissolve or disperse active in oil 2. Prepare aqueous or fat coating 3. Use an concentric nozzle, and press simultaneously the oil phase through the inner nozzle and the water phase through the outer one 4. Drop into gelling or cooling bath	Reservoir	70–90	150–8,000
Inclusion complexation	1. Mix carrier, active and water together 2. Incubate and dry if necessary	Molecular inclusion	5–15	0.001–0.01
Liposome entrapment	1. Disperse lipid molecules in water, with active agent in lipid or water phase 2. Reduce size by high shear or extrusion 3. Remove free active (option)	Various	5–50	10–1,000
Encapsulation by rapid expansion of supercritical fluid (RESS)	1. Create a dispersion of active and dissolved or swollen shell material in supercritical fluid 2. Release the fluid to precipitate the shell onto the active	Matrix	20–50	10–400
Freeze- or vacuum drying	1. Dissolve or disperse active agent and carrier material in water 2. Freeze the sample 3. Drying under low pressure 4. Grinding (option)	Matrix	Various	20–5,000
Preparation of nanoparticles	Various methods, see text	Various	Various	0.1–1

2.2.1 Spray-Drying and Agglomeration

Spray-drying is one of the oldest processes to encapsulate active agent. It is so common in foods that it is not always perceived as an encapsulate, e.g., aroma in a spray-dried form. Spray-drying of active agent is commonly achieved by dissolving, emulsifying, or dispersing the active in an aqueous solution of carrier material, followed by atomization and spraying of the mixture into a hot chamber (see Fig. 2.2 and Barbosa-Cánovas et al. 2005; Gharsallaoui et al. 2007). During this process a film is formed at the droplet surface, thereby retarding the larger active molecules while the smaller water molecules are evaporated. Optionally, one may also spray-dry active agent in organic solutions like acetone or ethanol; however, this is used much less for environmental and safety reasons (which also increase the costs).

Spray-dryers in the food industry are usually atomizing the infeed with a high-pressure nozzle or centrifugal wheel (also called rotary atomizer) and operate with a cocurrent flow of air and particles to give minimal overheating of the particle. This latter is important if the contents are heat sensitive or somewhat volatile (as is the case with aromas). However, cocurrently-dried particles are likely to be more porous than ones prepared in the counter-current mode.

The size of the atomizing droplets depends on the surface tension and viscosity of the liquid, pressure drop across the nozzle, and the velocity of the spray. The size of the atomizing droplets also determines the drying time and particle size.

The temperature of the droplet surface corresponds at any point in the dryer to the "wet bulb" temperature of the gas phase surrounding the droplet as long as the particle surface is wet. The wet bulb temperature under standard spray-drying conditions is of the order of 50°C. By controlling the air-inlet temperature

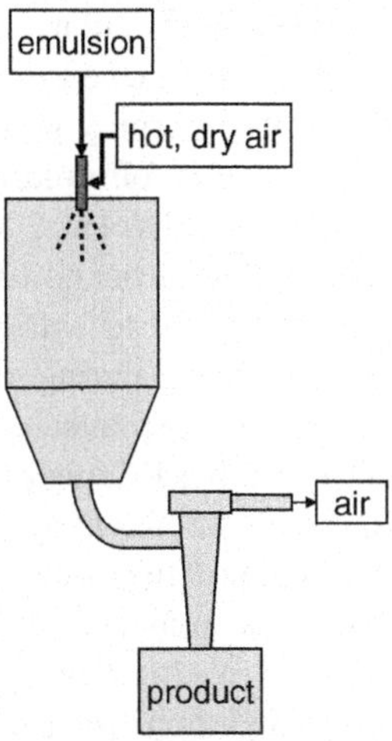

Fig. 2.2 Set-up of a spray-dryer with a cocurrent flow. The dried product is collected in a cyclone at the end

(typically 150–220°C), the flow rate, the feed rate, the feed temperature, and evaporative cooling, it should be ensured that the droplet temperature never exceeds 100°C. This temperature might be indicated by the air outlet temperature, which is typically 50–80°C. The larger the spray-dryer, the longer the residence time of the particle in the dryer (typically 5–100 s) and hence the larger the maximum size of the droplets that can be dried. Atomizing nozzles are usually mounted to spray downward, but it is also possible to spray upward like a fountain, which permits somewhat larger droplets to be dried because of the larger residence time of the droplet.

During the drying process a film is formed at the droplet surface and the concentration of ingredients in the drying droplet increases. Finally, a porous, dry particle is formed.

The carrier material used should meet many criteria, such as protection of active material, high solubility in water, molecular weight, glass transition, crystallinity, diffusibility, good film forming properties, good emulsifying properties, and low costs (Gharsallaoui et al. 2007). Examples from literature include natural gums (gum arabic, alginates, carrageenans, etc.), proteins (dairy proteins, soy proteins, gelatin, etc.), carbohydrates (maltodextrins and cellulose derivatives) and/or lipids (waxes, emulsifiers).

Conventional spray-dried encapsulates release their active agent immediately upon addition to water (which may also depend on the porosity of the particles). However, recent introductions of more hydrophobic and/or cross-linked carrier materials may provide a more gradual release upon dilution in water. Examples of these are denatured proteins, cross-linked proteins or cross-linked biopolymers.

Please see the reviews of Reineccius (2001, 2004), Gouin (2004), Barbosa-Cánovas et al. (2005), Desai and Park (2005), Gharsallaoui et al. (2007) and Jafari et al. (2008) for further, general information about encapsulation of food active agent by spray-drying.

For many applications, larger particles than those produced by spray-drying (in general about 10–150 μm) might be desirable. This might be achieved by agglomeration or granulation (Barbosa-Cánovas et al. 2005; Ortega-Rivas 2005). In general, this can be achieved in any equipment creating random movements. An option is fluidized bed spray granulation (also called spray-bed-drying), in which a spray-drying step is followed in one or two steps by a secondary agglomeration step in a fluid bed (Fuchs et al. 2006; see also the next section). Another option is to spray-dry onto another carrier powder (Fuchs et al. 2006). In both cases, the spray-dried particles are not fully dried after the first stage, and therefore remain sticky to facilitate agglomeration during the second phase. Alternatively, a binder solution (e.g., water) can be sprayed onto powder particles during high shear or tumbling (Litster 2003; Barbosa-Cánovas et al. 2005), or in a fluid bed (Uhlemann et al. 2002; Barbosa-Cánovas et al. 2005; Ortega-Rivas 2005).

An alternative process for the preparation of large particles is pressure agglomeration or compaction, in which spray-dried material is compressed under high pressure in extruders or presses, maybe together with additional maltodextrin, into lumps and then crushed into small pieces of about 0.7–3.0 mm (Barbosa-Cánovas et al. 2005; Ortega-Rivas 2005; Uhlemann et al. 2002). This process is useful for applications in which encapsulates should not segregate within food products.

2.2.2 *Fluid Bed Coating*

Fluid bed coating is a technique in which a coating is applied onto powder particles in a batch process (see Fig. 2.3) or a continuous set-up. The powder particles are suspended by an air stream at a specific temperature and sprayed with an atomized, coating material. With time, each particle will be gradually covered every time it is in the spraying zone. The coating material must have an acceptable viscosity to enable pumping and atomizing, must be thermally stable and should be able to form a film over a particle surface. In general, 5–50% of coating is applied, depending on the particle size of the core material and application of the encapsulate.

The coating material might be an aqueous solution of cellulose derivatives, dextrins, proteins, gums and/or starch derivatives, and the evaporation of its water content is then controlled by many factors such as the spray rate, the water content of the coating solution, the air flow, the humidity of the air inlet in the chamber, and the temperature of the coating solution, atomized air, and the material in the chamber (Dewettinck and Huyghebaert 1999; Guignon et al. 2002; Teunou and Poncelet 2002, 2005a). Often a so-called Würster set-up is used, in which the coating is sprayed in an inner column from the bottom (see Fig. 2.3, left picture). The air flow

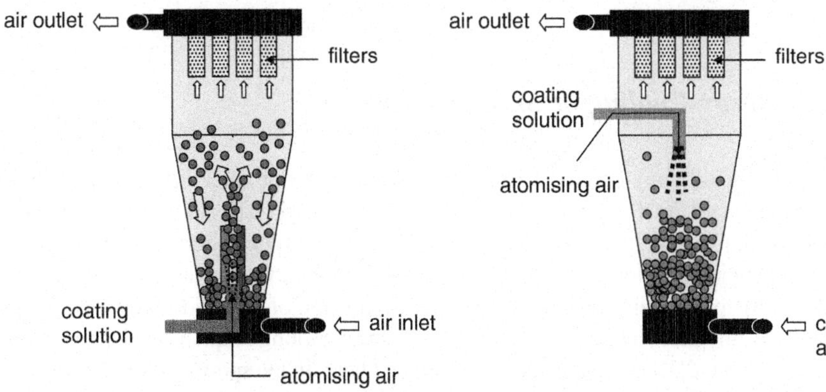

Fig. 2.3 Fluidized bed coating is achieved by an upwards air flow through a bed of particles and spraying a liquid with shell material. Here a so-called Würster coating set-up is shown at the *left*, in which process the coating material is sprayed onto powder particles within an inner column which brings the particles into circulation. On the *right*, a set-up is shown in which a coating solution is sprayed from the top onto powder particles. Other set-ups are also possible, which may include bottom-spraying without the inner column, side spraying with rotating disk and continuous configurations

rate is typically 80% in the center flow in the inner column and 20% in the periphery, which brings the powder particles into circulation. This increases the drying rate and reduces agglomeration. The bottom spray reduces the distance between the powder and the drops of coating solution, thereby reducing the risk of premature drying of the coating.

Alternatively, a molten lipid can be used as a coating material which can be either applied from the bottom or the top (see for the latter configuration the right picture in Fig. 2.3). Examples of lipids used are hydrogenated vegetable oils, fatty acids, emulsifiers and/or waxes. Care must be taken to prevent solidification of the lipid before it reaches the powder. This might be done by heating not only the storage vessel from which the molten lipid is pumped, but also the line, the nozzle, and atomizing air. Once in the chamber, the rate of congealing (solidification) is controlled by the application rate and the cooled, inlet air (often 10–20°C below its melting point). Product temperature too close to the melting temperature of the fat may result in sticky particles and thus agglomeration. At lower product temperature the congealing might occur before complete spreading so the coating might contain defects and pores.

The particles to be coated by fluid bed should ideally be spherical and dense, and should have a narrow particle size distribution and good flowability. Spherical particles have the lowest possible surface area and require less coating material for the same shell thickness than nonspherical ones. Sharp edges could damage the coating during handling. Fine and low-dense particles might face the risk of accumulating on the filter bags in the top of the machine.

Alternative air suspension coating technologies, e.g., pan coating, have been described by Teunou and Poncelet (2005b). In general, one applies a coating to make the powder more resistant to humidity. If desired, more than one coating can be applied on the powders (with increasing costs).

2.2.3 Spray-Cooling or Spray-Chilling

Spray-chilling or spray-cooling is another technology to produce lipid-coated active agent (Kjaergaard 2001; Uhlemann et al. 2002; Gouin 2004). The active agent might be soluble in the lipids, or be present as dry particles or aqueous emulsions. Firstly, droplets of molten lipid(s) are atomized into a chilled chamber (e.g., via nozzle, spinning disk or (centrifugal) co-extrusion), which results in solidification of the lipids and finally their recovery as fine particles. The initial set-up of spray cooling is quite similar to spray-drying (see Sect. 2.2.1), but no water is evaporated here. In the spray-chilling technique, the particles are kept at a low temperature in a set-up similar to the fluidized bed spray granulation (see Sect. 2.2.2), on which molten lipid droplets may adhere to already hard lipid particles before solidification. In general, the melting point of the lipid used is in the range of 34–42°C for spray-chilling, and higher for spray-cooling.

Rotating disk is another atomization method for the preparation of solid lipid particles (Sparks and Mason 1987). A suspension of particles in molten lipid is spread on the disk, followed by separation of coated particles by atomization at the edge of the rotating disk. The disk may be flat or bowl-shaped and can be heated. The drops solidify when falling from the disk. Depending on the droplet size and melt characteristics a certain falling height is required. The size of the particles depends on the core particles, melt viscosity, melt temperature, disk configuration and the rotational speed.

2.2.4 Melt Injection and Melt Extrusion

Carbohydrate materials can be mixed with an active when molten, at a temperature above 100°C, then pressed through one or more orifices (extrusion) and finally quenched to form a glass in which active agent have relatively little mobility. In general, the glass transition of encapsulates made by extrusion is between 30 and 70°C.

Basically, two processes to encapsulate active agent in a carbohydrate melt can be distinguished. One is melt injection, in which the melt (composed of sucrose, maltodextrin, glucose syrup, polyols, and/or other mono- and disaccharides) is pressed through one or more orifices (filter) and then quenched by a cold, dehydrating solvent. This is a vertical, screwless extrusion process. Generally isopropanol, and also liquid nitrogen, is used as the dehydrating solvent. The coating material hardens on contact with the dehydrating solvent, thereby encapsulating the active (Porzio 2004). The size of the extruded strands is reduced to the appropriate dimensions inside the cold solvent during vigorous stirring, thereby breaking up the extrudates into small pieces. Any residues of active agent on the outside will be washed away by the dehydrating solvent. Encapsulates made by melt injection are water-soluble and have particle sizes from 200 to 2,000 μm.

Encapsulation in a carbohydrate melt can also be achieved by using an extruder with one or more screws in a continuous process (see Fig. 2.4). This process is called

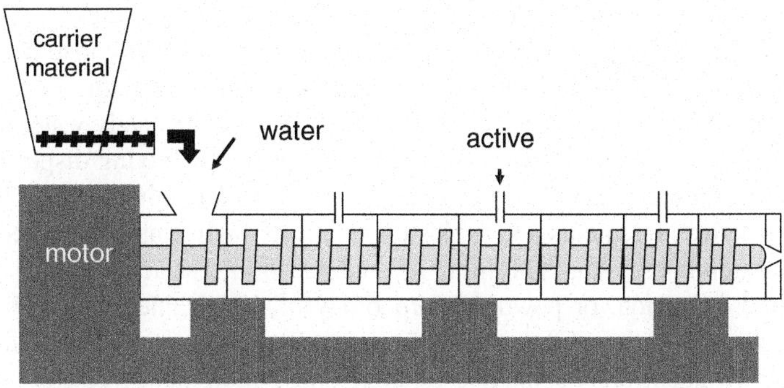

Fig. 2.4 Scheme of a melt extruder. Most often, extruders with two screws are preferred. Each section can be temperature controlled. The carrier material is commonly added via a twin screw feeder in the first section, and water and active can be added simultaneously or later

melt extrusion, and it can be regarded as a process very similar to melt injection; the main differences are that, in general, melt extrusion utilizes screws in a horizontal position and that the extrudates are not surface washed. Extruders are thermomechanical mixers that consist of one or more screws in a barrel. Most often, double screw extruders equipped with sinusoidal screws (self-wiping) are preferred for encapsulation. It is common to characterize the extrusion screws by the length/diameter (L/D) ratio, typically between 20 :1 and 40 :1. Transport of material within the extruder takes place by rotational, and sometimes oscillatory, movement of the screws. In the beginning (the feed zone), the screw design is such that a low pressure is generated to homogenize the feeding. In the subsequent zone(s), a gradual increase in pressure is achieved via the screw design to melt, further homogenize, and compress the exrudate. In the final part of the barrel, a constant screw design helps to maintain a continuous high pressure to ensure a uniform delivery rate of molten material out of the extruder. Most of the time, the barrel is also divided into sections to allow for section-controlled variation in temperature. At the end of the barrel, a "pre die" and "die head" determine the shape of the final product (e.g., sheets, ropes or threads). It can be equipped with a chopper/cutter to obtain granular extrudates. Alternatively, these can be obtained via postpreparation equipment like grinders or mills, see Ortega-Rivas 2005. Extrudates can be composed of starch, maltodextrins, modified starches, sugars, cellulose ethers (like hydroxypropyl cellulose or hydroxypropyl methyl cellulose), proteins, emulsifiers, lipids, and/or gums. Often, melt extrudates for use in food products are composed of "thermoplastic" starch (Yilmaz 2003; Yilmaz et al. 2005). Native starch is composed of semicrystalline granules with a size of 1–100 μm (see also Sect. 3.2.1.1). It consists of amylose and amylopectin, both containing only α-D-glucose units. Thermoplastic starch is obtained by destructuring the semicrystalline starch via simultaneous application of heat (around 110–120°C) and mechanical forces. The presence of plasticizers, such as water or glycerol/polyols, enables further processing. Plasticizers also influence other properties, such as glass transition temperature of the material, its solubility and morphology (Yilmaz et al. 1999 and references therein). Unmodified thermoplastic starch dissolves quickly in water. Modified starches that are tailor-made (using physical, chemical or enzymatic routes) might be used to make encapsulates more water resistant (Zasypkin and Porzio 2004). In addition, additives (such as plasticizers and other constituents in the formulation) and in situ modifications (e.g., heat treatment, surface modification, and induction of (local) crystallinity) might also be used to get relatively water-insoluble extrudates. Addition of the active ingredient might be in the mixing/dispersing zone of the extruder (at about halfway in the scheme of Fig. 2.4). This minimizes the residence time of the active ingredients and avoids the relatively high temperatures required to plasticize starch in case starch is still in its granular form. The active can be added as a gas, liquid, emulsion, or powder. Morphology of the obtained formulations will depend on the properties of the active agent as well as the matrix. In case the matrix and the active agent are compatible a single-phase morphology can be obtained, where the matrix behaves as a solvent. In case of incompatibility a two-phase morphology is likely to be obtained. For example, lipophilic compounds may mix well with starch modified with hydrophobic groups, in contrast to hydrophilic ones.

Alternatively, pre-encapsulation and surface modification are possible. The encapsulation efficiency (and the release kinetics) will depend on adequate mixing and dispersion of the encapsulant within the matrix. The use of emulsifiers may allow better control of these characteristics (Yilmaz et al. 2001). Unfortunately, the active load of extruded encapsulates is relatively low (typically less than 10%), which may have an impact on their cost-in-use.

2.2.5 Emulsification

Emulsions are kinetically rather than thermodynamically stable two-phase systems and ultimately, both the oil and water phase will separate. Proper formulation design of both phases and the interface, including choice of ingredients like emulsifiers, might prevent that (McClements 2005; Appelqvist et al. 2007). Emulsions are commonly made under high shear with, e.g., homogenizer, colloid mill, high shear mixer, or stirred vessel preferably equipped with baffles (see Fig. 2.5 for the latter).

Plain emulsions can be used as a delivery vehicle for either water soluble and/or lipophilic active agent in food products (Appelqvist et al. 2007). There are two considerations that must be taken into account when formulating an emulsion for controlled delivery. First, the emulsion system must be (storage) stable right up to the point

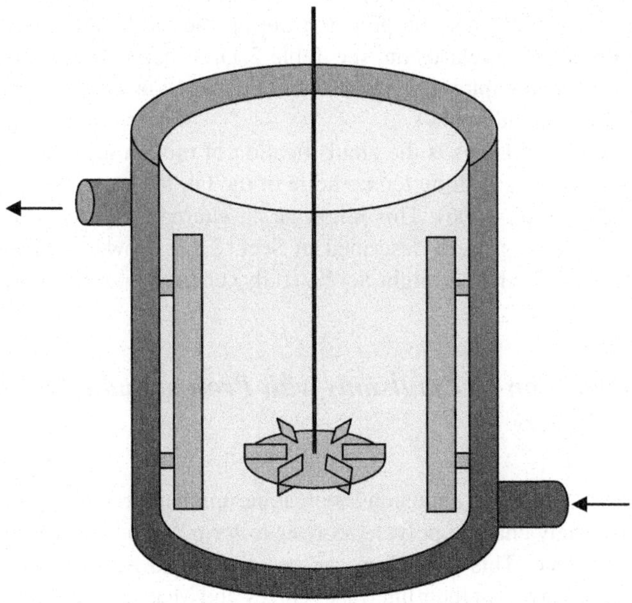

Fig. 2.5 Set-up of a stirred, double-wall vessel with 3–4 baffels and Rushton-impeller, which might be used for the preparation of emulsions or complex coacervates. This set-up can be used at both a lab scale and a factory scale

of application. Second, on application the emulsion should behave in a manner consistent for achieving the desired delivery. In many (but by no means all) cases this equates to the "making and breaking" of emulsions for stability and subsequent delivery.

Water soluble food active agent might be encapsulated in water-in-oil (w/o) emulsions or double emulsions of the type w/o/w (Appelqvist et al. 2007). Furthermore, oil-in-water (o/w) emulsions may affect taste (e.g., salt) by changing the aqueous phase volume and thus the concentration of taste molecules in water, and by suppressing contacts of salt with taste receptors. Lipophilic active agent (e.g., aroma, cartonenoids such as lycopene and beta-carotene, plant sterols, vitamin E, dietary fats) might be protected and delivered to consumers via o/w emulsions (Appelqvist et al. 2007; see also Chaps. 6 and 8).

Several technologies have been developed to produce highly uniform emulsion droplets (see Link et al. 2004; McClements 2005), such as reduction of polydispersity of already formed emulsions (including repeated fractionation and shearing immiscible fluids between uniformly separated plates; Mabille et al. 2003), or single-drop technologies like microfluidics. Monodispersed emulsions may have a more defined behavior and release pattern of entrapped active agent than polydispersed ones. This can be very important in pharmaceutics and when the emulsions are used as a template to make new materials for, e.g., electronics. Currently, it is not clear whether this would constitute a real advantage in food systems.

Oil-in-water emulsions might be dried by, e.g., spray-drying (see Sect. 2.2.1) or freeze-drying (see Sect. 2.2.13) to provide a powder. Such dry emulsions might be encapsulates or an instant formulation of beverages or other food products. Emulsion droplets might also be prepared during the processing of encapsulates (such as extrudates or co-extrusion, see Table 2.1), or act as templates for further processing (such as complex coacervates, microspheres or emulsions with multilayers; see Table 2.1 and below).

Another use of emulsions is the emulsification of molten fat or wax in water at a temperature above the melting temperature of the fat, followed by cooling during mixing (Mellema et al. 2006). This might be an alternative process to the spray-chilling/spray-cooling process described in Sect. 2.2.3. However, if the active is (partly) water-soluble, then it might not be (fully) encapsulated within the fat.

2.2.6 Preparation of Emulsions with Protein and/or Biopolymer Multilayers

A layer around "primary" emulsions with ionic emulsifier(s) can be formed by adsorbing oppositely charged polyelectrolytes to form "secondary" emulsions with a two-layer interface. This procedure can be repeated to form emulsion droplets with three or more layers at their interface (Guzey and McClements 2006). Removal of excess free polyelectrolytes by, e.g., centrifugation or filtration between the steps might be necessary.

This procedure is also called layer-by-layer (LBL) electrostatic deposition technique. It is a relatively new technique and its full potential is under investigation. It is a simple preparation technique at lab scale, but quite laborious at larger scales. Examples include emulsions with multilayers composed of β-lactoglobulin– ί-carrageenan, β-lactoglobulin–pectin, or sodium dodecyl sulfate (SDS)–chitosan–pectin.

2.2.7 Coacervation

Coacervates are made via a liquid–liquid phase separation mechanism of an aqueous solution into a polymer-rich phase (known as coacervate) and a polymer-poor phase. According to the number of polymer type(s) present, the process can be identified as (simple) coacervation when only one type of polymer is involved or complex coacervation when two or more types of polymers of opposite ionic charges are present. The coacervates used to encapsulate active agent are most often of the complex type. Their shell is frequently composed of gum arabic and gelatin. The technology was developed by National Cash Register Co. in the 1950s and was the basis of carbonless copy paper, the first commercial product with microencapsulates.

Complex coacervates are commonly made from an o/w emulsion with gelatin and gum arabic at a 1:1 w/w ratio and at a 2–4% w/w of each polymer dissolved in the water phase via adjusting the pH from neutral to about 4 under turbulent conditions in a stirred vessel (see Fig. 2.5) at >35°C, a temperature above the gelation temperature of gelatin (Gouin 2004; Lemetter et al. 2009). This creates three immiscible phases (oil, polymer-rich, and polymer-poor phase), and the polymer-rich phase droplets will deposit on the emulsion surfaces because of interfacial sorption. Alternatively, complex coacervation can be induced by dilution instead of pH adjustment; oil is emulsified in a 8–11% (w/w) gelatin solution, followed by addition of gum arabic and dilution water (Thies 2007). Upon cooling well below 35°C (Lemetter et al. 2009), the deposited gelatin and thus the shell will solidify. Factors like polymer concentrations, pH, turbulence of the system, emulsion size, ionic strength, and temperature affect the preparation process. After cooling, there is an option to crosslink the shell with, e.g., glutaraldehyde (Tabor et al. 1992; not allowed in Europe for food applications) or transglutaminase (Thies 2007). Finally, the coacervates are isolated and washed (if needed) via filtration or sedimentation (if their density is higher than the density of water, which depends on the relative amount of shell compared to the oil core) and might be dried by spray-drying or fluid bed drying. Optionally, gum arabic can be replaced by other negatively charged molecules like carboxymethylcellulose, pectin, carrageenan, alginate and alginate derivatives, or polyphosphate (Bakker et al. 1999; Gouin 2004; Thies 2007), or gelatin can be replaced by whey proteins (Weinbreck et al. 2003). Gelatin most often has a beef or pork origin, but as a Kosher or Halal alternative fish gelatin might be used. Each polymer combination operates at unique conditions in terms of pH, temperature, ionic strength, polymer levels, molecular weight, charge density, cooling rate, etc. Complex coacervates often have a very typical, oval shape.

Simple coacervation has been used less to encapsulate active agent. Examples are the encapsulation of o/w emulsions in gelatin where solubility is reduced by temperature or sodium sulfate, in 0.2 wt. % chitosan by increasing the pH with 0.1–1.5 wt. % sodium hydroxide (Hsieh et al. 2006), and in an aqueous solution of hydroxypropyl methylcellulose or methyl cellulose where simple coacervation was induced by addition of maltodextrin (Porzio and Madsen 1997; the maltodextrin also functioned as spray-drying carrier material for double coating).

2.2.8 Preparation of Microspheres by Extrusion or Emulsification

Microspheres are microbeads composed of a biopolymer gel network entrapping an active. The microspheres are commonly prepared in the presence of the active, but postloading of blank microspheres containing oil droplets with, e.g., aroma is also an option. Calcium-alginate gel is the best known gelling system used for the preparation of gel beads to encapsulate a wide variety of active agent, such as oil droplets containing aroma, cells, probiotics, yeast, or enzymes to name a few. These active agent are relatively large in size, as smaller ones will diffuse easily through the porous biopolymer network. Gelation of alginate in the presence of divalent cations can be easily controlled and does not require heating like other gelling biopolymers like agarose, agar, or carrageenan. Microspheres are commonly made via two different routes (Krasaekoopt et al. 2003; Gouin 2004):

(a) *The extrusion or dropping method*: This method consists of dropping droplets of an aqueous solution of 0.6–4 wt. % sodium alginate and active into a gelling bath of 0.05–1.5 M calcium-chloride solution. The dripping tool can be simply a pipette, syringe, vibrating nozzle, spraying nozzle, jet cutter, atomizing disk, coaxial air-flow, or electric field (see Fig. 2.6 and also Zhang et al. 2007 for dropping and spraying set-ups). In general, particles with a diameter between 0.2 and 5 mm can be made depending on the dripping tool and the viscoelasticity of the alginate solution. Alternatively, the extrusion or dropping method can be used with a concentric nozzle (co-extrusion), to prepare core-shell type of encapsulates with a lipophilic core and a shell of a gel network (see Sect. 2.2.9). In a recent study by Prüsse et al. (2008), different common bead production technologies were analyzed to check their ability to process fluids of different viscosities. Each of the technologies is suitable for the production of spherical microspheres (800 μm in diameter) from low-viscous sodium alginate solutions (up to 2% w/w), whereas high-viscous alginate solutions (≥3% w/w sodium alginate) cannot be processed with the vibration technology anymore. With the electrostatic, jet cutter, and coaxial air-flow technologies microsphere production was possible and a narrow size distribution was always achieved. However, the shape of the microspheres produced by coaxial air-flow was nonspherical and

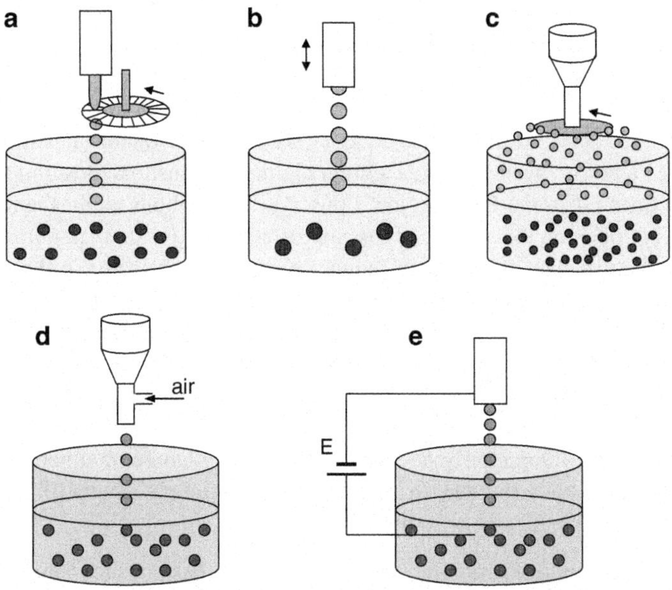

Fig. 2.6 Set-ups of three different ways of making microspheres. Aqueous solution of, e.g., sodium alginate and active are atomized by jet-cutter (**a**), pipette or vibrating nozzle (**b**), atomizing disk (**c**), coaxial air-flow (**d**), or electrostatic potential (**e**). The droplets fall into a batch of 0.05–1.5 M calcium chloride, resulting in instantaneous formation of calcium alginate microspheres

deformed egg-like or drop-like microspheres were obtained from 3 and 4% (w/w) sodium alginate solutions, respectively. In addition, extrusion technologies were compared with respect to productivity. The microsphere production rates of the coaxial air-flow and the electrostatic technology are very low. Thus, these technologies are limited to small/lab-scale applications when only a few grams of material have to be processed. The vibration technology exhibits up to 50 time higher production rates than the aforementioned technologies. Vibration systems, thus, are suitable for lab-scale as well as larger scale applications, assuming that multinozzle devices are used for larger scales. The JetCutter technology is suited both for lab-scale and large- up to industrial-scale microsphere production. Instead of calcium-alginate, one may prepare microspheres with other compositions, e.g., by dropping 4% κ-carrageenan into 0.3 M potassium chloride, or heated gelatin, agarose, or agar solution into a cold bath.

(b) *The emulsion method*: This technique utilizes emulsions to make microspheres. Several variants exist. One may add calcium chloride to an emulsion of water droplets of an alginate solution and active in vegetable oil. This results in the "break-up" of the emulsion and microbeads are formed by the gelation of the alginate droplets. Alternatively, both alginate and calcium (in an insoluble form such as calcium carbonate) can already be present in the water phase of the emulsion. Upon addition of an oil-soluble acid (such as acetic acid) the pH decreases, liberating free calcium ions in the system and initiating the gel formation of alginate droplet with calcium. Delta-glucono-lactone can also be employed for slower gelation kinetics, if needed. Another variant of the emulsion method is the preparation of a water-in-oil emulsion first, with calcium ions in the water phase, and second, addition of an aqueous alginate solution during stirring which produces a phase inversion, and calcium alginate begins to deposit on the newly formed drops (Casana Giner et al. 2006). Another colloid might then be added that will deposit on the surface of the microspheres (e.g., xanthan gum) and then a primary surfactant is added to reduce the size of the water in the oil drops. Agglomeration or deagglomeration may occur (depending on the process conditions) and finally the microspheres are hardened at an elevated temperature (75°C for 120 min). Gelling materials other than alginate can also be used in the emulsion technique, such as κ-carrageenan (gelation upon cooling with potassium ions), chitosan (crosslinking by addition of anions), gelatin (crosslinking by mixing with anionic polysaccharides, such as gellan gum, at neutral pH, followed by adjusting the pH to make gelatin positively charged), and pectin (chemically or physically crosslinked).

The emulsion method has the advantage that it can produce smaller microspheres (10 μm–1 mm) than the extrusion method (0.2–5 mm). It is also easier to scale-up. However, the emulsion method might be more expensive if vegetable oil has to be removed (and recycled), and the microspheres have to be washed sufficiently to eliminate the residual vegetable oil on the surface.

The presence of chelating agents (e.g., phosphate, lactate, citrate or bicarbonate) may interfere with the encapsulation process or alter the integrity of the calcium-alginate gels beads added into wet products. Posthardening for long periods of time in a solution with the crosslinker (e.g., storage of calcium-alginate microspheres for 1 day in 0.2 M calcium chloride solution), coating (e.g., with chitosan or poly-L-lysine, which are both not food grade), cross-linking with cationic polymers (e.g., chitosan), incorporation of additives (e.g., microcrystalline cellulose, hydrophobic starches) in the gel network, and/or modification of oil reservoir (if applicable) might be applied to modify the properties of the microspheres.

2.2.9 Co-extrusion

Co-extrusion is an extrusion technology which utilizes a concentric, multifluid nozzle, which may be stationary, rotating, or vibrating, It can be utilized to prepare spherical microbeads with a hydrophobic core of active agent and a hydrophilic or hydrophobic shell produced by interfacial gelling (e.g., with calcium-alginate or potassium-carrageenan) or cooling (e.g., gelatin or fat). Different set-ups are possible:

- Some equipment (e.g., from Inotech, Brace, Nisco) utilizes a vibrating multi-fluid nozzle to produce 80–1,500 micrometer particles. The technology is based on the principle that a laminar liquid jet is broken into equal-sized droplets by a superimposed vibration.

- Some other equipment is based on centrifugal co-extrusion, which leads to the formation of round beads at the edge of the nozzle due to Raleigh instabilities.
- The nozzle might also be submerged into a moving carrier and cooling fluid (see Fig. 2.7; Uhlemann et al. 2002). The submerged set-up prevents disruption of the shell upon contacting the cooling liquid. The capsules can be about 1–8 mm with typically a 70–95% load (aroma, fish oil, vitamins, freeze-dried probiotics dispersed in oil, etc.).
- Another option is to make use of a dual-feed spraying nozzle in combination with ultrasonic atomization (e.g., from Sono-Tek), which allows one to spray-dry immediately in the air after atomization takes place.

2.2.10 Inclusion Complexation

Molecular inclusion is the association of the active in a cavity-based material. The best known example is cyclodextrin (Hedges 1998; Szente and Szejtli 2004; Regiert 2008). Cyclodextrins are cyclic oligosaccharides of 6–8 D-glucose molecules, which are enzymatically joined through alpha 1–4 linkages in such a way that they to form a ring (see Fig. 3.7 in the next chapter). Some properties of cyclodextrins are listed in Table 2.2. Cyclodextrins containing six, seven, or eight glucose molecules are referred to as α-, β- and γ-cyclodextrin, respectively. Their diameters are about 14, 15 and 17 Å, respectively. Cyclodextrins have a lipophilic inner pocket of about 5–8 Å, in which an active molecule with the right size can be reversibly entrapped in an aqueous environment. However, this characteristic limits

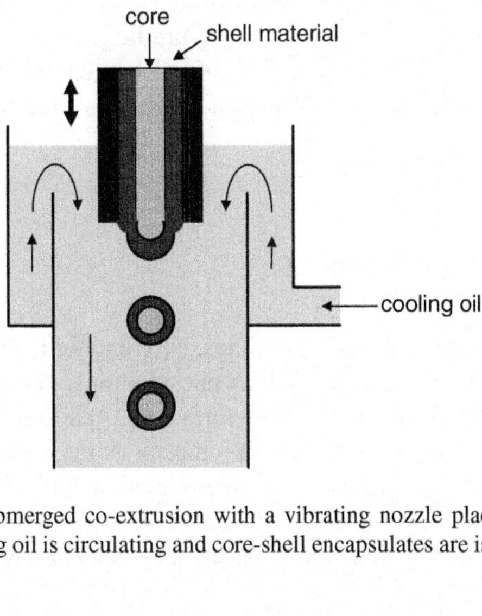

Fig. 2.7 Set-up of submerged co-extrusion with a vibrating nozzle placed into the carrier and cooling oil. The cooling oil is circulating and core-shell encapsulates are isolated from the cooling oil by, e.g., filtration

Table 2.2 Properties of cyclodextrins (Most data taken from Regiert (2008))

	α-cyclodextrin	β-cyclodextrin	γ-cyclodextrin
Number of glucose units	6	7	8
Molecular weight (g/mol)	973	1,135	1,297
Crystal water content (wt. %)	10.2	13.2–14.5	8.1–17.7
Molecule diameter (Å)	14.6	15.4	17.5
Cavity diameter (Å)	4.7–5.3	6.0–6.5	7.5–8.3
Solubility in water at 25°C (g/mol)	14.5	1.85	23.2
Hydrolysis of α-amylase	Negligible	Slow	Fast
Crystal water content	10.2	13.2–14.5	8.13–17.7

its loading capacity. Loading of cyclodextrins can be achieved by coprecipitation of the complex in aqueous solutions (essentially a laboratory method), by using a slurry of partially dissolved cyclodextrin (upto 45% w/w), by using a paste with 20–30% water, or by dry mixing (Hedges 1998). Temperature, time, the amount of water, and the particular active and cyclodextrin control the loading rate and efficiency. β-cyclodextrin is the most common one. The cyclodextrins might be branched enzymatically to increase their water solubility. Unfortunately, the use of cyclodextrin might be limited by regulatory rules. In Japan, cyclodextrins are regarded as a natural product. In the USA, α-, β- and γ-cyclodextrin have GRAS status. However, in the EU β-cyclodextrin is allowed in a limited number of products (chewing gum, potato, cereal, flour or starch based snacks, and in water-based flavored drinks; <1 g/kg) and α-cyclodextrin only has a regulatory status as a novel food since February 2007. Novel food status for the use of γ-cyclodextrin in the EU has been filed but not given as yet.

Other examples of molecular inclusion might be the entrapment of lipids by amylose (see Sect. 2.3) and the use of ligand-binding proteins (De Wolf and Brett 2000), such as the milk protein β-lactoglobulin. This protein belongs to the superfamily of lipocalins, together with retinol-binding protein and odor-binding proteins (Guichard 2006; Tromelin et al. 2006). β-lactoglobulin has a hydrophobic pocket, which binds fatty acids and aroma molecules in a pH and temperature dependent manner.

2.2.11 Liposome Entrapment

Liposomes consist of at least one closed vesicle composed of bilayer membranes which are made of lipid molecules, such as phospholipids (lecithin) and cholesterol (see also Sect. 3.2.3.4 and Fig. 3.19). They form when (phospho)lipids are dispersed in aqueous media and exposed to high shear rates by using, e.g., microfluidization or colloid mill. The underlying mechanism for the formation of liposomes is basically the hydrophilic–hydrophobic interactions between phospholipids and water molecules. Active agent can be entrapped within their aqueous compartment at a low yield, or within or attached to the membrane at a high yield. The particle size ranges from 30 nm to a few microns. Small vesicles tend to aggregate or fuse and may end up growing into micron-size particles during storage, which might be prevented by electrostatic repulsion (e.g., by addition of charged lipids in the membrane) or steric stabilization. Liposomes are currently mainly studied and used as advanced, pharmaceutical drug carriers (Torchillin and Weissig 2003), and their use in foods (Were et al. 2003; Gouin 2004; Taylor et al. 2005; Kosaraju et al. 2006; Mozafari et al. 2006; Takahashi et al. 2007) is quite limited due to its chemical and physical instability upon storage in especially emulsified food products, low encapsulation yield, leakage upon storage of liposomes containing water-soluble active agent, and the costs of raw materials (Zuidam et al. 2003). Liposomes are now used as drug delivery systems. For food applications, however, liposomes have mainly been studied to enhance ripening of hard cheeses and other applications in the food industry are very limited.

2.2.12 Encapsulation by Using Supercritical Fluid Technology

Supercritical fluids exist above a critical temperature and pressure at which the substance's liquid and gas phases are indistinguishable (Thies et al. 2003; Martin Del Valle and Galan, 2005). Their properties are intermediate to those of liquids and gases – liquid-like densities, gas-like viscosities, gas-like compressibility, and higher diffusivity and mass transfer than liquids. Many compounds can be brought into a supercritical state, such as water, propane, nitrogen, and carbon dioxide. The last one is probably the most interesting solvent for use in an encapsulation process, since it is environmentally friendly, it minimizes the use of organic solvent and water, and can be applied at reasonable pressures and temperatures (<30°C). It is actually applied to improve existing encapsulation processes:

- When supercritical fluid is released through a small nozzle, the abrupt pressure drop causes the supercritical fluid to evaporate or to transform into a much poorer solvent. Dissolved or swollen shell material will precipitate onto active agent dispersed in the supercritical fluid. This process is called Rapid Expension of Supercritical Solutions (RESS). Carbon dioxide is an apolar solvent, and therefore only shell materials like fat or wax will solubilize well in it. Hydrophilic proteins (e.g., gelatin) or polymers (e.g., cellulose, hydroxypropyl methylcellulose) can swell in it or might be solubilized by using cosolvents.
- Spray-drying of solvents containing supercritical carbon dioxide (also called supercritical assisted atomization, SAA) operates at relatively low temperatures, which might be beneficial for temperature-sensitive materials such as proteins or volatile flavors. Cosolvents might also be necessary here to dissolve active agent like proteins.
- Active agent and carrier material dissolved in organic solvent may be sprayed into supercritical fluid, thereby extracting the solvent from the incoming spray droplets and coprecipictating the active agent and carrier material. This process is called aerosol solvent extraction (ASES).

– Supercritical carbon dioxide might be used to dissolve lipids prior preparation of liposomes (see Sect. 2.2.11).

2.2.13 *Freeze-Drying and Vacuum Drying*

Active agent and carrier material dissolved in water can be freeze-dried to produce a porous, nonshrunken structure. Firstly, the sample is frozen at temperatures between −90 and −40°C and then dried by direct sublimation under low pressure and reduced temperature (between −90 and −20°C). After drying, the brittle cake obtained can be broken into smaller pieces by, e.g., grinding, if necessary. The use of relatively high amounts of cryoprotectants (like 10% milk proteins, 30% maltodextrin or 10% disaccharides) may help to stabilize sensitive active agent like probiotics (see Sect. 10.3.3) or sensitive encapsulates like liposomes (Zuidam et al. 2003).

The major disadvantages of freeze-drying are the high energy use, the long processing time, and the open porous structure obtained, which is in general not a very good barrier between the active and its surroundings. Compared to spray-drying, freeze-drying is upto 30–50 times more expensive (Gharsallaoui et al. 2007).

Vacuum-drying is very similar to freeze-drying, but operates at a temperature above the freezing point of the solvent (>0°C in case of water) and is therefore faster and cheaper.

2.3 Nanoparticles

The use of nanosized vehicles for the protection and controlled release of nutrients and bioactive food ingredients is a growing area of interest to the food science and technology community. The reasons for this are that nanoparticles might be incorporated into food products easily without sedimentation, without being noticed by the consumer, and/or with an enhanced bioavailability (Acosta 2009). The preparation of nanoparticles might be based on downsizing encapsulates prepared by "classical" technologies as discussed above, or by using new techniques. In this section, we only bring a few examples to demonstrate the potential of some of the techniques and concepts that evolved in recent years. For the sake of this section we will categorize them into lipid, protein, polysaccharide, and inorganic-based systems.

Lipid-based nanoencapsulation systems are among the most rapidly developing field of nanotechnology application in food systems. Lipid-based nanoencapsulation systems have several advantages, including the ability to entrap material with different solubilities and the use of natural ingredients on an industrial scale (Bummer 2004; Mozafari et al. 2006; Taylor et al. 2005). Lipid-based nanocarriers can also be used for targeted delivery of their contents to specific areas within the gastro-intestinal tract or food matrix. When referring to nanoscale lipid vesicles, the term nanoliposome has recently been introduced (Mozafari et al. 2006) to describe lipid vesicles whose diameter ranges tens of nanometres. These, so called, nanoliposomes have similar structural, physical, and thermodynamic properties as liposomes. The manufacture of nanoliposomes (as of liposomes) requires high energy for the dispersion of lipid/phospholipid molecules in the aqueous medium (see also Sect. 2.2.11). One of the most promising lipid-based nanodelivery systems for food applications is the development of nano-sized self-assembled liquids (NSSL) (Garti et al. 2005). NSSL vehicles tackle shortcomings of microemulsion systems. Mixtures of food-grade oils (in which two or more food-grade nonionic hydrophilic emulsifiers), cosolvent (polyol), and coemulsifiers that self-assemble to form mixed reverse micelles ("the concentrate") can be inverted into oil-in-water nano-droplets. This system is transformed into bicontinuous structures by dilution with an aqueous phase, progressively and continuously, without phase separation. These reversed micelles can solubilize compounds that are poorly soluble in water or in the oil phase. NSSLs can be used to solubilize hydrophobic substances several times their normal solubility. The use of solid lipid nanoparticles (SLN) and nano-structured lipid carriers (NLC, which are structured SLN) have been developed in the last two decades for mainly pharmaceutical purposes (Müller et al. 2000; Radtke et al. 2005), but may also find their way into foods. NLCs might be prepared by melting lipid(s), dissolving a lipophilic agent into the molten lipid, followed by hot high-pressure homogenization of the molten lipid phase in the presence of aqueous surfactant solution at 5–10°C above the melting temperature of the lipid. Alternatively, one may use a cold homogenization technique, in which the lipid with the agent is ground into 50–100 μm microparticles and then homogenized in an aqueous surfactant solution at a temperature well below the melting temperature of the lipid (<10–20°C below the melting temperature of the lipid). The cold homogenization procedure might be used to prepare SLN with temperature-sensitive or even hydrophilic agents. One may spray-dry such a suspension of SLN in the presence of a water-soluble carrier material to obtain a dried powder with a particle size between 20 and 100 μm (Shefer and Shefer 2003), with an option to entrap an extra water-soluble active in the water-soluble coating.

Protein-based nanoencapsulates: Yu et al. (2005) immobilized enzymes by incorporation into peptide nanotubes by enzyme engineering. Kamiya et al. (2006) used casein to form nano-sized protein micelles to hold hydrophobic substances. The group used transglutaminase to form ANS-encapsulated casein micelles with a particle size of 36 nm, which retained ≥50% ANS when treated with trypsin (ANS = 1-anilinonapthtalene-8-sulfonic acid, a fluorophore). This method is useful for manufacturing transparent supersaturated solutions by solubilization of hydrophobic substances in functional foods and pharmaceuticals. Semo et al. (2007) used self assembled casein micelles as nanocapsular vehicle. These authors realized that casein micelles are in effect nanocapsules created by nature to deliver nutrients, such as calcium, phosphate, and protein, to the neonate. Thus, they suggested using casein micelles, as a self assembled system for nanoencapsulation and stabilization of hydrophobic nutraceutical substances for enrichment food products. Vitamin D2 was used as a model for hydrophobic nutraceutical compounds. The reassembled micelles had average diameters of 146 and 152 nm without and with vitamin D2 respectively, similar to normal casein micelles, which are typically 150 nm on average. The vitamin concentration in the micelle was about 5.5 times more than in the serum. The use of protein–polysaccharide interaction to form encapsulation systems based on complex coacervation (see Sect. 2.2.7) was downsized to the nanoscale by Huang and Jiang (2004). They used coacervates formed by gelatin type A–carrageenan complexes as inexpensive encapsulation method for the green tea catechin epigallocatechin gallate (EGCG) in the micro- and nanoscale level. Nanoparticles can also be formed by combining molecular inclusion of lipophilic active agent like vitamin D or docosahexaenoic acid (DHA) by β-lactoglobulin (see Sect. 2.2.10) with complex coacervation (see Sect. 2.2.7) by electrostatic interactions between this β-lactoglobulin and polysaccharides like pectin at pH 4.5 (Zimit and Livney 2009, and references therein).

Polysaccharide-based nanoencapsulates: Chen and Wagner (2004) produced a 100 nm vitamin E nanoparticle product based on modified starch that was stable in a beverage and did not alter beverage appearance. Particles were produced by dissolving starch sodium octenyl succinate in distilled water with vitamin E acetate added slowly and homogenized with a high shear mixer until the emulsion droplet size was below 1.5 μm. The crude emulsion was then further homogenized until the emulsion droplets reach the target particle size, and spray-dried to yield a powder containing about 15% vitamin E acetate. Another interesting use of starch is molecular entrapment by amylose based on the interaction between amylose and lipids, which is characterized by amylose chains forming so-called V-crystalline forms (Lebail et al. 2000). In this form, the amylose chain forms a helix with a large cavity in which low molecular weight agents can be situated. The complexes have high melting temperatures and the complexed material is efficiently protected from oxidation. The digestibility of starch is decreased by the complex formation. The concept was examined using conjugated linoleic acid (CLA) as a model (Lalush et al. 2005). Thermal analysis of the complexes showed a transition temperature of the complex ranging from 88 to 95°C, suggesting stability of the complexes during food processing. Atomic force microscopy (AFM) scanning showed that the complexes had a globular structure of heterogeneous nature with an average diameter of 152 ± 39 nm. Stability tests showed that regardless of the complexation method and temperature, the complexes protect and inhibit the oxidation of the ligand. Enzymatic digestion of amylose-ligand complexes by pancreatic amylase showed that amylolytic enzymes could digest the complex, and the ligands were only released when the complexes were digested by amylases. This encapsulation method as outlined in Fig. 2.8 has been further developed to encapsulate a wide range of bioactive agent using continuous processes (Shimoni et al. 2007). By using dual feed jet homogenization, particles ranging from 400 nm to a few tens of microns can be produced.

Inorganic-based nanoencapsulates: Not many encapsulates are based on inorganic materials. One of the few examples might be silica nanoparticles obtained by sol-gel synthesis of silica gels in a w/o microemulsions to encapsulate enzymes (Barbé et al. 2008; Cellesi and Tirelli 2006). However, the process published was not food-grade, although this might be possible in theory. Another example is two-dimensional, layered double hydroxide nanohybrids, composed of vanillic acid,

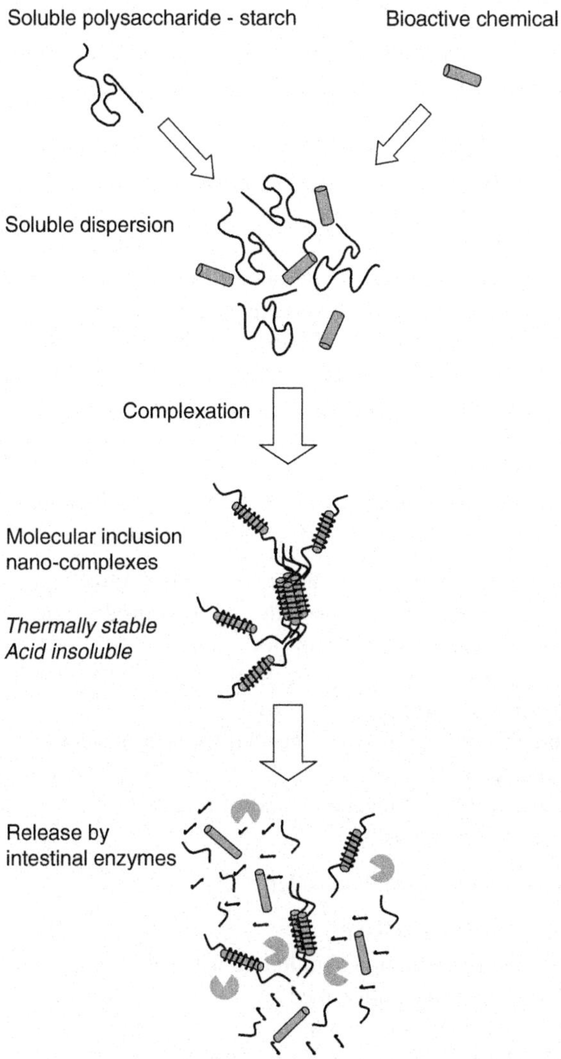

Fig. 2.8 Schematic presentation of the preparation of amylose-ligand complexes. The ligands release when complexes are digested by amylases

zinc, and aluminum oxides (molar ratios of 1.5:2:1–1.5:4:1; Hong et al. 2008). However, the vanillic acid release was a burst release in aqueous solutions.

2.4 Criteria to Select a Proper Encapsulation Technology

Ideally, the choice for a certain encapsulation technology should not be based on trial-and-error. So how does one choose?

The first question to answer is: Which benefit would one like to achieve? The desired properties as mentioned in Sect. 2.1 might be based on consumer insights, nutritional demands, or ideas to improve food processing and storage stability.

The second question should be whether encapsulation is the right technology to bring the desired benefit. Encapsulation should be considered as one of the last options to pursue in view of the negatives mentioned in Sect. 2.1. E.g., delivery of the active by proper design of the food matrix might be an easier way to deliver the functionality. Of course, this book would not exist if we believe that encapsulation is a not good option in many cases. Ubbink and Krüger (2006) introduced a retro-design approach to select from different technologies, by first systematically analyzing the physical, chemical, and biological properties of the active ingredient and the conditions in the food matrix. Based on this, the properties needed to realize the required performance are defined. Only then the technologies (which might be encapsulation) are selected to bring the desired benefit.

When one chooses encapsulation as a technology to deliver the desired benefit, one should consider carefully the design of the encapsulate:

- What are the physicochemical characteristics of the active?
- Which processing conditions are used during food production or processing?
- How will the encapsulates be stored prior to use?
- What will be the storage conditions of the food product containing the encapsulates prior to consumer use?
- Which particle size and density are needed to have it incorporated properly in the food product?
- What are the trigger(s) and mechanism(s) of release?
- What are the cost constraints?

Based on this analysis one should consider:

- Which coating should one select?
- Which process of encapsulation should one use?
- Which loading should one undertake?
- Are there any legal issues to consider?
- What is the freedom of use and IPR status?
- Can it be prepared at sufficient quantities, constant quality, and at the right time to ensure proper supply chain?

The next chapters of this book are aimed to answer these questions for specific active agent to be used for several product applications and food processes.

References

Acosta E (2009) Bioavailability of nanoparticles in nutrient and nutraceutical delivery. Curr Opin Colloid Interface Sci 14:3–15

Appelqvist IAM, Golding M, Vreeker R, Zuidam NJ (2007) Emulsions as delivery systems in foods. In: Lakkis JM (ed) Encapsulation and controlled release technologies in food systems. Blackwell Publishing, Ames, pp 41–81

Bakker MAE, Galema SA, Visser A (1999) Microcapsules of gelatin and carboxy methyl cellulose. Patent EP937496

Barbé CJ, Kong L, Finnie KS, Calleja S, Hanna JV, Drabarek E, Cassidy DT, Blackford MG (2008) Sol-gel matrices for controlled release: from macro to nano using emulsion polymerization. J Sol-Gel Sci Technol 46(3):393–409

Barbosa-Cánovas GV, Ortega-Rivas E, Juliano P, Yan H (2005) Food Powders. Physical properties, processing, and functionality. Kluwer Academic/Plenum Publishers, New York

Bummer PM (2004) Physical chemical considerations of lipid-based oral drug delivery – solid lipid nanoparticles. Crit Rev Ther Drug Carrier Syst 21:1–20

Casana Giner V, Gimeno Sierra M, Gimeno Sierra B, Moser M (2006) Continuous multi-micro-encapsulation process for improving the stability and storage life of biologically active ingredients. Patent EP1702675

Cellesi F, Tirelli N (2006) Sol-gel synthesis at neutral pH in w/o microemulsion: a method for enzyme nanoencapsulation in silica gel nanoparticles. Colloids Surf A Physicochem Eng Asp 288:52–61

Chen C-C, Wagner G (2004) Vitamin E nanoparticle for beverage applications. Chem Eng Res Des 82(A11):1432–1437

Desai KGH, Park HJ (2005) Recent developments in microencapsulation of food ingredients. Drying Technol 23:1361–1394

Dewettinck K, Huyghebaert A (1999) Fluidized bed coating in food technology. Trends Food Sci Tech 10:163–168

De Wolf FA, Brett GM (2000) Ligand-binding proteins: their potential for application in systems for controlled delivery and uptake of ligands. Pharmacol Rev 52(2):207–236

Frost & Sullivan (2005). Opportunities in the Microencapsulated Food Ingredients Market. London. http://www.frost.com/prod/servlet/report-brochure.pag?id=B716-01-00-00-00

Fuchs M, Turchiuli C, Bohin M, Cuvelier ME, Ordonnaud C, Peyrat-Maillard MN, Dumoulin E (2006) Encapsulation of oil in powder using spray drying and fluidized bed agglomeration. J Food Process Eng 75:27–35

Gharsallaoui A, Roudaut G, Chambin O, Voilley A, Saurel R (2007) Applications of spray-drying in microencapsulation of food ingredients: an overview. Food Res Intern 40:1107–1121

Garti N, Spernath A, Aserin A, Lutz R (2005) Nano-sized self-assemblies of nonionic surfactants as solubilization reservoirs and microreactors for food systems. Soft Matter 1:206–218

Gouin S (2004) Microencapsulation: industrial appraisal of existing technologies and trends. Trends Food Sci Tech 15:330–347

Guichard E (2006) Flavour retention and release from protein solutions. Biotechnol Adv 24:226–229

Guignon B, Duquenoy A, Dumoulin ED (2002) Fluid bed encapsulation of particles: principles and practice. Drying Technol 20(2):419–447

Guzey D, McClements DJ (2006) Formation, stability and properties of multilayer emulsions for application in the food industry. Adv Colloid Interface Sci 128–130:227–248

Hedges AR (1998) Industrial applications of cyclodextrins. Chem Rev 98:2035–2044

Hsieh WC, Chang CP, Gao YL (2006) Controlled release properties of chitosan encapsulated volatile citronella oil microcapsules by thermal treatments. Colloids Surf B Biointerfaces 53:209–214

Hong MM, Oh JM, Choy JH (2008) Encapsulation of flavour molecules, 4-hydroxy-3-methoxy benzoic acid, into layered inorganic nanoparticles for controlled release of flavor. J Nanosci Nanotechnol 8:5018–5021

Huang Q, Jiang Y (2004) Enhancing the stability of phenolic antioxidants by nanoencapsulation. Abstracts of Papers, 228th ACS National Meeting, Philadelphia, PA, United States, August 22–26

Jafari SM, Assadpoor E, He Y, Bhandari B (2008) Encapsulation efficiency of food flavours and oils during drying. Drying Technol 26:816–835

Kamiya T, Goto M, Shioashi Y, Noriki N (2006) Method for producing protein micelle structure having nano size to which hydrophobic substance is adsorbed and retained. Patent JP2006115751 A

Kjaergaard OG (2001) Multiple-core encapsulation: Prilling. In: Vilstrup P (ed) Microencapsulation of Food Ingredients. Leatherhead Publishing, Surrey, pp 197–214

Krasaekoopt W, Bhandari B, Deeth H (2003) Evaluation of encapsulation techniques of probiotics for yoghurt. Int Dairy J 13:3–13

Kosaraju SL, Tran C, Lawrence A (2006) Liposomal delivery systems for encapsulation of ferrous sulfate: preparation and characterization. J Liposome Res 16:347–358

Lalush I, Bar H, Zakaria I, Eichler S, Shimoni E (2005) Utilization of amylose-lipid complexes as molecular nanocapsules for conjugated linoleic acid. Biomacromolecules 6:121–130

Lebail P, Buleon A, Shiftan D, Marchessault RH (2000) Mobility of lipid in complexes of amylose-fatty acids by deuterium and ^{13}C solid state NMR. Carbohydr Polym 43(4):317–326

Lemetter CYG, Meeuse FM, Zuidam NJ (2009) Control of the morphology and size of complex coacervate microcapsules during scale up. AIChE Journal 55(6):1487–1496

Litster JD (2003) Scaleup of wet granulation processes: science not art. Powder Technol 130:34–40

Link DR, Anna SL, Weitz DA, Stone HA (2004) Geometrically mediated breakup of drops in microfluidic devices. Phys Rev Lett 92(5):054403-1–054403-4

Mabille C, Leal-Calderon L, Bibette J, Schmitt V (2003) Monodisperse fragmentation in emulsions: Mechanisms and kinetics. Europhys Lett 61(5):708–714

Martin Del Valle EM, Galan MA (2005) Supercritical fluid technique for particle engineering: drug delivery applications. Rev Chem Eng 21(1):33–69

McClements DJ (2005) Food Emulsions. Principles, practices and techniques. CRC Press, Boca Raton

Mellema M, Van Benthum WAJ, Boer B, Von Harras J, Visser A (2006) Wax encapsulation of water-soluble compounds for application in foods. J Microencapsul 23(7):729–740

Mozafari MR, Flanagan J, Matia-Merino L, Awati A, Omri A, Suntres ZE, Singh H (2006) Recent trends in the lipid-based nanoencapsulation of antioxidants and their role in foods. J Sci Food Agric 86(13):2038–2045

Müller RH, Mäder K, Gohla S (2000) Solid lipid nanoparticles (SLN) for controlled drug delivery – a review of state of the art. Eur J Pharm Biopharm 50:161–177

Ortega-Rivas E (2005) Handling and processing of food powders and particulates. In: Onwulata C (ed) Encapsulated and powdered foods. CRC Press, Boca Raton, USA, pp 75–144

Porzio MA, Madsen MG (1997) Double encapsulation process and flavorant compositions prepared thereby. Patent WO1997013416

Porzio M (2004) Flavor encapsulation: a convergence of science and art. Food Technology 58(7):40–47

Prüsse U, Bilancetti L, Bucko M, Bugarski B, Bukowski J, Gemeiner P, Lewinska D, Manojlović V, Massart B, Nastruzzi C, Nedović V, Poncelet D, Siebenhaar S, Tobler L, Tosi A, Vikartovska A, Vorlop K-D (2008) Comparison of different technologies for the production of alginate microspheres. Chem Pap 62(4):364–374

Radtke M, Souto EB, Müller RH (2005) Nanostructured lipid carriers: a novel generation of solid lipid drug carriers. Pharmaceut Tech Eur 17(4):45–50

Regiert M (2008) Molecular encapsulation in cyclodextrins. Speciality Chemicals Magazine, March, pp 22–24

Reineccius GA (2001) Multiple-core encapsulation: The spray drying of food ingredients. In: Vilstrup P (ed) Microencapsulation of Food Ingredients. Leatherhead Publishing, Surrey, pp 151–185

Reineccius GA (2004) The spray drying of food flavors. Drying Technol 22(6):1289–1324

Semo E, Kesselman E, Danino D, Livney YD (2007) Casein micelle as a natural nano-capsular vehicle for nutraceuticals. Food Hydrocolloids 21:936–942

Shefer A, Shefer SD (2003) Multi component controlled release system for oral care, food products, nutracetical, and beverages. Patent US20030152629

Shimoni E, Lesmes U, Ungar Y (2007) Non-covalent complexes of bioactive agents with starch for oral delivery. Patents WO2007122624 and US 2006794110

Sparks RE, Mason NS (1987) Method for coating particles or liquid droplets. Patent US4675140
Szente L, Szejtli J (2004) Cyclodextrins as food ingredients. Trends Food Sci Technol 15:137–142
Tabor BE, Owers R, Janus JW (1992) The crosslinking of gelatin by a range of hardening agents. J Photographic Sci 40(5–6):205–211
Takahashi M, Inafuku KI, Miyagi T, Oku H, Wada K, Imura T, Kitamoto D (2007) Efficient preparation of liposomes encapsulating food materials using lecithins by a mechanochemical method. J Oleo Sci 56(1):35–42
Taylor TM, Davidson PM, Bruce BD, Weiss J (2005) Liposomal nanocapsules in food science and agriculture. Crit Rev Food Sci Nutr 45:587–605
Teunou E, Poncelet D (2002) Batch and continuous fluid bed coating – review and state of the art. J Food Eng 53:325–340
Teunou E, Poncelet D (2005a) Fluid-bed coating. In: Onwulata C (ed) Encapsulated and powdered foods. CRC Press, Boca Raton, USA, pp 197–212
Teunou E, Poncelet D (2005b) Dry coating. In: Onwulata C (ed) Encapsulated and powdered foods. CRC Press, Boca Raton, USA, pp 179–195
Thies C, Ribeiro Dos Santas I, Richard J, Vandevelde V, Rolland H, Benoit J-P (2003) A supercritical fluid-based coating technology 1: Process considerations. J Microencapsul 20(1):87–96
Thies C (2007) Microencapsulation of flavors by complex coacervation. In: Lakkis JM (ed) Encapsulation and controlled release technologies in food systems. Blackwell Publishing, Ames, pp 149–170
Torchillin VP, Weissig V (2003) Liposomes, 2nd edn. Oxford University Press, A practical approach. Oxford
Tromelin A, Andriot I, Guichard E (2006) Protein-flavour interactions. In: Voilley A, Etiévant P (eds) Flavour in Food. CRC Press, Boca Raton, USA
Ubbink J, Krüger J (2006) Physical approaches for the delivery of active ingredients in foods. Trends Food Sci Technol 17:244–254
Uhlemann J, Schleifenbaum B, Bertram HJ (2002) Flavor encapsulation Technologies: an overview including recent developments. Perfumer & Flavorist 27:52–61
Weinbreck FCJ, De Kruiff CG, Schrooyen P (2003) Complex coacervates containing whey proteins. Patents WO03106014 and EP1371410
Were LM, Bruce BD, Davidson PM, Weiss J (2003) Size, stability, and entrapment efficiency of phospholipid nanocapsules containing polypeptide antimicrobials. J Agric Food Chem 51(27):8073–8079
Yilmaz G, Jongboom ROJ, Feil H, Hennink WE (2001) Encapsulation of sunflower oil in starch matrices via extrusion: effect of the interfacial properties and processing conditions on the formation of dispersed phase morphologies. Carbohydr Polym 45:403–410
Yilmaz G, Jongboom ROJ, Van Dijk C (2005) Thermoplastic starch as a biodegradable matrix for encapsulation and controlled release. In: Mallapragada SK, Narasimhan B (eds) Handbook of Biodegradable Polymeric Materials and Their Applications: volume 2 Applications. American Scientific Publishers, Stevenson Ranch, pp 58–76
Yilmaz G, Jongboom ROJ, Van Soest JJG, Feil H (1999) Effect of glycerol on the morphology of starch-sunflower oil composites. Carbohydr Polym 38:33–39
Yilmaz G (2003) Thermoplastic starch matrices for encapsulation and controlled release of volatile compounds. Ph.D. thesis, Utrecht University, The Netherlands
Yu L, Banerjee IA, Gao X, Nuraje N, Matsui H (2005) Fabrication and application of enzyme-incorporated peptide nanotubes. Bioconjug Chem 16(6):1484–1487
Zasypkin D, Porzio M (2004) Glass encapsulation of aromas with chemically modified starch blends. J Microencapsul 21(4):385–397
Zhang J, Li X, Zhang D, Xiu Z (2007) Theoretical and experimental investigations on the size of alginate microspheres prepared by dropping and spraying. J Microencapsul 24(4):303–322
Zimit P, Livney YD (2009) Beta-lactoglobulin and its nanocomplexes with pectin as vehicles for w-3 polyunsaturated fatty acids. Food Hydrocolloids 23:1120–1126
Zuidam NJ, Van Winden E, De Vrueh R, Crommelin DJA (2003) Stability, storage and sterilization of liposomes. In: Torchilin VP, Weissig V (eds) liposomes. Oxford University Press, Oxford, pp 149–165

Chapter 3
Materials for Encapsulation

Christine Wandrey, Artur Bartkowiak, and Stephen E. Harding

Abbreviations

Chemical Names

CMC	Carboxymethyl cellulose
EMC	Ethyl methylcellulose
GA	Gum arabic
GG	Gellan gum
GK	Gum karaya
GT	Gum tragacanth
HEC	Hydroxyethyl cellulose
HPC	Hydroxypropyl cellulose
HPMC	Hydroxypropyl methylcellulose
LBG	Locust bean gum
MC	Methylcellulose
MG	Mesquite gum
PVP	Polyvinylpyrrolidone
SSPS	Soluble soybean polysaccharide

Organizations/Services

CAS	Chemical Abstract Service

C. Wandrey (✉)
Ecole Polytechnique Fédérale de Lausanne, Laboratoire de Médecine Régénérative et de Pharmacobiologie, Station 15, CH-1015, Lausanne, Switzerland
e-mail: christine.wandrey@epfl.ch

A. Bartkowiak
Chair of Food Packaging and Biopolymers, The Westpomeranian University of Technology, Kazimierza Krolewicza 4, 71-550 Szczecin, Poland
e-mail: artur.bartkowiak@tz.ar.szczecin.pl

S.E. Harding
National Centre for Macromolecular Hydrodynamics, University of Nottingham, School of Biosciences, Sutton Bonington, UK
e-mail: steve.harding@nottingham.ac.uk

N.J. Zuidam and V.A. Nedović (eds.), *Encapsulation Technologies for Active Food Ingredients and Food Processing*,
DOI 10.1007/978-1-4419-1008-0_3, © Springer Science+Business Media, LLC 2010

EFSA	European Food Safety Authority
FAO	Food and Agriculture Organization
FDA	Food and Drug Administration (US)
JECFA	Joint Expert Committee Food and Agriculture
SCF	Scientific Committee on Food
WHO	World Health Organization

Characteristics

ADI	Acceptable daily intake
DA	Degree of acetylation
DE	Degree of esterification
DE	Dextrose equivalent
DP	Degree of polymerization
DR	Degree of reaction
DS	Degree of substitution
GRAS	Generally recognized as safe
GSFA	General standards for food additives
HM	High methoxylated
LM	Low methoxylated
MM	Molar mass
MMD	Molar mass distribution
WHO	World Health Organization

3.1 Introduction

A multitude of substances are known which can be used to entrap, coat, or encapsulate solids, liquids, or gases of different types, origins, and properties. However, only a limited number thereof have been certified for food applications as "generally recognized as safe" (GRAS) materials. It is worth mentioning that the regulations for food additives are much stricter than for pharmaceuticals or cosmetics. Consequently, some compounds, which are widely accepted for drug encapsulation, have not been approved for use in the food industry. Moreover, different regulations can exist for different continents, economies, or countries, a problem which has to be addressed by food producers who wish to export their products or who intend expanding their markets.

The aim of this chapter is to present and update the existing knowledge about materials either already in use or of potential use in the encapsulation of various ingredients or additives used by the food industry. This chapter summarizes the materials, their chemical and physical properties, as well as the principal methods available to measure and analyze these properties. It is further intended to outline some of the basic chemistry which facilitates the selection of the most appropriate materials for the encapsulation of a specific compound by a suitable technology. To a certain degree, some studies performed on specific encapsulation projects will be reviewed. The chapter will not deal with analyses of the economics involved and will not describe specific technologies. It will, however, refer to the legislation on food additives.

There are already a number of excellent books and abundant publications dealing comprehensively with biopolymers, polysaccharides, hydrocolloids, or gums, and their use in the food industry. Nevertheless, the correlation of data on materials with specific requirements for microencapsulation is limited. The requirement to improve this correlation is patently clear from the conclusions of three seminal reviews published in 1991, 1999, and 2005 dedicated to microencapsulation in the food industry (Jackson and Lee 1991; Gibbs et al. 1999; Desai and Park 2005a).

"Although microencapsulation has found applications in the food industry the technology remains far from being fully exploited… Exciting new techniques such as cocrystallization and liposome formation will improve the number and quality of encapsulated ingredients. Although cyclodextrins are discovered many years ago, their potential for flavor encapsulation was realized only recently. Cyclodextrins provide exceptional stability to oxidation and evaporative losses. Due to cost reduction and imminent FDA approval, cyclodextrins will likely find widespread use in foods." (Jackson and Lee 1991).

"Numerous developments have been made in the field of encapsulated food ingredients…. Limitations in many of the encapsulation techniques have occurred due to high cost of production and the lack of food-grade available materials. Research is necessary to eliminate these limitations. Encapsulation currently is an art that is difficult for the food scientist to master. The food scientist does not have the information available in databases to enable him to make informed choices concerning the most appropriate material and encapsulation process. For example, the appropriate blends of starches and maltodextrins as encapsulation material could prove highly beneficial. The development of cyclodextrins has led to new products with longer shelf-life, reduced volatility, and protection of heat-labile substances. Preliminary indications are that liposomes have many benefits for the food industry including protection of materials until desired release or delivery. There is a great deal of research that needs to be done concerning the use of liposomes in the food industry. Unlike the pharmaceutical industry, which can tolerate high costs, manufacturing costs will have to be reduced for food applications." (Gibbs et al. 1999).

"The use of microencapsulated food ingredients for controlled-release applications is a promising alternative to solve the major problem of food ingredients faced by food industries. The challenges are to select the appropriate microencapsulation technique and encapsulation material. Despite the wide range of encapsulated products that have been developed, manufactured, and successfully marketed in the pharmaceutical and cosmetic industries, microencapsulation has found a comparatively much smaller market in the food industry. The technology is still far from fully developed and has yet to become a conventional tool in the food technologist's repertoire for several reasons….development time is rather long…requires multidisciplinary cooperation…low margins…relative inertia of well-established corporations…understanding of the industrial constrains…" (Desai and Park 2005a).

The majority of materials used for microencapsulation in the food sector are biomolecules. In addition to carbohydrate polymers/polysaccharides, which are the most abundant of the four major classes of biomolecules, proteins and lipids are also biomolecules suitable for microencapsulation in the food sector. Dealing with the nomenclature of carbohydrates would by far exceed the purpose of this

Table 3.1 Materials suited for microencapsulation in the food industry

Origin	Carbohydrate polymer	Protein	Lipid
Plant	Starch – Derivatives Cellulose – Derivatives Plant exudates – Gum arabic – Gum karaya – Mesquite gum Plant extracts – Galactomannans – Soluble soybean Polysaccharide	Gluten (corn) Isolates (pea, soy)	Fatty acids/alcohols Glycerides Waxes Phospholipids
Marine	Carrageenan Alginate		
Microbial/animal	Xanthan Gellan Dextran Chitosan	Caseins Whey proteins Gelatin	Fatty acids/alcohols Glycerides Waxes Phospholipids (Shellac)

chapter. The chapter, therefore, refers the reader to the appropriate IUPAC publications (IUPAC-IUB 1980). Table 3.1 lists groups of biomolecules, arranged according to their origin, which are found to be most suitable either when used alone or when used in combination with others for microencapsulation in the food industry. These will be the major topic of this chapter. In addition, a few other compounds such as poly(vinylpyrrolidone) and inorganic molecules will be considered.

3.2 Materials

3.2.1 Carbohydrate Polymers

Carbohydrates comprise more than 90% of the dry mass of all biomass and more than 90% thereof are carbohydrate polymers – polysaccharides. These natural homo- and copolymers are composed of sugar residues and/or their derivatives. Many native polysaccharides contain a small percentage of peptide residues remaining from their biosynthesis. However, these are normally removed during processing. Native polysaccharides are of enormous varieties. Moreover, they also form valuable sources for chemically modified materials, thus expanding their applicability and usefulness. Overall, polysaccharides are of tremendous economical importance. Their world consumption – with one of the main sectors being food – exceeds by far, for example, the production of all synthetic polymers.

Different protocols for the classification/designation of carbohydrate polymers can be found in the literature. Sometimes, however, these protocols are not consistent and a combination of the classification related to the chemical structure and those related to the physical behavior exist. In particular, the designation as a "hydrocolloid" or a "gum" is not uniformly applied. Some authors use the terms as synonyms. Others understand gums as a subgroup of hydrocolloids. The names carbohydrate polymer and polysaccharide refer to the chemical structure; the designation as gum or hydrocolloid refers to the property that these polysaccharides hydrate in hot or cold water to form viscous solutions or dispersions at low concentration.

Overall, there are gums/hydrocolloids obtained by the chemical modification of native polysaccharides and others "harvested" from nature. The first group

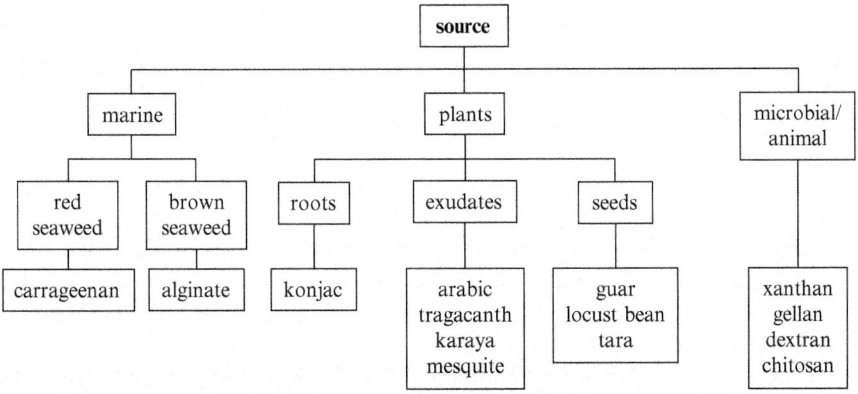

Fig. 3.1 Classification of natural carbohydrate polymers related to the source

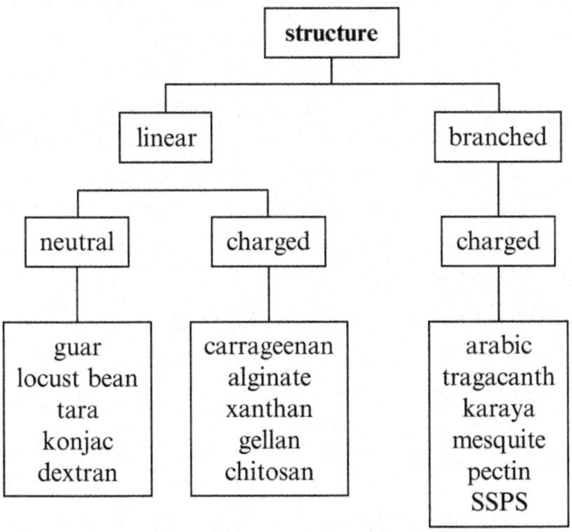

Fig. 3.2 Classification of carbohydrate polymers related to structural characteristics (Substances with only short side chains are considered here as principally linear. *SSPS* soluble soybean polysaccharide)

comprises derivatives of starch and cellulose. The second group includes a number of complex macromolecular structures such as plant exudates. Figure 3.1 presents a more detailed classification of natural food-grade materials with specification of their origin. Figure 3.2 provides the structural information. Carbohydrate polymers of both groups are subject of the subsequent sections.

3.2.1.1 Starch and Starch Derivatives

Chemical Description of Starch. (CAS# 9005-25-8) The polysaccharide starch is a polymer of α-D-glucose with the general chemical constitution $(C_6H_{10}O_5)_n$. Starch consists of two constitutionally identical but architecturally different polysaccharide molecules. These are essentially the linear amylose and the highly branched amylopectin. Normally, the amylose content is in the range of 20–30%, whereas the proportion of amylopectin is approximately 70–80%. The structure and composition varies with the source of the starch.

Amylose. (CAS# 2595-98-4) Figure 3.3 shows the principal chemical structure of amylose, poly[α-(1→4) anhydro-D-glucopyranose], consisting of almost exclusively linear molecules with α-(1→4)-linked D-glucose units, typically in the range 500–6,000 (Murphy 2000), forming a helix. The molar mass with a relatively narrow molar mass distribution depends on the botanical origin and the age of the plants, and also on the isolation procedure and technology. The measured molar mass can also depend on the sample preparation and the solubilization procedure used for the molar mass characterization. Some references even report molar masses up to one million (Parker and Ring 2001).

Amylopectin. (CAS# 9037-22-3) The principal structure of amylopectin molecules is shown in Fig. 3.4, side chains of about 30 D-glucose units are bonded with α-(1→6) linkages non-randomly at approximately every 20 to 30 glucose units along the chain. One amylopectin molecule may contain up to two million D-glucose units which is equivalent to a molar mass of approximately 4×10^8 g/mol.

Physico-chemical Characteristics of Starch. Starch is normally a white powder. Depending on its source, it may be odorless and tasteless. It is insoluble in cold water, ethanol, and most common solvents. It is naturally found radially packed in small grains of different shape (spherical or lentil-shaped) for which different ranges of size have been reported: 5–900 μm (Elias 1992), 1–100 μm (Murphy 2000). The size distribution determines the swelling behavior. The granules have been

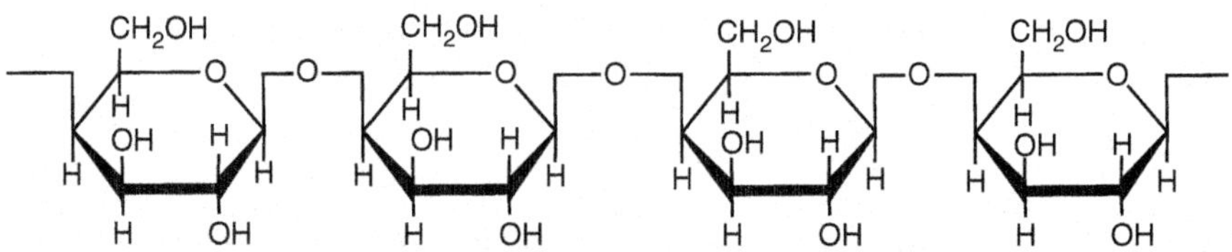

Fig. 3.3 Principal chemical structure of amylose

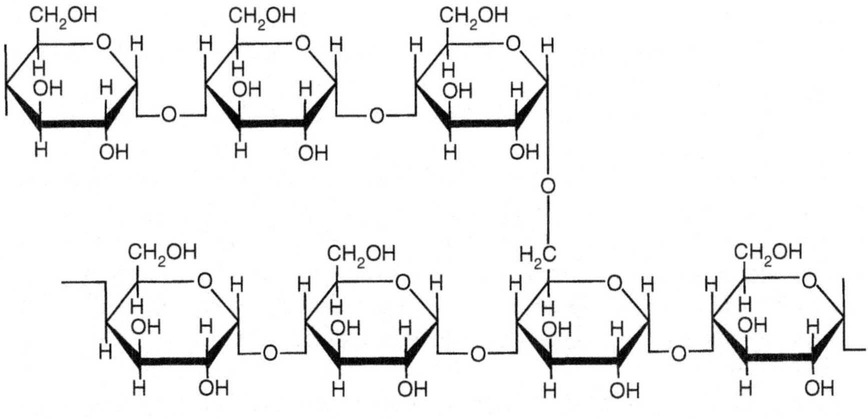

Fig. 3.4 Principal chemical structure of amylopectin

generally described as either larger and lenticular (lens-like, A-starch) or smaller and spherical (B-starch) (Ao and Jane 2007), with less swelling power. As the granules absorb water, they swell, lose crystallinity, and leach amylose. The higher the amylose content, the lower is the swelling power and the lower is the gel strength for the same starch concentration. However, a lower swelling power due to the high amylose content can be counteracted by a larger size of granule (Li and Yeh 2001; Singh et al. 2003; Chaplin 2007).

Both the amylose and amylopectin are complex macromolecules, with the capability of interacting intra- and intermolecularly to form ordered structures (Parker and Ring 2001). Amylose generally tends to wind up into a rather stiff left-handed single helix or forms even stiffer parallel left-handed double helical junction zones (Imberty et al. 1988). The helical structure is responsible for the specific solubility behavior. In starch, the amylose molecules are imbedded in a physical network of amylopectin molecules. This network prevents the amylose from forming a complete helical structure of the amylose.

Amylose has a very specific dissolution behavior. With hot water, amylose molecules dissolve as statistical coils of non-ordered segments and short helical parts whereas the amylopectin remains undissolved. In dilute solution, these short helical parts of the amylose grow relatively fast to longer helical parts, which organize laterally to form a double-stranded helix. Amylose slowly crystallizes from aqueous solution and becomes insoluble (retrogradation). During this process some of the water associated with the molecules is released. As a consequence, dried amyloses are no longer water soluble. In concentrated solutions of amylose, inter- and intramolecular association occurs with partial crystallization yielding a physical network – gelling of amylose (Elias 1992). The rate of retrogradation depends on the molar mass, concentration, temperature, and pH. The lower the temperature, concentration, and molar mass, the faster is the retrogradation. It is described as the fastest in the pH range 5–7, however, it can be retarded by salts (Rutenberg 1980).

The extended conformation of amylose causes the high viscosity of water-soluble starch and varies relatively little with temperature. Single helix amylose behaves similarly to the cyclodextrins. It has a relatively hydrophobic inner surface that holds water molecules, and these can be easily replaced with hydrophobic lipid or aroma molecules (Chaplin 2007).

The amylopectin molecules are oriented radially in the starch granule. As the radius increases so does the number of branches required to fill up the space, with the consequent formation of concentric regions of alternating amorphous and crystalline structure (Chaplin 2007). Some amylopectins (for example, from potato) have phosphate groups attached to some hydroxyl groups, which increase its hydrophilicity and swelling power. Intense stirring in cold water yields colloidal dispersions. Due to branching, no retrogradation and crystallization is observed. Amorphous powders are obtained by drying, which may be redispersed easily in water and do not really form gels.

Origin and Isolation of Starch. Starch is stored most abundantly in tubers, roots, seeds, and fruits, but can also be stored in leaves, stems, and pollen grains. It is usually possible to identify the plant source by the microscopic examination of grains (Buléon et al. 1998). Each grain typically contains several million amylopectin molecules and a much larger number of amylose molecules.

Starch is the second most abundant polysaccharide after cellulose. The sources and commercial isolation are shown in Fig. 3.5. Starch is supplied as comparably low cost products of controlled quality and ready availability. Pure amylopectin can be isolated from "waxy" maize starch whereas amylose is best isolated subsequent to specific hydrolysis of the amylopectin with pullulanase (Vorwerg et al. 2002). Genetic modification of starch crops has recently led to starches with both improved and targeted functionality (Jobling 2004; Chaplin 2007).

Starch Derivatives (Wurzburg 1995, 2006). Chemical, biochemical, and physical modifications of starch are known. Many functional derivatives of starch are marketed including cross-linked, oxidized, acetylated, hydroxypropylated, and partially hydrolyzed molecules. The aim of starch modification is to alter the structure and affect the hydrogen bonding in a controllable manner in order to enhance and extend the industrial applicability.

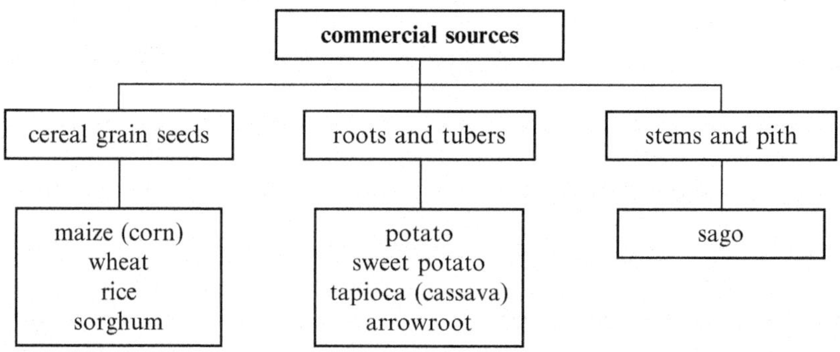

Fig. 3.5 Commercial sources of starch

Cross-linking replaces a limited number of hydrogen bonds between starch chains by permanent covalent bonds. As a result, swelling is inhibited and the starch offers higher heat and shear stability. In combination with the stabilization by introducing bulky groups, the shelf life can be improved. The starches modified in this way exhibit better tolerance to temperature fluctuations. Retrogradation can be prevented.

Hydrolysis (Blanchard and Katz 1995, 2006) is performed using 4 acids (chemical modification) or enzymes (selective biochemical modification) as catalysts. The resulting derivatives are assigned a Dextrose Equivalent (DE) value, which represents the degree of hydrolysis. Figure 3.6 illustrates a classification based on the DE.

Maltodextrins. (CAS# 9050-36-6) Maltodextrins are hydrolysates with a DE<20. They are creamy white hygroscopic polysaccharide powders, which are almost tasteless or only moderately sweet and easily digestible. In the US usually corn or potato starch is used for the modification whereas in Europe it is commonly wheat. The source of maltodextrins is important for coeliacs since the wheat-derived maltodextrin can contain traces of gluten. Overall, maltodextrin is an acid- or enzyme-catalyzed starch hydrolysate with M_w<4,000 g/mol.

Syrups. (CAS# 8029-43-4) Starch hydrolysates with a DE>20 are further subdivided into four corn syrup types (Fig. 3.4). These differ by their sweetness, which increases from type 1 to type 4 (Critical Data Tables 1975).

Dextrins. (CAS# 9004-53-9) The term dextrin, in its broadest sense, may refer to any product obtained by any method (e.g., heat, acid, enzyme) for degrading the starch. The physical properties of dextrins can cover a very wide range. A synonym used for dextrins is starch gums. The tensile strength of dextrin film is lower than that for starch and decreases with the degree of conversion. However, dextrin formulations can be prepared at higher concentrations than unmodified starch yielding films with higher proportions of solids, which dry faster and are thicker.

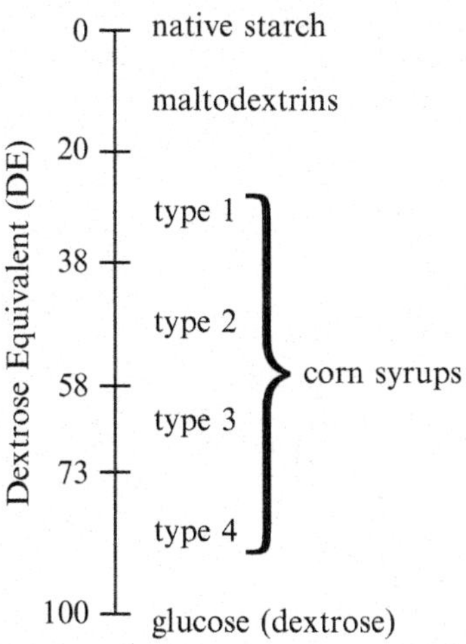

Fig. 3.6 Classification of starch hydrolysates based on the dextrose equivalent (DE) value

As an example, yellow corn dextrin is used to encapsulate water-insoluble flavors and oils by spray-drying (Wurzburg 2006).

Cyclodextrin. (CAS# 12619-70-4) Cyclodextrins are a family of cyclic oligosaccharides. Degradation of amylose, amylopectin, or glycogen by diluted acids or by enzymes of, e.g., *Bacillus macerans* yields these cyclic dextrins, which are non-branched oligomeric cycloamyloses composed of α-(1→4)D-glucopyranoside units. The ring structure results from the helical segments of the polysaccharide. Typical cyclodextrins contain six to eight glucose monomer units. The rings are cone-shaped with a cavity depth of 0.7–0.8 nm. However, much larger rings have also been derived. Figure 3.7a–c shows the chemical structures for typical cyclodextrins. They are denoted as follows (Szejtli 1998):

- α-cyclodextrin (CAS# 10016-20-3): six-sugar ring molecule, inner diameter 0.5 nm
- β-cyclodextrin (CAS# 7585-39-9) seven-sugar ring molecule, inner diameter 0.6 nm
- γ-cyclodextrin (CAS# 17465-86-0) eight-sugar ring molecule; inner diameter 0.8 nm

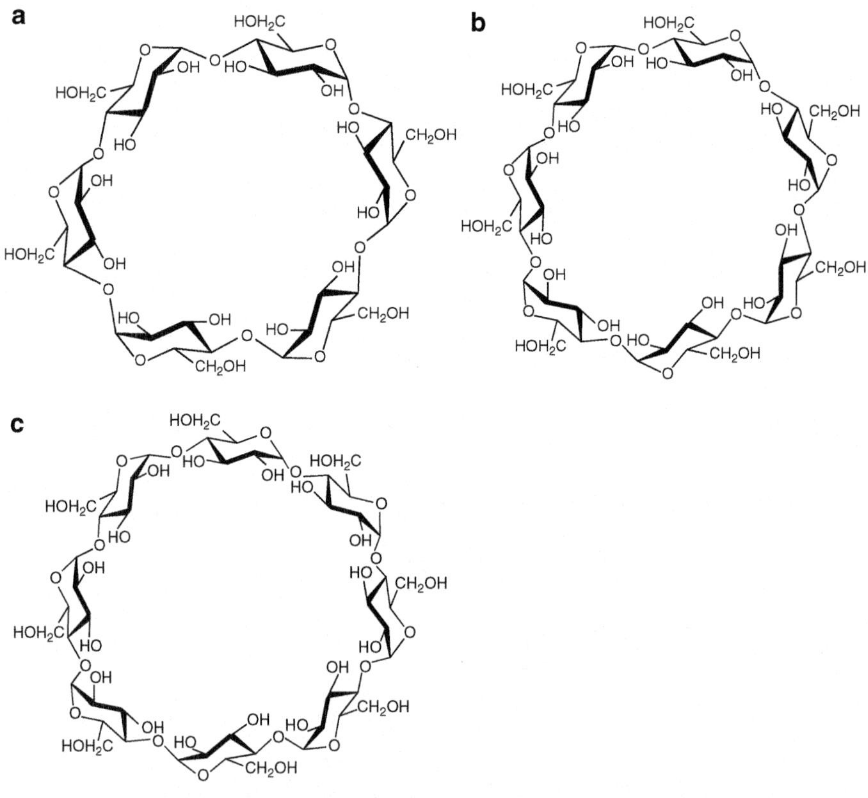

Fig. 3.7 Chemical structure of cyclodextrins. (**a**) α-cyclodextrin, (**b**) β-cyclodextrin, (**c**) γ-cyclodextrin

Cyclodextrins form inclusion complexes (Reineccius et al. 2005; see also Sect. 2.3.10). Their solubility in water at 25°C differs considerably, with γ-cyclodextrin being the best soluble (23.2 g/100 ml) followed by α-cyclodextrin (14.5 g/100 ml) and β-cyclodextrin (1.85 g/100 ml) (Blanchard and Katz, 2006).

Cyclodextrins are not adsorbed in the upper gastrointestinal tract. They are completely metabolized by the colon microflora (Szente and Szejtli 2004).

Polydextrose. It is also designated as (poly-D-glucose). It is a synthetic, highly branched polymer with many types of glycosidic linkages created by heating dextrose with an acid catalyst (acid-catalyzed condensation) and purifying the resulting water-soluble polymer. For example, melt-condensation of 89% D-glycose (dextrose), 10% sorbitol (D-glucitol), and 1% citric acid results in statistically branched polydextrose with primarily 1→6 linkage and molar masses up to 22,000 g/mol, which is water-soluble (Elias 1992). Polydextrose is preferably used as a bulking agent because it is tasteless and is similar to fiber in terms of its resistance to digestion.

Modified starch and maltodextrin have been used for coating and encapsulation by several technologies including spray-drying, fluidized bed spray-drying, fluidized bed granulation, compacting, and most importantly extrusion. Molecular inclusion is possible with cyclodextrins. Maltodextrins are cheaper than gum arabic and are, therefore, in some cases recommended as a partial substitute in encapsulation (Shahidi and Regg 1991).

3.2.1.2 Cellulose and Cellulose Derivatives

Chemical Description of Cellulose. (CAS# 9004-39-1) Cellulose (Coffey et al. 1995, 2006) is a polymer of β-D-glucose. The chain units are linked by β-(1→4)-glycosidic bonds. In contrast to starch, where all -CH_2OH groups are oriented along the same side of the molecular plane, the –CH_2OH groups in cellulose are oriented alternating above and below the plane thus producing long linear chains as is demonstrated in Fig. 3.8. Due to the absence of side chains, cellulose molecules can arrange close together and form rigid structures.

Only a few cellulose molecules are pure poly(β-(1→4)-glucopyranoses). The majority contains a low percentage of several other glucose units. As an example, cotton contains approximately 1.5% xylose as well as smaller amounts of mannose, galactose, and arabinose. Further, approximately one –COOH group per 500–1,000 glucose units can be found in native cellulose. These hydroxyl groups can react with acids to form esters and with alcohols to form ethers. Dependent on the origin, the degree of polymerization is in the range of 1,000–8,000 (Murray 2000), with a narrow molar mass distribution. Micro-crystalline cellulose is composed of 100–300 glucose units.

Physico-chemical Characteristics of Cellulose. Cellulose is insoluble in water and other ordinary solvents. It is a stiff polymer with an extended rod-like conformation. In microfibrils, the multiple hydroxyl groups on the glucose residues bond with each other, holding the chains firmly together. In contrast to starch, cellulose is

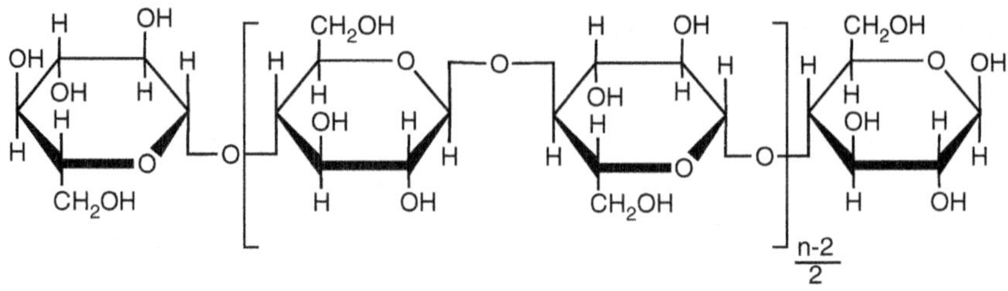

Fig. 3.8 Principal chemical structure of cellulose

much more crystalline. Whereas starch undergoes a crystalline to amorphous transition at 60–70°C in water, cellulose requires 320°C and 25 MPa to become amorphous in water (Chaplin 2007).

Origin and Isolation of Cellulose. Cellulose is the major structural material of plants. It is considered as the most abundant organic substance existing on the Earth. The primary cell wall of green plants is made of cellulose. Wood is largely cellulose, and cotton is almost (91%) pure cellulose. Other important natural sources are flax, hemp, jute, and straw. Acetic acid bacteria are also known to synthesize cellulose, as well as many forms of algae.

Derivatives of Cellulose. The hydroxyl groups of cellulose can partially or completely react with various reagents to produce derivatives with properties desired and/or required for food applications. Raw material used for the modification is cellulose pulp obtained from wood pulp or from cotton linters, the short fibers of cotton. The intended properties of the derivatives govern the selection of the raw material type. Theoretically, all three hydroxyl groups of the anhydroglucose unit can be subject to substitution. Cellulose ethers are the most important cellulose derivatives for food applications. Due to the limited solubility of cellulose, the majority of chemical modifications are performed in the heterogeneous media. Products with different degrees of substitution (DS) are obtained. The DS describes the average number of substituted hydroxy groups per sugar unit. Consequently, the maximum DS is three. GRAS cellulose ethers are summarized in Table 3.2. Selected structures are presented in Fig. 3.9.

Overall, the ultimate properties of modified cellulose depend on the substituent type, its frequency, and distribution along the polymer backbone, the molar mass, and the molar mass distribution. Practically, also the bulk properties, granular or powdered, can have an impact, for example, on the solution properties. In general, aqueous solutions of modified celluloses are odorless, colorless, and clear. Due to their capability of absorbing water appropriately to the relative humidity, storage without contact to humidity is strongly recommended.

Methylcellulose (MC). MC (Fig. 3.9a) is a hydrophilic white powder, which dissolves in cold water forming a clear viscous solution. Both the higher DS and the degree

Table 3.2 Food grade (GRAS) cellulose derivatives

Modification	CAS	E	DS[a] or DR[b]
Non	9004-39-1	460	
Methyl	9004-67-5	461	DS 1.3–2.6
Ethyl	9004-57-3	462	DS 2.1–2.6
Hydroxypropyl	9004-64-2	463	DR 4
Hydroxypropyl methyl	9004-65-3	464	DS (methyl) 0.9–1.8 DR (HP) 0.1–1.0
Ethyl methyl	9004-69-7	465	
Carboxy methyl (Na)	9004-32-4	466	DS 0.4–1.4
Ethyl hydroxyethyl	9004-58-4	467	
Crosslinked CMC	74811-65-7	468	

[a]*DS* Degree of substitution – average number of substituted hydroxyl groups per saccharide unit; different information from different sources, widest range taken

[b]*DR* Degree of reaction – average number of reagent molecules reacted with one saccharide unit

of polymerization (DP) result in lower solubility. The viscosity of MC solutions is reasonably stable over a wide pH range, pH 3–11. Three-dimensional gel formation occurs on heating above 50°C. MC of higher DS has a lower gelation temperature. The gels are reversible on cooling; however, a pronounced hysteresis is typical between heating and cooling. MC has good film forming properties and to a certain extent, surface activity. It is not digestible. The synthesis steps include heating of raw cellulose with NaOH and treatment with methyl chloride.

Hydroxypropyl methyl cellulose (HPMC). HPMC (Fig. 3.9b) is very similar to MC. It is supplied as white to off-white powder or granules, which swell and dissolve in water to form a viscous, non-ionic colloidal solution. It is also soluble in most polar solvents. Aqueous solutions are surface active and form films upon drying. They undergo reversible transformation from sol to gel upon heating and cooling, designated as reversible thermal gelation. However, the gel transition temperature depends on the ratio of methyl to hydroxypropyl derivatization. It can shift from 50 to 90°C. Moreover, the gel texture also changes with increasing hydroxypropyl substitution. In addition, the total DS influences the solubility. Typically, HPMC has a molar mass higher than 10,000 g/mol, $T_m = 220°C$, and a density of 1.6 g/ml (Greminger and Krumel 1980).

The solutions of both MC and HPMC exhibit pseudoplastic non-thixotropic flow properties, which are not a function of the DS within the range of commercial products. Deviation from the Newtonian behavior increases with the molar mass. Importantly, since flow properties are dependent on the molar mass and the molar mass distribution of the macromolecular substance, a blend of high and low molar mass polymers can have different flow properties compared to a polymer having the same solution viscosity as the blend but having a narrow molar mass distribution. This holds for modified celluloses and also for other biopolymers. Concentrations as low as 0.001–1% are able to reduce the surface tension and the interfacial tension (Greminger and Krumel 1980).

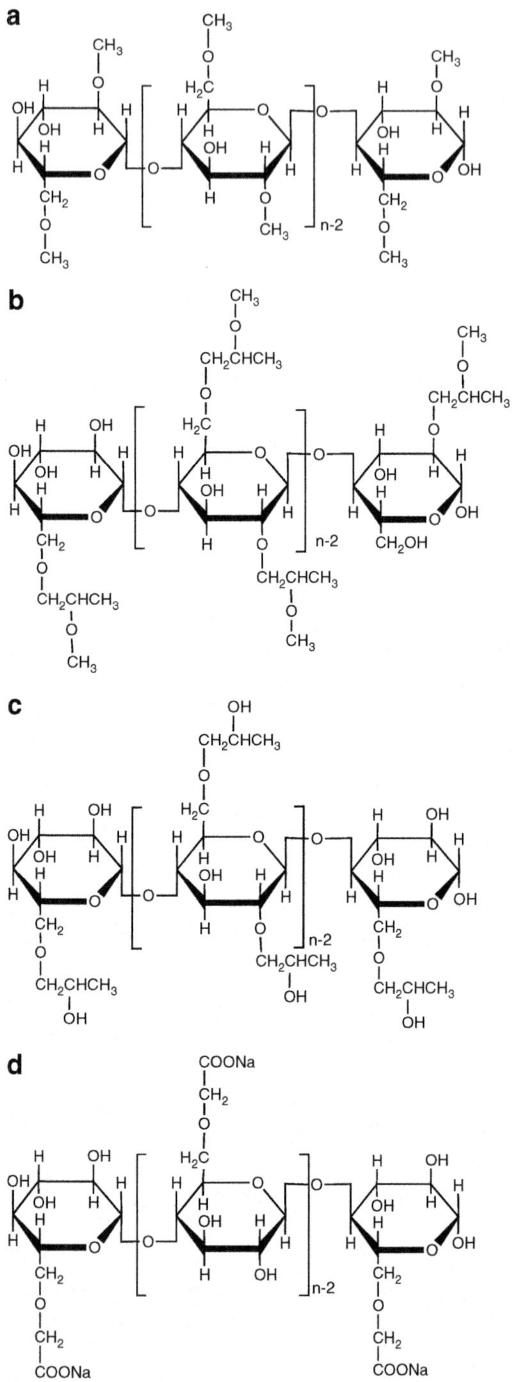

Fig. 3.9 Principal chemical structures of cellulose derivatives. (**a**) methyl (DS = 2), (**b**) methylhydroxypropyl, (**c**) hydroxypropyl, (**d**) carboxymethyl (DS = 1)

Hydroxypropyl cellulose (HPC). (Butler and Klug 1980; Picker-Freyer and Düring 2007) This non-ionic cellulose ether (Fig. 3.9c) is soluble in cold water. The viscosity of the solution can be adjusted by the degree of polymerization. Due to the presence of both the hydrophilic and hydrophobic groups, it becomes insoluble at temperatures above approximately 45°C. However, no gel is formed. Advantageous features include the solubility in ethanol and mixtures of ethanol and water, good film formation, and the high surface activity compared to most other hydrocolloids (Murray 2000). The films are flexible, glossy, and non-tacky (Butler and Klug 1980).

HPC is available in a wide range of molar mass from 6×10^4 to 1.2×10^6 g/mol. A high degree of reaction, DR=4, enhances the resistance to both acid hydrolysis and biological degradation. HPC is compatible with a number of high molar mass, high-boiling waxes and oils and can, therefore, be used to modify the properties of these materials. The addition of HPC to wax or oil systems will increase the system viscosity and improve the hardness and crack resistance of coatings made thereof (Butler and Klug 1980). It is compatible with most water-soluble gums and resins, and yields homogeneous solutions with carboxymethyl cellulose (CMC), hydroxyethyl cellulose (HEC), MC, gelatin, sodium caseinate, poly(ethylene oxide), carbowax, guar, alginate, and locust bean gum. With anionic polymers such as CMC and alginate, a synergistic viscosity increase is obtained, whereas with non-ionic polymers such as HEC, MC, guar gum, a lower than expected viscosity is obtained (Butler and Klug 1980).

Ethyl methylcellulose (EMC). It behaves like MC and HPMC. It is soluble in cold water and forms gels on heating. More interesting is the surface activity since the gels are comparably weak.

Ethylcellulose (EC). This derivative is a water-insoluble commercial thermoplast used for coating and controlled release applications, $T_m=135°C$.

Sodium carboxymethyl cellulose (CMC). CMC (Fig. 3.9d) is produced by reaction of cellulose with alkali and chloroacetic acid. The DP decreases slightly during the modification. The molar mass is typically in the range of 4×10^4–10^6 g/mol. The modification yields areas of high and low substitution. The purified product appears as a white to off-white, tasteless, odorless free-flowing powder.

CMC is an anionic linear polyelectrolyte, soluble in both cold and hot water. As typical for polyelectrolytes, the molecular conformation in aqueous solution strongly depends on the concentration, ionic strength, and pH. The polymer chains of CMC are most extended at low concentrations, low ionic strength, and high pH whereas with increasing concentration, ionic strength, and decreasing pH the molecules increasingly coil. At concentrations above the overlap concentration, thermo-reversible gels are formed in water. With heavy metals, three-valent cations, and the majority of polycation precipitation occurs. However, in the isoelectric range solubilization was observed with proteins such as soy protein and caseinate (Stelzer and Klug 1980).

CMC rapidly dissolves in cold water and gives clear and colorless solutions. The solution viscosity decreases during heating. Nevertheless, the solution behavior strongly depends on the molar mass and the DS. The more hydrophobic the lower substituted CMCs are thixotropic, the more hydrophilic the higher substituted ones exhibit pseudoplastic behavior. At low pH, CMC may form cross-links through lactonization between carboxylic acid and free hydroxyl groups (Chaplin 2007; Kästner et al. 1997).

3.2.1.3 Plant Exudates and Extracts

Polysaccharide plant exudates and extracts are complex macromolecular substances. Some of them are mixtures consisting of oligomers and polymers of different chemical structure and/or chain architecture. Several of the materials used for food encapsulation are products of gummnosis and are also therefore named plant gums in practice.

Gummnosis is widespread in the plant kingdom. It is the result of stress conditions such as wound, heat, and drought. Gums produced in response to the injury of the plant form a barrier at the lesion hindering the invasion by microorganisms. Only a few plant species are cultivated at present to provide gums for the food industry. Most of them belong to the Leguminosae family. Some examples of plants with the appropriate gum harvested thereof are listed below:

- *Acacia senegal* – gum arabic (gum senegal)
- *Astragalus spp.* – gum tragacanth
- *Cyamopsis tetragonolobos* – guar gum
- *Ceratonia siliqua* – locust bean gum

However, a wide range of exudates from other trees and shrubs are also harvested and consumed in their countries of origin, such as mesquite gum (Orozco-Villafuerte et al. 2005). These gums are generally harvested from wild growing trees.

Gum Arabic

Chemical Description of Gum Arabic (GA). (CAS# 9000-01-5) Gum arabic is a complex mixture of arabinogalactan oligosaccharides, polysaccharides, and glycoproteins. It is a branched neutral or slightly acidic substance. The chemical composition and the composition of the mixture can vary with the source, climate, season, age of trees, rainfall, time of exudation, and other factors. The backbone has been identified to consist of β-(1→3)-linked D-galactopyranosyl units. The side chains are composed of two to five β-(1→3)-linked D-galactopyranosyl units, joined to the main chain by 1,6-linkages. Both the main and the side chain contain units of α-L-arabinofuranosyl, α-L-rhamnopyranosyl, β-D-glucuronopyranosyl, and 4-O-methyl-β-D-glucuronopyranosyl. The latter two usually occur preferably as end-units (Verbeken et al. 2003). Depending on the source, the glycan components of GA contain a greater proportion of L-arabinose relative to D-galactose (*Acacia seyal*) or D-galactose relative to L-arabinose (*Acacia senegal*). The gum from *Acacia seyal* also contains significantly more 4-O-methyl-D-glucuronic acid but less L-rhamnose and unsubstituted D-glucuronic acid than that from *Acacia senegal* (Williams and Phillips 2000; Williams et al. 2006; Motlagh et al. 2006; Chaplin 2007).

GA has a complicated molar mass distribution. Some authors report a mixture of lower molar mass polysaccharide ($M_w \approx 2.5 \times 10^5$ for the major component) and higher molar mass hydroxyproline-rich glycoprotein ($M_w \approx 2.5 \times 10^6$ for the minor component) (Goodrum et al. 2000; Renard et al. 2006). Others describe three fractions: arabinogalactan (88.4%) with 3.8×10^5 g/mol, arabinogalactan-protein complex (10.4%) with 1.45×10^6 g/mol, and a low molar mass glycoprotein (1.2%) with 2.5×10^5 g/mol (Verbeken et al. 2003). Because it is a mixture, which varies with source, the exact chemical composition and molecular structures are still debated.

Physico-chemical Characteristics of Gum Arabic. Gum arabic is odorless, colorless, tasteless, and does not affect the odor, color, and taste of the system to which it is added. It is highly soluble in water and dissolves in both cold and hot water with concentrations up to 50 wt%. The solutions exhibit Newtonian behavior at concentrations up to 40 wt% and become pseudoplastic at higher concentration. The viscosity of solutions varies strongly with the GA type, pH, and ionic strength. Maximum viscosity is achieved between pH 6 and 7. GA acts as protective colloid and excellent emulsifier. The adhesive property is not related to the viscosity (Meer 1980a). However, and rather confusingly, the molecular aggregation can cause both shear thinning and time-dependent thickening behavior at low shear (Sanchez et al. 2002).

GA has the ability to create a strong protective film around oil droplets (Krishnan et al. 2005), which results from the highly branched arabinogalactan-protein structure containing both protein and polysaccharide moieties. The hydrophobic polypeptides anchor the polysaccharide onto the surface of the oil droplet and the hydrophilic carbohydrate chains prevent the aggregation by forming a thick charged layer. This is quite unique. Very few polysaccharide-protein systems have a stabilization mechanism comparable to GA (Nakamura et al. 2006; Kim and Morr 1996; Kim et al. 1996).

GA is compatible with most other plant hydrocolloids, proteins, carbohydrates, and modified starches. The viscosity of a solution of a mixture of GA and gum tragacanth tends to be lower than that of either constituent solution. Minimum solution viscosity was obtained for 80% gum tragacanth and 20% GA. Overall, when used as a fixative in the spray-drying of flavors, it forms a thin and impenetrable film around the flavor particle, protecting it from oxidation and evaporation and preventing it from absorbing moisture from the air (Meer 1980a; McNamee et al. 1998). GA was also found to be a useful prebiotic, which promotes beneficial physiological effects (Phillips et al. 2007).

Origin and Isolation of Gum Arabic. Gum arabic, also called gum acacia, is the oldest and best known of all natural plant gums. It is a dried exudate obtained from the stems and branches of *Acacia senegal* or *Acacia seyal* (FAO 1999). However, the name is also used for other gums produced by other *Acacia* species (Verbeken et al. 2003; FAO 1995). Despite there being more than 500 species of acacia trees, most commercial GA is produced from *Acacia senegal* and *Acacia seyal*, which are grown commercially throughout the Sahel from Senegal and Sudan to Somaliland.

Gum Tragacanth

Chemical Description of Gum Tragacanth. (CAS# 9000-65-1) Gum tragacanth is a complex mixture of highly branched, heterogeneous polysaccharides (Stauffer 1980). It occurs naturally as slightly acidic calcium, magnesium, or potassium salt with a molar mass of approximately 8.4×10^5 g/mol. Moreover, an elongated shape has been described with a length of 4,500 nm and a diameter of 19 nm (Gralen and Karrholm 1950). Gum tragacanth consists of two components: (a) a water-swellable component designated as bassorin, which contains the tragacanthic acid polymer with a molar mass of approximately 10^5 g/mol (Elias 1992), and (b) a water-soluble component, a colloidal hydrosol, tragacanthin with a molar mass of approximately 10^4 g/mol (Elias 1992). This is a neutral polysaccharide, which, at best, contains only small amounts (3%) of uronic acid (Aspinall and Baillie 1963) (sometimes there is confusion with the names). Due to the complexity of the substance, different molecular descriptions have been reported.

Depending on the source, the ratio of the water-swellable to the water-soluble fraction varies. Examples are 35:65 and 60:40. The tragacanthic acid fraction has a higher molar mass and a rod-like molecular shape (Stephen and Churms 1995). The main chain is formed by (1→4)-linked D-galactose residues with side chains of D-xylose units attached to the main chain by (1→3)-linkages. The water-soluble tragacanthin is a neutral, highly branched arabinogalactan with a spherical molecular shape. Its structure probably consists of a core composed of (1→6)- and (1→3)-linked D-galactose with attached chains of (1→2)-, (1→3)-, and (1→5)-linked L-arabinose (Verbeken et al. 2003; FAO 1995).

Physico-chemical Characteristics of Gum Tragacanth. Gum tragacanth is one of the most acid-resistant gums. It does not degrade at low pH. It is considered as a bifunctional emulsifier. First, it increases the viscosity of the aqueous phase and secondly it lowers the interfacial tension between the oil-in-water emulsion, and thus eliminates the need to incorporate surface-active agents. The suspending properties result from the negative charges. Gum tragacanth forms viscous aqueous solutions even at low concentration, which exhibits pseudoplastic flow properties. Similar to GA, it forms a protective film around oil droplets (Mohammadifar et al. 2006).

Origin and Isolation of Gum Tragacanth. Gum tragacanth is a dried exudate obtained from the stems and branches of *Astragalus gummifer* Labillardière and other Asiatic species of *Astragalus* (FAO 1992b). These plants grow in the highlands and deserts of Turkey, Iran, Iraq, Syria, Lebanon, Afghanistan, Pakistan, and Russia.

Gum Karaya

Chemical Description of Gum Karaya. (CAS# 9000-36-6) Gum karaya (Meer 1980b) is a complex, partially acetylated polysaccharide obtained as a calcium and magnesium salt. It has a branched structure and a molar mass up to 1.6×10^7 g/mol has been reported (Le Cerf et al. 1990). The backbone consists of α-D-galacturonic acid and α-L-rhamnose residues. Side chains are attached by (1→2)-linkage of β-D-galactose or by (1→3)-linkage of β-D-guluronic acid to the galacturonic acid of the main chain. Furthermore, half of the rhamnose residues of the main chain are (1→4)-linked to β-D-galactose units (Weiping 2000).

Physico-chemical Characteristics of Gum Karaya. Powdered gum karaya is white to grayish white. It is one of the least soluble of the exudate gums. Only 10% of the native gum solubilizes in cold water, increasing to 30% in hot water. After deacetylation, 90% dissolves in water (Verbeken et al. 2003). Gum karaya is compatible with other plant hydrocolloids, proteins, and carbohydrates.

Origin and Isolation of Gum Karaya. Gum karaya, sometimes known as Sterculia gum, is a dried exudate from the stems and branches of *Sterculia urens* Roxbourg and other species of *Stericula* or from the species of *Cochlospermum* (FAO 1992a). India is the largest producer of gum karaya.

Mesquite Gum

Chemical Description of Mesquite Gum. Mesquite gum is the neutral salt of a complex acidic branched polysaccharide. Its backbone is formed by (1→3)-linked β-D-galactose residues with (1→6)-linked branches, bearing L-arabinose (pyranose and furanose ring forms), L-rhamnose, β-D-glucuronate, and 4-O-methyl-β-D-glucuronate as single sugar or oligosaccharide side chains. It also contains a small amount of protein (0.7–5.8%) (Orozco-Villafuerte et al. 2003; Anderson and Farquhar 1982; Anderson and Weiping 1989; Goycoolea et al. 1997; Vernon-Carter et al. 2000). There is some chemical similarity with gum arabic (Islam et al. 1997).

Physico-chemical Characteristics of Mesquite Gum. Mesquite gum has a level of solubility comparable to GA. The color of the clear solutions is brown, slightly darker than those from GA. Hydrophobic affinity chromatography yielded five fractions differing in their molar masses (3.5×10^4–9.3×10^5 g/mol) and protein content (0.04% to about 30%). The two major fractions with >90% have protein contents much lower than 1% (Orozco-Villafuerte et al. 2003).

Solutions of up to 50% can be prepared. However, the increase of the viscosity with increasing concentration is steeper for mesquite gum in comparison to GA (Goycoolea et al. 1995). Mesquite gum has good film formation properties (Diaz-Sobac et al. 2002).

Origin and Isolation of Mesquite Gum. This gum is obtained from the mesquite tree (*Prosopis spp.*) or shrub, which grows in the southwest of the USA, Mexico, and other areas of the world. It is preferably harvested for local use and limited commercialized for local markets. Despite similar or even partially better application properties compared to GA, it finds either no use or limited use in industrial applications. The main reason for not using the mesquite gum in industrial applications is that its supply cannot be guaranteed, as all the gum is collected from wild growing trees (Orozco-Villafuerte et al. 2005).

Galactomannans

Chemical Description of Galactomannans. Locust bean gum (LBG) (Gidley and Reid 2006) (CAS# 9000-40-2), also called carob bean gum, tara gum (CAS# 39300-88-4), and guar gum (CAS# 9000-30-0, 9000-30-3, 9066-07-3) are galactomannans. They consist of linearly (1→4)-linked β-D-mannopyranosyl units with single α-D-galactopyranosyl units connected by (1→6) linkages as side branches. The three gum types differ in the ratio of D-mannosyl to D-galactosyl. On average, in LBG every fourth main chain unit bears a side unit whereas in tara gum it is every third, and in guar gum every second main chain unit, Fig. 3.10. The actual ratios reported in the literature vary slightly and differ from the average. For LBG it is between about 3.9:1 and 3.5:1 (Hoefler 2004; Chaplin 2007). The ratio corresponds to galactose weight contents of 17–26% for LBG, 25% for tara gum, and 33–40% for guar gum (Wielinga 2000). However, the side units are highly unevenly distributed along the backbone. More block polymer arrangement than statistical distribution occurs.

Physico-chemical Characteristic of Galactomannans. The three gum types differ in their solubility. The solubility increases with increasing number of side units. While guar is fully water soluble at room temperature, the solubility of tara is about 70% under these conditions but is completely soluble above 70°C. LBG has a limited water solubility at ambient temperature. It swells below 60°C but becomes soluble above 60°C and fully hydrated if heated for 10 min at 80°C (Hoefler 2004; Seaman 1980). Average molar masses ranging from about 1.5×10^5 to 1.5×10^6 g/mol have been reported for guar gum, having higher molar masses and producing higher viscosities of solutions than LBG. However, the isolation process, purification, solution preparation, and characterization method can have an impact on the molar masses obtained (Picout et al. 2001; Picout et al. 2002; Patel et al. 2006). The solutions of all three gum types exhibit pseudoplastic behavior.

Origin and Isolation of Galactomannan. LBG, tara, and guar are isolated from the endosperm of the seeds of the carob tree (*Ceratonia siliqua*), the tara shrub (*Cesalpinia spinosa*), and the guar plant (*Cyamopsis tetragonoloba*), respectively.

Pectins

Chemical Description of Pectins. (CAS# 9000-69-5) Pectins (May 2000; Lopez da Silva and Rao 2006) are high molar mass hetero-polysaccharides with at least 65 wt% of α-(1→4)-linked D-galacturonic acid-based units. These units may be present as free acid, salt (sodium, potassium calcium, ammonium), naturally esterified with methanol, or as acid amid in amidated pectins. Furthermore, a range of neutral sugars such as L-rhamnose, D-galactose, L-arabinose, D-xylose, and small amounts of others are part of the polymer chain. L-rhamnose units exist exclusively as (1→2)-linked in the main chain, whereas all other neutral sugar residues are bond preferably at the rhamnose and galactose units to the main chain (Elias 1992). Pectins exhibit a very complex, non-random structure with linear blocks of homo-poly(galacturonic acid), so-called smooth regions, and with highly branched blocks, so-called hairy regions (May 2000; Voragen et al. 1995).

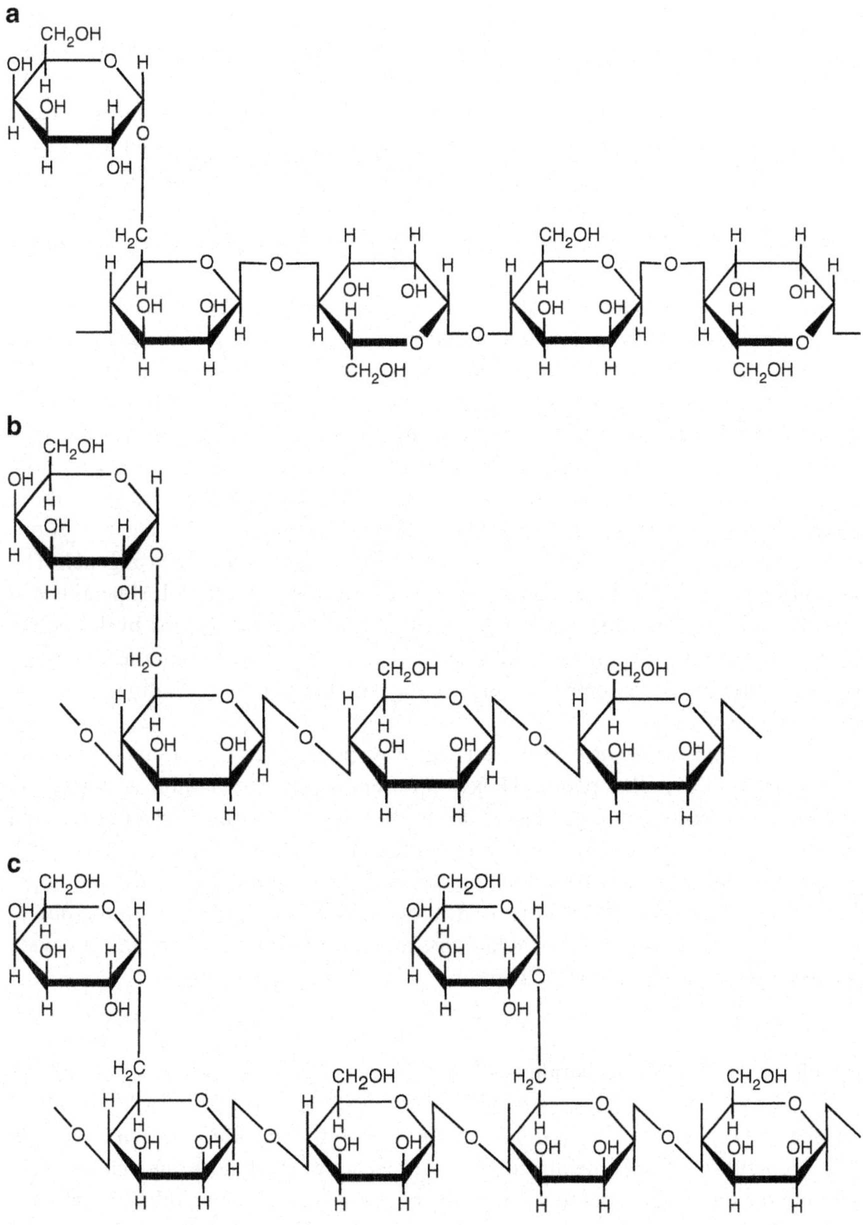

Fig. 3.10 Principal chemical structures of galactomannans. (**a**) locust bean gum, (**b**) tara gum, (**c**) guar gum

Pectins can differ by the degree of esterification of the carboxy groups of the galacturonic acid, which is in general in the range of 20–80%. Pectins with more than 50% esterification are designated as high-esterified (HM, high methoxylated) and distinguished from low-esterified pectins (LM, low methoxylated) with less than 50% ester groups. The molar mass depends on the pectin source and processing, and was reported to be in the range of 10^4–2×10^5 g/mol (Voragen et al. 1995). A small portion of the hydroxyl groups are acetylated in pectins from sugar beet, but not in those from citrus fruits.

Physico-chemical Characteristics of Pectins. Pectins are soluble in water but insoluble in most organic solvents. Aqueous solutions can be obtained up to concentrations between 6% and 12%, depending on the pectin type. The solutions have low viscosity compared to other plant gums. At low concentration, the rheological behavior is close to Newtonian behavior, but it shifts to pseudoplastic at higher concentration (Pedersen 1980).

Pectins are non-permanently negatively charged macromolecules and, therefore, behave as polyelectrolytes and their solution properties, and in particular gelation, are sensitive to pH variation and the presence of cations in the solution. Pectin solutions are most stable at pH 3–4 (Voragen et al. 1995). The gelation also depends on the nature and the quality of the raw material but primarily on the degree of esterification. Different gelation mechanisms have been identified for pectins with high methoxyl content (HM) and low (LM) methoxyl content. HM pectins gel only in the presence of sugars or other cosolutes in the low pH range, where the acid groups are not completely ionized. The gel strength increases with decreasing pH. The interaction between calcium ions and pectin governs the gelation of LM pectins and depends on the proportion and arrangement of the carboxyl groups in the pectin chain. The reactivity increases with decreasing degree of esterification. The presence of amide groups extends the range of the calcium ion concentration where the LM pectin forms gels.

Origin and Isolation of Pectins. Biologically, pectins serve as a kind of cement for the cell wall in higher plants. They are present in fruits in variable amounts and qualities. The main pectin producers are located in citrus fruit producing areas in Europe and Latin America, where the pectins are isolated from the citrus peel, orange peel, and residues from the extraction of citrus juice, citrus oil, and apple juice. Hot aqueous mineral acid is used to extract the raw material with the objective of obtaining pectins of high molar mass at high concentration. Pectins of a wide range of types are offered in the market.

Soluble Soybean Polysaccharide

Chemical Description of Soluble Soybean Polysaccharide. Soluble soybean polysaccharide (SSPS) is mainly composed of galactose, arabinose, and galacturonic acid, but the chain also contains many sugars at low quantities including rhamnose, glucose, xylose, and fucose. The exact structure is not fully understood. An average molar mass of several hundred thousand has been reported, though three components varying in their molar mass have also been identified (Nakamura et al. 2000). The negative charges, i.e., the galacturonic acid units, are located on the main backbone chain, whereas the shorter side chains contain neutral sugar units.

Physico-chemical Characteristics of Soluble Soybean Polysaccharide. Due to the negative charges in the backbone, SSPS is considered as an anionic polyelectrolyte. SSPS is soluble in both cold and hot water yielding solutions of relatively low viscosity. For comparable concentrations, the viscosity is considerably lower than that for guar gum or HM pectin, but up to concentrations of 30 wt% it is found to be higher than that for gum arabic. SSPS solutions do not gel. Moreover, the viscosity is only slightly affected by heat, salt, or by acidic pH variation (Furuta and Maeda 1999).

The adhesive and film forming properties of SSPS are excellent. Films processed without any additives show high resistance to tension. The films are generally colorless, transparent, water soluble, and edible and are suitable for coating the surfaces of food and food ingredients. SSPS can prevent oxidation of oils (Maeda 2000). SSPS has the ability to stabilize protein particles at low pH without increasing the viscosity. It can be used as an emulsifier and as a stabilizer for emulsions.

Origin and Isolation of Soluble Soybean Polysaccharide. SSPS is extracted from okara, the residue after oil and soy protein extraction from soybean (Maeda 2000). It is obtained as a powder of several varieties, which can be tailored for specific applications.

3.2.1.4 Marine Extracts

Seaweed can be regarded as another almost unlimited source of different types of polysaccharides for many industrial applications. Some of them have been found to be useful for encapsulation in the food sector.

Carrageenans

Chemical Description of Carrageenans. (CAS# 9000-07-1) Carrageenans (Guiseley et al. 1980; Piculell 1995, 2006; Imeson 2007) are a family of high molar mass sulfated polysaccharides, for which the individual structures strongly depend on the source and conditions during extraction and purification (Falshaw et al. 2001). Overall, the polymer chains comprise alternating (1→3)-linked β-D-galactopyranosyl and (1→4)-linked α-D-galactopyranosyl units. Some of the (1→3)-linked units occur as the 2- and 4-sulfates, whereas the (1→4)-linked units occur as the 2- and 6-sulfates, the 2,6-disufates, the 3,6-anhydride, and the 3,6-andydride-2-sulfate (Perceval 1972). Carrageenans are anionic polyelectrolytes. The half-ester content varies from 15% to 40%. Principally, there is a possibility of a continuous spectrum of carrageenan types including the biological precursors, which are modified by processing. After extraction and isolation, only the three types κ- (kappa), ι- (iota), and λ- (lambda) carrageenan are commercially available. They approach the respective ideal type in chemical composition (Guiseley et al. 1980). The idealized repeating units of the three limiting types are shown in Fig. 3.11. Deviating from the idealized structures, the practical sulfate group content/dimer repeating unit is 1.03 rather than 1.0 for κ, 1.49 rather than 2.0 for ι, and 2.09 rather than 3.0 for λ (Chaplin 2007).

Several analyses have shown that carrageenans are highly polydisperse with typical number average molar masses in the range of 10^5–2×10^5 g/mol and typical weight average molar masses in the range of 3×10^5–6×10^5 g/mol (Piculell 1995).

Physico-chemical Characteristics of Carrageenans. Some selected properties of carrageenans are summarized in Table 3.3. Due to the variation in their chemical structure, carrageenans exhibit a wide spectrum of the rheological behavior (Mangione et al. 2003). Dependent on the type, they can form viscous solutions and also thermally reversible gels with a texture varying from soft and elastic to firm

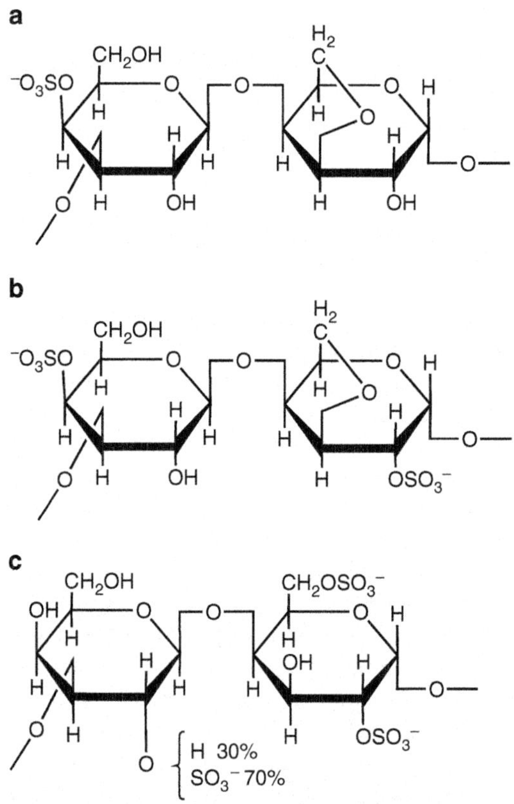

Fig. 3.11 Repeating dimer units of limit carrageenans. (**a**) κ-carrageenan: (1→3)-β-D-galactopyranose-4-sulfate-(1→4)-3,6-anhydro-α-D-galactopyranose-(1→3), (**b**) ι-carrageenan: (1→3)-β-D-galactopyranose-4-sulfate-(1→4)-3,6-anhydro-α-D-galactopyranose-2-sulfate-(1→3), (**c**) λ-carrageenan: (1→3)-β-D-galactopyranose-2-sulfate-(1→4)-α-D-galactopyranose-2,6-disulfate-(1→3)

Table 3.3 Selected properties of carrageenans (Davidson 1980; Imeson 2007)

Property	Carrageenan type		
	Kappa	Iota	Lambda
Solubility in hot water	T>70°C	T>70°C	Soluble
Solubility in cold water	Na-salt soluble; from limited to high swelling of K-, Ca-, and ammonium salts	Na-salts soluble, Ca-salts give thixotropic dispersions	All salts soluble; viscous pseudoplastic solutions
Gel formation	Thermo-reversible gels on cooling in the presence of appropriate counterions		Non-gelling
Gelation	Strongest with K^+	Strongest with Ca^{2+}	
Gel type	Brittle with syneresis; poor freeze-thaw stability	Elastic, no syneresis; good freeze-thaw stability	
pH stability	Stable at neutral alkaline pH; pH=3.5: hydrolysis of polymer solutions but gels are stable		Hydrolysis below pH 4.3
Salt tolerance	Poor	Good	Good

and brittle as specified for the three types in Table 3.3. κ- and ι-carrageenans have the ability to form elastic gels in the presence of certain cations such as K^+ and Ca^{2+}. The molecular mechanism for the conformal transitions in the solution that result in the aggregation and gelation is still not completely understood. Comprehensive NMR studies have contributed to some progress.

Synergistic effects are obtained when combining carrageenans with other gum types. Brittle κ-carrageenan gels may be softened with, for example, locust bean gum. ι-Carrageenan has less specific ionic binding but the increased ionic strength allows helices to form junction zones in soft elastic gels with good freeze-thaw stability. λ-Carrageenan is non-gelling but is the type with the highest anionic charge density, which causes the best solubility, extended chain conformation, and strong electrostatic interaction in solution (Janaswamy and Chandrasekaran 2002; Kara et al. 2006).

Origin and Isolation of Carrageenans. Carrageenans are prepared from red seaweed (*Rhodophycae*). Different seaweeds produce different carrageenans, for example, *Chondrus crispus* (κ and λ), *Eucheuma cottonii* (κ), *Eucheuma spinosum* (ι), *Gigartina* (κ and λ), *Furcellaria* (κ and λ), by extraction and subsequent treatment/modification of the biological precursors μ- and ν-carrageenan. Seaweed selection, processing, and blending of different extracts can control the variations of the carrageenan types (Piculell 1995, 2006; Imeson 2007).

Alginate

Chemical Description of Alginate. (CAS 9005-38-3 Na) Alginate (Draget 2000; Cottrell and Kovacs 1980b; Moe et al. 1995; Draget et al. 2006) is a family of linear anionic polysaccharides, which can be considered as copolymers of (1→4) linked α-L-guluronic acid (G) and β-D-mannuronic acid (M) residues. The linear chains are composed of homopolymeric regions of G- and M-blocks interspaced with regions of mixed sequences, MG-blocks (Grasdalen et al. 1979). Figure 3.12 shows a part of a sodium alginate chain. Depending on the seaweed extract, the isolation process, or the biotechnological procedure, the proportion and sequential arrangement of the two structural units vary widely. M and G blocks of various lengths can be present in the polymer chain. Consequently, for the description of the sequential structure not only the monomer composition (monad frequency) but also the diad and triad frequencies have to be known. The four diad (nearest neighbor) frequencies (F_{GG}, F_{GM}, F_{MG}, F_{MM}) and the eight possible triad frequencies (F_{GGG}, F_{GGM}, F_{MGG}, F_{MGM}, F_{MMM}, F_{MMG}, F_{GMM}, F_{GMG}) can be analyzed by NMR techniques (Grasdalen et al. 1979). The weight average molar mass of commercial sodium alginates varies from 4×10^4 to 5×10^5 g/mol. But some alginates with molar masses higher than 10^6 g/mol have also been isolated.

Physico-chemical Characteristics of Alginate. Alginate composition, structure, and molar mass govern the functional properties. The solubility of alginate in water is related to the rate of dissociation and the type of the counterion. At pH<3, both the M- and G-structures will precipitate as alginic acid. However, alternating structures precipitate at lower pH values compared to the alginates containing more homogeneous block structures. Neutralization of the alginic acid occurs at pH>4, where it is converted into its corresponding salt. Sodium alginate is an example of a water-soluble alginate.

The intramolecular electrostatic repulsion between the neighboring negative charges of each monomer unit forces alginate molecules into an extended random coil conformation (Smidsrød and Hang 1968). This results in highly viscous solutions even at relatively low alginate concentration. The dynamic viscosity increases exponentially with the molar mass, while the intrinsic flexibility of the alginate chains in solution increases in the order GG<MM<MG (Smidsrød et al. 1973). On the other hand, the selectivity for cation binding and gel forming properties strongly depend on the composition and sequence. Divalent cations preferably bind to the G-blocks. The ability to form ionotropic gels is based on this selective binding of cations. Gel-like networks are also formed with the polycation chitosan. As typical for linear homogeneously charged polyelectrolytes,

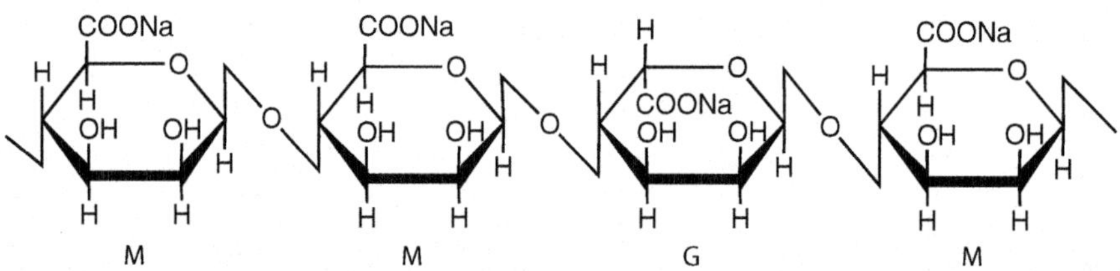

Fig. 3.12 Principal chemical structural units of a sodium alginate polymer chain

the chain conformation and, therefore, the solution viscosity strongly depend on the ionic strength.

The rheological behavior of sodium alginate solutions depends on the concentration and the shear rate. Higher concentrated solutions exhibit pseudoplastic flow even at low shear rates. At lower concentrations, the Newtonian behavior is observed at low shear rate whereas the solutions become pseudoplastic at higher shear rate (Draget 2000).

Alginate as a dry powder and, in particular as a solution, is subject to degradation leading to rapid decrease of the chain length detectable by decreasing viscosity. Many microorganisms digest alginate. In a deep freezer, sodium alginate may be kept for several years without significant degradation.

Origin and Isolation of Alginate. Alginates are quite abundant in nature. Commercial sodium alginates are generally produced from marine brown algae, though they may also be synthesized as an exocellular material by some bacteria (Chapman 1980). Moreover, biosynthesis is possible yielding alginates over a very wide range of tailored composition (Valla et al. 1996).

3.2.1.5 Microbial and Animal Polysaccharides

Polysaccharides produced biotechnologically by bacteria are biopolymers with novel and partially unique functional properties. Examples, which are interesting for food applications are xanthan, gellan, and curdlan, though the WHO has at the present time certified the latter only for a few countries.

Xanthan

Chemical Description of Xanthan. (CAS# 11138-66-2) Xanthan is a high molar mass anionic polyelectrolyte for which the principal chemical structure is shown in Fig. 3.13. It occurs as a mixed salt of sodium, potassium, and calcium. Its backbone consists of β-(1→4)-D-glucopyranosyl units with every second unit having a trisaccharide side chain attached at the C-3 position, one D-glucuronosyl unit between two D-mannosyl units. Therefore, xanthan may also be considered as consisting structurally of pentamers. Approximately, 40–50% of the terminal mannosyl units are 4,6-pyruvated. The non-terminal mannosyl units are mostly 6-acetylated.

The molar masses reported in the literature vary from 1.5×10^5 g/mol to several million, but are probably in the order of two million. Association phenomena observed for xanthan gum could explain these variations (Sworn 2000b; Morris 1995, 2006; Cottrell et al. 1980; Elias 1992). However, a consensus of the opinion exists that the polydispersity is relatively low.

Physico-chemical Characteristics of Xanthan. Xanthan is soluble in cold water. It hydrates rapidly in cold water without lumping if properly dispersed into the solvent. It is considered to be mainly non-gelling. The solutions show a very pronounced pseudoplastic behavior. The viscosity progressively reduces with increasing

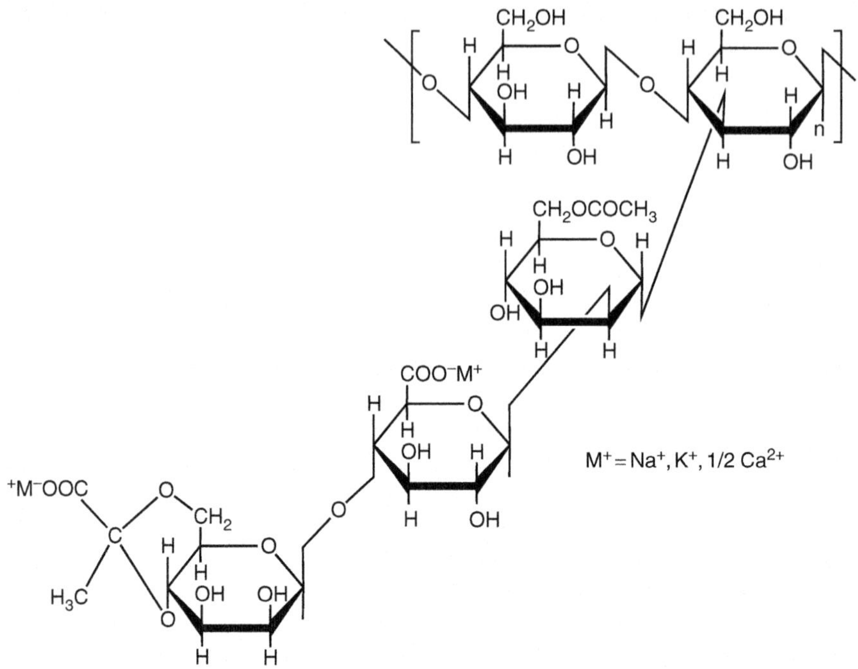

Fig. 3.13 Principal chemical structure of xanthan gum

shear stress but instantaneously recovers after removal of the shear stress, subsequently returning to the initial viscosity. In general, the solution viscosity has been found to be stable over a wide range of pH (2–12) and temperature. The high viscosity is assigned to the network formation due to the H-bonding and polymer chain entanglement. Two chains may be aligned to form a double helix, giving a rather rigid rod configuration. The conversion between the ordered double helical conformation and the single more-flexible extended chain may take place over hours between 40°C and 80°C. The shielding of the backbone by side chains could explain the extraordinary resistance to enzymatic degradation. Xanthan undergoes cryogelation (Giannouli and Morris 2003).

Contrary to the typical polyelectrolyte behavior, the addition of salt to a hitherto salt-free xanthan solution causes the viscosity to increase when the xanthan concentration is higher than 0.15%. An increase of viscosity has also been observed when heating salt-free solutions, but with salt (0.1% NaCl) only minor variation of the viscosity is observed in the temperature range 10–90°C (Cottrell et al. 1980a).

Synergistic interaction with guar gum leads to an enhanced viscosity, whereas with LBG and konjac mannan soft, elastic thermo-reversible gels are obtained at higher concentration (Sworn 2000b).

Origin and Isolation of Xanthan. The *Xanthomonas campestris* bacterium produces the xanthan gum in a pure culture of the bacterium by an aerobic fermentation process in a glucose medium. Use of different strains or fermentation conditions may give rise to differing degrees of actelylation and pyruvylation. Xanthan may contain cellulases, which prevents its use with cellulose derivatives.

Gellan

Chemical Description of Gellan. (CAS# 71010-52-1) Gellan (Sworn 2000a) is a high molar mass anionic polyelectrolyte. The principal chemical structure in Fig. 3.14 shows the tetrasaccharide repeating unit composed of one rhamnose, one glucuronic acid, and two glucose units. The 3-linked glucose unit is substituted with glyceryl at O(2) and with acetyl at O(6). It can, therefore, be designated as →4)-L-rhamnopyranosyl-α-(1→3)-D-glucopyranosyl-β-(1→4)-D-glucuronopyranosyl-β-(1→4)-D-glucopyranosyl-β-(1→ (Chaplin 2007). Molar masses are in the range up to 5×10^5 g/mol. The degree of esterification depends on the production process. Low acyl (deacetylated) and high acyl specifications are commercially available.

Physico-chemical Characteristics of Gellan. The solubility and solution properties depend on the degree of substitution and the type and concentration of ions present in the solution. Low acyl gellan is much more sensitive to ions and frequently a sequestrant has to be added for hydration, which generally takes place at temperatures above 90°C. The hydration of high acyl gellan gum is less dependent on the concentration of ions. Heating to 85–95°C is in general sufficient to fully hydrate the gum. Gellan is able to withstand heating to 120°C. The viscosity increases with the degree of acetylation. A gellan solution may invisibly hold particles in suspension but, unlike other gelling agents, without significantly increasing the solution's viscosity.

Thermo-reversible gelation takes place upon cooling of solutions in the presence of gelling cations in the case of low acyl and without gelling cations in the case of high acyl. The resulting gel texture depends on the degree of acetylation and can cover a wide range. Low acyl types form hard, non-elastic brittle gels, whereas high acyl types yield soft, elastic, transparent, and flexible gels. Comparison with other gelling agents manifests this wide range (Fig. 3.15). Nevertheless, the exact texture and quality of the gel is influenced by the concentration of divalent cations present.

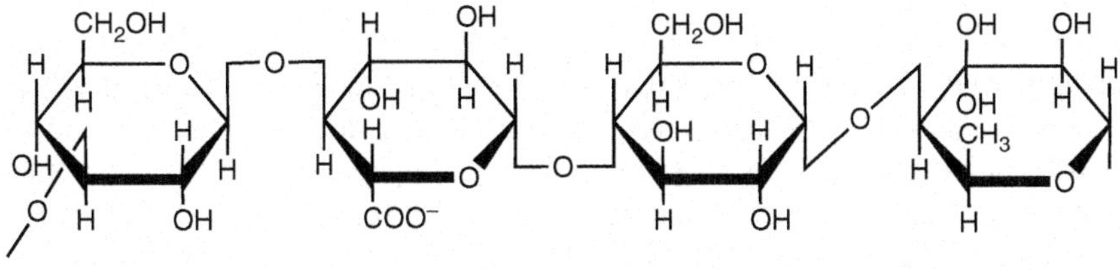

Fig. 3.14 Principal chemical structure of gellan gum

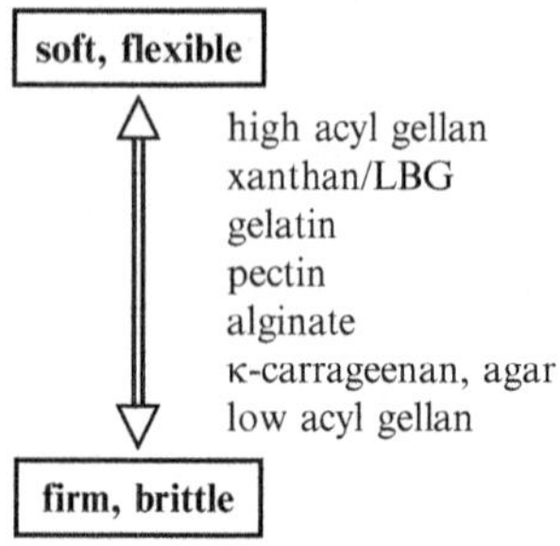

Fig. 3.15 Gel texture of different gelling macromolecular materials demonstrating the wide range of gel texture for gellan gum

Origin and Isolation of Gellan. Gellan is prepared commercially by aerobic submerged fermentation from the microorganism *Sphingomonas elodea* (Chandrasekaran and Radha 1995). High and low acyl types are available as off-white powders differing in their degree of decetylation. The latter is obtained by alkali treatment prior to isolation.

Dextran

Chemical Description of Dextran. Dextrans (CAS# 9004-54-0) are mainly linear neutral polymers of α-D-glucose linked by α-(1→6) glycosidic bonds, which can have variable amounts of α-(1→3) branches. Partially, the links in the main chain can be α-(1→4) or α-(1→2) links instead of α-(1→6) links. The fermentation process governs the chain architecture, molar mass, and polydispersity. The number - average of the molar mass of native dextrans is in the range of 2×10^5 g/mol. Due to the strong tendency to associate in the aqueous solution, apparent molar masses up to 5×10^8 have been obtained (Elias 1992).

Physico-chemical Characteristics of Dextran. Overall, dextrans are well soluble in water. Non-soluble portions result from higher degree of branching.

Origin and Isolation of Dextran. Microbial fermentation processes of sucrose yield dextrans, which are commercially supplied as powders or solutions. The bacteria used determine the chain architecture.

Chitosan

Chemical Description of Chitosan. (CAS# 9012-76-4) Chitosan (Winterowd and Sandford 1995; Vårum and SmidsrØd 2006) is a linear polysaccharide, which can be considered as a copolymer consisting of randomly distributed β(1→4) linked D-glucosamine and N-acetyl-D-glucosamine as illustrated in Fig. 3.16; block arrangement has also been reported (Inoue 1997). The composition is indicated by the degree of acetylation (DA), the fraction of acetyl-glucosamine units. The

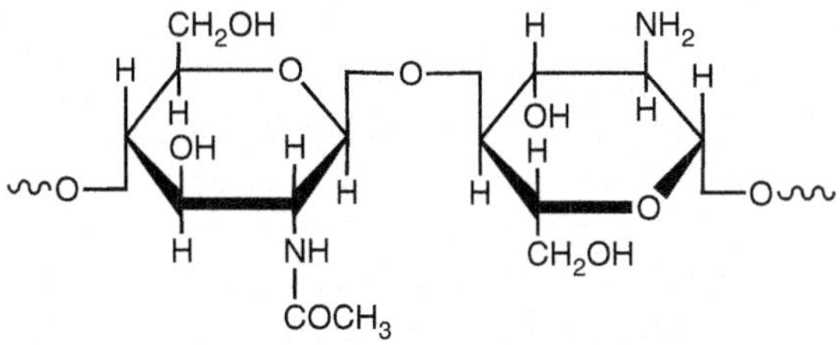

Fig. 3.16 Principal chemical structure of chitosan, deacetylated (*right*) and acetylated (*left*) chain units

molar mass depends on the source and the isolation technology, but it can reach values of about 5×10^5 g/mol. Lower molar masses and oligomers are obtained by the chain degradation under various conditions frequently yielding products, which differ from the composition and sequence arrangement of the parent molecules (Terbojevich 2000).

Physico-chemical Characteristics of Chitosan. Chitosan can be classified as a non-permanently charged cationic polyelectrolyte. Due to a pK_a value of approximately 6.5, chitosan is positively charged and soluble in acidic to neutral medium. The charge density and solubility depend on the DA. Only chitosan with a DA not exceeding 40% is soluble in acidic aqueous medium. Exceptions are oligomers, which have a higher solubility across a broader pH range. Chitosan forms gels with tripolyphosphate and alginate. Moreover, it has a very good film-forming ability (Domard and Domard 2002).

Origin and Isolation of Chitosan. Chitin, the main source of chitosan, has been evaluated to be as abundant as cellulose with an annual production of 10^{10}–10^{12} tons in biomass (Roberts 1992). Chitosan itself is much less present in nature. It has only been observed in some microorganisms and certain fungi. Only very recently, the commercial isolation from fungi has started. The main process of the alkaline deacetylation of crustacean chitins remains.

3.2.2 Proteins

Proteins are natural macromolecules composed of linear chains of amino acids. The possible sequences and frequency of 20 existing amino acids, more precisely 19 amino acids and one imino acid, in the chain give rise to their enormous variety. Proteins have a central role in all living organisms with a wide range of functions. Large amounts of proteins are directly used as food. The potential use for non-traditional applications is in the process of being extended.

3.2.2.1 Gluten

Chemical Description of Gluten. (CAS# 8002-80-0) Gluten is a complex mixture of gliadins (monomeric gluten proteins) with molar masses in the range of 3×10^4–8×10^4 g/mol (MacRitchie et al. 1990) and glutenins (polymeric gluten proteins) with molar masses in the range of 8×10^4 g/mol to several million (Kasarda 1989). Gliadins and glutenins comprise about 80% of the proteins contained in wheat seed. Both have unusually high levels of proline and glutamine and are therefore designated as "prolamins" (Shewry et al. 1986).

Gliadins represent a highly heterogeneous mixture of monomeric gluten proteins. Three structurally distinct groups, α-, γ-, and ω-types, can be distinguished (Shewry et al. 1986). The α- and γ-types are classified as sulfur-rich prolamins, whereas the ω-type gliandins are classified as sulfur-poor prolamins.

Wheat glutenin is a heterogeneous mixture of disulfide stabilized polymers of high-molar-mass glutenin subunits (HMM-GS) and low-molar-mass glutenin subunits (LMM-GS). More than 20 different HMM-GS have been identified so far in wheat varieties and new ones are being discovered frequently (Veraverbeke and Delcour 2002). Although LMM-GS show strong similarities with α- and γ-type gliadins, commonly classified as sulfur-rich prolamins, they differ in one very important characteristic: apart from intra-chain disulfide bonds, inter-chain disulfide bonds that lead to incorporation of LMM-GS in glutenin polymers also occur in LMM-GS. In these polymers, they form covalent links with the HMM-GS that form a separate third class of prolamins, the high-molar-mass prolamins, which typically have a higher content of glycine than the other two classes of prolamins (Shewry et al. 1986; Veraverbeke and Delcour 2002). Estimations of the size and structure of glutenin subunits and polymers are highly troubled by their low solubility in conventional buffers and the lack of crystallinity (Bottomley et al. 1982). Despite progress in unraveling the complexity, the structure is still largely a matter of speculation (Veraverbeke and Delcour 2002).

Physico-chemical Characteristics of Gluten. Glutenins show very low solubility in water due to a low content of amino acids with ionizable groups, the high content of non-polar amino acids and glutamine, and their high molar mass (Singh and MacRitchie 2001). Gliadins are soluble in distilled water but aggregate in salt solution (Van Vliet et al. 2002).

Surprisingly, few studies have been performed concerning the interfacial behavior of the wheat components (Örnebro et al. 2000). From studying the behavior of adsorbed layers of the four major wheat protein fractions at the air–water interface it was concluded that surface pressure increases in the order albumins < globulins < glutenins < gliadins (Keller et al. 1997). Further, a higher ability of gliadins to reduce the surface tension compared to glutenins has been found (Eliasson and Larsson 1993; Balla et al. 1997). Sprinkling the gluten powder on the water surface revealed that the film was almost completely elastic (Kokelaar et al. 1991).

Origin and Isolation of Gluten. Gliadin and glutenin exist, conjoined with starch, in the endosperm of some grass-related grains, notably wheat, rye, and barley. Being insoluble in water, they can be purified by washing away the associated starch.

Gluten proteins are strongly associated with baked products. However, the unique properties of wheat proteins have stimulated an interest in their use for other applications than traditional ones (Örnebro et al. 2000).

3.2.2.2 Milk Proteins

Bovine milk contains, in addition to water, lactose, fat, and other minor components, about 3.0–3.6 wt% proteins. Of this, caseins and whey proteins are the two major fractions. Detailed description of the milk proteins is presented in several reviews and book chapters (Ennis and Mulvihill 2000; Fox 1989; Fox 1992; Holt 1992; Creamer and MacGibbon 1996).

Caseins

Chemical Description of Caseins. Caseins (from Latin caseus "cheese") (CAS# 9000-71-9) are the most predominant phosphoproteins found in milk. The four principal primary proteins of the highly heterogeneous casein fraction are summarized in Table 3.4. Caseins can vary in their net charge, hydrophilicity, and metal binding. In milk they are present in the form of large approximately spherical micelles (molar mass about 10^8 g/mol), which can be separated from the molecularly dispersed whey proteins by ultracentrifugation.

Physico-chemical Characteristics of Caseins. Caseins are extremely heat-stable proteins. They do not coagulate by heat. They are insoluble at their isoelectric point, at about pH 4.6. However, the solubility behavior varies for casein fractions isolated from milk. Acid caseins are insoluble in the pH range 4–5, but they become soluble if the pH is increased above 5.5. Solutions of 10–15 wt% having high viscosity can be prepared at pH 6–7. They are also soluble at pH<3.5, but here the solutions are more viscous and become even gel-like. Rennet caseins are insoluble at pH 7 but soluble at pH exceeding 9. Calcium caseinate is the only milk protein

Table 3.4 Casein fractions and some characteristics typical for bovine milk (Ennis and Mulvihill 2000)

	Caseins			
Parameter	α_{s1}-casein	α_{s2}-casein	β-casein	κ-casein
Concentration wt%	0.9–1.5	0.3–0.4	0.9–1.1	0.3–0.4
Isoionic pH	4.94	5.45–5.23	5.14	5.61
Molar mass ($\times10^{-4}$ g/mol)	2.36	2.52–2.54	2.4	1.9

system reported to exhibit reversible thermal gelation. Overall, milk proteins have good surface active properties ranked in the order β-casein > monodispersed casein micelles > serum albumin > α-lactalbumin > α_s-casein = κ-casein > β-lactoglobulin. Surface films of sodium caseinate or β-casein have the best flexibility and lowest viscoelasticity compared with the films of other materials. Caseins are good fat emulsifiers. The water solubility of films formed from caseins/caseinates depends on the pH conditions used in their preparation. The water vapor permeability of the films depends on the protein type.

Origin and Isolation of Caseins. Caseins are generally manufactured from the skim milk by destabilizing the micelles. Numerous processes are known. The main products obtained are mineral acid casein, lactic acid casein, and rennet casein. Fractionation is possible and other caseinates may be obtained by different manufacturing processes. Spray-dried sodium caseinate is the most commonly used water-soluble caseinate.

Whey Proteins

Chemical Description of Whey Proteins. Whey is a by-product of cheese or casein production and has several commercial uses. Whey proteins primarily include α-lactalbumin, β-lactoglobulin, immunoglobulins, and serum albumin but also numerous minor proteins. The four principal proteins of the highly heterogeneous whey protein fraction are summarized in Table 3.5. α-Lactalbumin is a calcium metalloprotein with four intramolecular disulfide cross-links, for which genetic variants exist. β-Lactoglobulin has two intramolecular disulfide cross-links and one free SH group. Many genetic variants are known which vary in their degree of glycosylation.

Physico-chemical Characteristics of Whey Proteins. Whey proteins are globular proteins (Chen et al. 2006), soluble in their native forms in the ionic environment of milk, almost independent of pH. But they become insoluble at their isoelectric point (pH about 5) at very low ionic strength. In contrast to caseins, whey proteins

Table 3.5 Whey protein fractions and some characteristics typical for bovine milk (Ennis and Mulvihill 2000)

	Whey proteins			
Parameter	α-lactalbumin	β-lactoglobulin	Immunoglobulins	Serum albumin
Concentration (wt%)	0.07–0.15	0.2–0.4	0.06–0.1	0.01–0.04
Isoionic point	4.2–4.5	5.2		5.3
Molar mass ($\times 10^{-4}$ g/mol)	1.4	1.8	15-90	6.6

denature at temperatures above 70°C and become insoluble. Then they form thermally irreversible gels of different quality. Solutions of non-denatured whey proteins are much less viscous than caseinate solutions. They exhibit Newtonian flow at concentrations in the range of 4–12 wt% but become pseudoplastic in the range of 18–29 wt% concentration. As already mentioned for caseins, milk proteins exhibit good surface-active properties. Films, formed when thermally induced disulfide cross-linking takes place, are excellent gas barriers (Madene et al. 2006), while the incorporation of lipids reduces the water vapor permeability. Tensile strengths were found to be similar to synthetic films. The films were generally flavorless, and transparent to translucent depending on the protein source (Chen 1995; Lee and Rosenberg 2000).

Cold induced gelation of globular whey protein (Barbut and Foegeding 1993; Chen et al. 2006) can be achieved by adding calcium to a preheated mixture. A mechanism of cross-linking carboxyl groups with calcium has been suggested for the gelation of prenaturated whey proteins at ambient temperature (Roff and Foegeding 1996; Chen et al. 2006).

Because of their flexibility and amphiphilic nature, globular proteins rapidly adsorb on the emulsion interface, where they self-aggregate and form continuous and homogeneous membranes around oil droplets through intermolecular beta-sheet interactions (Lefèvre and Subirade 2003; Chen et al. 2006). By coating oil droplets with charged layers, protein films provide an electrostatic barrier against flocculation and coalescence.

Various procedures can yield microcapsules from whey proteins (Rosenberg and Lee 2004).

Origin and Isolation of Whey Proteins. Whey, the source of whey proteins, remains after removal of fat and caseins from milk. The following products are isolated: whey powders of different quality (demineralized, delactosed, and demineralized-delactosed), whey protein concentrate (WPC), whey protein isolate (WPI), lactalbumin, and individual whey protein fractions. WPC contains a low level of fat and typically 35–80 wt% protein. During the isolation of WPI, the removal of fat and lactose is intended. The protein content becomes > 90 wt%. β-Lactoglobulin is the most important whey protein (Madene et al. 2006).

It is likely that the range of proteins recovered from milk will extend in the future. Tailoring of milk protein products to meet specific functional requirements for individual applications will become increasingly important. The use of milk protein products in emerging technologies, such as the manufacture of edible films (Chen 1995) and the microencapsulation of ingredients (Rosenberg and Yong 1993) will benefit from further development (Ennis and Mulvihill 2000).

3.2.2.3 Gelatin

Chemical Description of Gelatin. (CAS# 9000-70-8) Gelatins (Ledward 2000) vary widely in their size and charge distribution. However, they generally have a characteristic primary structure determined by the parent collagen. Almost every third unit in all chains is glycine, followed by proline and 4-hydroxyproline as next frequently occurring residues. A typical amino acid composition of type A gelatin derived from pigskin is (residues per 100 units): glycine 33, prolin 13, alanine 11, hydroxyproline 9, arginine 5, serine 3.5, aspartic acid 3, lysine 3, glutamic acid 2.5, and further leucine, valine, phenylalanine, threonine, isoleucin, hydroxylysine, methionine, histidine, and tyrosine. Especially the minor components can vary, depending on the source of the raw material and processing technique. In type B gelatins, aspargine and glutamine are missing due to their conversion into aspartic acid and glutamic acid (Ledward 2000).

Gelatin is a heterogeneous mixture of single- or multi-stranded polypeptides, each with extended left-handed proline helix conformations and containing between 300 and 4,000 amino acids (Chaplin 2007).

Commercial gelatins are highly heterogeneous in size. Individual molecules may be single chains of molar masses less than 5×10^4 g/mol or multi-strand polymers with molar masses of over one million. The gelatin molecules are subdivided into several molar mass ranges corresponding to the most commonly occurring sizes, with the highest molar mass range covering 3.4×10^5–5.4×10^5 g/mol and the lowest one covering 1×10^4–2.5×10^4 g/mol. Nevertheless, each of these fractions is also heterogeneous with regard to the size and shape (Ledward 2000).

Physico-chemical Characteristics of Gelatin. Gelatins in their pure form are translucent, brittle solid substances, which are colorless or slightly yellow, almost tasteless, and odorless. Gelatins melt when heated and solidify when cooled again. Mammalian gelatins dissolve in hot water forming solutions of high viscosity, which gel on cooling below 35–40°C. Alternatively, the fish skin gelatin has a lower gelation temperature, typically around 5°C for cold-water fish skin gelatin. However, a gelation temperature of about 12°C was reported for warm-water fish skin gelatin (Avena-Bustillos et al. 2006). Gelatin is also soluble in most polar solvents. The aqueous solutions show viscoelastic flow and streaming birefringence. The solubility of the gelatin is determined by the source and the method of manufacture. If gelatins are put into contact with cold water, only portions dissolve, however, they can take up water up to 5–10 times its own weight while swelling to an elastic mass.

Gelatins are amphiphilic and can behave as polyelectrolytes in aqueous solution, depending on the pH. The isoelectric points of type A gelatins are in the range of pH 7–9.4, whilst type B gelatins have isoelectric points in the range of pH 4.8–5.5. Apart from these differences, type A gelatins usually have lower intrinsic viscosities for a given molar mass than type B gelatins. The amphiphilic nature provides good emulsifying property.

Gelatin gels formed on cooling of solutions with concentrations above about 1 wt%, depending on the quality of the gelatin and the pH, are clear, elastic, transparent, and thermo-reversible. On warming to 35–40°C, the gels will dissolve again. Whereas from dilute solutions gels with a "melt in mouth" texture are obtained, the gels from higher concentrated solutions exhibit elastic gum-like textures. Gelatin gels from aqueous solutions exist over only a limited temperature range, the upper limit being the melting point of the gel, which depends on the gelatin grade and concentration, and the lower limit, the ice point at which ice crystallizes (Ledward 1986).

In solution, gelatins undergo a coil-helix transition followed by the aggregation of helices to form collagen-like right-handed triple-helical proline/hydroxyproline rich junction zones. Gelatin films with higher triple-helix content swell less in water and are therefore much stronger (Bigi et al. 2004; Chaplin 2007).

The amphotheric hydrocolloid gelatin forms complex coacervates with anionic polysaccharides such as gum arabic. At low pH, gelatin becomes positively charged and also forms coacervates with negatively charged gellan gum (Madene et al. 2006).

Fish skin gelatins (Avena-Bustillos et al. 2006) of low pyrolidine content are far poorer gelling agents than gelatins of similar molar mass derived from warm-blood mammals (Ledward 2000). Nevertheless, the low water vapor permeability observed for warm-water fish gelatin, and particularly for cold-water fish gelatin films (Avena-Bustillos et al. 2006), can be advantageous for encapsulation applications to reduce water loss at low temperatures such as in refrigerated or frozen food systems.

Origin and Isolation of Gelatin. Gelatins do not occur naturally but are manufactured from collagens by processes that destroy the secondary and higher structures of collagen. The latter is the major constituent of all white fibrous connective tissue occurring in animals (cartilage, sinews, skin, ossein). The process of manufacture is very complex using hides, skins, or bones from domesticated cattle, pigs, and horses as preferred sources, but also fish skin is used as a source. Two principal processes are distinguished finally yielding two principally different types of gelatin: type A gelatins by acid treatment (pH 1.5–3.0) and type B by alkaline treatment (pH 12) of collagen. Overall, the source and the process determine the final gelatin properties. Numerous commercial gelatin types are available in the market.

3.2.3 Lipids

Characteristic for lipids is their general insolubility in water; they are hydrophobic. Lipids involve molecules and substances of large diversity and structural variety such as oils, fats, waxes, and phospholipids. These are widely distributed in nature.

3.2.3.1 Fatty Acids and Fatty Alcohols

Chemical Description of Fatty Acids and Fatty Alcohols. Two subgroups of fatty acids may be distinguished, saturated and unsaturated acids, which in addition can be of variable length. Saturated fatty acids are linear monocarboxylic acids of the overall formula $CH_3(CH_2)_nCOOH$. Alternatively, unsaturated fatty acids have one or more double bonds in their chain, giving rise to different chain configurations, namely cis and trans. Further, short-, medium-, and long-chain fatty acids are distinguished, which differ in the number of carbons in the aliphatic tail, having less than 8, 8–14, and more than 16 carbons, respectively. In fatty alcohols, a hydroxyl group has replaced the carboxylic group.

Physico-chemical Characteristics of Fatty Acids and Fatty Alcohols. Short chain aliphatic acids are miscible in water; however, the water-solubility rapidly decreases with increasing chain length. The melting points of fatty acids vary over a wide range, with the unsaturated acids having a much higher melting point. Examples are (Beyer 1968): palmitic acid ($n=14$) 62.6°C, stearic acid ($n=16$) 69.4°C, behenic acid ($n=20$) 74–78°C. At room temperature, fatty acids tend to undergo auto-oxydation.

Fatty alcohols behave as non-ionic surfactants and have emulsifying properties.

Origin and Isolation of Fatty Acids and Fatty Alcohols. Fatty acids are produced by the hydrolysis of the ester linkage of naturally occurring fats and oils, which are in general triglycerides. Glycerol is obtained as the byproduct from this process. Reduction of fatty acids yields fatty alcohols.

3.2.3.2 Glycerides

Chemical Description of Glycerides. Triglyceride (triacylglycerol, triacylglyceride), diglyceride (diacylglycerol), and monoglyceride (monoacylglycerol) belong to the family of glycerides. The principal chemical structures are shown in Fig. 3.17. Three, two, or only one fatty acid chains are covalently bonded to a glycerol molecule by ester linkages. In triglyceride, the three fatty acids may be the same, and may differ by one or all three. Thereof, the acids may be saturated or unsaturated. The most common residues are those with 16, 18, and 20 carbons. The same holds for diglycerides. Here, the two fatty acids can be located at any one of the three carbon positions. In monoglyceride, the fatty acid can be located either at the C-1 or C-2 position.

Physico-chemical Characteristics of Glycerides. Glycerides are not soluble in water. Di- and monoglycerides have emulsifying properties. The melting points of the glycerides strongly depend on the chemical nature but also on the symmetry of the fatty acid residues, their distribution over the carbon positions. For example, the highest melting point of milk fats has the fully symmetric tristearate with $T_m=72$°C (Walstra 1999).

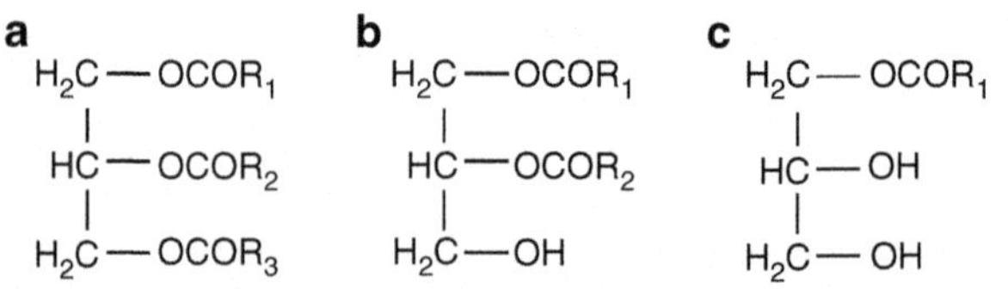

Fig. 3.17 Glycerides. (**a**) triglyceride, (**b**) diglyceride, (**c**) monoglyceride

Origin and Isolation of Glycerides. Triglyceride is the main constituent in animal fats and plant/vegetable oils. Most natural fats contain a mixture of different triglycerides. The heterogeneity of the mixture determines the span of the melting range. The commercial source of mono- and diglycerides is the same as for triglycerides, but total synthesis is also possible.

3.2.3.3 Waxes – Beeswax, Carnauba Wax, Candellia Wax

Chemical Description of Waxes (Parish et al. 2002). Waxes are esters of fatty acids. In contrast to fats and oils, the fatty acids are not esters of glycerol but of higher primary monovalent alcohols. The main components of beeswax (CAS# 8012-89-3) are triacontylpalmitate (about 75%), triacontylcerotinate (about 10%), and paraffin (about 15%) (Beyer 1968).

Physico-chemical Characteristics of Waxes. Waxes are practically insoluble in water. The color of beeswax varies from nearly white to brownish. It melts in the range of 62–64°C. Beeswax (Ross Waxes 2008a) is compatible with most other waxes and oils, fatty acids, glycerides, and hydrocarbons.

Carnauba wax (CAS# 8015-86-9) (Ross Waxes 2008b; Strahl and Pitsch 2008a) is one of the hardest natural waxes. The melting point is in the range 78–85°C (typically 83°C). The compatibility with other materials is similar as for beeswax.

Candelilla wax (CAS# 8006-44-8) (Ross Waxes 2008c; Strahl and Pitsch 2008b) is soluble in many organic solvents. It is light brown to light yellow, and melts in the range of 67–79°C. It is not as hard as carnauba wax. It is compatible with all vegetable and animal waxes, fatty acids, a large variety of natural and synthetic resins, glycerides, and hydrocarbons in certain proportions,

Origin and Isolation of Waxes. Waxes are isolated from animal and plant products. Beeswax is secreted by young honey bees to construct the honeycomb. Carnauba wax is obtained from the leaves of palm trees preferably in Brazil. Candellila wax is derived from the leaves of the Candelilla shrub, which grows in the northern Mexico.

3.2.3.4 Phospholipids – Liposomes

Chemical Description of Phospholipids. The general chemical structure of one of the main phospholipids or phosphodiglycerides, i.e., phosphatidylcholine, is shown in Fig. 3.18. Phospholipids contain two long chain fatty acids. The third hydroxyl group of the basic glycerol is modified with the phosphoric acid linked to a base such as choline or ethanolamine. Depending on the position of the phosphor-containing group, it is linked to the primary or secondary hydroxyl group of the glycerin, α and β phospholipids are distinguished (Beyer 1968). As with triglycerides, numerous species are possible by variation of the different head groups and fatty acyl substitution at the first and second position of the glycerol backbone (Weiner 2002). The most

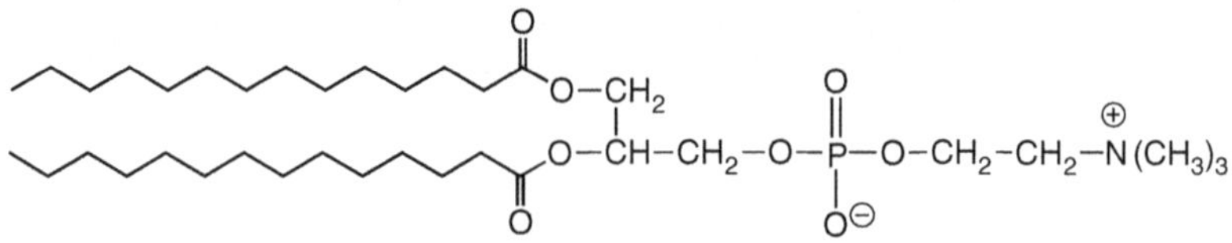

Fig. 3.18 Phospholipid – phosphatidylcholine

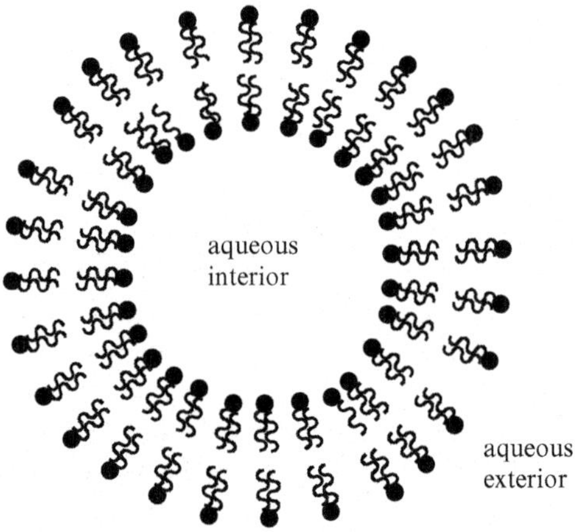

Fig. 3.19 Schematic structure of a liposome

abundant phospholipids are lecithins (phosphatidylcholine) (Taylor et al. 2005), where the base is choline. The fatty acid components may be saturated or unsaturated, for example palmitic, stearic, oleic, linoleic, or linolenic acid. Other polar lipids are kephalins (phosphatidyl-ethanolamines) where the base is ethanolamine.

It has to be mentioned that the name lecithin is also used for the naturally occurring mixture of glycolipids, triglycerides, and phospholipids.

Physico-chemical Characteristics of Phospholipids. Phospholipids are ionic amphiphiles (betains). Diacyl phospholipids generally form liquid crystalline suspensions as long as the temperature is held at or above the phase transition. Due to the amphiphilic character they function well as emulsifying and dispersing agents (Weiner 2002). When mixed with water, they aggregate or self-assemble into well organized and defined structures, and bilayers. Applying energy during the mixing process forces the bilayer to form liposomes, which in general do not form spontaneously (Taylor et al. 2005). In liposomes, an aqueous interior is separated by one or more phospholipid bilayers from the aqueous exterior (Fig. 3.19) (Lasic

1998). The manufacturing techniques used and the intensity of mixing employed, govern the size of the liposomes. The mesoscale shape taken by the liposome is a balance between applied mechanical forces and the tendency to assume specific self-assembled structures (Winterhalter and Lasis 1993). Generally, liposomes are only stable for a defined period of time; i.e., they are kinetically but not thermodynamically stable (Reineccius 1995).

One of the key parameters for liposomal systems is the so-called gel to liquid crystalline transition temperature, where the bilayer loses much of its ordered packing structure due to the "melting" of hydrocarbon chains of the lipids. Typically, longer chains, saturation of the acids, and strong head group interactions translate into higher phase transition temperatures. Liposomal content may leak out fast, especially around or above the gel to liquid crystalline temperature. In addition, liposomal properties and functionality depend on external parameters, such as pH and ionic strength of the medium (Taylor et al. 2005, see also Sect. 2.3.11).

Origin and Isolation of Phospholipids. Phospholipids are present in all animal and plant cells. Commercially, they are isolated from the egg yolk (lecithin, CAS# 93685-90-6) and soybean oil (lecithin, CAS# 8002-43-5) (Hsieh et al. 2002), but are also produced from milk fat globular membrane isolated from buttermilk (Thompson and Singh 2006). Lecithin is commercially available at high purity.

3.2.4 Others

3.2.4.1 Polyvinylpyrrolidone (PVP)

PVP, 1-Ethenyl-2-pyrrolidoinone homopolymer, (CAS# 9003-39-8), for which the chemical structure is shown in Fig. 3.20, is a synthetic neutral polymer with molar masses in the range of 10^4–5×10^5 g/mol. PVP powder is well soluble in water and organic solvents. The viscosity of the solution depends on the molar mass and the concentration. Aqueous solutions exhibit Newtonian behavior. The good film forming ability makes it an excellent polymer for coatings, which requires good temperature stability (Blecher et al. 1980).

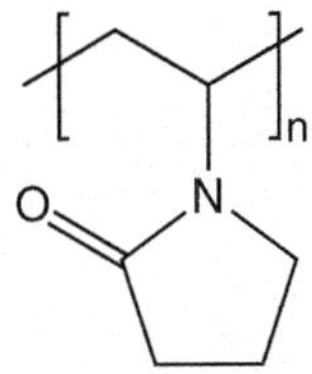

Fig. 3.20 Chemical structure of the polyvinylpyrrolidone

3.2.4.2 Paraffin

Paraffin (CAS# 8002-74-2) (Ross Waxes 2008d; Strahl and Pitsch 2008c) is a family of linear hydrocarbons with the general formula C_nH_{2n+2}. Dependent on the number of carbon atoms *n*, which can vary from 1 to about 40, they behave at ambient temperature as gas, liquid, or solid. Paraffin wax, the solid form, has $n>20$. It is a white, odorless, tasteless waxy material, which is also available at food-grade quality. Paraffin is edible but not digestible. Melting points ranging from 48 to 95°C are reported by suppliers and in table books (INCHEM 2008). The source is petroleum but synthesis is also possible.

3.2.4.3 Shellac

Shellac (CAS# 9000-59-3) is secreted by the lac insect *Laccifer lacca (Kerria lacca)*, found in the forests of northeast India, Thailand, Bangladesh, Indochina, and China. The secretion is collected from trees and processed in several steps to form shellac, which is soluble in alkaline media (ammonia, sodium carbonate, sodium hydroxide) and in several organic solvents. The exact chemical composition of shellac is unknown. It appears to be composed of a network of hydroxy fatty acid esters and sesquiterpene acid esters with a molar mass of about 1,000 g/mol. The composition is a function of the source and time of harvest. The physical properties of shellac may also vary. For example, the reported melting point ranges from 77°C to 120°C. Shellac is soluble in ethanol, methanol, glycols, glycol ethers, and alkaline water (Yates and Field 1960). Coatings obtained from alcoholic solutions are of superior durability and hardness. It is edible.

India, Thailand, and China are the main producers. The production volume is difficult to estimate but is assumed to be of the order of 10,000–20,000 tonnes per year, and increasing (Sengupta 1972; FAO 2008)

3.2.4.4 Inorganic Materials

There are several food-grade inorganic materials, which have been described as being useful for coatings or microencapsulation in food applications. They can be utilized alone or in combination with other materials. They include tripolyphosphate (CAS# 7758-29-4) (Desai and Park 2006), silicon oxides, or aluminum oxides (Amberg-Schwab et al. 2006).

3.3 Analytical and Characterization Methods

The suitability of a GRAS chemical substance for the encapsulation basically results from its molecular characteristics, i.e., the chemical composition, the molecular architecture, the molar mass, as well as the homogeneity/heterogeneity/

polydispersity of these characteristics. Exclusively, these molecular parameters govern all substance/material properties including solid material properties, behavior of melts and solutions, and molecular interactions. Therefore, despite the fact that industrial interest is primarily on material properties and their correlation with the properties of the final products, the knowledge of molecular characteristics and their correlation with material properties is a prerequisite for correct material selection and process optimization so as to obtain the desired texture and functionality in food products. Table 3.6 lists material properties, which are important for encapsulation materials, and molecular characteristics from which they result.

Carbohydrate polymers, proteins, and lipids are isolated and manufactured from natural resources. For the majority thereof, the chemical composition, structure, molecular dimensions, and homogeneity vary more or less, as discussed in Sect. 2 of this chapter. Such variation can complicate and render more difficult their application since sophisticated quality control procedures have to be performed continuously.

Due to the different chemical nature, molecular architecture, size, and physical properties of carbohydrates, proteins, and lipids, the three groups require rather different characterization methods and analysis techniques. Common to all is that separation and purification has to precede all analysis and characterization.

This section is not intended to provide comprehensive information about polysaccharide, protein, and lipid analysis and characterization. The reader is referred instead to appropriate specialist literature. Each of the materials discussed in Sect. 3.2 has its own peculiarities. Therefore, focus will be on practical aspects of methods and techniques and will be limited to polysaccharides, the dominating materials of this chapter.

The most frequent situation encountered is that the materials supplier provides more or less comprehensive basic information/specification about the particular substance that has been ordered. This information can be sufficient in many cases for established industrial applications though can be insufficient for research as well as product and technology development. The basic information is usually restricted to the chemical structure/composition, degree of purity, and the range of macromolecular characteristics.

Table 3.6 Substance and molecular characteristics of macromolecular encapsulation materials

Substance characteristics	Molecular characteristics
Solubility	Chemical composition
Rheology of melt and/or solution	– Sequence of monomer units
Transition temperatures (T_m, T_g)	– Homopolymer or heteropolymer
Stability (pH, T)	– Neutral or charged
Surface activity	Molecule architecture
Film forming ability	– Linear
Gelation ability	– Branched
Crystallinity	– Cross-linked
Density	– Linkage of monomer units
	Molar mass/molar mass distribution

Only for these basic characteristics, which are subject to variability and have practical relevance, will the appropriate analytical and characterization methods and procedures be considered here. Some characteristics, such as the type of monomer units and their linkage, do not change for a given polysaccharide even if the percentage or the molar mass varies. They can be taken as known without the need for repetitive examination.

3.3.1 Isolation and Purification

Polysaccharides suitable for encapsulation have to be isolated and purified prior to quantitative analysis and characterization. Impurities, which do not interfere with the practical application, can falsify analytical results. In addition, some of the materials presented here are mixtures of two or more components, which have to be analyzed separately. Also, the percentage of the components is of practical interest. Its variation may affect the rheological behavior, surface activity, or gelation behavior.

Purification methods are mainly based on physical characteristics of the macromolecules such as molecular size, charge, polarity, solubility, or specific interactions. Commonly used purification and separation techniques are: filtration, centrifugation, dialysis, preparative size exclusion chromatography, ion exchange chromatography, affinity chromatography, or electrophoresis.

3.3.2 Composition Analysis

The dramatic improvement of analytical methods during the last two decades, the increase of the sensitivity, precision, accuracy, and complexity, has contributed to a more comprehensive and better identification and quantitative determination of biopolymers. Nevertheless, the analysis of biopolymers still remains a challenge.

The composition analysis of polysaccharides (and also of proteins) requires chain degradation (chemical or enzymatic hydrolysis) as a first step, followed by separation of the monosaccharides released and their derivatives. Finally, the detection and quantification complete the analysis. Frequently, the separation, detection, and quantification are combined (Aspinall 1982), for example gas chromatography with mass spectrometry (GC-MS). Further applicable separation methods are: high-performance liquid chromatography (HPLC), capillary zone electrophoresis, and some methods already listed for purification above. Preferably, spectroscopic methods are used for the detection and quantification including UV-VIS, spectroscopy, infrared spectroscopy (IR, FTIR), and nuclear magnetic resonance (NMR) spectroscopic techniques. For specific composition analysis colorimetry, circular dichroism (CD), and polarimetry can be used. NMR has become the most powerful and non-invasive physicochemical technique for determining polysaccharide structure (Brummer and Cui 2006).

Electrochemical methods such as the acid-base titration, potentiometry, complex titration, and conductometry (Wandrey and Hunkeler 2002) are useful in both identifying and quantifying charges in polysaccharides. These methods do not require degradation of the macromolecules.

Overall, the type of the biopolymer under investigation governs the selection of separation, detection, and quantification methods. Figure 3.21 illustrates the principal procedure of composition analysis.

3.3.3 Macromolecular Characterization

Macromolecular characterization methods serve to primarily determine the molar mass (MM) and/or the molar mass distribution (MMD). Commonly used absolute and relative methods are listed in Table 3.7, together with the parameters, which they deliver. Absolute methods, such as osmometry, analytical ultracentrifugation, light scattering techniques, or mass spectrometry directly yield the MM and/or the MMD. Relative methods, such as dilution viscometry, gel permeation chromatography/size exclusion chromatography (GPC/SEC), or field-flow fractionation (FFF) require previous calibration in order to obtain the MM or MMD. However, frequently the molecular separation techniques GPC/SEC and FFF are combined with detectors which yield directly information about the MM and MMD. Such coupled, hyphenated, and multidimensional procedures have become the state-of-the-art for the separation and analysis of macromolecules. In addition to information about the molecular dimensions, hydrodynamic, thermodynamic, and conformational information also become available. Analytical ultracentrifugation (AUC) is particularly suitable for the macromolecular characterization of polysaccharides (see e.g., Harding 2005).

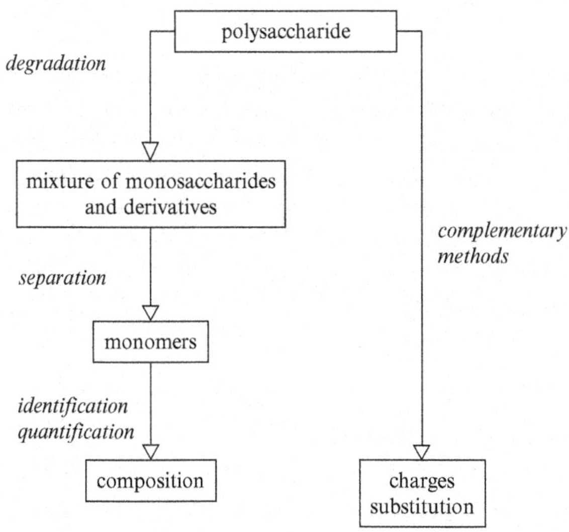

Fig. 3.21 Scheme for polysaccharide composition analysis

Table 3.7 Macromolecular characterization methods

Method	Direct	With calibration/fitting
Absolute methods		
Membrane osmometry	M_n	
Vapor pressure osmometry	M_n	
Analytical ultracentrifugation		
Sedimentation velocity	s, r_h	M_n, M_w, M_z, MMD
Sedimentation equilibrium	M_n, M_w	
Static (classic) light scattering	M_w, $<s^2>_z^{1/2}$	
Dynamic light scattering	D, r_h	M_d, MMD
Mass spectrometry	M_n, M_w, M_z, MMD	
Relative methods		
Dilution viscometry	$[\eta]$	M_v
GPC/SEC	V_e	r_h, MMD
Super critical fluid chromatography	V_e	MMD
Field-flow fractionation	V_e	r_h, MMD

GPC/SEC Gel permeation chromatography/size exclusion chromatography
M_n Number average molar mass
M_w Weight average molar mass
M_z *z*-average molar mass
MMD Molar mass distribution
s Sedimentation coefficient
r_h Hydrodynamic radius
D Diffusion coefficient
$<s^2>_z^{1/2}$ z-average root mean square radius
$[\eta]$ Intrinsic viscosity
V_e Elution volume

The macromolecular characterization of neutral polysaccharides can be performed, in general, in a two-component mixture, containing only the macromolecules and solvent. By contrast, the characterization of charged polysaccharides (polyelectrolytes) and proteins requires more efforts. Low molar mass salt/buffer has to be added to the aqueous solution for sufficient electrostatic screening, allowing the application of the usual characterization methods as for neutral macromolecules. Nevertheless, such multi-component mixtures remain more complicated. The electrostatic interactions, which interfere with the characterization, are suppressed but not removed.

3.3.4 Rheological Characterization

Since the rheological behavior of polysaccharides, proteins and lipids is of particular importance for encapsulation processes, some consideration of this is included in this chapter. Very often material suppliers specify macromolecular characteristics in terms of the dynamic viscosity of a solution of defined concentration. This is critical since a solution containing a mixture of two narrowly distributed fractions, differing in the average molar mass, can have the same dynamic viscosity as a solution containing only one broadly distributed fraction. However, the rheological behavior will differ considerably.

Rheological measurements can also be very useful for facilitating the acquisition of knowledge about the molecular size, shape, solvent and macromolecular interactions, and intermolecular network formation. Certain polysaccharides even exhibit a very characteristic rheological behavior due to their particular molecular structure and composition. Nevertheless, there are also certain generalities of the rheological behavior.

The rheological behavior of polymer solutions or melts depends on both the molecular characteristics and the concentration. There is a strong interrelation. Of course, the medium conditions (temperature, pH, ionic strength, type of salt ions present) have an impact too, however, they can be manipulated for comparison. Different concentration ranges have to be distinguished: first, concentrated solutions where polymer molecules are entangled, not separated, and form a network; secondly, dilute solutions where polymer molecules are separated from each other and molecular dimensions depend on the chemical structure, the molar mass, the temperature, and the quality of the solvent. The transition occurs at the so-called critical polymer concentration or overlap concentration c^*. Below c^*, neutral polymers in very good solvents can still slightly expand, however, retaining their coil conformation. In the case of polyelectrolytes, such as charged polysaccharides the coil expansion is much pronounced and also the electrostatic atmosphere around polyelectrolyte molecules becomes more extended. Under certain conditions, the molecules approach a rod-like conformation. Overall, and under otherwise comparable conditions, the higher the molar mass, the stiffer the molecules, and the higher the charge density of the polymer, the lower is c^*. Polysaccharide solutions normally exhibit Newtonian behavior at concentrations below c^*, whereas above c^* non-Newtonian behavior is usually observed.

To study the rheological behavior, dilution viscometry yielding the intrinsic viscosity and different techniques of rheometry are employed. The intrinsic viscosity, $[\eta]$, can be correlated with the molar mass in a Mark–Houwink relationship $[\eta]=KM^a$. The exponent a of such a correlation provides additional information about the macromolecular conformation.

Many polysaccharide solutions exhibit similar flow behavior at higher concentration. Frequently, Newtonian properties with a constant zero-shear viscosity over a limited shear range are observed at low shear rates. With increasing shear rate shear thinning occurs (pseudoplastic behavior), i.e., the solution viscosity decreases non-linearly. There are several rheological models to describe the flow behavior (Rao 1999).

Overall, to understand the correlation between chemical composition and molar mass on the one hand and the rheological behavior on the other hand, well purified and fractionated polysaccharide samples of low polydispersity have to be the subject of correlation studies. Even if such samples will never be used practically, their study is necessary for identifying reliable basic correlations, which then will contribute to the product development and process optimization.

3.4 Regulatory Aspects

3.4.1 Safety Evaluation of Existing and Potential Encapsulation Materials as Food Additives

All food additives including microcapsules must have demonstrated useful purpose and undergo a rigorous scientific safety evaluation before they can be approved for use (Saltmarsh 2000). At an international level there is a Joint Expert Committee, from the Food and Agriculture Organization (FAO) and the World Health Organization (WHO), on Food Additives (JECFA). The JECFA has been in existence since 1955 and serves as a scientific advisory committee to FAO, WHO, Member Governments and the Codex Alimentarius Commission. Its principal role is to assess the human health risks associated with the consumption of additives to food and to recommend Acceptable Daily Intake (ADI) levels. The acceptable daily intake (ADI) approach to toxicological evaluation was initiated by the Joint FAO/WHO Expert Committee on Food Additives in 1961. The procedure involves collecting all relevant data, ascertaining the completeness of the available data, determining the no-effect level using the most sensitive indicator of the toxicity, the so-called "no-observed-adverse-effect level" (NOAEL), and applying an appropriate safety factor to arrive at the ADI for man. The ADI provides a large safety margin and is the amount of a food additive that can be consumed daily over a lifetime without any adverse effect on health.

The Codex Alimentarius Commission, a joint FAO/WHO activity which develops guidelines for food safety globally, is also drawing up new "General Standards for Food Additives" (GSFA), with the aim of establishing a harmonized, workable, and indisputable international standard for world trade. Only those additives that have been evaluated by the JECFA can be applied as food additives. Thanks to such strict regulation and thorough testing, food additives can be considered safe components in human diet that are contributing to the rapid evolution of the food supply throughout the world.

Each listed and accepted food additive specified by JECFA has its own specification card in the Compendium of Food Additive and Flavoring Agents Specifications (JEFCA 2008a), where all information and definitions related to the chemical composition, structure and properties are listed. Moreover, each card describes in detail common standardized identification methods and, more importantly, identifies purity requirements on an acceptable level of dry content, other chemicals including heavy metals, and specific microbiological criteria. Furthermore, from the 61st JECFA meeting even more specific information including the chemical and technological aspects of new or existing food additives are available to the public as Chemical and

Technical Assessment (CTA). The CTA reflects and emphasizes the role that the chemical characterization plays in the risk assessment of food additives. These documents provide basic information related to identity, purity, and use of the food additive, as related to its risk assessment. The CTAs are available at the FAO JECFA website (JEFCA 2008b) although there is no intention at present to publish in print. Furthermore, useful information on characterization methods and techniques provide the Combined Compendium of Food Additive Specifications "Analytical Methods, Test Procedures and Laboratory Solutions, which is referenced in the JECFA specifications" (JEFCA 2006).

In the EU, up until the creation of the European Food Safety Authority (EFSA), the safety evaluation of additives in Europe had been done by the Scientific Committee on Food (SCF). At present, it is the EFSA Panel on Food Additives, Flavorings, Processing Aids and Materials in Contact with Food (AFC Panel), who is in charge of this task. EU legislation requires studies investigating the range of intakes across a population and to address any changes in consumption patterns. Occasional intakes over the ADI are unlikely to cause any harm because of the 100-fold safety margin. In Europe, the market for food products is harmonized by several rules for authorization and conditions for the use of additives. In 1989, the European Community adopted a Framework Directive (89/107/EEC) (European Parliament and Council Directive 87/107/EEC 1988), which set out the criteria by which additives would be assessed and provided for the adoption of three specific technical directives: Directive 94/35/EC on sweeteners; Directive 94/36/EC on colors, and Directive 95/2/EC on additives other than sweeteners and colors (European Parliament and Council Directive 95/2/EC 1995). These three directives establish the list of additives, which can be used (to the exclusion of others), the foods in which they can be used and any maximum levels that are appropriate. The purity required for these additives is laid down in directives defining specific purity criteria.

Regulatory information associated with already established or potential materials, which can be used for encapsulation in the food sector (Table 3.8), is continuously updated on the web and provided in standard bibliographies (e.g., General Standard for Food Additives 1995; Codex Alimentarius 2008).

3.5 Strategies for the Selection of Materials

Besides energy and water supply, food is, and will remain, one of the three major fundamental topics for the continuation of mankind. However, the problems of fast-multiplying populations in developing countries causing lack of resources on the one hand and serious health problems in industrially developed countries due to wrong nutrition on the other, require innovative solutions. Moreover, increasing consumer health concerns associated with growing demand for healthier food is stimulating innovative and new product development in the food industry internationally. This development is an important factor for the expanding worldwide

Table 3.8 List of food additives with INS Numbers and ADIs already applied or with potential in encapsulation technologies (JECFA-Reviewed Food Additives) (General Standard for Food Additives 1995)

Name	INS/E[a] number	ADI[b]
Alginic acid	400	NS
Sodium alginate	401	NS
Potassium alginate	402	NS
Ammonium alginate	403	NS
Calcium alginate	404	NS
Propylene glycol alginate	405	70 mg/kg
Agar	406	NL
Carrageenan	407	NS
Locust/carob bean gum (LBG)	410	NS
Guar gum	412	NS
Gum tragacanth	413	NS
Gum arabic	414	NS
Xanthan gum	415	NS
Karaya gum	416	NS
Tara gum	417	NS
Gellan gum	418	NS
Curdlan	424	NS
Konjac flour glucomannan	425	NS
Pectins (amidated and nonamidated)	440	NS
Cyclodextrin, alpha	457	NS
Cyclodextrin, gamma	458	NS
Cyclodextrin, beta	459	5 mg/kg
Powdered cellulose	460ii	NS
Methyl cellulose	461	NS
Ethyl cellulose	462	NS
Hydroxypropyl cellulose	463	NS
Hydroxypropyl methyl cellulose	464	NS
Methyl ethyl cellulose	465	NS
Sodium carboxymethyl cellulose	466	NS
Ethyl hydroxyethyl cellulose	467	NS
Cross-linked sodium carboxymethyl cellulose	468	NS
Sodium carboxymethyl cellulose, enzymatically hydrolyzed	469	NS
Beeswax, white and yellow	901	A
Candelilla wax	902	A
Carnauba wax	903	7 mg/kg
Shellac	904	A
Microcrystalline wax	905ci	20 mg/kg
Gum benzoic	906	ND
Polydextrose A and N	1,200	NS
Insoluble polyvinylpyrrolidone	1,202	NS
Dextrins, white and yellow roasted starch	1,400	NS

(continued)

Table 3.8 (continued)

Name	INS/E[a] number	ADI[b]
Acid treated starch	1,401	NS
Alkaline treated starch	1,402	NS
Bleached starch	1,403	NS
Oxidized starch	1,404	NS
Enzyme treated starch	1,405	NS
Monostarch phosphate	1,410	NS
Distarch glycerol	1,411	ND
Distarch phosphate	1,412	NS
Phosphated distarch phosphate	1,413	NS
Acetylated distarch phosphate	1,414	NS
Starch acetate	1,420	NS
Starch acetate esterified with vinyl acetate	1,421	NS
Acetylated distarch adipate	1,422	NS
Hydroxypropyl starch	1,440	NS
Hydroxypropyl distarch phosphate	1,442	NS
Starch sodium octenyl succinate	1,450	NS
Acetylated oxidized starch	1,451	NS

[a]*INS/E* International Numbering and E-number systems for food additives. The INS was prepared by the Codex Committee on Food Additives and Contaminants for the purpose of providing an agreed international numerical system for identifying food additives in ingredient lists as an alternative to the declaration of the specific name. An E-number signifies approval of an additive by the EU. To obtain an E-number, the additive must have been fully evaluated for safety by the SCF or the European Food Safety Authority. E stands for edible
[b]*ADI* acceptable daily intake estimated by JECFA/OFAS based on animal toxicology or human studies; expressed as either a numerical value (mg/kg bw/day – body weight per day), or in general as *A* acceptable, *NL* not limited, *NS* not specified – very low toxicity, *ND* not determined

interest in the functional food (Lopez-Rubio et al. 2006). Therefore, the creation of novel functionalities of active ingredients in complex food matrices is of increasing importance for the food industry.

Traditional active ingredients are flavors, vitamins, and minerals. Relatively novel ones are probiotic microorganisms and various classes of bioactive compounds (Ubbink and Krüger 2006). Moreover, the food industry is aware of the potential of particle coating technology and has identified numerous potential applications for its use (Werner et al. 2007). Nevertheless, compared to the pharmaceutical industry and medical fields, the microencapsulation, coating and embedding of active ingredients is as yet much less used.

By analyzing the publications related to materials used for microencapsulation in the food industry, which are able to entrap/encapsulate ingredients in homogeneous matrices and/or capsules surrounded by protecting walls, one will realize that many, if not the majority, of these publications have been based on the selection of materials on empirical grounds. Several materials and combinations are explored on a trial and error basis in order to identify the most suitable one. A strictly scientific selection starting with the analysis of the targeted application or product has rarely been reported. Nevertheless, principal strategies become obvious, namely selection of the material related to the technological requirement, selection considering the properties of the ingredient, or selection primarily oriented to the final product quality.

Of course, the application of microencapsulation in the food industry can have various goals such as the creation of a totally new product, improvement of an existing product, protection of a known ingredient, improvement of an existing process, replacement of an existing technology, along with others. All these aspects will also affect the strategies for the selection of a matrix/wall material. Nevertheless, the fundamental knowledge of the chemistry and physico-chemical properties of the materials will remain the prerequisite for successful product development. There is no substitute for the study and detailed analysis of the properties of potential matrix/wall material in order to conclude or predict its behavior under conditions present in food formulations. However, one has to be aware of the specific problems that may result from the use of natural materials, which can vary, within certain limits, from batch to batch. Serious restrictions may also result from economical aspects.

3.5.1 Use of Established Materials

A number of well-established industrial technologies for the microencapsulation of food ingredients including appropriate material description have been summarized and discussed in various review papers. A representative but not comprehensive list of such reviews is given here: Air-suspension particle coating in the food industry (Werner et al. 2007), Bioactive packaging: turning foods into healthier foods through biomaterials (Lopez-Rubio et al. 2006), Food protein-based materials as neutraceutical delivery systems (Chen et al. 2006), Physical approaches for the delivery of active ingredients in food (Ubbink and Krüger 2006), Flavor encapsulation and controlled release (Madene et al. 2006), Recent development in microencapsulation of food ingredients (Desai and Park 2005a), Liposomal nanocapsules in food science and agriculture (Taylor et al. 2005), Chances and limits of microencapsulation in modern food processing (Kruckeberg et al. 2003), Microencapsulation: industrial appraisal of existing technologies and trends (Gouin 2004), Cyclodextrins as food ingredients (Szente and Szejtli 2004), Technological challenges for future probiotic food (Mattila-Sandholm et al. 2002), Encapsulation in the food industry (Gibbs et al. 1999), Novel applications of liposomes (Lasic 1998), Fundamental aspects of controlled release in foods (Pothakamury and Barbosa-Cánovas 1995), Microencapsulation and the food industry (Jackson and Lee 1991).

Some of these papers include recommendations for the selection of materials. Tables 3.9 and 3.10 list recent microencapsulation studies with materials-oriented

Table 3.9 Examples of application of materials, which have been published recently

Active	Material for encapsulation	Technology	Comment	Reference
Vitamins				
Vitamin C, acerola, synthetic ascorbic acid	(a) Maltodextrin of DE20Mixture DE20/gum arabic (GA);		DE20/GA (3:1) most effective for vitamin C protection	(a) Righetto and Netto (2006)
	(b) Maltodextrin DE25, GA, or mix of both	(b) Spray-drying		(b) Righetto and Netto (2005)
Vitamin C	Tripolyphosphate/chitosan	Spray-drying	Process optimization	Desai and Park (2005b, 2006)
Vitamin E	Starch	High-pressure homogeniza-tion	Beverage fortification, nanoparticles	Chen and Wagner (2004)
Vitamin A acetate dissolved in coconut oil	hi-CAP 100 (starch octenylsuccinate, OSA-starch)	O/w emulsion, spray-drying	Effect of humidity	Xie et al. (2007)
Folic acid	Alginate-pectin	Coacervation	High efficiency: improved stability of folic acid during cheese ripening	Madziva et al. (2006)
Aroma/flavor				
European pear aroma	α-cyclodextrin (CD), GA, soybean soluble polysaccharide (SSPS), highly branched cyclic dextrin (HBCD)	Spray-drying, freeze-drying	Aroma content variation from 1.35 g/100 g with HBCD to 14.1 g/100 g with GA; MCs with α-CD and GA stable to 120°C	Tobitsuka et al. (2006)
	Cyclodextrins	Inclusion complexation	Fundamental data to obtain powdered pear flavor; best: α-CD	Tobitsuka et al. (2005)
Oregano, citronella, and marjoram flavors	Milk-protein based matrices: WPC, skimmed milk powder (SMP)	Spray-drying	Efficiency (% of flavoring entrapped in MC): 54.3% (marjoram in WPC) to 80.2% (oregano in SMP); particle size: 6–280 μm for SMP, 2–556 μm for WPC	Baranauskiene et al. (2006)

(continued)

Table 3.9 (continued)

Active	Material for encapsulation	Technology	Comment	Reference
Limonene	Gum arabic-sucrose-gelatin	Freeze-drying	1:1:1 most efficient	Kaushik and Roos (2007)
Cardamom oil	Mesquite gum	Spray-drying	High flavor retention of 83.6% for an oil: gum ratio of 1:4	Beristain et al. (2001)
Probiotics				
Bifidobacterium lactis	Hydrated gellan, xanthan gums	Extrusion	Suitable means for supplying viable probiotics to the food	McMaster et al. (2005)
Bifidobacterium PL1	Starch	Spray coating, spray-drying	Modified starch might not be suitable for use as an encapsulation material for probiotic strains	O'Riordan et al. (2001)
Lactobacillus acidophilusBifidobacterium. lactis	Alginate/Hi-Maize™ starch	Emulsion	Microencapsulation enhanced the survival of probiotic cultures compared to free cells in yogurts stored over 7 weeks, but negative influence on the textural properties	Kailasapathy (2006)
Lactobacillus. acidophilus 547B. bifidum ATCC 1994, Lactobacillus casei 01	Alginate/$CaCl_2$/chitosan		Survival of encapsulated bacteria was by about 1 log cycle higher	Krasaekoopt et al. (2006)
Bifidobacterium. breve R070Bifidobacterium. longum R023	Milk fat, denaturated whey proteins	Emulsion, spray-drying		Picot and Lacroix (2004)
Lactobacillus sp.	Gum arabic (GA), gellan gum (GG), mesquite gum (MG), and binary mixtures thereof	Interfacial polymerization	Highest viability in microcapsules of GA/MG mixtures	Yáñez-Fernández et al. (2008)
Lipids				
Conjugated linoleic acid (CLA)	Whey protein concentrate (WPC), gum arabic (GA), blend WPC/maltodextrin 10DE (1:1, w/w)	Spray-drying	WPC: best morphology and encapsulation efficiency, lowest CLA degradation	Jimenez et al. (2006)
Linoleic acid	Gum arabic	Spray-drying	Analysis of the oxidation process	Fang et al. (2005)

			best wall materials	(2006)
Lipids: oleic acid, linoleic acid, stearic acid	Potato starch, waxy maize starch, tapioca starch	Microwave heating	Facile process; formation of an amorphous matrix; no lipid interaction/reaction with starch matrix	Kapusniak and Tomasik (2006)
Oil				
Vegetable oil	Maltodextrin/GA, 3/2, w/w	Spray-drying + agglomeration in an air fluidized bed	Good oxidation protection; suitable floatability and wettability of the powders	Fuchs et al. (2006)
Fish oil	(a) Modified cellulose, skim milk powder, mixture of fish gelatin/corn starch	Spray-drying	(a) Changes in fish oil sensory quality; resulting powder was not stable; oxidized in the presence of air; stability may be improved when stored under vacuum	(a) Kolanowski et al. (2007)
	(b) Methylcellulose (MC), hydroxypropyl methylcellulose (HPMC), maltodextrin		(b) High oil retention; especially MC improved the stability and concentration in the powder	(b) Kolanowski et al. (2004)
	Modified starch	Spray-drying	Optimization of loading	Tan et al. (2005)
Tuna o/w emulsion	Lecithin-chitosan, corn syrup	Layer-by-layer deposition + spray-drying	Result: tuna oil droplets in a carbohydrate wall matrix; good physico-chemical properties	Klinkesorn et al. (2006)
Mixtures				
Ferrous sulfate + ascorbic acid	Liposomes	Pro-liposome and micro-fluidization	Size: about 5 μmSize: 150–200 nm Demonstration of the feasibility of microfluidization-based liposomal delivery systems for large scale food/nutraceutical applications	Kosaraju et al. (2006)
Iron (ferric pyrophosphate), iodine (potassium iodate), Vitamin A (retinyl palmitate)	Hydrogenated palm fat	Spray cooling	Microcapsules added to local salt in Morocco; highly stable microcapsules for salt fortification in Africa	Wegmuller et al. (2006)

(continued)

Table 3.9 (continued)

Active	Material for encapsulation	Technology	Comment	Reference
Others				
Chito-oligo saccharide	Polyglycerol monostearate (PGMS)		Application to commercial milk	Choi et al. (2006)
Anthocyanin pigments of black carrot	Maltodextrins: DE 10, DE 20-23, DE 28-31	Spray-drying	Best powder with DE 20–23	Ersus and Yurdagel (2007)
Isoflavone, β-galactosidase	Medium-chain triacylglycerol (MCT), polyglycerol monostearat (PGMS)		Both could be microencapsulated with fatty acid esters and released effectively in simulated intestinal condition	Kim et al. (2006)
allyl isothiocyanate – AIT (pathogens inhibitor)	Gum arabic		AIT microencapsulated in GA could be used in chopped refrigerated beef to reduce/eliminate *E. coli*	Chacon et al. (2006)
IgY	Whey protein concentrates (WPC-34, 50 and 80); whey protein isolate (WPI)	Emulsification, heat gelation	Effective for controlled IgY release to food systems	Lee et al. (2004)

Table 3.10 Examples of general applications of materials

Material	Application	Technology	Comment	Reference
Hybrid polymer coated with inorganic oxide layer (SiOx, AlOx)	Barrier against oxygen, water vapor, flavor permeation	Sol-gel	For food packaging	Amberg-Schwab et al. (2006)
Starch – linear amylose	Inclusion complexes with a wide variety of flavor compounds; controlled favor release	Complexation	Build-up and breakdown of starch flavor structures studied	Heinemann et al. (2005)
Whey protein concentrate 75	Soy oil	Spray-drying	Powders with oil content 20–75% (w/w)	Hogan et al. (2001a)
Sodium caseinate				Hogan et al. (2001b)
Beeswax, carnaubawax	Water-soluble compounds	Emulsion, solid or liquid preparation	Colorants	Mellema et al. (2006)

aspects. Table 3.11 reports about which materials can be used for which technology. This table may be regarded as an extension of Table 2.1 in Chap. 2. Often several materials are used in mixtures with other polymers. Therefore, some of the technologies listed in Table 3.11 may be applicable only if the appropriate material is mixed with others.

3.5.2 Identification and Definition of Criteria for Selection of a Material

The definition of the purpose of encapsulation is an important criterion for the selection of a material. The purpose could be, for example, the increase of shelf life, masking of taste, simplification of handling, guarantee of controlled and/or targeted release, or improvement of appearance. Answers to questions such as (Desai and Park 2005a):

- What functionality should the encapsulate provide to the final product?
- Are there restrictions for the coating material?
- Which concentration of the encapsulate has to be guaranteed?
- Which type of release is intended?
- What are the stability requirements?
- Are there cost constrains?

Table 3.11 Summary of the suitability of carbohydrate polymers, proteins, lipids, and a few other materials for principal technologies used for microencapsulation in the food industry

Material	Technologies
Starch and derivatives	Spray-drying, fluidized bed coating, extrusion, freeze-drying, microwave-assisted heating
– Maltodextrins	Spray-drying, fluidized bed coating, extrusion
– Syrups	Fluidized bed coating, extrusion, co-crystallization
– Cyclodextrin	Inclusion complexation (molecular inclusion)
Cellulose and derivatives (MC, HPMC, HPC, EMC, EC)	Spray-drying, fluidized bed coating, extrusion, emulsification/precipitation
– CMC	Coacervation, microwave-assisted encapsulation
Plant exudates (GA, GT, GK, MG)	Spray-drying, fluidized bed coating, extrusion, coacervation, freeze-drying
Plant extracts	
– Guar, LBG, tara	Extrusion, phase separation
– Pectins	Spray-drying, coacervation,
– SSPS	Spray-drying, freeze-drying
Marine extracts (carrageenans, alginate)	Spray-drying, extrusion, coacervation, emulsification
Microbial/animal extracts (xanthan, gellan, dextran, chitosan)	Spray-drying, coacervation, emulsification
Proteins	Fluidized bed coating
– Gluten	Spray-drying, coacervation, emulsification
– Caseins	Spray-drying, extrusion
– Whey proteins	Spray-drying, emulsification
– Gelatin	Spray-drying, extrusion, coacervation, freeze-drying
Lipids	
– Fatty acids and alcohols	Fluidized bed coating, spray chilling/cooling, extrusion, centrifugal suspension separation
– Glycerides	Spray chilling/cooling, extrusion, centrifugal suspension separation
– Waxes	Fluidized bed coating, extrusion, emulsification
– Liposomes	Liposomal entrapment
Others	
– PVP	Spray-drying
– Paraffin	Fluidized bed coating, spray chilling/cooling
– Shellac	Spray-drying
Inorganic	Sol-gel transition
– CO_2, N_2, water	Supercritical fluid technology

will further support the identification of a matrix or wall material.

Overall, materials for microencapsulation have to fulfill all or some of the following requirements (based on Desai and Park 2005a):

– Have good rheological properties at high concentration (if needed) and easy work ability during the encapsulation

- If applicable, disperse or emulsify the active material and stabilize the emulsion produced
- Do not react with the material to be encapsulated
- Seal and hold the active material within its structure during processing or storage
- If applicable, completely release solvent or other material used during encapsulation under drying or other desolvating conditions
- Provide maximal protection of the active against environmental conditions
- Are inexpensive
- Are food grade and legally allowed
- Are available at large quantities and constant quality

Depending on the ingredient, the process, and the application the requirements for the matrices may be different, requiring various shapes (films, spheres, irregular particles), or various structures such as compact or porous, amorphous or crystalline, rubbery or glassy. As an example, requirements for flavor encapsulation are: no reactivity with the core material, a form that is easy to handle (low viscosity at high concentration), complete elimination of solvent, maximum protection of the active ingredient, good emulsion stabilization properties, and effective redispersion behavior (Madene et al. 2006).

Ultimately, the approval of matrix/coating material by, for example, the FDA (US) or the European Food Safety Authority will govern the choice of material for microencapsulation in the food industry.

3.5.3 Strategies

After criticizing that the choice of technology and encapsulation material is often based on trial and error and not on a fundamental understanding of the physical and chemical phenomena determining the stability, release, perception, and digestion, Ubbink and Krüger (2006) proposed a strategy focusing on the desired functionality in the food product. The target application at the outset needs to be analyzed using scientific principles including materials science, physical chemistry, and biophysics. It is recommended to first precisely define the functionality and performance of the encapsulate in the final application, then to analyze the physical, chemical, and biological properties of the encapsulate and the conditions prevailing in the matrix/capsule. This includes identifying both the conditions required for maintenance of the performance of the ingredient and the limits to the conditions set by the properties of the matrix/capsule. It is emphasized that the functionality is defined solely based on the analysis of the interaction of the ingredient with the matrix/capsule and it does not relate at this stage to a specific technology. This concept/strategy of functionality postpones the selection of a technology to a later stage.

Werner et al. (2007) pointed out that the specific core and the ability of the coating material to impart the desirable characteristics to the product mainly influence the choice of appropriate coating material. It is their opinion that, although the performance of the coating in the final application is crucial, the matching of the material to the process technology and process conditions is likely to be of equal importance, and yet is almost certainly overlooked in practice. Typically, a large number of coating materials must be tested in order to determine their suitability. There is a lack of knowledge about matching of encapsulate and matrix/capsule/coating properties. It was, therefore, concluded that in order to speed up the product development it is important to establish guidelines for polymer selection, and based not only on their performance in the final application, but also on their behavior in a coating process. There are attempts to predict coating properties based on physico-chemical properties of the coating materials, their hygroscopicity, viscosity, surface tension, and glass transition data. However, wet film properties rather than solution properties determine the coating quality. To understand the coating process, knowledge is required about the physico-chemical properties of both the coating solution and the films that form on these solutions as they dry (Dewettinck et al. 1998).

An example of a quantitative method for selecting the most suitable biopolymer has been given by Pérez-Alonso et al. proposing blends (Pérez-Alonso et al. 2003). The estimation of the activation energy of carbohydrate polymer blends provided a quantitative discrimination parameter for selecting the most suitable materials for protecting microencapsulated lipids against oxidative deterioration. However, no correlation with molecular parameters was considered.

Despite proposing various strategies/concepts for the appropriate selection of materials, which differ in some aspects, as demonstrated by the examples considered above, most of these strategies have in common the need for detailed analysis and correlation of the properties of ingredients or materials. There is no disagreement with this point. Interdisciplinary scientific developments at the interfaces between biomaterials science, physical chemistry, biophysics, and encapsulation technology will increasingly be incorporated (Ubbink and Krüger 2006) and will contribute to the progress of using microencapsulation in the food industry. In the biomedical field, the development of matrices for controlled release of bioactive substances (drugs) is already a fact and an active research area with constant improvement. This extensive knowledge gathered in the pharmaceutical and medical fields can be further exploited or rethought for the development of novel functional foods (Lopez-Rubio et al. 2006). Some limitations may, however, exist due to the use of not always food grade materials. But there is no doubt about the technological aspects.

3.6 Concluding Remarks

This chapter demonstrates that a variety of food-grade materials is in principle available for use for microencapsulation of food ingredients. The previous Chap. 2 reported a number of technologies that are useful and applicable for this purpose. The challenge consists in adopting the microcapsules containing the active ingredients with a maximum of functionality to the whole food matrix. To achieve an optimum of the key functionality, which is usually determined by the barrier properties of the coating/film/matrix, a detailed knowledge of the chemical and physical properties of the food-grade encapsulation material is required. Due to the natural origin of the majority of these materials, a certain variability of the molecular characteristics and composition has to be taken into account. This requires defined specification controlled by sophisticated analysis and characterization. Further, the interaction between encapsulation material and ingredient deserves attention. Knowing the chemical nature of both may allow a priori conclusions. Nevertheless, for the final application, the properties of the coating/capsule/barrier formed by the carbohydrate polymers, proteins, lipids, their mixtures, or complexes are of particular importance. Because the microencapsulation technology and the final environment in the food may govern these properties, interdisciplinary collaboration is required for optimal solutions. Conclusions from studies, which establish basic correlations, will be favorable to support the development of food innovations. Apart from this, economical considerations will remain crucial for the selection of the most appropriate encapsulant materials.

Acknowledgments The authors acknowledge the support of Ingrid Margot during the preparation of the manuscript. We also thank Redouan Mahou for designing the chemical structures.

References

IUPAC-IUB Joint Commission on Biochemical Nomenclature (JCBN), Polysaccharide nomenclature (Recommendations 1980) European Journal of Biochemistry 126:439–441 (1982); Pure and Applied Chemistry 54:1523–1526 (1982); Journal of Biological Chemistry 257:3352–3354 (1982); Archives of Biochemistry and Biophysics 220:330–332 (1983); ref. 2, pp. 174–176

Amberg-Schwab S, Weber U, Burger A, Nique S, Xalter R (2006) Development of passive and active barrier coatings on the basis of inorganic–organic polymers. Monatsh Chem 137:657–666

Anderson DMW, Farquhar JGK (1982) Gum exudates from the genus *Prosopis*. Int Tree Crops J 2:15–24

Anderson DMW, Weiping W (1989) The characterization of proteinaceous *Prosopis* (mesquite) gums which are not permitted food additives. Food Hydrocolloids J 3:235–242

Ao Z, Jane J (2007) Characterization and modeling of the A- and B-granule starches of wheat, triticale, and barley. Carbohydr Polym 67:46–55

Aspinall GO, Baillie J (1963) Gum tragacanth. Part II. The Arabinogalactan. J Chem Soc 318:1714–1721

Aspinall GO (1982) Chemical characterization and structure determination of polysaccharides. In: Aspinall GO (ed) The polysaccharides, Vol. 1. Academic Press, New York, p 35

Avena-Bustillos RJ, Olsen CW, Olson DA, Chiou B, Yee E, Bechtel PJ, McHugh TH (2006) Water vapor permeability of mammalian and fish gelatin films. J Food Sci 71:E202–E207

Balla A, Blecker C, Razafindralambo H, Paquot M (1997) Interfacial properties of wheat gluten films from flours with different breadmaking qualities. Sci Aliments 17:271–278

Baranauskiene R, Venskutonis PR, Dewettinck K, Verhe R (2006) Properties of oregano (Origanum vulgare L.), citronella (Cymbopogon nardus G.) and marjoram (Majorana hortensis L.) flavors encapsulated into milk protein-based matrices. Food Res Int 39:413–425

Barbut S, Foegeding EA (1993) Ca^{2+}-induced gelation of pre-heated whey protein isolate. J Food Sci 58:867–871

Beristain CI, Garcia HS, Vernon-Carter EJ (2001) Spray dried encapsulation of cardamom (*Elettaria cardamomum*) essential oil with mesquite (*Prosopis juliflora*) gum. Food Science Technol – Lebenson Wiss Technol 34:398–401

Beyer H (1968) Lehrbuch der Organischen Chemie. S Hirzel Verlag, Leipzig

Bigi A, Panzavolta S, Rubini K (2004) Relationship between triple-helix content and mechanical properties of gelatin films. Biomaterials 25:5675–5680

Blanchard PH, Katz FR (1995) Starch hydrolysates. In: Stephen AM (ed) Food Polysaccharides and their applications. Marcel Dekker, Inc, New York, pp 99–122

Blanchard PH, Katz FR (2006) Starch hydrolysates. In: Stephen AM, Phillips GO, Williams PA (eds) Food polysaccharides and their applications, 2nd edn. Taylor & Francis, Boca Raton, FL, pp 119–145

Blecher L, Lorenz DH, Lowd HL, Wood AS, Wyman DP (1980) Polyvinylpyrrolidone. In: Davidson RL (ed) Handbook of water-soluble gums and resins. McGraw-Hill, New York, pp 21-1–21-17

Bottomley RC, Kearns HF, Schofield JD (1982) Characterisation of wheat flour and gluten proteins using buffers containing sodium dodecyl sulfate. J Sci Food Agric 33:481–491

Brummer Y, Cui SW (2006) Detection and determination of polysaccharides in food. In: Stephen AM, Phillips GO, Williams PA (eds) Food polysaccharides and their applications, 2nd edn. Taylor & Francis, Boca Raton, FL, pp 675–712

Buléon A, Colonna P, Planchot V, Ball S (1998) Starch granules: structure and biosynthesis. Int J Biol Macromol 23:85–112

Butler RW, Klug ED (1980) Hydroxypropylcellulose. In: Davidson RL (ed) Handbook of water-soluble gums and resins. McGraw-Hill, New York, pp 13-1–13-6

Chacon PA, Buffo RA, Holley RA (2006) Inhibitory effects of microencapsulated allyl isothiocyanate (AIT) against Escherichia coli O157:H7 in refrigerated, nitrogen packed, finely chopped beef. Int J Food Microbiol 107:231–237

Chandrasekaran R, Radha A (1995) Molecular architectures and functional properties of gellan gum and related polysaccharides. Trends Food Sci 6:143–148

Chaplin MF (2007) Water structure and Science. http://www.lsbu.ac.uk/water/. Accessed on October 18, 2007

Chapman DJ (1980) Seaweeds and their uses. Chapman and Hall, London and New York

Charalampopoulos D, Wang R, Pandiella SS, Webb C (2002) Application of cereals and cereal components in functional foods: a review. Int J Food Microbiol 79:131–141

Chen H (1995) Functional properties and applications of edible films made of milk proteins. J Diary Sci 78:2563–2583

Chen CC, Wagner G (2004) Vitamine E nanoparticles for beverage applications. Chem Eng Res Des 82:1432–1437

Chen LY, Remondetto GE, Subirade M (2006) Food protein-based materials as nutraceutical delivery systems. Trends Food Sci Technol 17:272–283

Choi HJ, Ahn J, Kim NC, Kwak HS (2006) The effects of microencapsulated chitooligosaccharide on physical and sensory properties of the milk. Asian-Australas J Anim Sci 19:1347–1353

Code Alimentarius (2008) http://www.codexalimentarius.net/web/standard_list.jsp and http://www.codexalimentarius.net/gsfaonline/additives/index.html

Coffey DG, Bell DA, Henderson A (1995) Cellulose and cellulose derivatives. In: Stephen AM (ed) Food Polysaccharides and their applications, 2nd edn. Marcel Dekker, Inc., New York, pp 123–153

Coffey DG, Bell DA, Henderson A (2006) Cellulose and cellulose derivatives. In: Stephen AM, Phillips GO, Williams PA (eds) Food polysaccharides and their applications, 2nd edn. Taylor & Francis, Boca Raton, FL, pp 147–179

Cottrell IW, Kang KS, Kovacs P (1980a) Xanthan Gum. In: Davidson RL (ed) Handbook of water-soluble gums and resins. McGraw-Hill, New York, pp 24-1–24-30

Cottrell IW, Kovacs P (1980b) Alginates. In: Davidson RL (ed) Handbook of water-soluble gums and resins. McGraw-Hill, New York, pp 2-1–2-43

Creamer LK, MacGibbon AKH (1996) Some recent advances in the basic chemistry of milk proteins and lipids. Int Dairy J 6:539–568

Critical data tables (1975) 3rd ed. Corn Refiners Association, Washington, DC

Davidson RL (1980) Handbook of water-soluble gums and resins. McGraw-Hill, New York

Desai KGH, Park HJ (2005a) Recent developments in microencapsulation of food ingredients. Drying Technol 23:1361–1394

Desai KGH, Park HJ (2005b) Encapsulation of Vitamin C in triphosphate cross-linked chitosan microspheres by spray drying. J Microencapsul 22:179–192

Desai KGH, Park HJ (2006) Effect of manufacturing parameters on the characteristics of Vitamin C encapsulated tripolyphosphate-chitosan microspheres prepared by spray-drying. J Microencapsul 23:91–103

Dewettinck K, Huyghebaert A (1998) Top-spray fluidized bed coating: Effect of process variables on coating efficiency. Food Sci Technol – Lebenson Wiss Technol 31(6):568–575

Dewettinck K, Deroo L, Messens W, Huyghebaert A (1998) Agglomeration tendency during top-spray fluidized bed coating with gums. Food Sci Technol – Lebenson Wiss Technol 31(6):576–584

Diaz-Sobac R, Garcia H, Beristain CI, Vernon-Carter EJ (2002) Morphology and water vapor permeability of emulsion films based on mesquite gum. J Food Process Preserv 26(2):129–141

Domard A, Domard M (2002) Chitosan: Structure-Properties Relationship and Biomedical Application. In: Dumitriu S (ed) Polymeric Biomaterials. Marcel Dekker, New York, pp 187–212

Draget KI (2000) Alginates. In: Phillips GO, Williams PA (eds) Handbook of hydrocolloids. Woodhead Publishing Limited, Cambridge, England, pp 379–395

Draget KI, Moe ST, Skjåk-Bræk G, Smidsrød O (2006) Alginates. In: Stephen AM, Phillips GO, Williams PA (eds) Food polysaccharides and their applications, 2nd edn. Taylor & Francis, Boca Raton, FL, pp 289–334

Elias H-G (1992) Makromoleküle. Band 2, Technologie. Hüthig & Wepf Verlag, Basel, Heidelberg, New York

Eliasson A-C, Larsson K (1993) Cereals in breadmaking: A molecular colloidal approach. Marcel Dekker, Inc., New York

Ennis MP, Mulvihill DM (2000) Milk proteins. In: Phillips GO, Williams PA (eds) Handbook of hydrocolloids. Woodhead Publishing Limited, Cambridge, England, pp 189–217

Ersus S, Yurdagel U (2007) Microencapsulation of anthocyanin pigments of black carrot (Daucuscarota L.) by spray drier. J Food Eng 80(3):805–812

European Parliament and Council Directive 87/107/EEC (1988) on the approximation of the laws of the Member States concerning food additives authorized for use in foodstuffs intended for human consumption. Official Journal of the European Communities L40:27–33; 11.2.89

European Parliament and Council Directive 95/2/EC (1995) on food additives other than colors or sweeteners. Official Journal of the European Communities L61: 1–40; 18.3.95

Falshaw R, Bixler HJ, Johndro K (2001) Structure and performance of commercial kappa-2 carrageenan extracts I. Struct Anal, Food Hydrocolloids 15:441–452

Fang X, Shima M, Adachi S (2005) Effects of drying conditions on the oxidation of linoleic acid encapsulated with gum arabic by spray-drying. Food Sci Technol Res 11(4):380–384

FAO (1992a) Karaya gum. (Food and nutrition paper 52). FAO, Rome

FAO (1992b) Tragacanth gum. (Food and nutrition paper 53). FAO, Rome

FAO (1995) Gums, resins and latexes of plant origin. (Non-wood forest products 6). FAO, Rome

FAO (1999) Gum arabic. (Food and nutrition paper 52, addendum 7). FAO, Rome

FAO (2008) Non-wood forest products for rual income and sustainalble forestry insect dyes-5. http://www.fao.org. Accessed January 28, 2008

Fox PF (1989) The milk protein system. In: Fox PF (ed) Developments in Dairy Chemistry – 4. Functional Milk Proteins. Elsivier Applied Science Publishers, London, pp 1–53

Fox PF (1992) Advanced Dairy Chemistry – Volume 1. Elsevier Applied Science Publishers, London

Fuchs M, Turchiuli C, Bohin M, Cuvelier ME, Ordonnaud C, Peyrat-Maillard MN, Dumoulin E (2006) Encapsulation of oil in powder using spray drying and fluidised bed agglomeration. J Food Eng 75(1):27–35

Furuta H, Maeda H (1999) Rheological properties of water-soluble soybean polysacccharides extacted under weak acidic conditions. Food Hydrocolloids 13:267–274

General Standard for Food Additives (1995) Codex General Standard 1a, CODEX STAN 192 (1995/rev. 8-2007)

Giannouli P, Morris ER (2003) Cryogelation of xanthan. Food Hydrocolloids 17:495–501

Gibbs BF, Kermasha S, Alli I, Mulligan CN (1999) Encapsulation in the food industry: a review. Int J Food Sci Nutr 50:213–224

Gidley MJ, Reid JSG (2006) Galactomannans and other cell wall storage polysaccharides in seeds. In: Stephen AM, Phillips GO, Williams PA (eds) Food polysaccharides and their applications, 2nd edn. Taylor & Francis, Boca Raton, FL, pp 181–215

Goodrum LJ, Patel A, Leykam JF, Kieliszewski MJ (2000) Gum arabic glycoprotein contains glycomodules of both extensin and arabinogalactan-glycoproteins. Phytochemistry 54:99–106

Gouin S (2004) Microencapsulation: industrial appraisal of existing technologies and trends. Trends Food Sci Technol 15(7–8):330–347

Goycoolea FM, de la Barca AM Calderón, Balderrama JR, Valenzuela JR (1997) Immunological and functional properties of the exudate gum from northwestern Mexican mesquite (Prosopis spp.) in comparison with gum arabic. Int J Biol Macromol 21:29–36

Goycoolea FM, Morris ER, Richardson RK, Bell AE (1995) Solution rheology of mesquite gum in comparison with gum arabic. Carbohydr Polym 27:37–45

Gralen N, Karrholm M (1950) The physicochemical properties of solutions of gum tragacanth. J Colloid Sci 5(1):21–36

Grasdalen H, Larsen B, Smidsrød O (1979) A p.m.r. study of the composition and sequence of uronate residues in alginates. Carbohydr Res 68(1):23–31

Grasdalen H, Larsen B, Smidsrød O (1981) ^{13}C-n.m.r. studies of monomeric composition and sequence in alginate. Carbohydr Res 89(2):179–191

Greminger GK Jr, Krumel KL (1980) Alkyl and hydroxyalkylalkylcellulose. In: Davidson RL (ed) Handbook of water-soluble gums and resins. McGraw-Hill, New York, pp 3-1–3-25

Größl M, Harrison S, Kaml I, Kenndler E (2005) Characterisation of natural polysaccharides (plant gums) used as binding media for artistic and historic works by capillary zone electrophoresis. J Chromatogr 1077:80–89

Guiseley KB, Stanley NF, Whitehouse PA (1980) Carrageenan. In: Davidson RL (ed) Handbook of water-soluble gums and resins. McGraw-Hill, New York, pp 5-1–5-30

Harding SE (2005) Analysis of polysaccharide size, shape and interaction. In: Scott DJ, Harding SE, Rowe AJ (eds) Analytical Ultracentrifugation – Techniques and Methods. RCS Publishing, Cambridge, pp 231–252

Heinemann C, Zinsli M, Renggli A, Escher F, Conde-Petit W (2005) Influence of amylose-flavor complexation on build-up and breakdown of starch structures in aqueous food model systems. Food Sci Technol – Lebenson Wiss Technol 38:885–894

Hoefler AC (2004) Hydrocolloids. Eagan press, St. Paul, MN

Hogan SA, McNamee BF, O'Riordan ED, O'Sullivan M (2001a) Microencap-sulating properties of whey protein concentrate 75. J Food Sci 66:675–680

Hogan SA, McNamee BF, O'Riordan ED, O'Sullivan M (2001b) Microencapsulating properties of sodium caseinate. J Agric Food Chem 49:1934–1938

Holt C (1992) Structure and stability of bovine casein micelles. Adv Protein Chem 43:63–151

Hsieh YF, Chen TL, Wang YT, Chang JH, Chang HM (2002) Properties of liposomes prepared with various lipids. J Food Sci 67(8):2808–2813

Imberty A, Chanzy H, Pérez S (1988) The double-helical nature of the crystalline part of A-starch. J Mol Biol 201:365–378

Imeson A (2007) Carrageenan. In: Phillips GO, Williams PA (eds) Handbook of hydrocolloids. Woodhead Publishing Limited, Cambridge, England, pp 87–102

INCHEM (2008) http://www.inchem.org Accessed March 3, 2008

Inoue Y (1997) NMR determination of the degree of acetylation. In: Muzzarelli RAA, Peter MG (eds) Chitin Handbook. Atec, Grottamare, Italy, pp 133–136

Islam AM, Phillips GO, Sljivo A, Snowden MJ, Williams PA (1997) A review of recent developments on the regulatory, structural and functional aspects of gum arabic. Food Hydrocolloids 11(4):493–505

Jackson LS, Lee K (1991) Microencapsulation and the Food Industry. Food Sci Technol – Lebenson Wiss Technol 24:289–297

Janaswamy S, Chandrasekaran R (2002) Effect of calcium ions on the organization of iota-carrageenan helices: an X-ray investigation. Carbohydr Res 337:523–535

JEFCA (2006) Combined Compendium of Food Additives Specifications, vol. 4, FAO JECFA Monographs 1, Rome

JEFCA (2008a) Compendium of Food Additive and Flavoring Agents Specifications: JECFA specifications for food additives and flavoring agents, on-line edition http://www.fao.org/ag/agn/jecfa-additives/search.html?lang=en

JEFCA (2008b) Chemical and Technical Assessments, available on-line on FAO JECFA website: http://www.fao.org/ag/agn/agns//jecfa_archive_cta_en.asp

Jimenez M, Garcia HS, Beristain CI (2006) Spray-dried encapsulation of conjugated linoleic acid (CLA) with polymeric matrices. J Sci Food Agric 86:2431–2437

Jobling S (2004) Improved starch for food and industrial applications. Curr Opin Plant Biol 7:210–218

Kailasapathy K (2006) Survival of free and encapsulated probiotic bacteria and their effect on the sensory properties of yoghurt. Food Sci Technol – Lebenson Wiss Technol 39:1221–1227

Kapusniak J, Tomasik P (2006) Lipid microencapsulation in starch. J Microencapsul 23:341–348

Kara S, Arda E, Kavzak B, Pekcan Ö (2006) Phase transitions of κ-carrageenan gels in various types of salts. J Appl Polym Sci 102:3008–3016

Kasarda DD (1989) Glutenin structure in relation to wheat quality. In: Pomeranz Y (ed) Wheat is unique. American Association of Cereal Chemists, Inc, St. Paul, MN, pp 277–302

Kästner U, Hoffmann H, Dönges R, Hilbig J (1997) Structure and solution properties of sodium carboxymethyl cellulose. Colloids Surf A: Physicochem Eng Asp 123:307–328

Kaushik V, Roos YH (2007) Limonene encapsulation in freeze-drying of gum arabic-sucrose-gelatin systems. Food Sci Technol – Lebenson Wiss Technol 40:1381–1391

Keller RCA, Orsel R, Hamer RJ (1997) Competitive adsorption behaviour of wheat flour components and emulsifiers at an air-water interface. J Cereal Sci 25:175–183

Kikuchi S, Fang X, Shima M, Katano K, Fukami H, Aiachi S (2006) Oxidation of arachidonoyl glycerols encapsulated with saccharides. Food Sci Technol Res 12:247–251

Kim NC, Jeon BJ, Ahn J, Kwak HS (2006) In vitro study of microencapsulated isoflavone and beta-galactosidase. J Agric Food Chem 54(7):2582–2586

Kim YD, Morr CV (1996) Microencapsulation properties of gum Arabic and several food proteins: Spray-dried orange oil emulsion particles. J Agric Food Chem 44(5):1314–1320

Kim YD, Morr CV, Schenz TW (1996) Microencapsulation properties of gum Arabic and several food proteins: Liquid orange oil emulsion particles. J Agric Food Chem 44(5):1308–1313

Klinkesorn U, Sophanodora P, Chinachoti P, Decker EA, McClements DJ (2006) Characterization of spray-dried tuna oil emulsified in two-layered interfacial membranes prepared using electrostatic layer-by-layer deposition. Food Res Int 39(4):449–457

Kokelaar JJ, Prins A, de Gee M (1991) A new method for measuring the surface dilational modules of a liquid. J Colloid Interf Sci 146:507–511

Kolanowski W, Jaworska D, Weissbrodt J, Kunz B (2007) Sensory assessment of microencapsulated fish oil powder. J Amer Oil Chem Soc 84(1):37–45

Kolanowski W, Laufenberg G, Kunz B (2004) Fish oil stabilisation by microencap-sulation with modified cellulose. Int J Food Sci Nutr 55(4):333–343

Kosaraju SL, Tran C, Lawrence A (2006) Liposomal delivery systems for encapsulation of ferrous sulfate: Preparation and characterization. J Liposome Res 16:347–358

Krasaekoopt W, Bhandari B, Deeth HC (2006) Survival of probiotics encapsulated in chitosan-coated alginate beads in yoghurt from UHT- and conventionally treated milk during storage. Food Sci Technol – Lebenson Wiss Technol 39(2):177–183

Krishnan S, Kshirsagar AC, Singhal RS (2005) The use of gum arabic and modified starch in the microencapsulation of a food flavoring agent. Carbohydr Polym 62(4):309–315

Kruckeberg S, Kunz B, Weissbrodt J (2003) Chances and limits of microencapsulation in progressive food industry. Chemie Ingenieur Technik 75(11):1733–174

Lasic DD (1998) Novel applications of liposomes. Trends Biotechnol 16:307–321

Le Cerf D, Irinei F, Muller G (1990) Solution properties of gum exudates from *Sterculia urens* (karya gum). Carbohydr Polym 13:375–386

Ledward DA (1986) Gelation of gelatin. In: Mitchell JR, Ledward DA (eds) Functional properties of food macromolecules. Elsevier Applied Science Publishers Ltd, London, pp 171–201

Ledward DA (2000) Gelatin. In: Phillips GO, Williams PA (eds) Handbook of hydrocolloids. Woodhead Publishing Limited, Cambridge, England, pp 67–86

Lee JJ, Park IB, Cho YH, Huh CS, Baek YJ, Park J (2004) Whey protein-based IgY microcapsules prepared by multiple emulsification and heat gelation. Food Sci Biotechnol 13(4):494–497

Lee SJ, Rosenberg M (2000) Whey protein-based microcapsules prepared by double emulsification and heat gelation. Food Sci Technol – Lebenson Wiss Technol 33(2):80–88

Lefèvre T, Subirade M (2003) Formation of intermolecular β-sheet structures: A phenomenon relevant to protein film structure at oil-water interfaces of emulsions. J Colloid Interf Sci 263:59–67

Li J-Y, Yeh A-I (2001) Relationships between thermal, rheological characteristics and swelling power for various starches. J Food Eng 50:141–148

Lopez da Silva JA, Rao MA (2006) Pectins: Structure, functionality, and uses. In: Stephen AM, Phillips GO, Williams PA (eds) Food polysaccharides and their applications, 2nd edn. Taylor & Francis, Boca Raton, FL, pp 353–411

Lopez-Rubio A, Gavara R, Lagaron JA (2006) Bioactive packaging: turning foods into healthier foods through biomaterials. Trends Food Sci Technol 17(10):567–575

MacRitchie F, du Cross DL, Wrigley CW (1990) Flour polypeptides related to wheat quality. In: Pomeranz Y (ed) Advances in Cereal Science and Technology, Vol. X. American Association of Cereal Chemists, Inc, St. Paul, MN, pp 79–145

Madene A, Jacquot M, Scher J, Desobry S (2006) Flavour encapsulation and controlled release - a review. Int J Food Sci Technol 41:1–21

Madziva H, Kailasapathy K, Phillips M (2006) Evaluation of alginate-pectin capsules in Cheddar cheese as a food carrier for the delivery of folic acid. Food Sci Technol – Lebenson Wiss Technol 39:146–151

Maeda H (2000) Soluble soybean polysaccharide. In: Phillips GO, Williams PA (eds) Handbook of hydrocolloids. Woodhead Publishing Limited, Cambridge, England, pp 309–320

Mangione MR, Giacomazza D, Bulone D, Martorana V, San Biagio PL (2003) Thermoreversible gelation of k-carrageenan: relation between conformational transition and aggregation. Biophys Chem 104:95–105

Mattila-Sandholm T, Myllarinen P, Crittenden R, Mogensen G, Fonden R, Saarela M (2002) Technological challenges for future probiotic foods. Int Dairy J 12(2–3):173–182

May CD (2000) Pectins. In: Phillips GO, Williams PA (eds) Handbook of hydrocolloids. Woodhead Publishing Limited, Cambridge, England, pp 169–188

McMaster LD, Kokott SA, Slatter P (2005) Micro-encapsulation of Bifidobacterium lactis for incorporation into soft foods. World J Microbiol Biotechnol 21:723–728

McNamee BF, O'Riordan ED, O'Sullivan M (1998) Emulsification and microencapsulation properties of gum arabic. J Agric Food Chem 46:4551–4555

Meer W (1980a) Gum Arabic. In: Davidson RL (ed) Handbook of water-soluble gums and resins. McGraw-Hill, New York, pp 8-1–8-21

Meer W (1980b) Gum Karaya. In: Davidson RL (ed) Handbook of water-soluble gums and resins. McGraw-Hill, New York, pp 10-1–10-13

Mellema M, Van Benthum WAJ, Boer B, Von Harras J, Visser A (2006) Wax encapsulation of water-soluble compounds in food. J Microencapsul 23:729–740

Moe ST, Draget KI, Skjåk-Bræk G, Smidsrød O (1995) Alginates. In: Stephen AM (ed) Food Polysaccharides and their applications. Marcel Dekker, Inc, New York, pp 245–286

Mohammadifar MA, Musavi SM, Kiumarsi A, Williams PA (2006) Solution properties of targacanthin (water-soluble part of gum tragacanth exudate from *Astragalus gossypinus*). Int J Biol Macromol 38:31–39

Morris VJ (1995) Bacterial Polysaccharides. In: Stephen AM (ed) Food Polysaccharides and their applications. Marcel Dekker, Inc, New York, pp 341–376

Morris VJ (2006) Bacterial Polysaccharides. In: Stephen AM, Phillips GO, Williams PA (eds) Food polysaccharides and their applications, 2nd edn. Taylor & Francis, Boca Raton, FL, pp 413–454

Motlagh S, Ravines P, Karamallah KA, Ma Q (2006) The analysis of Acacia gums using electrophoresis. Food Hydrocolloids 20:848–854

Murphy P (2000) Starch. In: Phillips GO, Williams PA (eds) Handbook of hydrocolloids. Woodhead Publishing Limited, Cambridge, England, pp 41–65

Murray JCF (2000) Cellulosics. In: Phillips GO, Williams PA (eds) Handbook of hydrocolloids. Woodhead Publishing Limited, Cambridge, England, pp 219–230

Nakamura A, Furuta H, Maeda H, Nagamatsu Y, Yoshimoto A (2000) The structure of soluble soybean polysaccharide. In: Nishinari V (ed) Hydrocolloids, Part 1. Elsevier Science, Amsterdam, pp 235–241

Nakamura A, Yoshida R, Maeda H, Corredig M (2006) Soy soluble polysaccharide stabilization at oil-water interfaces. Food Hydrocolloids 20(2–3):277–283

O'Riordan K, Andrews D, Buckle K, Conway P (2001) Evaluation of microencapsulation of a Bifidobacterium strain with starch as an approach to prolonging viability during storage. J Appl Microbiol 91:1059–1066

Örnebro J, Nylander T, Eliasson A-C (2000) Interfacial behavior of wheat proteins. J Cereal Sci 31:195–221

Orozco-Villafuerte J, Buendía-González L, Cruz-Sosa F, Vernon-Carter EJ (2005) Increased mesquite gum formation in nodal explants cultures after treatment with a microbial biomass preparation. Plant Physiol Biochem 43:802–807

Orozco-Villafuerte J, Cruz-Sosa F, Ponce-Alquicira E, Vernon-Carter EJ (2003) Mesquite gum: fractionation and characterization of the gum exuded from *Prosopis laevigata* obtained from plant tissue culture and from wild trees. Carbohydr Polym 54:327–333

Parish EJ, Boos TL, Li S (2002) The chemistry of waxes and sterols. In: Akoh CC, Min DB (eds) Food lipids: chemistry, nutrition, and biotechnology, 2nd edn. Marcel Dekker, New York, pp 103–132

Parker R, Ring SG (2001) Aspects of the physical chemistry of starch. J Cereal Sci 34:1–17

Patel TR, Picout DR, Harding SE, Ross-Murphy SB (2006) Pressure cell assisted solution characterization of galactomannans. 3. Application of analytical ultracentrifugation techniques. Biomacromolecules 7:3513–3520

Pedersen JK (1980) Pectins. In: Davidson RL (ed) Handbook of water-soluble gums and resins. McGraw-Hill, New York, pp 15-1–15-21

Perceval E (1972) Chemistry of Agaroids, Carrageenans and Furcellarans. J Sci Food Agric 23(8):933–940

Perez-Alonso C, Baez-Gonzalez JG, Beristain CI, Vernon-Carter EJ, Vizcarra-Mendoza MG (2003) Estimation of the activation energy of carbohydrate polymers blends as selection criteria for their use as wall material for spray-dried microcapsules. Carbohydr Polym 53(2):197–203

Phillips GO, Ogasawara T, Ushida K (2007) The regulatory and scientific approach to defining gum arabic (*Acacia Senegal* and *Acacia seyal*) as a dietary fibre. Food Hydrocolloids 22:24–35

Phillips GO, Williams PA (2001) Tree exudat gums: natural and versatile food additives and ingredients. Food Ingredients Anal Int 23:26–28

Picker-Freyer KM, Düring T (2007) Physical Mechanical and Tablet formation properties of hydroxypropylcellulose: In pure form and in mixtures. AAPS Pharma Sci Tech 8(4):E92

Picot A, Lacroix C (2004) Encapsulation of bifidobacteria in whey protein-based microcapsules and survival in simulated gastrointestinal conditions and in yoghurt. Int Dairy J 14(6):505–515

Picout DR, Ross-Murphy SB, Errington N, Harding SE (2001) Pressure cell assisted solution characterization of polysaccharides. 1. Guar gum. Biomacromolecules 2:1301–1309

Picout DR, Ross-Murphy SB, Jumel K, Harding SE (2002) Pressure cell assisted solution characterization of polysaccharides. 2. LBG and tara gums. Biomacromolecules 3:761–767

Piculell L (1995) Gelling Carrageenans. In: Stephen AM (ed) Food Polysaccharides and their applications. Marcel Dekker, Inc, New York, pp 205–244

Piculell L (2006) Gelling Carrageenans. In: Stephen AM, Phillips GO, Williams PA (eds) Food polysaccharides and their applications, 2nd edn. Taylor & Francis, Boca Raton, FL, pp 239–287

Pothakamury UR, Barbosa-Canovas GV (1995) Fundamental aspects of controlled release in foods. Trends Food Sci Technol 6(12):397–406

Rao MA (1999) Rheology of fluid and semisolid foods. Aspen Publishers, Gaitherburg

Reineccius CA (1995) Liposomes for contolled-release in the food-industry. Encapsulation and Controlled Release of Food Ingredients. ACS Symp Ser 590:113–131

Reineccius TA, Reineccius GA, Peppard TL (2005) The effect of solvent interactions on alpha-, beta-, and gamma-cyclodextrin flavor molecular inclusion complexes. J Agric Food Chem 53:388–392

Renard D, Lavenant-Gourgeon L, Ralet M-C, Sanchez C (2006) Acacia senegal gum: continuum of molecular species differing by their protein to sugar ratio, molecular weight, and charges. Biomacromoleccules 7:2637–2649

Righetto AM, Netto FM (2005) Effect of encapsulating materials on water sorption, glass transition and stability of juice from immature acerola. Int J Food Prop 8(2):337–346

Righetto AM, Netto FM (2006) Vitamin C stability in encapsulated green West Indian cherry juice and in encapsulated synthetic ascorbic acid. J Sci Food Agric 86:1202–1208

Roberts GAF (1992) Chitin Chimistry. Macmillan, London

Roff CF, Foegeding EA (1996) Dicationic-induced gelation of pre-denatirated whey protein isolate. Food Hydrocolloids 10:193–198

Rosenberg M, Lee SJ (2004) Water-insoluble, whey protein-based microspheres prepared by an all-aqueous process. J Food Sci 69:E50–E58

Rosenberg M, Yong SL (1993) Whey proteins as microencapsulating agents. Microencapsulation of anhydrous milkfat-structure evaluation. Food Struct 12:31–41

Ross Waxes, Frank B. Ross Co. Inc. (2008a) Data tables for Beeswax: http://www.frankbross.com/beeswax.htm

Ross Waxes, Frank B. Ross Co. Inc. (2008b) Data tables for Carnauba Wax: http://www.frankbross.com/carnauba_wax.htm

Ross Waxes, Frank B. Ross Co. Inc. (2008c) Data tables for Candelilla Wax: http://www.frankbross.com/candelilla_wax.htm

Ross Waxes, Frank B. Ross Co. Inc. (2008d) Data tables for Paraffin Wax: http://www.frankbross.com/paraffin_wax.htm

Rutenberg MW (1980) Starch and its modifications. In: Davidson RL (ed) Handbook of water-soluble gums and resins. McGraw-Hill, New York, pp 22-1–22-83

Saltmarsh M (ed) (2000) Essential Guide to Food Additives. Leatherhead Food RA Publishing, Surrey, Leatherhead, pp 1–322

Sanchez C, Renard D, Robert P, Schmitt C, Lefebvre J (2002) Structure and rheological properties of acacia gum dispersions. Food Hydrocolloids 16:257–267

Seaman JK (1980) Locust Bean Gum. In: Davidson RL (ed) Handbook of water-soluble gums and resins. McGraw-Hill, New York, pp 14-1–14-16

Sengupta SC (1972) Twenty-five years of research in lac. Indian Farming 22:13–23

Shahidi FR, Regg RB (1991) Encapsulation of the pre-formed cooked cured-meat pigment. J Food Sci 56:1500

Shewry PR, Tatham AS, Forde J, Kreis M, Miflin BJ (1986) The classification and nomenclature of wheat gluten proteins: a reassessment. J Cereal Sci 4:97–106

Singh H, MacRitchie F (2001) Application of polymer science to properties of gluten. J Cereal Sci 33:231–243

Singh N, Singh J, Kaur L, Singh Sodhi N, Singh Gill B (2003) Morphological, thermal and rheological properties of starches from different botanical sources. Food Chem 81:219–231

Smidsrød O, Glover RM, Whittington SG (1973) The relaive extension of alginates having different chemical compositions. Carbohydr Res 27(1):107–118

Smidsrød O, Hang A (1968) Dependence upon uronic acid composition of some ion.exchange properties of Alginates. Acta Chem Scand 22(6):1989–1997

Stauffer KR (1980) Gum Tragacanth. In: Davidson RL (ed) Handbook of water-soluble gums and resins. McGraw-Hill, New York, pp 11-1–11-30

Stelzer GI, Klug ED (1980) Carboxymethylcellulose. In: Davidson RL (ed) Handbook of water-soluble gums and resins. McGraw-Hill, New York, pp 4-1–4-28

Stephen AM, Churms SC (1995) Gums and Mucilages. In: Stephen AM (ed) Food Polysaccharides and their applications. Marcel Dekker, Inc, New York, pp 377–440

Strahl & Pitsch. Quality waxes (2008a) Data tables Carnauba Wax from http://www.spwax.com/spcarnau.htm

Strahl & Pitsch. Quality waxes (2008b) Data tables Candelilla Wax from http://www.spwax.com/spcandel.htm

Strahl & Pitsch. Quality waxes (2008c) Data tables Paraffin and Microcrystalline Wax from http://www.spwax.com/spparaff.htm

Sworn G (2000a) Gellan gum. In: Phillips GO, Williams PA (eds) Handbook of hydrocolloids. Woodhead Publishing Limited, Cambridge, England, pp 117–135

Sworn G (2000b) Xanthan gum. In: Phillips GO, Williams PA (eds) Handbook of hydrocolloids. Woodhead Publishing Limited, Cambridge, England, pp 103–115

Szente L, Szejtli J (2004) Cyclodextrins as food ingredients. Trends Food Sci Technol 15:137–142

Szejtli J (1998) Introduction and general overview of cyclodextrin chemistry. Chem Rev 98:1743

Tan LH, Chan LW, Heng PWS (2005) Effect of oil loading on microspheres produced by spray drying. J Microencapsul 22(3):253–259

Taylor TM, Davidson PM, Bruce BD, Weiss J (2005) Liposomal nanocapsules in food science and agriculture. Crit Rev Food Sci Nutr 45:587–605

Terbojevich M (2000) Cellulosics. In: Phillips GO, Williams PA (eds) Handbook of hydrocolloids. Woodhead Publishing Limited, Cambridge, England, pp 367–378

Thompson AK, Singh H (2006) Preparation of liposomes from milk fat globule membrane phospholipids using a Microfluidizer. J Dairy Sci 89:410–419

Tobitsuka K, Miura M, Kobayashi S (2005) Interaction of cyclodextrins with aliphatic acetate esters and aroma components of La France pear. J Agric Food Chem 53:5402–5406

Tobitsuka K, Miura M, Kobayashi S (2006) Retention of a European pear aroma model mixture using different types of saccharides. J Agric Food Chem 54:5069–5076

Ubbink J, Krüger J (2006) Physical approaches for the delivery of active ingredients in foods. Trends Food Sci Technol 17:244–254

Valla S, Ertesvåg H, Skjåk-Bræk G (1996) Genetics and biosynthesis of alginates. Carbohydr Eur 14:14–18

Van Vliet T, Martin AH, Bos MA (2002) Gelation and interfacial behavoiur of vegetable proteins. Curr Opin Colloid Interface Sci 7:462–468

Vårum KM, Smidsrød O (2006) Chitosans. In: Stephen AM, Phillips GO, Williams PA (eds) Food polysaccharides and their applications, 2nd edn. Taylor & Francis, Boca Raton, FL, pp 497–520

Veraverbeke WS, Delcour JA (2002) Wheat protein composition and properties of wheat glutenin in relation to breadmaking functionality. Critical Rev Food Sci Nutr 42(3):179–208

Verbeken D, Dierckx S, Dewttlinck K (2003) Exudate gums: occurrence, production, and applications. Appl Microbiol Biotechnol 63:10–21

Vernon-Carter EJ, Beristain CI, Pedroza-Islas R (2000) Mesquite gum (Prosopis gum). In: Doxastakis G, Kiosseoglou V (eds) Novel macromolecules in food systems, developments in food science 41. Elsevier, The Netherlands, pp 217–238

Voragen AGJ, Pilnik W, Thibault J-F, Axelos MAV, Renard CMGC (1995) Pectins. In: Stephen AM (ed) Food Polysaccharides and their applications. Marcel Dekker, Inc, New York, pp 287–339

Vorwerg W, Radosta S, Leibnitz E (2002) Study of a preparative-scale process for the production of amylose. Carbohydr Polym 47:181–189

Walstra P (1999) Diary Technology: Principles of milk properties and processes. Marcel Dekker, New York, p 63

Wandrey C, Hunkeler D (2002) Study of polyion counterion interaction by electrochemical methods. In: Tripathy S, Kumar J, Nalwa HS (eds) Handbook of polyelectrolytes and their applications, vol. 2. American Scientific Publishers, Stevenson Ranch, pp 147–172

Wegmuller R, Zimmermann MB, Buhr VG, Windhab EJ, Hurrell RE (2006) Development, stability, and sensory testing of microcapsules containing iron, iodine, and vitamin a for use in food fortification. J Food Sci 71(2):S181–S187

Weiner AL (2002) Lipid excipients in phamaceutical dosage forms. In: Encyclopedia of Pharmaceutical Technology EDs.: J. Swarbrick and J.C. Boylan. Marcel Dekker, New York, pp 1659–1672

Weiping W (2000) Tragacanth and karaya. In: Phillips GO, Williams PA (eds) Handbook of hydrocolloids. Woodhead Publishing Limited, Cambridge, England, pp 231–246

Werner SRL, Jones JR, Paterson AHJ, Archer RH, Pearce DL (2007) Air-suspension particle coating in the food industry: Part I - State of the art. Powder Technol 171(1):25–33

Wielinga WC (2000) Galactomannans. In: Phillips GO, Williams PA (eds) Handbook of hydrocolloids. Woodhead Publishing Limited, Cambridge, England, pp 137–154

Williams PA, Phillips GO (2000) Gum arabic. In: Phillips GO, Williams PA (eds) Handbook of hydrocolloids. CRC Press, Cambridge, England, pp 155–168

Williams PA, Phillips GO, Stephen AM, Churms SC (2006) Gums and Mucilages. In: Stephen AM, Phillips GO, Williams PA (eds) Food polysaccharides and their applications, 2nd edn. Taylor & Francis, Boca Raton, FL, pp 455–495

Winterhalter M, Lasis DD (1993) Liposome stability and formation: Experimental parameters and theories on the size distribution. Chem Phys Lipids 64:35–43

Winterowd JG, Sandford PA (1995) Chitin and Chitosan. In: Stephen AM (ed) Food Polysaccharides and their applications. Marcel Dekker, Inc, New York, pp 441–462

Wurzburg OB (1995) Modified starches. In: Stephen AM (ed) Food Polysaccharides and their applications. Marcel Dekker, Inc, New York, pp 67–97

Wurzburg OB (2006) Modified starches. In: Stephen AM, Phillips GO, Williams PA (eds) Food polysaccharides and their applications, 2nd edn. Taylor & Francis, Boca Raton, FL, pp 87–118

Xie YL, Zhou HM, Zhang ZR (2007) Effect of relative humidity on retention and stability of vitamin A microencapsulated by spray drying. J Food Biochem 31:68–80

Yáñez-Fernández J, Ramos-Ramirez EG, Salazar-Montoya JA (2008) Rheological characterization of dispersions and emulsions used in the preparation of microcapsules obtained by interfacial polymerization containing Lactobacillus sp. Eur Food Res Technol 226(5):957–966

Yates P, Field GF (1960) Shellolic acid, a cedrenoid sesquiterpene from shellac. J Am Chem Soc 82(21):5764–5765

Zohuriaan MJ, Shokrolahi F (2004) Thermal studies on natural and modified gums. Polymer Test 23:575–579

Chapter 4
Characterization Methods of Encapsulates

Zhibing Zhang, Daniel Law, and Guoping Lian

4.1 Introduction

Food active ingredients can be encapsulated by different processes, including spray drying, spray cooling, spray chilling, spinning disc and centrifugal co-extrusion, extrusion, fluidized bed coating and coacervation (see Chap. 2 of this book). The purpose of encapsulation is often to stabilize an active ingredient, control its release rate and/or convert a liquid formulation into a solid which is easier to handle. A range of edible materials can be used as shell materials of encapsulates, including polysaccharides, fats, waxes and proteins (see Chap. 3 of this book). Encapsulates for typical industrial applications can vary from several microns to several millimetres in diameter although there is an increasing interest in preparing nano-encapsulates. Encapsulates are basically particles with a core-shell structure, but some of them can have a more complex structure, e.g. in a form of multiple cores embedded in a matrix. Particles have physical, mechanical and structural properties, including particle size, size distribution, morphology, surface charge, wall thickness, mechanical strength, glass transition temperature, degree of crystallinity, flowability and permeability. Information about the properties of encapsulates is very important to understanding their behaviours in different environments, including their manufacturing processes and end-user applications. E.g. encapsulates for most industrial applications should have desirable mechanical strength, which should be strong enough to withstand various mechanical forces generated in manufacturing processes, such as mixing, pumping, extrusion, etc., and may be required to be weak enough in order to release the encapsulated active ingredients by mechanical forces at their end-user applications, such as release rate of flavour by chewing. The mechanical strength of encapsulates and release rate of their food actives are related

Z. Zhang (✉) and D. Law
School of Chemical Engineering, University of Birmingham, Edgbaston, Birmingham, B15 2TT, UK
e-mail: z.zhang@bham.ac.uk

G. Lian
Unilever Discover, Colworth, Shanrnbrook, Bedford, MK44 1LQ, UK

N.J. Zuidam and V.A. Nedović (eds.), *Encapsulation Technologies for Active Food Ingredients and Food Processing*,
DOI 10.1007/978-1-4419-1008-0_4,

to their size, morphology, wall thickness, chemical composition, structure etc. Hence, reliable methods which can be used to characterize these properties of encapsulates are vital. In this chapter, the state-of-art of these methods, their principles and applications, and release mechanisms are described as follows.

4.2 Physical Characterization Techniques

4.2.1 Characterization by Microscopy, X-ray and Laser Light Scattering

Physical properties of encapsulates include their size, size distribution, shape, morphology and surface charge, which can be measured by commercially available instruments based on principles of microscopy, X-ray and laser light scattering.

4.2.1.1 Microscopy

Optical and electron microscopy have been widely used to provide important information about size, surface topography, shell thickness and sometimes porosity of the shell material of encapsulates in food product. The performance and dissolution properties of encapsulates with active ingredients are often related to the physical properties.

Conventional Optical Microscopy

A standard upright optical microscope is capable of characterising structures that are 0.2 μm or greater as restricted by the wavelength of light. Often digital images or video images of a specimen are taken, followed by quantitative analysis using image analysis software. The size of single encapsulates can be measured. If encapsulates are not spherical, their shape may be characterized by a shape parameter circularity, which is sometimes called "roundness (RN)", and corresponds to a minimum value of unity for a circle. It is defined by the following equation:

$$RN = P^2 / \left(4\pi A\right) \tag{4.1}$$

where P is the perimeter of single encapsulates and A is their cross-sectional area.

Confocal Laser Scanning Microscopy

Confocal Laser Scanning Microscopy (CLSM) can be used to produce in-focus images of a fluorescent specimen by optical sectioning. The fluorescent specimen is illuminated by a point laser source, and each volume element of the specimen is associated with discrete fluorescence intensity. Images of a thick object can be obtained point-by-point and its three-dimensional structure can be reconstructed with a computer software package. For example, gelatin and gum arabic have been used to encapsulate an oil-based active ingredient by a complex coacervation process. The polymers were labelled with fluorescent markers. The spatial distribution of both gelatin and arabic gum throughout the encapsulate shell was identified by CLSM (Lamprecht et al. 2000). Moreover, when fluorescently labelled casein was added as a macromolecular model compound to the coacervation process, a gradiental distribution of casein in the shell materials was observed. It was found that casein had the highest concentration at the oil-wall interface. In another study, CLSM was demonstrated to be effective in studying the distribution of polymers and cross-linking ions in alginate-poly-L-lysine (PLL)-alginate encapsulates made by fluorescent-labelled polymers (Strand et al. 2003).

Transmission Electron Microscopy

Due to the limitation of optical microscopy, transmission electron microscopy (TEM) that is capable of resolving structures with smaller dimensions than optical microscopy has often been used (Mathiowitz 1999). TEM focuses electrons emitted by a heated filament to provide images restricted only by wavelength of electrons, which are approximately 0.003 nm. TEM must be operated under high vacuum to prevent collisions of electrons with air molecules, which can cause losses in resolution. Electromagnets are also used to provide sufficient energy to penetrate samples. TEM has been used as reliable microscopy to quantify the morphology and shell thickness of encapsulates (Xu and Du 2003; Chiu et al. 2005). Figure 4.1 shows a TEM micrograph of an encapsulate with a core of chamomile oil. However, this technique is time consuming as sample preparation can be quite tedious.

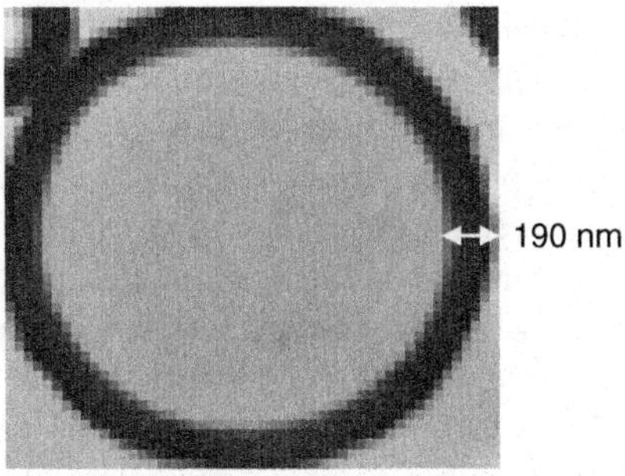

Fig. 4.1 TEM micrograph of a melamine formaldehyde micro-capsule encapsulating chamomile oil (Courtesy of Ms Yue Long, University of Birmingham, UK)

Scanning Electron Microscopy

Scanning electron microscopy (SEM) has a resolution of approximately 3 nm, which has the feature of simplicity in sample preparation and ease of operation. It has a great advantage because of its capability to analyse samples ranging in size from nano-metre to centimetre scale (Weiss et al. 1995; Shu et al. 2006; Yan-yu et al. 2006; Roueche et al. 2006). The spacious chamber and goniometer of a scanning electron microscope can accommodate relatively large samples as compared to a transmission electron microscope and provide nearly unlimited points of viewing with the assistance of translational, tilting, and rotary movements. Nonetheless, SEM does not distinguish colours as optical microscopy does and has lower resolution compared to TEM.

Environmental Scanning Electron Microscopy

Environmental Scanning Electron Microscopy (ESEM) offers several important advances in scanning electron microscopy. While a conventional scanning electron microscope requires a relatively high vacuum in the specimen chamber to prevent atmospheric interference with primary or secondary electrons, an environmental scanning electron microscope may be operated with a low vacuum (up to 10 Torr of vapour pressure, or one 76th of atmosphere pressure) in the specimen chamber and allows "wet mode" imaging, which is essential to investigate the surface tomography of wet encapsulates (Ren et al. 2007).

4.2.1.2 X-Ray Micro-computed Tomography

X-ray micro-computed tomography (CT) is an increasingly popular method to image complex three-dimensional structures with a spatial resolution in the micrometre range. The schematic diagram of an X-ray micro-CT system is shown in Fig. 4.2. The principle of this technique is by firstly acquiring X-ray images of an object placed between the X-ray source and the detector. Then the X-ray images are reconstructed and analysed using a commercially available software package. X-ray micro-CT offers the advantage of being a non-invasive and non-destructive testing method. Unlike some other testing methods, defects (including delamination, micro-cracking, fibre fracture, fibre pullout, matrix cracking, inclusions, voids, and impact damage) created during the preparation of the sample, which is required to adapt to a particular testing environment, can be avoided (Schilling et al. 2005). In addition, X-ray micro-CT allows unobstructed visual access to a sample's inner architecture in a completely non-destructive way and reconstructs interior structural details with a resolution on a scale of interest for such evaluation. It provides accurate details of variation of X-ray absorption within an object regardless of its structure properties and density gradients (Stock 1999). Previously X-ray micro-CT has been used successfully for quantitative measurements of localised density variations in cylindrical

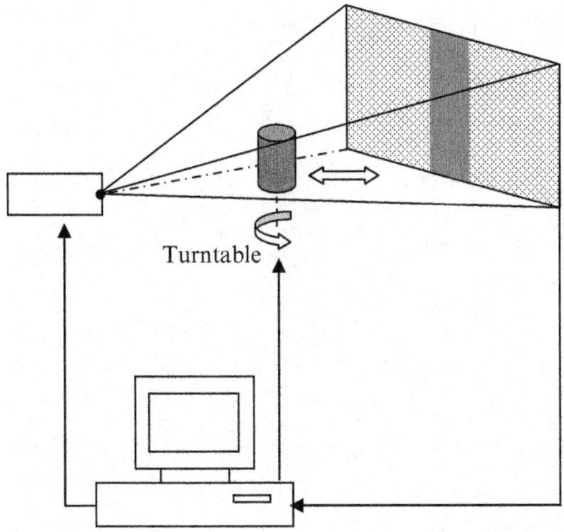

Fig. 4.2 Schematic diagram of the X-ray micro-computed tomography system

tablets. In all cases, an important heterogeneity in density was observed with higher densities concentrated in the peripheral region and lower densities in the middle (Busignies et al. 2006). This technique was also proven to be feasible to characterize the irregular interconnected pore structure of actual particle beds in order to study the fluid transport phenomena in a filter cake (Lin and Miller 2004). Recent work has shown that the internal structure of a calcium shellac bead can be characterized (see Fig. 4.3). However, the X-ray micro-CT cannot detect fine pores such as gel pores less than 0.01 μm (Rattanasak and Kendall 2005) and pores in granules less than 4 μm in diameter (Farber et al. 2003).

4.2.1.3 Laser Light Scattering

Although microscopy combined with image analysis can be used to measure the size of single encapsulates, it is a relatively slow technique particularly if a large number of encapsulates need to be measured. For spherical encapsulates, in wet or dry samples, with particles in the size range of 0.02–2,000 μm, measurement of their size distribution can be carried out using a laser light scattering technique, such as Malvern particle sizing, which can measure thousands of encapsulates within a couple of minutes. However, the refractive indexes of the shell material and the suspending medium are required. A similar technique to laser light scattering called single particle optical sensing (SPOS) has also been used to measure particle size distribution (O'Hagan et al. 2005). SPOS is based on measurement of the magnitude of pulse generated by single particles passing though a small photozone, illuminated by light from a laser diode or incandescent bulb, which can be correlated with the size of the particles. For non-spherical encapsulates or encapsulates which tend to form aggregates, microscopy combined with image analysis is

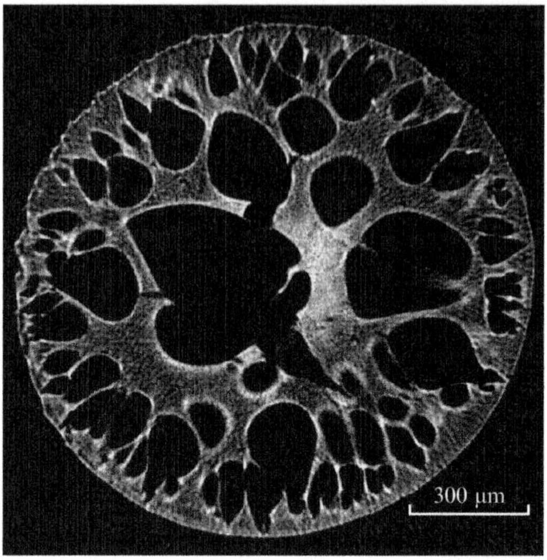

Fig. 4.3 Two-dimensional image of the central layer of a calcium shellac bead, which was reconstructed from pictures taken by X-ray micro-computed tomography (Law 2007)

preferred. Recently, a technique based on focused beam reflectance measurement (FBRM) has been used to provide in situ/on-line characterization of non-spherical particles by measuring chord lengths of particles, which are converted to particle size and shape of the particles by a mathematical model (Li et al. 2005). However, the data obtained by this technique require validation.

The electro kinetic potential in a suspension with encapsulates may be characterised by their zeta potential. It is a measure of the magnitude of the repulsion or attraction forces between encapsulates, which has a unit of mV. The value may be related to the storage stability of the encapsulates in suspension. In general, a particulate system with zeta potential smaller than −35 mV or greater than 35 mV tends to be stable (Legrand et al. 1999). Zeta potential can be measured by applying an electric field across the suspension. Encapsulates within the suspension with a zeta potential will migrate toward the electrode of opposite charge with a velocity proportional to the magnitude of the zeta potential. This velocity can be measured using laser Doppler anemometry. The frequency shift or phase shift of an incident laser beam caused by these moving encapsulates is measured as the particle mobility, which can be converted to the zeta potential by inputting the dispersant viscosity. Typical instruments for such measurement include Malvern Zetasizer Nano-Z (Malvern Instruments, UK).

It has been reported that the zeta-potential of polyamide encapsulates in suspension depends on the pH, concentration of electrolytes and surfactants in the suspension. Increasing the ionic strength led to a decrease in the zeta potential (Yan et al. 1992). Labhasetwar and Dorle (1991) measured the zeta potential of gelatin, methylcellulose and agar encapsulates at regular intervals during ageing at 45°C. It was found that there was an initial sharp rise in Zeta potential followed by a progressive decrease. This demonstrates that zeta potential is a useful parameter to investigate the changes occurring in the encapsulating material of encapsulates during ageing. Some encapsulates can be made using a layer-by-layer deposition, e.g. poly-L-lysine (P-Lys) and poly-L-aspartic acid (P-Asp) onto a negatively charged liposome. The zeta potential of such encapsulates changed between positive and negative at each layer deposition (Fujimoto et al. 2007).

4.2.2 Mechanical Characterization

Different techniques have been developed to characterize the mechanical properties of encapsulates, which can be classified into indirect and direct methods. A cone and plate shearing apparatus was established to measure the mechanical strength of encapsulates by determining their resistance to fluid shear force (Peirone et al. 1998). Alternatively, encapsulate strength was evaluated by determining the percentage of intact encapsulates remaining after being agitated with glass beads for a period of time (Leblond et al. 1996), or exposing them to bubble disengagement in a bubble column (Lu et al. 1992). The results from these indirect methods are difficult to interpret since the damage to the encapsulates depends not only on their mechanical strength, but also on the hydrodynamics of the process equipment, which is poorly understood. For semi-permeable encapsulates, an osmotic pressure test was implemented providing a rapid means of assessing their mechanical stability (Van Raamsdonk and Chang 2001).

Direct methods include compression of a layer of encapsulates between two glass plates. Their mechanical strength was determined based on the number of encapsulates broken under a given applied weight on the top plate, but it was observed that the weight concentrated onto the largest encapsulates first caused them to break, followed by the smaller ones (Ohtsubo et al. 1991). Although this method is practically useful, it conceals (any) difference in mechanical strength between encapsulates within a sample. Despite being more time consuming, results attained from the assessment of single encapsulates is more accurate and reliable as compared with a sample of encapsulates because the inhomogeneity of encapsulates in the sample due to their variations in size and structure can be identified (Schuldt and Hunkeler 2000). Direct methods on single encapsulates include the use of a micropipette aspiration technique or an atomic force microscope probe to measure the elastic properties of single encapsulates. Unfortunately, the former technique cannot be used to determine the force required to rupture the encapsulates (Grigorescu et al. 2002), whilst the latter relies on compression of single encapsulates between a rigid spherical bead and a flat surface (Lulevich et al. 2003), which is difficult to implement. A uniaxial compression of single encapsulates was attempted by means of a texture analyser consisting of a penetrometer with a stress gauge (Edwards-Levy and Levy 1999). It provided a measure of a particle's resistance to compressive force. However, assessment of single encapsulates was often prevented when their size went down to μm range (Martinsen et al. 1989). The limitation may be overcome by using a novel micro-manipulation technique that offers the capability to obtain the force vs. deformation data for compressing single encapsulates to rupture. The schematic diagram of a micro-manipulation rig is given in Fig. 4.4. The principle of this technique is to compress single encapsulates (dry or wet) between a probe connected with a force transducer and surface of a glass slide or glass chamber. The force transducer is connected with a fine micro-manipulator and its moving speed and distance can be controlled accurately. When single micro-encapsulates are compressed, the force imposed on them and their displacement are measured simultaneously. Moreover, the three-dimensional deformation of the encapsulates are also measured by two video cameras (for side view and bottom view respectively) connected to a video recorder or a computer with two image capture cards. This technique was initially used to measure the bursting strength of single animal cells, but later explored to mechanically characterize various biological and non-biological particles, including micro-spheres, encapsulates, etc. (Zhang et al. 1991; Zhang et al. 1992; Zhang et al. 1999; Stenekes et al. 2000; Sun and Zhang 2001; Zhao and Zhang 2004; Chung et al. 2005).

The micro-manipulation technique has been used to measure the mechanical strength of encapsulates of different size, shell thickness and shell composition (Zhang et al. 1999; Sun and Zhang 2001, 2002; Zhao and Zhang 2004). For example, the mechanical properties of single melamine formaldehyde (MF) encapsulates with diameters of 1–12 μm were determined, including their viscoelastic and elastic-plastic properties (Sun and Zhang 2001). It was found that the encapsulates were mainly elastic up to a deformation of 19 ± 1%. Beyond this point, the encapsulates underwent plastic deformation and were ruptured at a deformation of 70 ± 1%. However, the corresponding deformations at the yield point and at the rupture of

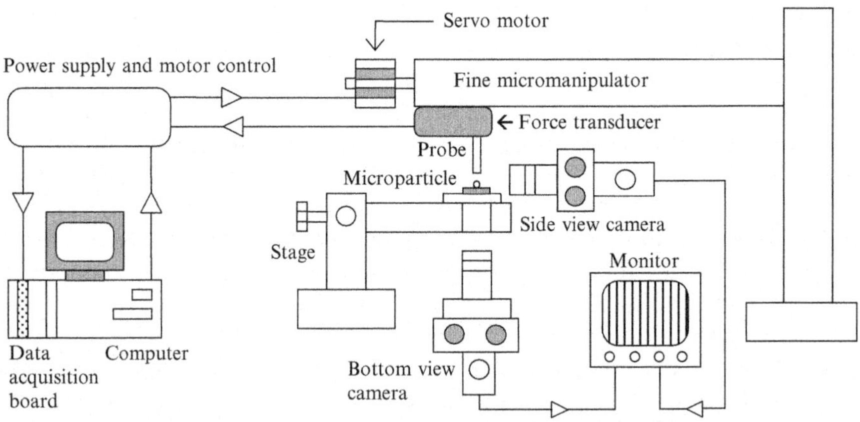

Fig. 4.4 Schematic diagram of the micro-manipulation rig. Permitted by Taylor & Francis

urea-formaldehyde encapsulates were 17 ± 1% and 35 ± 1% respectively, which implies that urea-formaldehyde encapsulates were more brittle than melamine formaldehyde (Sun and Zhang 2002). Besides the shell composition, the rupture strength of these encapsulates depended on their size and shell thickness. From the micromanipulation measurements the rupture force of different encapsulates can be compared by extracting information from a force-displacement curve up to the rupture. The curve can also be fitted with theoretical equations derived from a theoretical model, e.g. Lardner and Pujara's model (Lardner and Pujara 1980; Liu et al. 1996). This was used to determine the intrinsic mechanical property parameters, for example Young's modulus, and for further mathematical modeling to determine viscous-elastic and plastic parameters when appropriate. Such information is essential for quantifying the mechanical properties of encapsulates for a given sample and comparing the mechanical properties between different formulations.

Some encapsulates, e.g. calcium alginate beads, have certain permeabilities. When they are compressed, there can be a loss of liquid from the encapsulates and consequently their rupture force depends on the compression speed (Zhao and Zhang 2004; Wang et al. 2005). In order to minimise the speed-dependent behavior such encapsulates can be compressed at a high speed, e.g. 1,000 μm/s, which can be achieved using a newly developed micro-manipulation rig (Wang et al. 2005). This new high speed micro-compression tester, with two complementary high-speed videos, is a powerful tool for investigating the mechanical properties of hydrated encapsulates at the micro-scale.

The mechanical properties of single hydrated dextran encapsulates (<10 μm in diameter) with a model protein drug embedded were also measured by the micromanipulation technique, and the information obtained (such as the Young's modulus) was used to derive their average pore size and further to predict the protein release rate (Stenekes et al. 2000).

For encapsulates at the submicron scale, the above described micro-manipulation technique based on using conventional microscopy is not adequate to image single encapsulates clearly. Very recently, a novel ESEM-based nano-manipulation technique has been developed to characterize encapsulates at nano-scale. ESEM is primarily used to visualise materials on nano-scales under wet mode (Liu et al. 2005). To enable the mechanical properties of single encapsulates on such small scales to be measured, a nano-manipulation device with a force transducer was placed in the chamber of an environmental scanning electron microscope, and was used to compress single encapsulates and then measure the force imposed on them simultaneously. The nano-manipulation technique has been applied to investigate the mechanical properties of melamine formaldehyde (MF) encapsulates including their fracture mode (see Fig. 4.5) (Ren et al. 2007).

Understanding the mechanical strength of encapsulates is essential to a wide range of applications in controlled release of active ingredients including food encapsulates. Although not much work has been done on single food encapsulates, it is believed that the micro-/nano-manipulation technique described is a very powerful and unique tool to determine the mechanical properties of encapsulates made of different materials and structures.

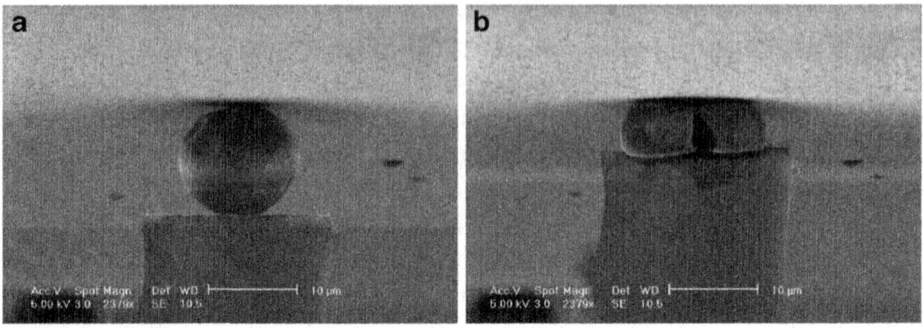

Fig. 4.5 A single melamine formaldehyde micro-capsule held between a force probe and a slide in the chamber of an environmental scanning electron microscope: (**a**) before compression; (**b**) after rupture. Permitted by Maney Publishing

4.2.3 Determination of Glass Transition and Degree of Crystallinity

The structure of encapsulates defines how the molecules in encapsulates are organised. Two parameters are often used to characterise the fine structure of encapsulates: glass transition temperature of encapsulates and crystallinity of encapsulate shell materials. These parameters may be used in designing encapsulates with desirable mass transfer properties (Le Meste et al. 2002).

4.2.3.1 Glass Transition Temperature

Glass transition temperature (T_g) is a property parameter of an amorphous material above which it behaves like a liquid (rubbery state) and many of the physical properties of the material change dramatically, including increase in heat capacity. Therefore, T_g can be determined by the measurement of a sudden increase in heat capacity with temperature, often using differential scanning calorimetry (DSC) (Bhandari and Howes 1999). DSC is a thermo analytical technique for measuring the energy required to maintain zero temperature difference between a sample and a reference (Hu et al. 2007). A similar technique to DSC is thermo gravimetric analysis (TGA), which is based on measuring weight changes associated with transformations of a material on heating (Foster and Clifford 1966), and has been used to identify the thermal decompositions of a copolymer (Shukla and Srivastava 2004).

Below the T_g the diffusion through carrier material is limited in a manner similar to those of a crystalline phase (glassy state). Above the T_g the carrier material is in rubbery state and the diffusion of molecules is relatively fast. This phenomenon can be used to design encapsulates that are storage stable below the T_g, which triggers release above a certain temperature.

Understanding the glass transition temperature may also be used to control the pore size of encapsulate shells. For example, encapsulates can be formed by close packing of a layer of polymer particles, e.g. poly(methyl methacrylate) (PMMA) at the interface of the oil/water (O/W) droplets and such systems have been called "colloidosomes" (Dinsmore et al. 2002). The adsorbed polymer particles were then swollen with further monomer and an initiator to effect further polymerisation in the particle shell layer, causing the particles therein to first bridge and then to grow into a more continuous film. This process was aided by heating the system at 105°C for 5 min, which is above the glass transition temperature ($T_g = 100$°C) of the polymer. It was found that after heating; the particles coalesced slightly, creating approximately 150-nm-diameter bridges between them. The resultant colloidosome therefore contained a precise array of uniform holes in an elastic shell. Increasing the sintering time led to smaller pores: after 20 min, the particles coalesced completely and the holes were fully closed.

Besides, measurement of the glass transition temperature can help to understand interactions between the molecules of core and shell materials. Sasaki et al. (2006) used DSC to measure T_g of encapsulates consisting of calcium carbonate whisker as a core and crosslinked polystyrene as a shell. The crosslinked shell was found to show higher glass transition temperatures (T_g) than the corresponding bulk values. It was also revealed that a thicker shell exhibited a lower T_g than a thinner shell, and that encapsulates without the core (hollow encapsulates) exhibited lower T_g than the corresponding core/shell encapsulates.

4.2.3.2 Degree of Crystallinity

The degree of crystallinity of the encapsulate shell materials can be characterised by DSC, X-ray diffraction (XRD) or nuclear magnetic resonance (NMR).

In addition to glass transition temperature, DSC can also be used to characterise crystalline melt temperatures. From knowledge of both the glass transition temperature and crystalline melt properties, information regarding the degree of crystallinity can be obtained. XRD relies on measurement of a diffraction pattern generated when an encapsulate specimen is irradiated with a parallel beam of monochromatic X-rays and the atomic lattice of the specimen acts as a three-dimensional diffraction grating causing the X-ray beam to be diffracted to specific angles. The angles and intensities of the diffracted beam can be used to infer the degree of crystallinity. NMR is a technique based on monitoring how spinning nuclei with magnetic dipoles interact with an applied magnetic field and absorb radiation (Seamus 2003; Jonathan 2003). It was found that the spectrum of NMR can be separated into two components: a broad component associated with the rigid crystalline region and a narrow component associated with mobile non-crystalline regions (Montes de Oca et al. 2004). NMR is designed for structural determination and analytical applications.

Ricciardi et al. (2004) used DSC, XRD and NMR to measure the degree of crystallinity of poly(vinyl alcohol) (PVA) hydrogels, and demonstrated that these three methods had different accuracies, which depended on the complexity of their structure.

The degree of crystallinity of polyamide capsules was estimated by XRD (Yoshioka et al. 2007), which were prepared from reacting diamines and diacid chlorides in an acetone or dioxane solution that included water, using the precipitation polymerization method employing ultrasonic irradiation. They found that the degree of crystallinity greatly depended on the combination of the diamine compound, diacid chloride compound, and reaction solvent.

The permeabilities of the polyurea micro-capsules for encapsulated cyclohexane were determined in conjunction with the degree of crystallinity of the polymer forming the membranes by XRD (Yadav et al. 1997). For cyclohexane diffusion through the water-swollen polyurea membranes used in their work, the permeability ranged from 6.5×10^{-9} to 3.75×10^{-8} m/s. It was shown that the product of the permeability and membrane thickness varied over an order of magnitude when the degree of crystallinity was changed from 21% to 33%.

In another study, nano-capsules based on polyureas and polyamides were prepared based on polycondensation reaction of two complementary monomers and spontaneous formation of oil in water emulsion. It has been found that the permeability of the polymeric wall of polyureas and polyamides was related to its crystallinity (Montasser et al. 2007).

4.2.4 Flowability of Dry Encapsulates

Dry encapsulates are like powders and their flowability is important in handling and processing operations, such as flow from hoppers and silos, transportation, mixing, compression, processing and packaging (Fitzpatrick et al. 2004a, b). The standard techniques to characterize the flowability of powders can also be applied to encapsulates although attention should be paid to handling encapsulates in order to minimise their mechanical damage.

Regarding the methods to characterize flowability of powders, Schulze (1996a, b) compared 16 flowability test methods of powders and concluded that the Jenike shear test was the most accurate method. Shear cell techniques for measuring powder flow properties were pioneered by Jenike (1964). In short, the horizontal shear forces needed to make a powder flow are measured as a function of different vertical pressures. A comprehensive explanation of the fundamental of shear cell techniques and schematic diagram can be found in Fitzpatrick et al. (2004a) and Carson and Wilms (2006). In conjunction with the data obtained from a shear cell technique, Jenike developed a theoretical background for the flow of solids from silos, a tester to determine the relevant flow properties, and a design procedure consistent with theory, flow properties, and industrial practice. Jenike test has been used to determine the influence of relative humidity and temperature on the flowability of different food powders (Teunou and Fitzpatrick 1999). However, the main disadvantage of Jenike test is that it is difficult to conduct (Zou and Brusewitz 2002). So far, there has not been much work in the public domain on the flowability of dry encapsulates.

4.3 Mass Transfer

In this section, the mechanisms of mass transfer that underpin the application of encapsulation are discussed. The emphasis has been placed on the mathematical modelling and experimental characterization of the fundamental factors that affect mass transfer, including solute partition and equilibrium, interfacial mass transfer and rate limiting steps.

4.3.1 Mass Transfer Mechanisms

Release of active ingredients from encapsulates can be induced or triggered by various methods including diffusion, mechanical rupture, melting, dissolution, hydration, enzyme attack, chemical reaction, hydrolysis, disintegration and so forth. In the food industry, the most commonly used method to trigger controlled release is by hydration. For example, the active is released from dry products when water is added. The release of the active from encapsulates depends on several mutually interacting processes such as diffusion of the flavor compound through the matrix, type and geometry of the particles, transfer from the matrix to the environment, and degradation/dissolution of the matrix material (Pothakamury and Barbosa-Canovas 1995). The dynamics of hydration and enzymatic reactions of the matrix with saliva also have important effects.

The physical process of controlled release of active ingredients from encapsulates can be described by diffusion and/or convection or a combination of both. The driving force of diffusion is the gradient in the chemical potential of the active ingredients between encapsulates and the surrounding environment. Convective mass transfer can be triggered by hydration, phase change, pH change, reaction and mechanical fracture of encapsulates. A good example of convective mass transfer is the mechanical rupture of micro-encapsulated flavor by chewing.

Other methods used to trigger controlled release include changes in pH and temperature. Karel and Langer (1988) reported a study on controlled release of enzymes from liposomes through the change in pH. Release of actives can also be triggered by the melting of wall material composed of lipids or waxes. Components such as salts, leavening agents, flavorings and nutrients are released when the shell is destroyed by melting. Encapsulates made of hardened fats are insoluble in water and the content can be released when they are subjected to shear or increased temperature which melts the fat. This type of encapsulates is widely used in soup mixes, bakery products or high-fat products (Dziezak 1988). The release of active ingredients depends strongly on the heat transfer that governs the melting process. Yeo et al. (2005) reported a study on flavor encapsulation in complex coacervate encapsulates using gelatin and gum arabic. Solubility of their encapsulate shell was temperature dependant.

In drug delivery, osmotic trigger has also been used. Molecules such as glucose, hydroxyethylcellulose, glycerol, poly (ethylene glycol)s, and dextrans of different molecular weights are extensively employed as osmotic agents (Santus and Baker 1995; Verma et al. 2002). The osmotic agents form the expandable core contained within a semi-permeable compartment. Pharmaceutical agent held within is released during the expansion and disintegration of the core material. Hydrogels have also attracted wide research interest as controlled release devices due to their tunable chemical and three-dimensional physical structure, high water content, good mechanical properties, and biocompatibility (Langer and Peppas 2003). Bioresponsive, "intelligent" or "smart" hydrogels can regulate drug release through responding to environmental stimuli by swelling and deswelling. Various bioresponsive and thermoresposive hydrogels have been developed for drug delivery (Zhang and Chu 2002; Huang et al. 2004).

For encapsulation in food industry, much of the challenge is balancing the need in protecting active molecules during processing and storage (from oxidation, leaking, and loss by evaporation) and controlling the rate of release to achieve optimal effect. Drying of the encapsulates/granules to a glassy state is a well-established technology and various drying processes have been developed including spray drying, fluidized bed drying, freeze drying, extrusion and microwave drying. In a dry glassy state, diffusion of active molecules is prohibited. The release of the active is triggered when dry encapsulates are hydrated. Diffusion of active molecules in hydrated matrix and/or dissolved solution becomes orders of magnitudes faster.

4.3.2 Mass Transfer Properties

A major challenge in controlled release is to understand the complex interactions of various dynamic processes accompanying the mass transfer. In particular, the dynamics of heat transfer, phase change, reaction, hydrolysis and hydration all have profound impact on the mass transfer properties of encapsulates. Understanding such complex interplay that governs the active release from encapsulates is critical.

Many researchers have sought a better understanding of the effects that govern the release of active ingredients from complex matrices of encapsulates as this represents an important target in many fields, including the food industry (Guichard 2000). An overview of physical chemistry relevant to flavor release has been presented previously (Taylor 1998). De Roos (2000) showed that two factors control the rate of flavor release from products, the comparative volatility of the aroma compounds in the food matrix and air phase under equilibrium conditions (thermodynamic factor) and the resistance to mass transport from product to air (kinetic factor).

Most encapsulates have complex heterogeneous structures. A wide range of materials can be selected for encapsulation. Micro-encapsulates can be made of sugars, gums, proteins, natural and modified polysaccharides, lipids and synthetic polymers (see Chap. 3 of this book). For food applications, the selection is somewhat limited because food safety is an additional consideration.

Mass transfer in heterogeneous encapsulates is exceedingly complex. A number of studies have been reported on the diffusion in complex networks of synthetic and/or biological gels (e.g. Kosto and Deen 2005; Pluen et al. 1999; Amsden 1998; Fatin-Rouge et al. 2003; Johnson et al. 1996; Phillips et al. 1989; Odijk 2000; Phillips 2000). Theoretical models proposed include hindered diffusion by hydrodynamic interaction, trapped diffusion by excluded structure volume and depletion theory. Diffusion and partition of solutes in lipid has received considerable attention.

Mass transfer in heterogeneous encapsulates depends strongly on the environmental condition under consideration. A good knowledge of the physico-chemical interactions occurring between compounds and the main constituents of foods such as lipids, polysaccharides, and proteins, is required for food active control. Mass transfer in complex encapsulates depends not only on the diffusion property but also on the thermodynamic properties such as gas-liquid equilibrium (for volatiles) and liquid-solid equilibrium.

4.3.2.1 Gas-Liquid Equilibrium for Release of Volatiles

Gas-liquid equilibrium is an important property of mass transfer of volatiles from encapsulates with a liquid core and/or liquid foods. When a volatile substance is added to a gas-liquid system, the substance will distribute itself in the liquid and gas phases. At equilibrium, the partial pressure of a volatile compound in the gas phase is related to its concentration in the liquid solution. In dilute solution, this relationship is described by Henry's law

$$p_{\mathrm{i}} = Hx_{\mathrm{i}} \tag{4.2}$$

where p_{i} is the partial pressure of a volatile compound in headspace, x_{i} is the mole concentration in the liquid solution, and H is Henry's constant.

For ideal gas, the partial pressure of volatile can be related to its mole fraction in the headspace and the atmosphere pressure by the relationship $y_i = p_i/P_0$. The other form of Henry's law is written as

$$y_{\mathrm{i}} = \frac{H}{P_0} x_{\mathrm{i}} = mx_{\mathrm{i}} \tag{4.3}$$

Net flux of mass transfer between gas and liquid takes place if the gas-liquid equilibrium is violated.

4.3.2.2 Gas-Solid Equilibrium for Release of Volatiles and Water Sorption Isotherm

Another important mass transfer property of encapsulates might be gas-solid equilibrium. The relationship is characterised by the sorption isotherm. The sorption of oxygen on encapsulates is related to the oxidation of lipids and other labile materials. The sorption of water vapour by encapsulates has received much attention because of its important effect on the storage and controlled release properties.

Water sorption isotherm is described by water activity (a_w), which is defined as the ratio of the partial pressure of water vapour at equilibrium, p_w, to the corresponding saturated vapour pressure, p_{sat}

$$a_w = \frac{p_w}{p_{sat}} \tag{4.4}$$

In other words, the water activity of a solid material is related to the relative humidity.

Water sorption isotherm of a given encapsulate system can be determined experimentally. Several empirical equations have been proposed. The Langmuir isotherm relates the water activity to adsorbed water content (θ_w) by

$$a_w\left(\frac{1}{\theta_w} - \frac{1}{\theta_0}\right) = \frac{1}{\beta\theta_0} \tag{4.5}$$

where θ_0 is the water content corresponding to the monolayer adsorption and β is a constant.

The other commonly used water sorption isotherm is the BET equation

$$\frac{a_w}{(1-a_w)\theta_w} = \frac{1+(\beta-1)a_w}{\beta\theta_0} \tag{4.6}$$

Water sorption isotherm has a number of effects on the mass transfer of active ingredient in encapsulates. Firstly, for many applications, hydration triggers the release of active ingredients. Secondly, water sorption isotherm has strong interaction with solute diffusivity of matrix materials. Thirdly, water vapor adsorption by encapsulates changes the quality of encapsulates during storage and transportation.

4.3.2.3 Solute Partition and Adsorption

Emulsions and dispersions are common structures of fabricated foods. When an active molecule such as flavor is added to a system of two immiscible phases, the solute may distribute itself in the two phases to reach equilibrium. The ratio of the concentrations of a solute in the two phases is called the partition coefficient. The partition coefficient of a solute in two immiscible fluids is relatively insensitive to temperature and concentration. The relationship is expressed by the following equation

$$C_y = KC_x \tag{4.7}$$

where K is the partition coefficients, and C_x, C_y are the concentrations of a solute in phase X and Y respectively.

Liquid-liquid equilibrium can also be extended to describe the partition of a solute binding and absorption to dispersed particles of complex fluids. The binding of volatile compounds to starch has been classified into two types. On the one hand, the flavor compound surrounded by the amylose helix through hydrophobic bonding is known as an inclusion complex. On the other hand, polar interactions have been determined which involve hydrogen bonds between the hydroxyl groups of starch and aroma compounds.

Of the many immiscible liquid-liquid systems, octanol-water equilibrium is widely studied and has generated great interest. Octanol-water partition coefficient is noted as P_{ow} or K_{ow}. Experimental data of octanol-water partition coefficients are available for a large number of chemicals. Methods have also been proposed for calculating octanol-water partition coefficients based on the molecular structures of solute. The fragment or group contribution method and atomistic method are the most commonly applied methods. The reader is referred to the monograph of Sangster (1997) for more details on the fundamental physical chemistry of octanol-water partition, including the measurement and prediction methods.

4.3.2.4 Solute Diffusion in Complex Media

The theory of Fick's diffusion applies to homogeneous media. The use of Fick's diffusion equation to describe the mass transfer of encapsulates is an oversimplified view. For emulsions and dispersions, where the size of the dispersed particles is sufficiently small compared to that of the encapsulates, it is possible to use the effective diffusion coefficient which is related to the diffusion and partition properties of the continuous and dispersed phases as follows (Lian 2000)

$$D_e = D_c \frac{D_c(1-\varphi)\xi + P_{dc}D_d(1+\xi\varphi)}{D_c(\xi+\varphi) + P_{dc}D_d(1-\varphi)} \tag{4.8}$$

where φ is the volume fraction of the dispersed phase, P_{dc} the partition coefficient of solute between dispersed phase and continuous phase, D_c and D_d are the diffusion coefficients of the solute in the continuous phase and dispersed phase respectively.

Thus, even for relatively simple structures of dispersions and emulsions, the effective diffusion property depends not only on the diffusion coefficients of the two phases, but also on the partition coefficient. The above effective diffusion coefficient was used to predict aroma release from oil-containing gel particles dispersed in aqueous solution and there was a good agreement with the experimental data (Lian et al. 2004).

Solute diffusion in complex fluids is affected by both temperature and molecular size. The following Einstein-Stokes equation can be used to estimate the diffusion coefficient of the active (D) in carrier material:

$$D = \frac{KT}{6\pi\eta r} \tag{4.9}$$

where K is the Boltzman constant, T is the temperature (K), η is the viscosity (Pa s), and r is the hydrodynamic radius of the active. This equation shows that the diffusion and thus the active release out of encapsulates increases with increasing temperature, decreasing viscosity of the carrier material and decreasing molecular size of the entrapped active.

Solute diffusion in lipids can also be estimated using the following equation (Mitragotri 2002)

$$D_{\mathrm{lip}} \approx D_0 \exp\left(-0.4r^2\right) \tag{4.10}$$

where D_0 is the diffusion coefficient in isotropic model hydro-carbon.

Diffusion of solute in complex networks of synthetic and/or biological gels has received significant attention (e.g. Kosto and Deen 2005; Pluen et al. 1999; Amsden 1998; Fatin-Rouge et al. 2003). Theoretical models proposed include hindered diffusion by hydrodynamic interaction, trapped diffusion by excluded structure volume and depletion theory (Amsden 1998).

4.3.3 *Characterizing Mass Transfer Properties*

Many of the physical-mechanical-structural properties of importance to the mass transfer of encapsulates have been discussed in the forgoing sections, for example, various microscopic methods for microstructure, DCS method for phase change and NMR for degree of crystallinity. Methods for characterizing the equilibrium properties (liquid-gas, solid-gas, and liquid-liquid) can be also found elsewhere (Rao and Rizvi 1994). Here, only methods for characterizing the dynamic mass transfer process of encapsulates are discussed.

4.3.3.1 Diffusion Cell

Diffusion cell can be used for measuring the diffusion properties of composite wall materials of encapsulates. The system is made of a membrane of the tested composite material separating two liquid-filled, well-stirred compartments. One compartment contains solute of a known concentration, whilst the other, known as the recipient side, is initially free of the solute. Measuring the concentration change with time allows the effective diffusion coefficient, D_e, to be obtained. The review by Westrin et al. (1994), concludes that the diffusion cell is a good method. However, the method is limited to encapsulate materials that can form thin membranes.

4.3.3.2 Release from Encapsulate Dispersion

The widely used method is to measure the release of active ingredients from encapsulates dispersed in a liquid media. Gentle agitation is applied to ensure good mixing and suspension of the dispersed encapsulates. The increase in solute concentration in the liquid phase is measured to determine the release kinetics and effective diffusion coefficient, D_e. Measurement can be automated and continuous, if an online analysis method for the active ingredient is available. The method has been generally regarded as an established method for measuring the release kinetics from particular systems, such as the flavor release from tea leaf particles (Lian and Astill 2002).

4.3.3.3 MS-Breath

MS-breath is a relative new method for in-mouth flavor release measurement. The method involves the use of atmospheric pressure chemical ionisation or proton transfer-reaction mass spectroscopy (APCI-MS or PTR-MS) to measure the real time release of volatile compounds from the mouth and into the nasal headspace during eating. Exhaled air from the nose is gently sucked into the mass spectrometer where the concentrations of the volatile compounds are continuously detected as protonated ions. The main advantage of the MS-breath method is that it obtains real time release profile that can be directly correlated to sensory properties. However, the release of the volatile from encapsulates to the nasal headspace has several rate-limiting steps and a complicated mathematical model is required to obtain the mass transfer properties of encapsulates (Lian et al. 2004).

4.3.4 Modeling Release Kinetics

To characterize mass transfer properties of encapsulates, reliable mathematical models are required to predict the release kinetics and fit with experimental data. Modeling release kinetics is also important for designing encapsulate systems (material, structure, size and loading) to achieve the desired release rate. When a simple diffusion model is used to describe the release kinetics of solutes from encapsulates, mathematical equations can be derived for various shaped encapsulates. For a homogeneous spherical encapsulate under infinite sink condition, the amount of released material is related to the size by the following Crank equation

$$\frac{M_t}{M_0} = 1 - \sum_{n=0}^{\infty} \frac{6}{\pi^2 n^2} \exp\left(-\frac{D_e \pi^2}{R^2} n^2 t \right) \tag{4.11}$$

where M_t is the amount of material released from the encapsulate, M_0 is the initial loading, D_e is the effective diffusion coefficient of solute through the encapsulate matrix and R is the radius of the encapsulate.

Corresponding to the early stage of release, the equation can be approximated by

$$\frac{M_t}{M_0} = 6\left(\sqrt{\frac{D_e t}{\pi R^2}} - 3\frac{D_e t}{R^2} \right) \tag{4.12}$$

This forms the so-called penetration theory for modeling flavor release with repetitive surface renewal (Overbosch et al. 1991).

At longer time, (4.11) can be approximated by the following equation

$$\frac{M_t}{M_0} = 1 - \exp\left(-1.2\pi^2 \frac{D_e}{R^2} t\right) \tag{4.13}$$

This is the first-order release kinetics which has been widely used to describe solute release from dispersions of encapsulates. The half-time at which 50% of the contents of the encapsulates is released can be calculated from (4.13) as follows:

$$\frac{M_t}{M_0} = 1 - \exp\left(-1.2\pi^2 \frac{D_e}{R^2} t_{1/2}\right) = 0.5 \tag{4.14}$$

which results in

$$t_{1/2} = \frac{\ln(0.5)R^2}{1.2\pi^2 D} \tag{4.15}$$

Equation (4.13) might also be described as:

$$\frac{M_t}{M_0} = 1 - \exp[-kt] \tag{4.16}$$

The term first-order equation has historically been derived from the equation:

$$\frac{\partial(M_0 - M_t)}{\partial t} = k(M_0 - M_t)^n \tag{4.17}$$

where n is the reaction order. Solving this latter equation for $n = 1$ gives the first-order release equation. For zero-order release, one assumes that the release is not dependent on its concentration and therefore $n = 0$:

$$\frac{\partial(M_0 - M_t)}{\partial t} = k \tag{4.18}$$

which can also be integrated to obtain the following equation

$$\frac{M_t}{M_0} = kt \tag{4.19}$$

In the latter equation, $t < 1/k$. Please note zero-order kinetic provides constant release of actives with time (see Fig. 4.6). When the active ingredient loaded to an encapsulate system is over saturated, the release rate of encapsulated solute to an infinite sink environment becomes constant.

The temperature dependence of k might be described by the Arrhenius equation:

$$k = Ae^{-E_a/R_g T} \tag{4.20}$$

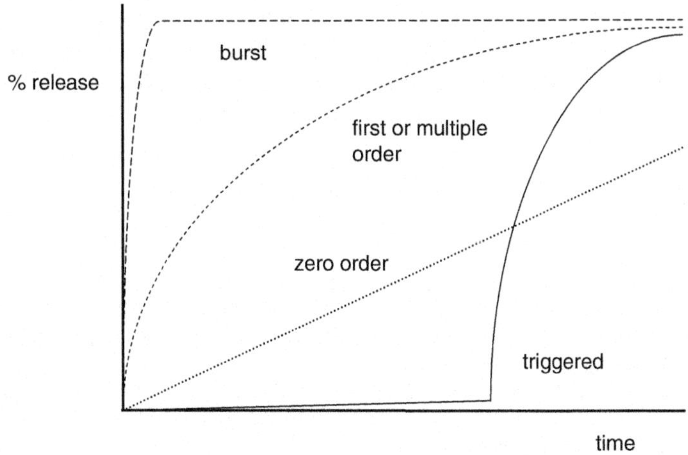

Fig. 4.6 Release profiles from encapsulates. Burst release is typical for soluble or broken ones. Release might also be triggered after a certain period of time by, e.g. pH change or addition of enzymes. Zero-order release is obtained with over saturated amounts of active in the core or with micro-capsules where the thin shell is the rate limiting step. First- or multiple-order release is common for matrix type of encapsulates and release due to other interactions. See also text

where A is the frequency factor, E_a is the activation energy, R_g is the gas constant.

Generally, encapsulates exhibit more complex release kinetics due to the interactions with other rate limiting steps such as hydration, structural breakdown and enzyme attack. Even for relatively simple systems such as in-mouth flavor release from oil-containing gel particles, there are two rate limiting steps, the release of aroma from gel particle to the saliva and the release of aroma from saliva to the retro nasal head space (Lian et al. 2004). The release kinetics exhibits two-components, corresponding to a fast component and a slow component. The gel effect of particle size interacts with the partition properties. Only for highly hydrophobic aroma molecules, the change of gel particle size has the expected effect.

4.4 Conclusions

The success of a formulation/encapsulation process depends on whether the formed encapsulates have desirable properties, including physical, mechanical, structural and mass transfer properties. In this chapter, the state-of-art methods which are currently used for characterization of these properties are reviewed. However, it seems that most of the published work on the characterization of the encapsulate properties focused on physical and mass transfer properties. Less attention has been paid to mechanical characterization, and a very limited amount of work has been done on structural characterization, and the relationship between these properties. In the future, more efforts should be made toward understanding the mechanical and structural properties, and how they are related to physical and mass transfer properties.

References

Amsden B (1998) Solute diffusion in hydrogels. An examination of the retardation effect. Polym Gels Networks 31:13–43

Bhandari BR, Howes T (1999) Implication of glass transition for the drying and stability of dried foods. J Food Eng 40:71–79

Busignies V, Leclerc B, Porion P, Evesque P, Couarraze G, Tchoreloff P (2006) Quantitative measurements of localized density variations in cylindrical tablets using X-ray microtomography. Eur J Pharm Biopharm 64:38–50

Carson JW, Wilms H (2006) Development of an international standard for shear testing. Powder Technol 167:1–9

Chiu GNC, Abraham SA, Ickenstein LM, Ng R, Karlsson G, Edwards K, Wasan EK, Bally MB (2005) Encapsulation of doxorubicin into thermosensitive liposomes via complexation with the transition metal manganese. J Control Release 104:271–288

Chung JT, Vlugt-Wensink KDF, Hennink WE, Zhang Z (2005) Effect of polymerization conditions on the network properties of dex-HEMA microspheres and macro-hydrogels. Int J Pharm 288:51–61

De Roos KB (2000) Physicalchemical models of flavor release from foods. In: Roberts DD, Taylor AJ (eds) Flavor release, ACS Symposium Series 763. American Chemical Society, Washington, DC

Dinsmore AD, Hsu MF, Nikolaides MG, Marques M, Bausch AR, Weitz DA (2002) Colloidosomes: Selectively permeable capsules composed of colloidal particles. Science 298:1006–1009

Dziezak JD (1988) Microencapsulation and encapsulated ingredients. Food Technol 42:136–151

Edwards-Levy F, Levy M-C (1999) Serum albumin-alginate coated beads: mechanical properties and stability. Biomaterials 20:2069–2084

Farber L, Tardos G, Michaels JN (2003) Use of X-ray tomography to study the porosity and morphology of granules. Powder Technol 132:57–63

Fatin-Rouge N, Milon N, Buffle J, Goulet RR, Tessier A (2003) Diffusion and partitioning of solutes in agarose hydrogels: the relative influence of electrostatic and specific interactions. J Phys Chem B 107:12126–12137

Fitzpatrick JJ, Barringer SA, Iqbal T (2004a) Flow property measurement of food powders and sensitivity of Jenike's hopper design methodology to the measured values. J Food Eng 61:399–405

Fitzpatrick JJ, Iqbal T, Delaney C, Twomey T, Keogh MK (2004b) Effect of powder properties and storage conditions on the flowability of milk powders with different fat contents. J Food Eng 64:435–444

Foster DS, Clifford LH (1966) Encyclopedia of industrial chemical analysis. Interscience, New York

Fujimoto K, Toyoda T, Fukui Y (2007) Preparation of bionanocapsules by the layer-by-layer deposition of polypeptides onto a liposome. Macromolecules 40:5122–5128

Grigorescu G, Rosinski S, Lewinska D, Ritzén LG, Viernstein H, Teunou E, Poncelet D, Zhang Z, Fan X, Serp D, Marison I, Hunkeler D (2002) Characterization of microcapsules: recommended methods based on round-robin testing. J Microencapsul 19:641–659

Guichard E (2000) Interaction of food matrix with small ligands influencing flavour and texture – foreword. Food Chem 71:299–300

Hu Y, Zhang J, Sato H, Noda I, Ozaki Y (2007) Multiple melting behavior of poly(3-hydroxybutyrate-co-3-hydroxyhexanoate) investigated by differential scanning calorimetry and infrared spectroscopy. Polymer 48:4777–4785

Huang G, Gao J, Hu ZB, John JVS, Ponder BC, Moro D (2004) Controlled drug release from hydrogel nanoparticle networks. J Control Release 94:303–311

Jenike AW (1964) Storage and flow of solids. Bulletin 123, Engineering Experiment Station, University of Utah, USA
Johnson EM, Berk DA, Jain RK, Dean WM (1996) Hindered diffusion in argarose gels: test of effective medium model. Biophys J 70:1017–1026
Jonathan AI (2003) NMR spectroscopy in inorganic chemistry. Oxford University Press, New York
Karel M, Langer R (1988) Controlled release of food-additives. ACS Symp Ser Am Chem Soc 370:177–191
Kosto KB, Deen WM (2005) Hindered convection of macromolecules in hydrogels. Biophys J 88:277–286
Labhasetwar VD, Dorle AK (1991) A study on the zeta potential of microcapsules during aging. J Microencapsul 8:83–85
Lamprecht A, Schafer UF, Lehr CM (2000) Characterization of microcapsules by confocal laser scanning microscopy: structure, capsule wall composition and encapsulation rate. Eur J Pharm 49:1–9
Langer R, Peppas NA (2003) Advances in biomaterials, drug delivery, and bionanotechnology. AIChE J 49:2990–3006
Lardner TJ, Pujara P (1980) Compression of spherical cells. Mechanics Today 5:161–176
Law NGD (2007) Stabilisation and targeted delivery of enzyme nattokinase by encapsulation. Doctoral Dissertation, The University of Birmingham, UK
Le Meste M, Champion D, Roudaut G, Blond G, Simatos D (2002) Glass transition and food technology: a critical appraisal. J Food Sci 67:2444–2458
Leblond FA, Tessier J, Halle JP (1996) Quantitative method for the evaluation of biomicrocapsule resistance to mechanical stress. Biomaterials 17:2097–2102
Legrand P, Barratt G, Mosqueira V, Fessi H, Devissaguet JP (1999) Polymeric nanocapsules as drug delivery systems. A review. STP Pharma Sci 9:411–418
Li MZ, Wilkinson D, Patchigolla K (2005) Determination of non-spherical particle size distribution from chord length measurements. Part 2: Experimental validation. Chem Eng Sci 60:4992–5003
Lian G (2000) Modeling flavor release from oil-containing gel particles. In: Roberts DD, Taylor AJ (eds) Flavor release. ACS Symposium Series 763. American Chemical Society, Washington, DC, pp 201–211
Lian G, Astill C (2002) Computer simulation of the hydrodynamics of teabag infusion. Food Bioproducts Processing 80:155–162
Lian GP, Malone M, Homan JE, Norton IT (2004) A mathematical model of volatile release in mouth from the dispersion of gelled emulsion particles. J Control Release 98:139–155
Lin CL, Miller JD (2004) Pore structure analysis of particle beds for fluid transport simulation during filtration. Int J Mineral Process 73:281–294
Liu KK, Williams DR, Briscoe BJ (1996) Compressive deformation of a single microcapsule. Phys Rev E 54:6673–6680
Liu T, Donald AM, Zhang Z (2005) Novel manipulation in environmental SEM for measuring the mechanical properties of single nano-particles. Mater Sci Technol 21:289–294
Lu GZ, Thompson FG, Gray MR (1992) Physical modelling of animal cell damage by hydrodynamic forces in suspension cultures. Biotechnology and Bioengineering 40:1277–1281
Lulevich VV, Radtchenko IL, Sukhorukov GB, Vinogradova OI (2003) Deformation properties of nonadhesive polyelectrolyte. J Phys Chem B 107:2735–2740
Martinsen A, Skjakbraek G, Smidsrød O (1989) Alginate as immobilization material. 1. Correlation between chemical and physical properties of alginate gel beads. Biotechnol Bioeng 33:79–89
Mathiowitz E (1999) Encyclopedia of controlled drug delivery. Wiley, New York
Mitragotri S (2002) A theoretical analysis of permeation of small hydrophobic solutes across the stratum corneum based on scaled particle theory. J Pharm Sci 91:744–752
Montasser I, Briancon S, Fessi H (2007) The effect of monomers on the formulation of polymeric nanocapsules based on polyureas and polyamides. Int J Pharm 335:176–179
Montes de Oca H, Ward IM, Klein PG, Ries ME, Rose J, Farrar D (2004) Solid state nuclear magnetic resonance study of highly oriented poly(glycolic acid). Polymer 45:7261–7272

O'Hagan P, Hasapidis K, Coder A, Helsing H, Pokrajac G (2005) Particle size analysis of food powders. In: Onwulata C (ed) Encapsulated and powdered foods. CRC, Boca Raton, FL, pp 215–245
Odijk T (2000) Depletion theory of protein transport in semi-dilute polymer solutions. Biophys J 79:2314–2321
Ohtsubo T, Tsuda S, Tsuji K (1991) A study of the physical strength of fenitrothion microcapsules. Polymer 32:2395–2399
Overbosch P, Agterhof WGM, Haring PMG (1991) Flavour release in the mouth. Food Reviews International 7(2):137–184
Peirone MA, Delaney K, Kwiecin J, Fletch A, Chang PL (1998) Delivery of recombinant gene product to canines with nonautologous microencapsulated cells. Hum Gene Ther 9:195–206
Phillips RJ (2000) Hydrodynamic model for hindered diffusion of proteins and micelles in hydrogels. Biophys J 79:3350–3353
Phillips RJ, Dean WM, Brady JF (1989) Hindered transport in fibrous memberanes and gels. AIChE J 35:1761–1769
Pluen A, Netti PA, Jain RK, Berk DA (1999) Diffusion of macromolecules in agarose gels: Comparison of linear and globular configurations. Biophys J 77:542–552
Pothakamury UR, Barbosa-Canovas GV (1995) Fundamental aspects of controlled release in foods. Trends Food Sci Technol 6:397–406
Rao M, Rizvi SSH (1994) Engineering properties of foods. Marcel Dekker, New York
Rattanasak U, Kendall K (2005) Pore structure of cement/pozzolan composites by X-ray microtomography. Cement Concrete Res 35:637–640
Ren Y, Donald AM, Zhang Z (2007) Investigation of the radiation damage to microcapsules in an Environmental SEM. Mater Sci Technol 23:857–864
Ricciardi R, Auriemma F, Gaillet C, De Rosa C, Laupretre F (2004) Investigation of the crystallinity of freeze/thaw poly(vinyl alcohol) hydrogels by different techniques. Macromolecules 37:9510–9516
Roueche E, Serris E, Thomas G, Perier-Camby L (2006) Influence of temperature on the compaction of an organic powder and the mechanical strength of tablets. Powder Technol 162:138–144
Sangster J (1997) Octanol-water partition coefficients: fundamentals and physical chemistry. Wiley Series in Solution Chemistry, New York
Santus G, Baker RW (1995) Osmotic drug-delivery. A review of the patent literature. J Control Release 35:1–12
Sasaki T, Kawagoe S, Mitsuya H, Irie S, Sakurai K (2006) Glass transition of crosslinked polystyrene shells formed on the surface of calcium carbonate whisker. J Polym Sci B Polym Phys 44:2475–2485
Schilling PJ, Karedla BR, Tatiparthi AK, Verges MA, Herrington PD (2005) X-ray computed microtomography of internal damage in fiber reinforced polymer matrix composites. Compos Sci Technol 65:2071–2078
Schuldt U, Hunkeler D (2000) Characterization methods for microcapsules. Minerva Biotechnologica 12:249–264
Schulze D (1996a) Measuring powder flowability: a comparison of test methods. Part I. Powder Bulk Eng 10:17–28
Schulze D (1996b) Measuring powder flowability: a comparison of test methods. Part II. Powder Bulk Eng 10:45–61
Seamus PJH (2003) Analytical chemistry. Oxford University Press, New York
Shiu C, Zhang Z, Thomas CR (1999) A novel technique for the study of bacterial cell mechanical properties. Biotechnol Tech 13:707–713
Shu B, Yu W, Zhao Y, Liu X (2006) Study on microencapsulation of lycopene by spray-drying. J Food Eng 76:664–669
Shukla A, Srivastava AK (2004) Synthesis and characterization of functional copolymer of linalool and vinyl acetate: a kinetic study. J Appl Polym Sci 92:1134–1143
Stenekes RJH, De Smedt SC, Demeester J, Sun GZ, Zhang ZB, Hennink WE (2000) Pore sizes in hydrated dextran microspheres. Biomacromolecules 1:696–703
Stock SR (1999) X-ray microtomography of materials. Int Mater Rev 44:141–164

Strand BL, Morch YA, Espevik T, Skjak-Braek G (2003) Visualization of alginate-poly-L-lysine-alginate microcapsules by confocal laser scanning microscopy. Biotechnol Bioeng 82:386–394

Sun G, Zhang Z (2001) Mechanical properties of melamine-formaldehyde microcapsules. J Microencapsul 18:593–602

Sun G, Zhang Z (2002) Mechanical strength of microcapsules made of different wall materials. Int J Pharm 242:307–311

Taylor AJ (1998) Physical chemistry of flavour. Int J Food Sci Technol 33:53–62

Teunou E, Fitzpatrick JJ (1999) Effect of relative humidity and temperature on food powder flowability. J Food Eng 42:109–116

Van Raamsdonk JM, Chang PL (2001) Osmotic pressure test: a simple, quantitative method to assess the mechanical stability of alginate microcapsules. J Biomed Mater Res 54:264–271

Verma RK, Krishna DM, Garg S (2002) Formulation aspects in the development of osmotically controlled oral drug delivery systems. J Control Release 79:7–27

Wang C, Cowen C, Zhang Z, Thomas CR (2005) High-speed compression of single alginate microspheres. Chem Eng Sci 60:6649–6657

Weiss G, Knoch A, Laicher A, Stanislaus F, Daniels R (1995) Simple coacervation of hydroxypropyl methylcellulose phthalate (HPMCP). II. Microencapsulation of ibuprofen. Int J Pharm 124:97–105

Westrin BA, Axelsson A, Zacchi G (1994) Diffusion measurements in gels. J Control Release 30:189–199

Xu Y, Du Y (2003) Effect of molecular structure of chitosan on protein delivery properties of chitosan nanoparticles. Int J Pharm 250:215–226

Yadav SK, Khilar KC, Suresh AK (1997) Release rates from semi-crystalline polymer microcapsules formed by interfacial polycondensation. J Memb Sci 125:213–218

Yan NX, Zhang MZ, Ni PH (1992) Size distribution and zeta potential of polyamide microcapsules. J Memb Sci 72:163–169

Yan-yu X, Yun-mei S, Zhi-peng C, Qi-neng P (2006) Preparation of silymarin proliposome: a new way to increase oral bioavailability of silymarin in beagle dogs. Int J Pharm 319:162–168

Yeo Y, Bellas E, Firestone W, Langer R, Kohane DS (2005) Complex coacervates for thermally sensitive controlled release of flavor compounds. J Agric Food Chem 53:7518–7525

Yoshioka Y, Asao K, Yamamoto K, Tachi H (2007) New method for fabricating aromatic polyamide particles with a narrow particle size distribution. Macromol Reaction Eng 1:222–228

Zhang YL, Chu CC (2002) Biodegradable dextran-polylactide hydrogel networks: their swelling, morphology and the controlled release of indomethacin. J Biomed Mater Res 59:318–328

Zhang Z, Ferenczi MA, Lush AC, Thomas CR (1991) A novel micromanipulation technique for measuring the bursting strength of single mammalian-cells. Appl Microbiol Biotechnol 36:208–210

Zhang Z, Ferenczi MA, Thomas CR (1992) A micromanipulation technique with theoretical cell model for determining mechanical properties of single mammalian cells. Chem Eng Sci 47:1347–1354

Zhang Z, Saunders R, Thomas CR (1999) Mechanical strength of single microcapsules determined by a novel micromanipulation technique. J Microencapsul 16:117–124

Zhao L, Zhang Z (2004) Mechanical characterization of biocompatible microspheres and microcapsules by direct compression. Artif Cells Blood Substit Immobil Biotechnol 32:25–40

Zou Y, Brusewitz GH (2002) Flowability of uncompacted marigold powder affected by moisture content. J Food Eng 55:165–171

Chapter 5
Encapsulation of Aroma

Nicolaas Jan Zuidam and Emmanuel Heinrich

5.1 Introduction

Flavor is one of the most important characteristics of a food product, since people prefer to eat only food products with an attractive flavor (Voilley and Etiévant 2006). Flavor can be defined as a combination of taste, smell and/or trigeminal stimuli. Taste is divided into five basic ones, i.e. sour, salty, sweet, bitter and umami. Components that trigger the so-called gustatory receptors for these tastes are in general not volatile, in contrast to aroma. Aroma molecules are those that interact with the olfactory receptors in the nose cavity (Firestein 2001). Confusingly, aroma is often referred to as flavor. Trigeminal stimuli cause sensations like cold, touch, and prickling. The current chapter only focuses on the encapsulation of the aroma molecules.

Aroma consists of many volatile and odorous organic molecules. Most of them are in a gas or liquid state, but also some solid materials may have a distinct smell (e.g. vanillin and menthol). In general, aroma molecules have a low MW (often between 100 and 250) and can be classified as hydrocarbons, alcohols, aldehydes, ketones, esters, acids, sulphides, etc. Typical examples of aroma molecules are shown in Fig. 5.1. Most of the aroma molecules are lipophilic, but some are hydrophilic (their logarithmic values of the octanol–water partition coefficient, log *P*, ranges from −1 to 7). Aroma molecules can be either added to food products, produced during processing of the food product (the so-called processing or reaction flavor) or are formed during cooking of the food product. Aroma can reach the nose cavity directly when the food is not yet in the mouth (ortho-nasally: direct smell) or via the oral cavity (retro-nasally). Gas and thus aroma is transferred from the oral cavity to the nasal cavity via nasal airflow, which is influenced by mouth action like saliva production, mastication and swallowing. Aroma release from food products before and after eating is controlled by both thermodynamic and kinetic parameters,

---

N.J. Zuidam (✉) and E. Heinrich
Unilever R&D Vlaardingen, Olivier van Noortlaan 120, 3133 AT Vlaardingen, The Netherlands
e-mail: klaas-jan.zuidam@unilever.com

N.J. Zuidam and V.A. Nedović (eds.), *Encapsulation Technologies for Active Food Ingredients and Food Processing*,
DOI 10.1007/978-1-4419-1008-0_5, © Springer Science+Business Media, LLC 2010

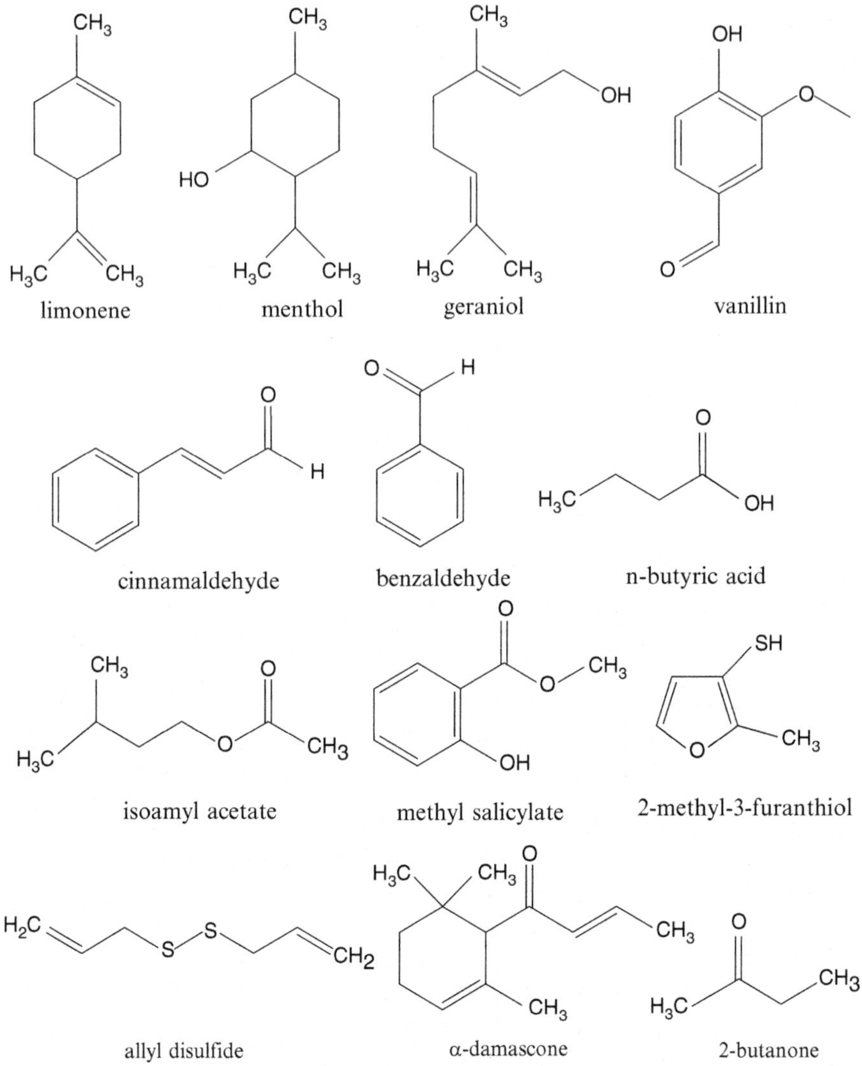

Fig. 5.1 Molecular structures and odour of typical aroma molecules. The hydrocarbon limonene can be found in citrus fruits and has lemon-like odour. The alcohols menthol and geraniol have a mint-like and rose-like smell, respectively. In addition, menthol gives a fresh, cooling sensation. The aldehydes vanillin, benzaldehyde and cinnamic aldehyde have a vanilla, bitter almond-like, and pungent, spicy note, respectively. The acid *n*-butyric acid has a rancic, butter-like odour. The ester isoamylacetate is found in fruit and has a banana or pear odour, and the ester methyl salicylate is found in many plant cells (like wintergreens). The sulphide 2-methyl-3-furanthiol has a roasty, meaty note, and the sulphide diallyl disulphide has a characteristic garil odour. The ketones α-damascone have a pleasant fruity, rosy odour and 2-butanone has an acetone-like smell. The molecular structures and odour information have been obtained from Burdock 1994

which depends on the aroma characteristics and on the composition and the physical state of the matrix. These parameters will determine the volatility of the flavor compounds and their resistance to mass transfer between different phases, especially from the product to air (Druaux and Voilley 1997; Van Ruth and Roozen 2002; De Roos 2003). Proper choice of food composition and food microstructures may thus control aroma release during food product preparation and consumption. Encapsulation might be one of the tools in such a design (Porzio 2007a).

Aroma can be encapsulated to improve aroma functionality and stability in products. The possible benefits of encapsulated aromas are (Uhlemann et al. 2002; Angelich 2005):

- Superior ease of handling (conversion of liquid aroma oil into a powder, which might be dust free, free flowing, and might have a more neutral smell)
- Improved stability in the final product and during processing (less evaporation, degradation or reaction with other components in the food product)
- Improved safety (e.g. reduced flammability, no concentrated aroma oil handling)
- Creation of visible and textural effects (visual cues)
- Adjustable aroma properties (particle size, structure, oil- or water-dispersable)
- Controlled aroma release (aroma differentiation, or aroma retention)

Exact information about the annual sales of aroma delivery systems is not available. The flavor and fragrance industry had a turnover of about US$20.5 billion in 2008 (see http://www.leffingwell.com/top_10.htm). Ubbink and Schoonman (2003) assumed that about half of this turnover is flavor, and that about 20–25% of all flavors are estimated to be sold in an encapsulated form (US$1.4–1.7 billion in 2001 and 2–2.5 billion in 2008). However, Campanile (2004) estimated that the total market for aroma delivery systems was only half of this value, i.e. at about $750 m in 2002. He also claims that it is growing faster than the aroma market. About 80–90% of the encapsulates are spray-dried ones, 5–10% are prepared by spray-chilling, 2–3% are prepared by melt extrusion and 2% are prepared by melt injection (Ubbink and Schoonman 2003; Porzio 2004, 2007b, 2008). Other type of encapsulates contribute <1% to all encapsulates.

The current chapter first reviews the general concept and benefits of aroma encapsulation, then the existing encapsulation technologies and finally provides recommendations for the use of the different encapsulation technologies for several types of food products.

5.2 Mechanisms of Aroma Release Out of Encapsulates

Aroma is often diluted in a co-solvent, which may affect the encapsulation process (Porzio 2004). Propylene glycol and ethanol are hydrophilic solvents used with, e.g., fruit aromas (low molecular weight esters, aldehydes and alcohols), and fractionated coconut oil or medium-chain triglycerides are used with more hydrophobic aromas.

Morphology of the encapsulates will depend on the aroma properties as well as the carrier material. The majority of aroma molecules are lipophilic and may not mix well with hydrophilic carrier material; they will then be present as oil droplets. In case a more hydrophobic carrier material is used, aroma may mix well with the carrier material and then single-matrix morphology is obtained. The latter morphology might also be obtained when entrapping water-soluble aroma molecules (like vanillin) in hydrophilic carrier materials.

In general, the diffusion through carrier material decreases with increasing molecular size (steric hindrance, related to molecular weight), decreasing vapor pressure (volatility) and increasing log *P* of the aromas (Goubet et al. 1998). In carbohydrate carrier materials, the order of retention was in the order of alcohols > ketones = esters > acids, but many exceptions to this order were found (Goubet et al. 1998). This indicates that not only the presence of chemical groups is influencing the retention, but also other characteristics like polarity may play a role here; the higher the polarity the lower the retention.

The molecular weight of the coating material, its conformation, its chemical composition and its physical state can all influence the barrier properties. Higher molecular weight of the carrier material will limit aroma diffusion as was, e.g., found for maltodextrins (Goubet et al. 1998). However, for the highest degree of polymerization lower aroma retention was found, probably due to lower degrees of interactions between aroma and carrier material. When an amorphous coating material is in a glassy state, molecules may have little relative mobility, similar to those of a crystalline phase (Goubet et al. 1998; Ubbink and Schoonman 2003; Soottitantawat et al. 2004; Carolina et al. 2007). The glassy state is characterised by a so-called glass transition temperature (T_g). Below the T_g the diffusion through carrier material is limited in a manner similar to those of a crystalline phase (glassy state), in contrast to temperatures above the T_g when the carrier material is in rubbery state. The T_g increases with higher MW of carbohydrate (Ubbink and Schoonman 2003). However, low MW matrices provide better storage stability in the glass transition than high MW matrices, since the first have a lower residual open and a closed porosity (Ubbink and Schoonman 2003). Usually, a compromise has to be made between a high T_g of the carrier material (which is favored by high MW material) and low porosity (which increases with the use of low MW materials). The T_g decreases with increasing percentages of plasticisers and exposure to relatively high humidity (the T_g of amorphous solid water is at −135°C). When the water content of the carrier material increases and the holding temperature is for some time above the T_g, the carrier material may have sufficient mobility to associate and crystallise and this results in forcing out of aroma and thus aroma loss (Goubet et al. 1998). Benczédi (2002) stated that mass transport of aromas is also affected by polymer cross-linking or plasticization, as induced by temperature or partial pressure of water. Hildebrand's solubility parameters for aromas and polymers could be useful to predict their compatibility. The lack of compatibility between lipophilic aromas and hydrophilic carrier material may indicate the chances of success of hydrophilic carriers as encapsulation carriers.

Release rate of each aroma compound might be fitted to Avrami's equation (Desai and Park 2005):

$$R = \exp(-(kt)^n) \tag{5.1}$$

where R is the retention of the aroma compounds during release, t is the time, n is the parameter representing the release mechanism, and k is the release rate constant. The Avrami's equation has originally been used to describe crystal growth of polymers. The parameter n has a value of 1 for first-order kinetics and 0.54 for diffusion-limiting reaction kinetics. Taking a logarithm of both sides of (5.1) twice gives:

$$\ln(-\ln R) = n \ln k + n \ln t \tag{5.2}$$

Using (5.2), one can determine the parameter n as the slope by plotting $\ln(-\ln R)$ against $\ln t$, and the release rate constant k from the interception at $\ln t = 0$. Please see Chap. 4 for more information on release kinetics.

5.3 Encapsulation Technologies for Aroma

Table 5.1 summarises most common aroma encapsulation technologies currently available in the open literature. This section will discuss each technology in more details below. Final recommendations for their use in several types of application in food products are provided in the next paragraph.

5.3.1 Spray-Drying

Spray-drying is one of the oldest processes to encapsulate aroma. It is so common in foods to use aroma in a spray-dried, fine powder form that one often tends to forget it is a form of encapsulation. About 80–90% of the encapsulated aroma is prepared by this method (Ubbink and Schoonman 2003; Porzio 2007b). This is, however, not so obvious since aroma compounds evaporate faster than water. Spray-drying of aroma is achieved by dispersion of the aroma in an aqueous solution of carrier material, followed by atomization and spraying of the emulsion into a hot chamber (see Fig. 2.2). The co-current, hot air stream has a predetermined temperature within the 160–220°C range, which heats the droplets almost instantaneously to 100°C (Porzio 2007b). During this process a film is formed at the droplet surface and the concentration of ingredients in the drying droplet keeps increasing, thereby retarding the larger aroma molecules while the smaller water molecules continue to diffuse to the exterior at a considerable rate (Ré 1998; Ubbink and Schoonman 2003; Porzio 2004, 2007b; Jafari et al. 2008). The final powder consists of dry particles containing a dispersion of fine aroma droplets and having a particle size of 10–150 μm.

Table 5.1 Aroma encapsulates and their load, particle size, morphology and release mechanism

Technology	Load (%)	Particle size (μm)	Morphology	Release mechanism based upon
(Fluidised) Spray-drying	10–50	10–400	Droplets in matrix	Dissolution
Granulation/agglomeration/compaction	5–40	200–3,000	Droplets in matrix	Dissolution
Fluid bed coating	10–50	200–5,000	Droplets in matrix	Dissolution
Spray-chilling/cooling	10–20	20–200	Matrix	Diffusion, heat
Melt injection	5–25	200–2,000	Droplets in matrix	Dissolution
Melt extrusion	5–40	300–5,000	Droplets in matrix	Swelling + dissolution, heat
Complex coacervates	20–90	10–800	Core with wall	Diffusion
Microspheres	20–40	10–800	Matrix of droplets	Diffusion, dissolution
Co-extrusion	70–95	800–8,000	Core with wall	Diffusion, heat
Yeast cells	1–70	25–30	In cell membrane	Diffusion, heat
Cyclodextrin	8–10	0.002	Molecular inclusion	Dissolution
Silica particles	5–50	10–1,000	Matrix	Diffusion, heat
Co-crystallization/co-precipitation	Various	Various	Crystals	Dissolution

Please see the end of Sect. 5.1 for estimations of the relative contributions of each technology to the commercial market

The carrier material of spray-dried aroma should ideally have (Ubbink and Schoonman 2003; Reineccius 2004; Porzio 2004; Jafari et al. 2008):

- A high degree of solubility
- Limited viscosity at 35–45% (w/w) in water
- Good emulsifying properties
- Good film forming and drying properties
- A non-hydroscopic character
- A bland taste
- Should be non-reactive
- Should protect the aroma from oxidation
- Should be available at low cost

Examples of carrier material used for spray-drying are mono- and disaccharides, maltodextrin, corn syrup solids, modified starches (such as *n*-octenylsuccinic anhydride-substituted (OSA or OSAN) starches), gums (such as gum arabic or larch gum), milk or soy proteins, hydrolysed gelatin, and combinations thereof. Often, a blend of carrier material is used to obtain both emulsifying properties (e.g. with modified starches or gum arabic) and prevention of oxidation via the formation of a glassy state (e.g. with sugars or hydrolysed starches; see also Sect. 5.2). The carrier material is mixed with aroma in preferably deionised water and then emulsified, prior to spray-drying. One should use the highest infeed solids level possible (35–40% range); this will result in fast formation of the semi-permeable film around the droplet surface during atomization (Reineccius 2004). Especially highly volatile aroma components are much better retained with increasing infeed solid level (Charve and Reineccius 2009). However, increasing the solid level too much may result in non-dissolved material and increased viscosity, which may slow down the formation of the film at the droplet surface during atomization and thus promote aroma losses. So there might be an optimum solids level. One may encounter relatively low pH's of about 3 when using OSA-starches, due to associated *n*-octenylsucinnic acid group, which may cause degradation of acid-sensitive flavors like citral or acetal (Porzio 2004, 2007b).

The retention of aroma upon spray-drying highly depends on the type of aroma. In general, the less volatile, the larger the molecular size (related to molecular weight), and the lower the polarity, the higher the retention of aroma components upon spray-drying (Goubet et al. 1998; Jafari et al. 2008). Volatile aroma compounds like acetaldehyde, diacetyl, dimethylsulphide and ethanol might partially be lost upon spray-drying. One may compensate for these losses by using higher initial levels of these molecules before spray-drying or by post-blending them upon spray-drying at optimal process and carrier conditions (Porzio 2007b). The retention of lipophilic aroma upon spray-drying increases with decreasing emulsion size (optimal below 1–2 μm; Soottitantawat et al. 2003), increasing emulsion stability during the spray-drying process (Liu et al. 2000, 2001; Jafari et al. 2007a, 2008), and larger powder particle size (Jafari et al. 2007b, 2008). For hydrophilic aromas there might be an optimal emulsion size at 2–3 μm, depending on the type of aroma (Soottitantawat et al. 2003). Hydrophilic aromas might also be entrapped by spray-drying water-in-oil-in-water double emulsions, with aroma trapped within the inner water droplets prior to spray-drying (Brückner et al. 2007).

The drying process should bring the carrier material in a glassy state, which properties might be improved by addition of low MW carbohydrates (such as sugars, corn syrup solids and/or polyols) at 10–35% of the total carrier formula (Porzio 2004; see also Sect. 5.2 for more information about glassy state and about the effect of low MW carbohydrates on porosity). The glassy state will also prevent sticking of encapsulates to the spray-dry chamber. In general, the ratio carrier material to aroma is about 4:1 (w/w). Higher ratios generally results in better aroma retention during spray-drying, which is related to an increased diffusion path length to the surface of the droplet/particle. The use of cyclodextrin might be another way to enhance aroma retention during spray drying (Shiga et al. 2004; Yoshii et al. 2005) (see also Sect. 5.3.11). Most of the aroma loss during spray-drying is probably occurring during the stage when water in the droplet exceeds its boiling point and bubbles are bursting from the interior of the droplet to the outside just before the particle structure is set (Ré 1998; Reineccius 2004; Jafari et al. 2008). Water solubility of aroma may also influence aroma loss, due to its ability to diffuse with the water through the droplet during the spray-drying process (Ré 1998).

The storage stability of spray-dried aromas is typically between 6 and 12 months, and generally shows a semi-logarithmic decay with time and a dependence on moisture and temperature. The glassy state of the carrier material limits caking and enhances aroma retention upon storage (Soottitantawat et al. 2004; Carolina et al. 2007). This latter may also be improved by the use of lipid coatings or OSA-modified starches like Capsul and especially Hi-Cap 100 (Soottitantawat et al. 2005a, b; Baranauskienė et al. 2005, 2007; Krishnan et al. 2005; Vaidya et al. 2006; Kanakdande et al. 2007). Vaidya et al. (2006) found that a 4:1:1 blend of gum arabic/maltodextrin/Hi-Cap 100 offered a better protection of cinnamon oleoresin upon storage for 6 weeks at 25°C than gum arabic alone. However, blends of gum arabic and maltodextrin or gum arabic and Hi-Cap 100 could not provide the protection offered by gum arabic alone. The same result was found for cumin oleoresin by the same research group (Kanakdande et al. 2007). When using gum arabic with or without maltodextrin as the carrier, Soottitantawat et al. (2005a) observed that the aroma stability upon storage improved when the powder particle size (typically between 10 and 150 μm) and emulsion size increased, probably due to a reduced surface area. However, this was not found with Hi-Cap 100. Shiga et al. (2001) found that limonene in a spray-dried powder with β-cyclodextrin was more stable than without, but this was not found for ethyl *n*-hexanoate.

Charve and Reineccius (2009) found better retention over storage time with higher infeed solids of gum arabic or modified starch, which was attributed to changes in particle structure (e.g. less porous). However, this was not found when using soy or whey protein isolate as the carrier material. The use of proteins (soy or whey protein isolate, or sodium caseinate) provided better limonene protection upon storage than gum arabic or modified starch, most likely due to limited oxygen permeability and thus prevention of oxidation. However, these carrier materials turned brown upon storage within 1 week, which was probably caused by Maillard reaction between (*E*)-2-hexanal and the protein.

The advantages of spray-drying are

- Continuous production: the dried particles can be collected continuously at the bottom of the spray-dryer
- Easy operation: constant quality is possible when drying conditions are held constant and it can be automated
- Low cost
- Historically, a lot of experience with this technique
- Wide choice of carrier material and equipment

A disadvantage of spray-drying might be that very volatile aromas (fresh top-notes like ethyl acetate, for example) can be (partially) lost during the process (Reineccius 2004), thereby changing the balance of some flavor. Furthermore, some aromas might be oxidised during the spray-drying process. Due to their small size, spray-dried aroma powders might fall through a tea bag and are therefore not preferred for this application (Porzio 2004). In general, spray-dried aromas are water-soluble, which might either be desirable or not. For example, in chewing gum the aroma is spray-dried to allow faster release of the aroma upon hydration in the mouth than when it is in the gum matrix (De Roos 2003). However, spray-dried aroma powders dissolves quickly in water-containing food products and can then not provide a controlled release benefit over non-encapsulated aroma during cooking or eating. A delayed release of spray-dried aroma in water might only be achieved when the carrier material has been irreversibly changed during the preparation process. An example is protein denaturation (Porzio 2007b).

5.3.2 *Agglomeration or Granulation*

For some applications, larger aroma particles than those produced by spray-drying (in general about 10–150 μm) might be desirable. An option is fluidised bed spray granulation (also called spray-bed-drying), in which a spray-drying step is followed by a secondary agglomeration step to produce larger particles of about 0.3–0.9 mm in one (Schleifenbaum et al. 2000) or in two steps (De Roos et al. 2001). Another option is to spray-dry onto another carrier powder (Trophardy 2007a). In all cases, the spray-dried particles are not fully dried after the first stage, and therefore remain sticky to facilitate agglomeration during the second phase.

Alternatively, a binder solution (e.g. water) can be sprayed onto spray-dried product within a fluid bed (Uhlemann et al. 2002). The high moisture levels that remain for a relatively long period of time can result in some volatile losses. The somewhat larger particles will have less oil exposed to the surface of the particles. (This does not necessarily correlate with better shelf-life (Buffo et al. 2002).)

An alternative process for the preparation of larger aroma particles is pressure agglomeration or compaction, in which spray-dried aroma is compressed under high pressure together with additional maltodextrin into lumps and then crushed into small pieces of about 0.7–3.0 mm (Uhlemann et al. 2002; Barnekow and Fexer 2007). These type of particles are useful for applications in which the aroma encapsulates should not be separated from the food products (e.g. in tea bags).

The effect of granulation/agglomeration/compaction on aroma release in use is very small and will be comparable to spray-dried particles prior this treatment (see Sect. 5.3.1).

5.3.3 *Fluid bed Coating*

Fluidised bed coating is a technique in which a coating is applied onto powder particles suspended in air (see Fig. 2.3 and Sect. 2.2.2). The powder particles might be aroma encapsulates obtained via another encapsulation technique (e.g. spray-drying). The coating material can be a fat or a wax, which will then solidify on the surface of the aroma encapsulates (Uhlemann et al. 2002; Cho and Park 2002). This might be useful to provide aroma release upon heating. An alternative way of applying a fat coating was disclosed in patent by Popplewell and Porzio (1998), which states that fat-coated encapsulates can be prepared by (1) mixing the encapsulates with molten fat to obtain a slurry, and (2) cooling the slurry to obtain a solid mass in which the active agent is dispersed in solid fat. Ethylcellulose might be present in the molten fat and the solid mass can be milled or grinded to reduce its particle size.

Fluid bed coating might also be achieved by spraying powder particles with aqueous polymer solutions, which will leave a polymer film onto the powder particles after water evaporation. Examples of such coatings are cellulose derivatives (e.g. ethyl cellulose, methyl cellulose, hydroxypropyl cellulose, hydroxypropyl methyl cellulose 8), gums, proteins, or shellac. In general, these coatings will solubilise in water, unless they react with the powder to form an insoluble film. An example of this might be spraying of an aqueous mixture of limonene, starch and alginate or pectin onto a powder mixture of starch and calcium gluconate in a fluid bed (Trophardy 2007b; see also Sect. 5.3.8). Daragh and Stone (1975) disclosed in their patent double encapsulation of volatile flavors by coating flavor in lipid particles (made by spray-chilling, see next section) with a water-soluble material (gum arabic or dextrin). These encapsulates were added to butter or margarine, and the inventors claimed superior flavor when one of these two products was used to bake cookies or cake.

The primary advantage of fluid bed coating is that it can modify the surface properties of the aroma encapsulates, e.g., thermal release properties or additional aroma protection. Disadvantages are the relatively long coating times (which may easily take 2–20 h in a batch process) and the substantial amount of coating materials needed to ensure complete covering (depending on the application and particle size of the core material). These two disadvantages may make fluid bed coated encapsulates relatively costly.

5.3.4 Spray-Chilling/Spray-Cooling

Spray-chilling or spray-cooling is a technology to produce lipid particles with aroma. About 5–10% of all aroma encapsulates are prepared by this technique (Porzio 2004, 2007b, 2008). It is a similar process as spray-drying (see Sect. 2.2.1 and 5.3.1), but no water is evaporated here and the air used is cold or refrigerated. First, a molten lipid–aroma mixture is prepared, atomised as fine droplets into a chilled chamber (e.g. via nozzle, spinning disk or (centrifugal) co-extrusion), then solidified and finally recovered as fine particles (Uhlemann et al. 2002; Gouin 2004; Porzio 2004). The resulting particles consist of a fine dispersion of liquid aroma in fat crystals. Optionally, the aroma might be spray-dried (Bush Boake Allen Ltd 1973) or encapsulated in a microsphere prior spray-chilling (Coyne et al. 2005). In general, the melting point of the lipid used is in the range of 34–42°C for spray-chilling, and higher for spray-cooling. The aroma is released when the lipid material becomes liquid above its melting point. This trigger is often utilised during processing or cooking. One should realise that lipophilic aroma might diffuse relatively easily through fat crystals; therefore it is usually not recommended for applications that require long shelf lives (Popplewell et al. 2003).

An alternative process to spray-chilling/spray-cooling might be emulsification of molten fat or wax in water at a temperature above the melting temperature of the fat, followed by cooling during mixing. However, the part of aroma that is in the water phase will not be encapsulated within the fat. One may spray-dry this suspension of fat particles in the presence of water-soluble matrix material to obtain a free-flowing powder (see Sect. 5.3.1), which is also claimed in a patent by Shefer and Shefer (2003).

5.3.5 Melt Injection

Carbohydrate materials (like sucrose, maltodextrin, glucose syrup, polyols, modified starches, other mono- and disaccharides) can be mixed with aroma and emulsifiers when molten at 110–140°C, then pressed through one or more orifices (at 1–7 bar) and finally quenched to form a glass. This process is based upon the confectionary hard candy procedure, and is called melt injection or vertical, screwless extrusion. The extrusion is usually carried out into a cold, dehydrating organic solvent, commonly isopropanol (IPA). The coating material hardens on contacting the isopropanol, thereby encapsulating the aroma as finely dispersed droplets in a solid matrix (see Fig. 5.2). The size of the extruded strands is reduced to the appropriate dimensions inside the cold solvent during vigorously stirring, thereby breaking up the extrudates into small pieces. Any residual aroma is removed from the surface of the encapsulate by the isopropanol, thereby preventing oxidation of the non-encapsulated aroma (which is especially important when for example citrus oils are to be encapsulated (Ubbink and Schoonman 2003; Porzio 2004, 2008)). If the aroma components are sensitive towards oxidation, the melt injection process can be carried out under nitrogen. Isopropanol is by far the most used cooling solvent. However, it is flammable and its vapor is regarded as a hazardous waste requiring specialised equipment for handling and storage. An alternative might be the use of liquid nitrogen as the cooling solvent (Subramaniam et al. 2006).

About 2% of all aroma encapsulates are made by melt injection (Porzio 2004, 2007b, 2008). This technology is used by, e.g., Firmenich as 'Durarome®'. Interestingly, the same technology has been used by them to encapsulate fish oil (Valentinotti et al. 2006; see also Chap. 6). Durarome® products from Firmenich are rod-shaped with a typical particle size between 200 and 2,000 μm, and the aroma loading is between 5 and 25%. The carrier material used in melt injection is commonly water-soluble, and can be sensitive to humidity (depending on composition and particle size). The use of low MW carbohydrate for the preparation of Durarome® products results in relatively low porosity (see Sect. 5.2) and enhanced barrier properties and thus decreases the oxidation of aroma compounds upon storage. One may add anti-caking agents as tricalcium phosphate or pyrogenic silica to keep the powder free-flowing during storage.

Bohn et al. (2005) found that the average glass transition temperature (T_g; see also Sect. 5.2) of a rapid dissolving Durarome® with 4–5 wt% water decreased rapidly from 46°C to about −50°C upon humidification. Durarome® products' T_g responded similarly to pure freeze-dried amorphous sucrose upon humidification. The release of benzaldehyde at 25°C was triggered when the T_g was below −5°C

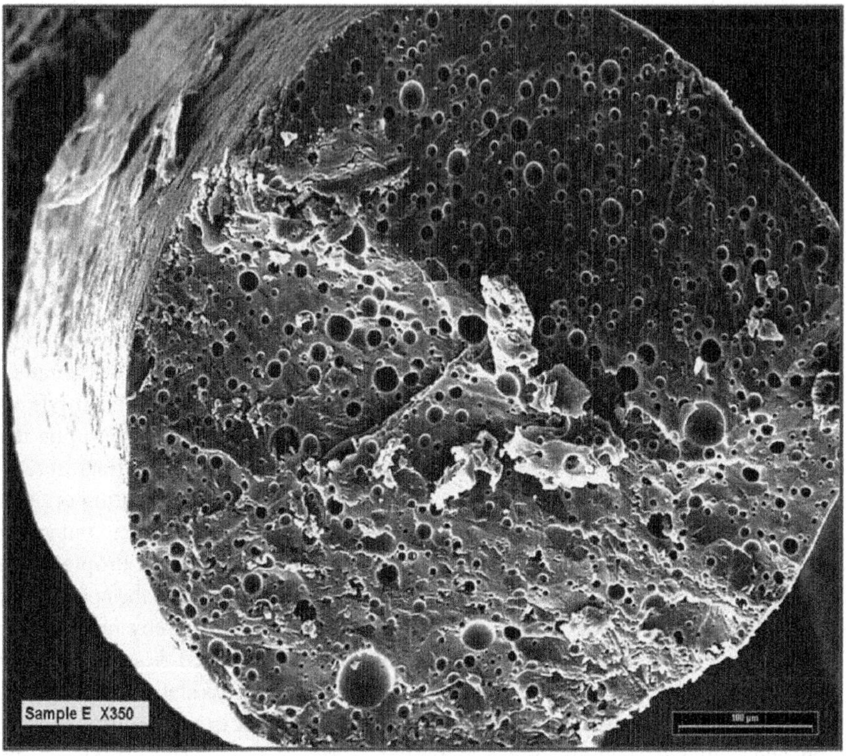

Fig. 5.2 Scanning electron micrograph of a Durarome® product in which the aroma is encapsulated as finely dispersed droplets in a solid matrix. This picture has been kindly provided by Firmenich S.A., Geneva

after exposure at a relative humidity of 60% or higher (this value might be different for other aroma components). As seen in other studies, the release occurred at a temperature above the T_g, but not at that temperature. Please note that other Durarome® products may behave differently upon exposure to high relative humidity, since the stability towards humidity and the dissolution rate of Durarome products can be modulated through a selection of the encapsulation matrix and particle characteristics.

Bouquerand et al. (2004) from Firmenich described that incorporation of 3.3 wt% agar agar in a Durarome composed of 57 wt% maltodextrin, 30 wt% aroma and 8 wt% water (patented by McIver et al. 2002) resulted in Fickian diffusion-controlled, first-order aroma release due to the formation of a hydrogel. If no agar was used, a faster, zero-order release was found due to gradual dissolution/erosion. The authors could model the aroma release when the diffusion of water into the encapsulates was taken into account.

5.3.6 Melt Extrusion

Aroma can also be encapsulated in a carbohydrate melt by using an extruder with one or more screws in a continuous process (see Sect. 2.2.4 and Porzio 2004, 2008). Melt extrusion can be regarded as a very similar process as melt injection; the main differences are that in general melt extrusion utilises screw(s) in a horizontal position and that the extrudates are not surface washed. First, a dispersion of aroma in a melt of carbohydrates with water as a plasticizer is formed within such an extruder, then expelled under high pressure through a die to form sheets, ropes or threads of different dimensions (depending on the die geometry), and finally an amorphous or solid glass is formed upon rapid cooling by air or cold bath. The extrudate can be broken into smaller pieces by cutting with a rotating knife immediately after the die or by grinding after the cooling step. The extruder has screws which are designed to melt, further homogenise and compress the extrudate, and may have section-controlled variation in temperature (melt temperatures are generally about 105–125°C). In general, the aroma melt extrudates have a dense structure and their glass transition is between 30 and 70°C. Extrudates can be composed of starch, maltodextrins, modified starches, sugars, cellulose ethers (like hydroxypropyl cellulose or hydroxypropyl methyl cellulose), proteins, emulsifiers, lipids and/or gums. One of the most widely used food polymer material is starch. So-called thermoplastic starch is obtained by destructuring starch via the simultaneous application of heat (around 110–120°C) and mechanical forces in the extruder (Yilmaz 2003; Yilmaz et al. 2005). The presence of plasticisers, such as water or glycerol/polyols, enables further processing and properties of the extrudates, such as the control of the glass transition temperature of the material, its solubility and morphology (Yilmaz et al. 1999 and references therein; Ubbink and Schoonman 2003). If one would like to delay the dissolution of melt extrudates in water, one may use relatively high molecular weight polymers and process these at higher temperature and/or plasticiser content (Ubbink and Schoonman 2003). Modified starches that are tailor-made (using physical, chemical or enzymatic routes) might also be utilised to make extrudates more water-resistent (Zasypkin and Porzio 2004). In addition, additives (such as plasticisers and other constituents in the formulation) and in situ modifications (e.g. heat treatment, surface modification, and induction of (local) crystallinity) might be used (Yilmaz et al. 2005). Polymers which form a hydrogel upon hydration (like cross-linked gelatin or gluten, instant starch and sodium alginate in the presence of calcium) might also be used in melt extrudates to provide a controlled release of the flavor, e.g., upon eating a lemon-flavored chewy candy (Henson et al. 2006).

Encapsulation of aroma via melt extrusion can be achieved by adding it in the mixing/dispersing zone of the extruder (at about halfway the screw(s), see Fig. 2.4). This minimises the residence time of aroma and avoids too long exposure to the relatively high temperatures required. The aroma can be added as an oil, an o/w emulsion or in a spray-dried form. Alternatively, pre-encapsulation and surface modification are possible. Morphology of the obtained formulations will depend on the properties of the ingredients as well as the matrix (see Sect. 5.2). The encapsulation efficiency (and the release kinetics) will depend on adequate mixing and dispersion of the aroma within the matrix. The use of emulsifiers may allow better control of these characteristics (Yilmaz et al. 2001).

Some aroma loss during encapsulation may happen when the extrudate is expelled under high pressure through the die out of the extruder (flashing due to the pressure drop) and during the grinding of the extrudates. One may prevent this by cooling at high pressures (20–30 atmospheres) till the carrier material is in the glassy state (Fulger and Popplewell 1994). Due to their highly dense structure, extruded aroma encapsulates have in general very good stability and are therefore often used in dry beverage systems (e.g. tea bags) to prevent volatile losses and oxidation during storage. Carrier modifications might improve aroma survival in aqueous products. Unfortunately, the aroma load of such extruded encapsulates is relatively low (typically less than 10%), which may have an impact on their cost in use.

Aroma release from melt extrudates will occur upon water uptake or dissolution of the melt extrudates. As seen above with melt injection, the aroma release from melt extrusion occurs at a temperature well above the T_g of the carrier material. Blake et al. (2003) even stated that aroma loss did not correlate particularly well with the T_g when melt extrudates took up moisture dynamically (within a few hours). At 20°C the bulk of aroma only released when the moisture content became 20% and the T_g of the extrudates had dropped far below 20°C.

Several major flavor houses commercialise aroma encapsulates made by melt extrusion technologies. About 2–3% of all aroma encapsulates are produced by melt extrusion (Porzio 2004, 2007b). Aroma melt extrudates are in general water-soluble. Therefore, such extrudates are usually stable in dry food products but not in water-containing ones.

5.3.7 Coacervates

Coacervates are made via a liquid–liquid phase separation mechanism of an o/w emulsion into a polymer-rich phase (known as coacervate) and a polymer-poor phase (see Sect. 2.2.7, Soper 1995 and Thies 2007). According to the number of polymer(s) present, the process can be identified as (simple) coacervation when only one polymer is involved or complex coacervation when two or more polymers of opposite electric charges are present. The coacervates used to encapsulate aroma are most often of the complex type.

The shell of complex coacervates is commonly composed of gelatin and gum arabic, and can be crosslinked with, e.g., glutaraldehyde (not allowed in Europe for food applications) or transglutaminase (Soper and Thomas 1998). Optionally, the gum arabic can be replaced by other negatively charged molecules like carboxymethylcellulose, pectin, alginate, alginate derivatives or polyphosphate (Meyer 1992; Gouin 2004; Bakker et al. 1999). Gelatin might be replaced by whey proteins (Weinbreck et al. 2003). Fish gelatin might also be applied for Halal use

(Subramaniam and Reilly 2004). Each system has its own optimal operating conditions. For example, complex coacervates might be made from an aroma in an o/w emulsion with gelatin and gum arabic at a 1:1 w/w ratio and at a 2% w/w of each polymer dissolved in the water phase. The pH is then adjusted from neutral to about 4 under turbulent conditions in a stirred vessel at >35°C, a temperature above the gelation temperature of gelatin (see Fig. 2.5; Gouin 2004; Lemetter et al. 2009). This creates three immiscible phases (oil, polymer-rich and polymer-poor phase), and the polymer-rich phase droplets will deposit on the emulsion surfaces due to interfacial sorption. Alternatively, complex coacervation can be induced by dilution instead of pH adjustment from a 8 to 11% (w/w) gelatin solution (Thies 2007). Upon cooling well below 35°C (Lemetter et al. 2009), the deposited gelatin and thus the shell will solidify and can be crosslinked by glutaraldehyde or transglutaminase (Thies 2007). Complex coacervates might also be postloaded with aroma by solvent (Thies 2007). Aroma components to be encapsulated are dissolved in ethanol and then mixed with in water swollen complex coacervates under agitation for several hours, followed by ethanol removal.

An example of simple coacervation has been given by Hsieh et al. (2006). They emulsified citronella oil in the presence of 0.2 wt% chitosan and increased the pH with 0.1–1.5 wt% sodium hydroxide to form a shell of chitosan around the emulsion droplets. After washing twice with distilled water, the dispersion of the encapsulates was improved by incubation with coconut oil amphoteric surfactant for 10 days. The resulting microcapsules were dried in a vacuum oven. The aroma released out of dried microcapsules within a few hours (depending on the temperature). Interestingly, the authors suggest that the structure of the chitosan wall shrinked once the microcapsules were thermally pre-treated at 80°C. Another example of simple coacervation is disclosed in a patent by Porzio and Madsen (1997); coacervation of hydroxypropyl methylcellulose or methyl cellulose was induced by addition of maltodextrin, which also functioned as a spray-drying carrier material.

Less than 1% of aroma encapsulates are composed of complex coacervates (Porzio 2007b). Release out of complex coacervates might be achieved by pressure, heat or water dissolution. It is more the oil (the co-solvent) than the aroma that is encapsulated in complex coacervates, and aroma is in equilibrium with the entrapped oil and the outside (Malone and Appelqvist 2003; see also the next section). The load of aroma can be very high (20–90%), and might be adjusted to match the density of a liquid food product. (The shell is heavier than water, and the oil is lighter.)

5.3.8 Microspheres

Microspheres are biopolymer gel microbeads which may have embedded oil droplets. These matrix-type of encapsulates are obtained by gelling a biopolymer forming a network around oil droplets in water. In general, microspheres are made via the extrusion/dropping method or via the emulsion method (see Sect. 2.2.8 and Fig. 5.3).

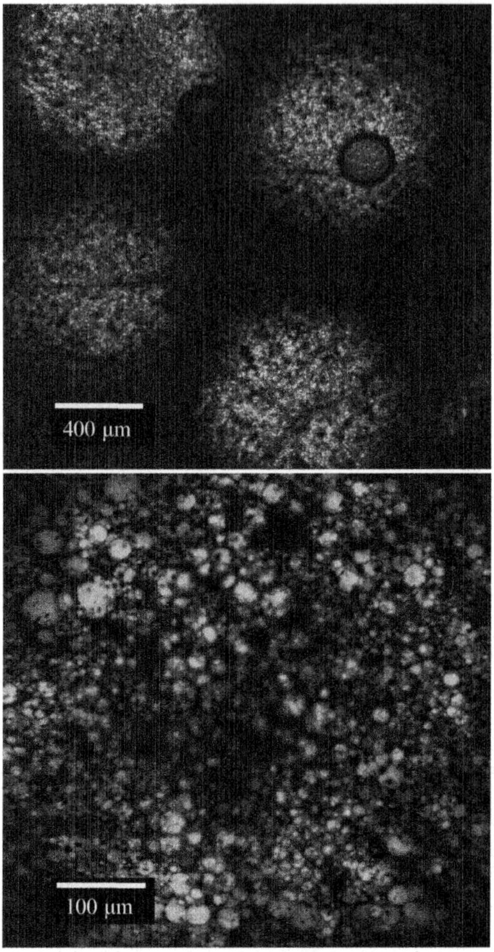

Fig. 5.3 Fluorescence confocal laser scanning micrograph (CLSM) of calcium alginate microspheres containing oil droplets with aroma. The pictures have been kindly provided by our colleague Han Blonk from the Advanced Measurement and Imaging Department of Unilever R&D, Vlaardingen

In the latter case, one has to start with a double emulsion (o/w/o) before gelation of the polymer. The main gelling system used is based on calcium alginate gels (see for example Chang and Dobashi 2003; Bouwmeesters and De Roos 1998; Soper et al. 2004), but other systems might be applied as well: potassium – κ-carrageenan, agarose, gelatin, gellan gum, agar, etc. Aroma-containing microspheres are prepared either in the presence of aroma in oil droplets (which may partition partly in the continuous oil phase when using the emulsion method) or by post-loading of dry microspheres containing blank oil droplets in the presence of trace amounts of water (Bouwmeesters and De Roos 1998; Ubbink and Schoonman 2003). The aroma load by post-loading usually ranges from 10 to 30 wt%.

The emulsion method offers the advantage of providing smaller microspheres (25 μm to 1 mm) than the extrusion method (1–5 mm). It is also easier to scale-up. However, the emulsion method is also more expensive especially when scaling-up, since the vegetable oil has to be removed (and possibly recycled), and the microspheres have to be washed sufficiently to eliminate the residual oil on the surface.

Recently, Trophardy (2007b) disclosed a method to prepare microspheres by fluid bed coating of biopolymers on reactive powder. For example, an aqueous mixture of limonene, starch and alginate or pectin was sprayed onto a powder mixture of starch and calcium gluconate in a fluid bed (see also Sect. 5.3.3). This resulted in insoluble encapsulates with an aroma load of about 10–50% and a particle size of 100–2,000 μm.

Aroma itself is not encapsulated into microspheres, but it is more the oil emulsion droplets inside the microspheres that are retained (Malone and Appelqvist 2003; the same is true for aroma in complex coacervates, see above). Aroma is interacting with the oil emulsions and is in equilibrium with hydrophobic environments outside (if present). The retention of aromas within the microspheres is controlled by both their initial lipophilic medium, consisting of triglyceride oil, and the porosity of the gel matrix surrounding the oil droplets. In a water-containing food application, for example, the (swollen) gel network increases the path-length through which the aromas must diffuse before coming under the influence of advective conditions existing in the environment outside the microspheres. The higher the density of the gel network, the slower the release. This explains why slower aroma release is found with increasing alginate and calcium concentrations and with increasing cross-linking time (see, e.g., Chang and Dobashi 2003). Unfortunately, the biopolymer concentration cannot be increased a lot since a too high viscosity of this solution is hindering the preparation process of microspheres.

The aroma release performance of microspheres described above is usually not better than coacervates capsules in terms of retention during cooking. Nevertheless, microspheres are in general more robust than coacervates and can sustain harsh food processing conditions. In low water content food applications, one may expect a good retention of aroma since the biopolymer gel might not be in swollen state (reduced network porosity); increased aroma retention has been shown for example with alginate microspheres when baking crackers (De Roos 2003).

Aroma release out of microspheres might be delayed by coating with lipids (by spray-chilling, see Coyne et al. 2005), or by the use of microcrystalline cellulose (Soper et al. 2004) or hydrophobic starches in the gel network (Bouwmeesters and De Roos 1998).

5.3.9 Co-extrusion

Reservoir type of encapsulates (see Fig. 2.1) can also be made via co-extrusion with the use of a multifluid nozzle consisting of concentric tubes (see Sect. 2.2.9 and Angelich 2005). Some equipment (e.g. from Inotech, Brace, Nisco) utilises a vibrating nozzle to produce 150–8,000 μm particles. The technology is based on the principle that a laminar liquid jet is broken into equally sized droplets by a superimposed vibration (see Fig. 2.6b). Another option is to make use of a dual-feed nozzle in combination with ultrasonic atomization (from, e.g., Sono-Tek) or centrifugal co-extrusion.

Co-extrusion can be utilised to prepare microbeads with a core of aroma and a shell produced by interfacial gelling (e.g. calcium alginate, potassium – κ carrageenan) or cooling (e.g. gelatin or fat) (Schleifenbaum et al. 2006; Mangos et al. 2007). If gelatin is used as the shell (typically at 70–90 wt%) one may use 30–10 wt% plasticiser, e.g., glycerol or propylene glycol (Schleifenbaum et al. 2006).

Cooling is often done via so-called submerged co-extrusion (Morishita et al. 1984; Schleifenbaum et al. 2006). This technique allows one to co-extrude an aqueous polymer solution that gels on cooling at the outer-side and an (aroma) oil to be encapsulated at the inner-side through a vibrating, multi-fluid nozzle that is submerged into a moving carrier and cooling fluid (see Fig. 2.7). The submerged set-up prevents disruption of the shell upon contacting the cooling liquid. The capsules can be 0.8–8 mm with 70–95% load. Optionally, the capsules can be coated with sugar, chocolate or fat (Kamaguchi et al. 2001; Wonschik et al. 2005).

Co-extruded aroma encapsulates may have a similar aroma retention performance as complex coacervates and microspheres. However, they commonly have a thin shell and are relatively large, which may make them more sensitive to shear than the other two types of encapsulates. Last but not least, the processing costs are quite high, which may limit their use for food applications.

5.3.10 Yeast Cells

Yeast cells (e.g. baker's yeast) can be laden with aroma essential oil by simply incubating these two compounds in water under gentle stirring (Bishop et al. 1998; Normand et al. 2005; Heinrich 2006; Dardelle et al. 2007). An example is shown in Fig. 5.4. The passive diffusion of aroma across the cell wall and the cytoplasmic membrane of yeast cells probably relies on its relative affinity for the internal lipid phase of the yeast cell (Bishop et al. 1998). The internal cell matter is simultaneously displaced with the uptake of aroma, which is emulsified by internal lipids. Yeast cell capsules can be seen as relatively natural aroma encapsulates since only natural ingredients are used and the process itself is also fully natural.

The plasma membrane appears to be the major permeability barrier to the diffusion of aroma essential oil into the yeast cells. The requirements for the aroma molecules to pass through the hydrated phospholipid bilayer are to be lipophilic enough ($\log P$ between 0.5 and 4) as well as small enough in size (Crothers and Nelson 2006). Active agent with a MW less than 1,000 can generally be encapsulated. The chain length cut-off is about 12 atoms of carbon in a straight chain; this was also verified by us when investigating yeast-encapsulated ginger aroma compounds

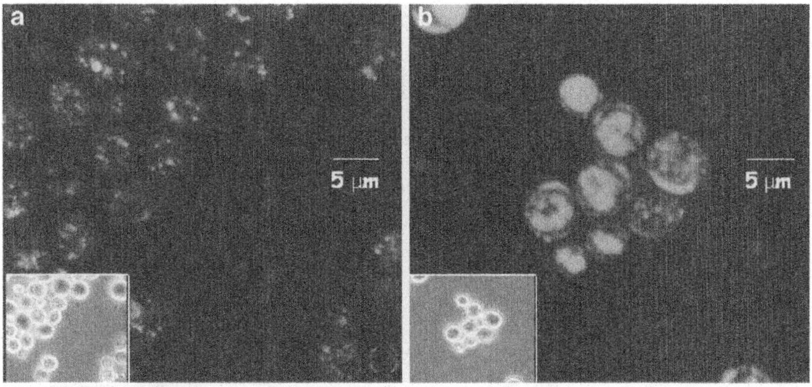

Fig. 5.4 Fluorescence confocal laser scanning (Nile Blue staining at 0.1% w/v)/phase contrast micrographs of yeast cells with (**a**) no added aroma (original baker's yeast) and (**b**) after encapsulation of limonene essential oil during 4-h incubation at 40°C (loading ca. 32% w/w) (Heinrich 2006). The pictures have been kindly provided by our colleague Han Blonk from the Advanced Measurement and Imaging Department of Unilever R&D, Vlaardingen

(Heinrich 2006, 2007). Furthermore, the fluidity of the plasma membrane can be manipulated simply by changing the temperature to alter the encapsulation rates of aroma compounds as well as the encapsulation kinetics (Bishop et al. 1998). For example encapsulating aroma essential oils at 60°C is more efficient than at 40°C, but also more costly especially since incubation has to last for a few hours for sufficient aroma loading. Equally the release of aroma compounds, also diffusion-controlled through the cell membrane, is faster at 60°C than at 20°C (Heinrich 2006, 2007).

Under dry conditions aroma release rates are considerably low, and at up to A_w of around 0.7 hardly any aroma is released; release appears to be largely limited by the mass transfer in the yeast cell wall (Bishop et al. 1998). Dardelle et al. (2007) even claim that aroma encapsulated in yeast cells in a dry state leaked out only at temperatures above 243°C (no proof provided). Upon rehydration this limitation no longer holds and release out of yeast cells is no longer limited by their cell walls but only by their phospholipid membranes. So aroma release from yeast is only extremely slow when the yeast is not hydrated (in dry state or in oil). When the yeast capsules are fully hydrated the aroma release is still very slow in purely aqueous systems and depends on the applied temperature to the system (see, e.g. Fig. 5.5). It has been shown with baker's yeast cells in water that encapsulated limonene or ginger aromas, for example, are releasing faster (higher aroma release rate) at higher temperatures (temperature was increased from 10 to 60°C; Heinrich 2006, 2007). Surprisingly, the aroma release rate was found to be constant and equal to the release rate at 60°C for temperatures of 60–82°C (higher temperatures were not studied). Flavored rice and pasta are typical examples of applications that can benefit from the yeast encapsulation technology (Dardelle et al. 2007). However, it was also shown that, as soon as triglyceride fats and/or emulsifiers are

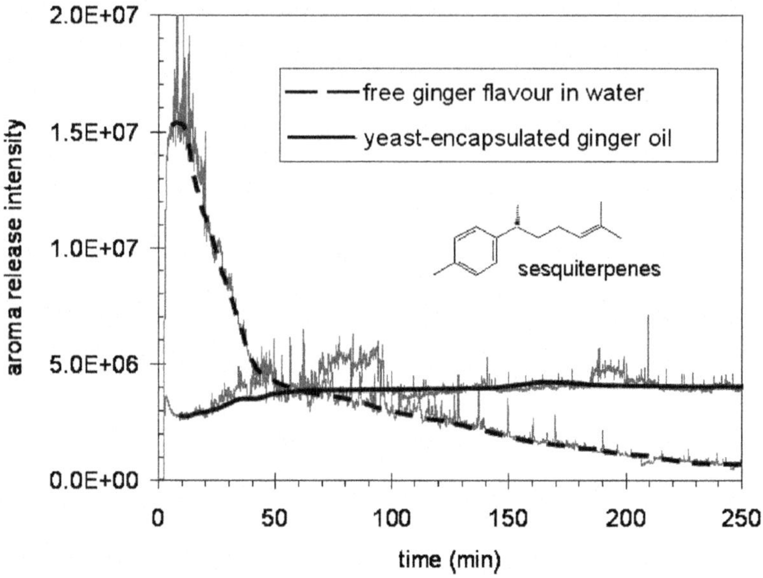

Fig. 5.5 Sesquiterpenes aroma release intensity profiles, measured by APCI-MS, of yeast-encapsulated ginger essential oil (*continuous line*) and free ginger oil (*dotted line*), both dispersed in pure water and heated at a constant temperature of 65°C. Aroma concentration was 100 ppm in both samples

present in the final application, aroma stability in yeast capsules (during storage or during cooking) is severely impacted. Therefore in many food applications this technology has some performance limitations. Another area of applications where yeast capsules are expected to perform well is dry food products (low A_w) that do not require water for cooking.

It is also believed that yeast cell capsules have an additional unique property of specific in-mouth delivery (Dardelle et al. 2007). The mechanisms that are believed to be involved here are a first adhesion of yeast capsules into the oral mucosa (via strong interactions between residual chitosan on the surface of the cells and the mucin protein from the mucus layer), followed by a fusion of the lipophilic aroma compounds onto the lipophilic surface of the keratinised areas of the mucosa.

Aroma in yeast cells is commercialised by Firmenich under the trade name 'Thermarome' under exclusive license from Micap Ltd., and often provided in a spray-dried form.

5.3.11 Cyclodextrins

Cyclodextrins are cyclic oligosaccharides of 6–8 glucose molecules, which are formed by enzymatic modification of partially hydrolysed starches. β-cyclodextrin is the most common one (Kant et al. 2004; Reineccius et al. 2004, 2005; Szente and

Szejtli 2004; see also Sect. 2.2.10). The cyclodextrins might be branched enzymatically to increase their water solubility (Ajisaka et al. 2000). Cyclodextrins have a lipophilic inner pocket, in which an aroma molecule with the right size can be entrapped via hydrophobic interactions (so-called molecular inclusion). The entrapment is usually carried out by dissolving cyclodextrin in limited amount of water, followed by addition of aroma under vigorous stirring. A temperature of 60–80°C might be used to facilitate dissolution of cyclodextrin and complex formation. Unfortunately, the loading of aroma into cyclodextrins is limited. For example, only 97 mg of lemon oil could be entrapped by 1 g of β-cyclodextrin after spray- or vacuum-drying of a paste (Bhandari et al. 1999). The load of other aromas varies in most cases between 8 and 10% (Szente and Szejtli 2004). Yoshii et al. (2006) found for several aromas an aroma/cyclodextrin molar ratio between 0.4 and 1.4. The binding is reversible and release of hydrophobic volatiles from cyclodextrins in aqueous systems is similar to that observed with emulsions (Kant et al. 2004), which indicates that the effect of cyclodextrin is diminished in the presence of an excess of emulsions. Cyclodextrin may protect aroma against oxidation, light- or heat-induced degradations and evaporation (Szente and Szejtli 2004). Freeze-dried flavor – β-cyclodextrin complexes retained flavor better and had a higher oxidative stability upon storage than spray-dried flavors using a maltodextrin, gum arabic, N-Lok, and gelan gum mixture at a 30/26.4/39.6/4 (w/w) ratio (Cho and Park 2009). The oxidative stability might be due to the fact that the flavor – β-cyclodextrin powder was less hygroscopic than the spray-dried flavor powder. Yoshii et al. (2006) found that addition of cyclodextrins for the retention of aromas in boiling water might only have a small effect in comparison to that at room temperature.

About 1% of all aroma encapsulates are based on the use β-cyclodextrin (Porzio 2004, 2007b, 2008). A typical application includes protection of unstable aromas in beverages (e.g. lemon flavoured teas). Cyclodextrins are also used to improve aroma retention during spray-drying (Shiga et al. 2004; Yoshii et al. 2005) or manufacturing of extruded snack food (Yuliani et al. 2004; 2006a). Unfortunately, the use of cyclodextrins might be limited by regulatory rules. In the US β-cyclodextrin is on the GRAS list since 1998, at a 2% level in numerous food products (Szente and Szejtli 2004), and also α- and γ-cyclodextrins have a GRAS status. Within the EU β-cyclodextrin is only allowed in chewing gum, potato and water-based flavoured drinks (see http://www.codexalimentarius.net/gsfaonline/additives/index.html). Only in February 2007, α-cyclodextrin got a novel food status and the one for γ-cyclodextrin has not been granted yet.

5.3.12 Silica Particles

Amorphous silicas are used to adsorb aroma prior to further encapsulation, e.g., within fluid bed coating (see Sect. 5.3.3 and Redd and Sell 1997) or extrusion (Rasp et al. 2000).

A new technology to encapsulate aroma is the use of porous sol-gel made silica (Veith et al. 2004a, b, 2005). Hydrolysis of tetraethyl orthosilicate was performed for 48 h at ambient temperature, and then neutralised to induce gelation. At pH 6 an aroma mixture was added and after further neutralization gelation took place. As the porosity decreased, aroma molecules were entrapped more efficiently in the silica particles. The retention performance at ambient temperature decreased from alcohols > aldehydes ≥ esters > terpenes as with polar organic matrices (Veith et al. 2004b). Open sol-gel made silica particles showed an increased retention with increasing aroma load, while denser silica matrices had a maximum retention with increasing load. The aroma release was within 5–20 min in absolute ethanol at 25°C (Veith et al. 2004a). Further research will be necessary to estimate the full potential of food grade silicas in food products.

5.3.13 Co-crystallization/Co-precipitation

Co-crystallization or co-precipitation involves spontaneous cryztallization or precipitation, in which aroma (or other actives) can be entrapped. This technology is not often used (much less than 1% of all aroma encapsulates; Porzio 2004, 2007b, 2008), but discussed here to give a complete overview of aroma encapsulation possibilities.

For instance, supersaturated sucrose syrup at a temperature above 120°C can form spontaneous crystals upon cooling. Aroma can be added at the time of crystallization, thereby being encapsulated (see Madene et al. (2006) and references therein). The crystal structure of sucrose can be modified to form aggregates of very small, water soluble crystals that incorporate the aromas either within the crystals or by entrapment.

Recently, water-insoluble aroma complexes were made by co-precipitation of an emulsion of aroma oil in an aqueous solution of sodium caseinate by bringing the system to the isoelectric point (Yuliani et al. 2004, 2006b; Begum 2005). Lemon oil was used as a model aroma compound at 5, 10, 15, 20 and 25% (w/w). It was emulsified in a sodium caseinate solution, followed by precipitation of the emulsion at pH 4.5. The wet precipitate granules were fluid bed dried to obtain the encapsulated product in a dry granular form. Grinding provides a free-flowing powder. The capsules are edible and water-insoluble. The retention of lemon aroma in dried capsule was maximum (77.5% total oil) at 15% oil level. The total oil retained in the encapsulated powder granules from precipitation method were comparable to that retained by the conventional spray drying method which produces water-soluble encapsulated powder. The same group also studied the used of these encapsulates in extruded model food products composed of corn starch. They found improved aroma retention upon extrusion of the food product (Yuliani et al. 2006b), but it was lower than that obtained with cyclodextrin (see Sect. 5.3.11 and Yuliani et al. 2006a). These capsules might be more sensitive to shear during extrusion than aroma–cyclodextrin complexes, and the presence of sodium caseinate material also influenced the properties of the extrudates and process parameters.

Another example from the literature is the co-precipitation of aroma and zein, a water-insoluble protein. Zein nanospheres have been proposed as novel drug delivery systems, and only recently as novel encapsulates for the delivery of essential oil (Parris et al. 2005). Zein nanospheres contained 10–20% essential oil and were prepared by dissolving 250 mg of oil and 1.0 g of zein in 15 ml of 85% ethanol, followed by rapid dispersion of this solution into 40 ml of water containing 0.01% silicone fluid. The resulting opaque solution was then freeze-dried. Fifty per cent release was obtained within 4–5 h at 37°C and at pH 7.4 (with or without pepsin). Nanoparticles of 2-dimensional, layered double hydroxide nanohybrids composed of vanillic acid, zinc and aluminium oxides were recently made by co-precipitation of aqueous solutions containing vanillic acid, zinc nitrate 6-hydrate and aluminium nitrate 9-hydrate at molar ratios of 1.5:2:1 to 1.5:4:1 via titration of sodium hydroxide to a pH of 8 (Hong et al. 2008). The entrapment of vanillic acid was based on electrostatic interaction and about 33–39% of vanillic acid could be entrapped. However, the vanillic acid release was a burst one in aqueous solutions.

5.4 Applications of Aroma Encapsulates

Aroma might be encapsulated to provide superior handling during production, improved safety, creation of visible and/or textural effects, proper incorporation of aroma into the food product, improved stability and/or controlled aroma release (see also Sect. 5.1). Selecting the best encapsulation system for a specific aim(s) depends on various criteria, as already outlined in Sect. 2.4. In general, efficacy of aroma encapsulates is a delicate balance between shelf-life stability and release at the desired moment and place. Stability of the aroma encapsulates during preparation of the food product prior use, costs of the technology, scaling up, IPR and legal aspects are important aspects to consider as well.

5.4.1 Aroma Retention During Production and Storage in Food Products

Unfortunately, the studies about incorporation and storage stability of aroma encapsulates in food products are limited. Most studies focus on aroma release upon application or on storage stability of aroma encapsulates as such (not incorporated into food products). Nevertheless, this paragraph will discuss some general guidelines.

Basically all encapsulation technologies discussed in the previous section may provide superior handling and improved safety by the conversion of liquid aroma into a powder. If these are the only benefits obtained, one may choose the cheapest option (i.e. spray-drying).

If one would like to create visible and/or textural effects (e.g. coloured particles), the particle size should be large enough to be noticed by consumers by eye or in the mouth. In general, particles above 50 μm are noticeable in the mouth and below they may not, depending on the hardness of the capsules and the contrast between this property and the one of the food matrix. For example, large particles will be easily noticed by eye or in the mouth in a drink, but not in a food bar.

Of course, aroma encapsulates should be properly incorporated into the food products. For food powders, this means that the aroma encapsulates are homogeneously distributed. The particle size of aroma encapsulates might be adjusted to avoid segregation in a food powder by proper choice and design of the encapsulation process or by post-treatment via agglomeration or compaction. For example, spray-dried aroma might fall out of tea bags. To avoid this, one may utilise melt extrudates or (pressure) agglomeration to get larger aroma particles that will stay in the tea bags. For food liquids, one may utilise small encapsulates that are not noticeable by consumers and that do not sedimentate. An example is the use of cyclodextrins in beverages. However, cyclodextrins will not be useful when a fat phase is present in the beverage (e.g. milk), since then aroma will migrate quickly into the fat.

Stability of aroma during processing of food products might be another benefit of using encapsulates. Most encapsulates are relatively stable upon mixing with dry food powders and storage in closed packages ($A_w < 0.3$), although care might be taken for high shear during powder streams and processing steps in the factory. Large particles may, e.g., be ruptured by the rolling process used in making stick gum (Angelich 2005). Smaller particles are more shear resistant.

In principle, aroma encapsulates can be added to liquid food products as a dry powder or as a slurry. In juices, encapsulates might not be noticed due to the presence of natural dispersed material. Sedimentation of encapsulates might be prevented by matching the density of the liquid food product. In addition, care must be taken not to destroy encapsulates (especially large capsules and/or reservoir type of encapsulates) during processing of liquid foods, especially during emulsification steps requiring high shear.

If the food products contain a reasonable amount of water, encapsulates with water soluble carrier material (such as (poly-)saccharides, maltodextrin, (modified) starches, gums, and proteins) will dissolve. Encapsulates with such a carrier material (prepared by, e.g., spray-drying) cannot deliver any further benefit of aroma retention. (Cyclodextrins might be the exception if no fat is present, see above.) The use of coatings that are water-insoluble, e.g. with hard fat or wax coatings, might be an alternative, although osmotic forces may lead to rapid aroma release if the core is water-soluble and cracks in the coating exist (Ubbink and Schoonman 2003). Another challenge here is that aqueous food products will generally be pasteurised to guarantee microbiological safety and then the hard fat or wax coatings will usually melt during such a heat treatment. In such cases, post-addition of such encapsulates after the preparation process of food products (which increases costs) or encapsulates with coating that do not dissolve in water and do not melt upon heating might be an option (e.g. cyclodextrins if no fat is present or encapsulates with a matrix based on cellulose derivatives). The following literature findings exemplify what kind of improved aroma stability by encapsulation can be achieved during processing:

- β-Cyclodextrin (see Sect. 5.3.11) or sodium casein precipitates (see Sect. 5.3.13) have been used to retain D-limonene upon extrusion cooking with starch, which is widely used to manufacture snack foods (Yuliani et al. 2004, 2006a, b).
- Reineccius et al. (2004) found a flavour retention benefit with β-cyclodextrin as well upon preparation of hard candy, fruit leathers, and angel food cake.
- Flavor in yeast cells survived the processing conditions of granular cooking aids and kept their controlled release benefit (Bonnet 2006).
- Campanile (2004) claimed that fluid bed coated encapsulates can resist processing conditions in typical sugar confectionery (10 min at 120–130°C for, e.g., hard boiled candies) or gum systems (up to 40 min at 80°C for, e.g., wine gums or aerated chews).
- Improved aroma retention was found when using aroma in complex coacervates upon coating of cereal samples with berry or citrus flavored molten sugar solution of 100°C, followed by 10 or 20 min drying in a convection oven at 79°C (Soper 1995).
- The use of melt extrudates with a high glass transition temperature in a hard candy production showed better aroma retention than melt extrudates composed of a standard blend of dextrins and sugars (Popplewell et al. 2003).

Aroma encapsulates in dry food powders ($A_w < 0.3$) may increase aroma stability by preventing reactions with other food components in the products, degradation or evaporation. For example, Soper (1995) showed improved berry aroma retention when using complex coacervates upon storage of sugar coated ready-to-eat cereal for 10 weeks at 40°C. Oxidation-sensitive aroma like citrus oils might have a better shelf-life in encapsulates prepared by melt injection or melt extrusion than prepared by other methods because of their barrier properties to oxygen (Ubbink and Schoonman 2003). Aroma evaporation might occur via sorption or permeation through the packaging material into the open atmosphere (López-Carballo et al. 2005) or by opening the pack. Encapsulation may prevent such evaporation or migration processes.

Upon storage of dry food products in closed packages, aroma might migrate from encapsulates via the gas phase into other hydrophobic pockets of the food matrix (e.g. oil, starch, etc.). Encapsulates which do not have aroma present on their surface by design and carrier material with a high glass transition temperature (T_g), prepared for example by melt extrusion or fluid bed coating, may be especially useful for stabilization of labile aroma molecules under dry conditions. The glassy state of carrier material limits caking and enhances aroma retention upon storage. Moisture uptake ($A_w > 0.3$) through the package or by opening might increase leakage and/or oxidation of aroma in, e.g., spray-dried form, since water will decrease the T_g of the carrier material (see Sect. 5.2). The increase of aroma release might occur at a temperature well above the T_g (the difference depends on the aroma molecule), but probably not at that T_g temperature (Bohn et al. 2005). This increase will depend on the physico-chemical characteristics of both the carrier material and the aroma. Moisture damage may be prevented by using encapsulates with more hydrophobic carrier materials (including spray-dried ones and lipid coated ones), with complex mixtures of carbohydrates which do not easily crystallise (Ubbink and Schoonman 2003) and/or with highly dense structures (e.g. melt extrudates, see Popplewell et al. 2003).

Aroma encapsulates in liquid food products might be less storage stable. Encapsulates with water-soluble coatings dissolve and are therefore not storage stable. Complex coacervates, microspheres, cyclodextrins or yeast cells with aroma are only storage stable in aqueous food products if no fat and/or emulsifiers are present. For example, aroma in dried coacervates composed of gelatin and gum arabic was completely recovered in the fat phase upon storage in margarine (Unilever, unpublished results). Encapsulates with coating that are water-insoluble might be an alternative for aqueous food products if they can be post-added or survive pasteurization (see above). Furthermore, lipophilic aroma may diffuse through the fat, therefore lipid-coated encapsulates are not recommended for applications that require long storage. Encapsulates in oil-based products without water might use oil-insoluble coatings, which are often water-soluble ones (such as (poly-)saccharides, maltodextrin, (modified) starches, gums, and proteins). Within Unilever one has found that simple spray-dried aroma is very well retained in the powder particles upon storage in sunflower oil for over 1 year at ambient temperatures (unpublished results). Only very volatile aroma molecules like acetaldehyde leaked partly out of the encapsulate. So, volatile aroma might migrate from encapsulates into the oil phase upon storage. Small additions of water to the oil resulted in immediate release of the aroma. Dardelle et al. (2007) found no significant aroma release from dry yeast capsules after 15 days of incubation in oil (in contrast to wet yeast cells).

5.4.2 Aroma Retention During Cooking

Only encapsulates with a good stability during production and storage until use (see previous section) will be able to deliver a controlled release benefit in application. Although this seems very obvious, it is often overlooked. Aroma release benefits from encapsulates in food products are often claimed without checking this processing and storage stability requirement. Possible controlled release benefits during cooking (baking, boiling, frying, grilling, etc.) are improved aroma differentiation, aroma burst or aroma retention.

Aroma differentiation by encapsulation means that the aroma perception of a product might change during cooking. In theory this might be used during cooking (e.g. to signal that the meal is ready), but this is difficult to achieve in practice. Most often, aroma differentiation is used to give an aroma burst when the food product is dissolved, e.g., upon addition of a tea bag with tea and aroma encapsulates into hot water.

Improved aroma retention can be achieved by delayed release from encapsulates during the cooking process. This means that (1) encapsulates should not dissolve quickly during cooking and (2) that the aroma diffuses only slowly with time out of encapsulates. Ideally, encapsulates should be of the reservoir type (see Fig. 2.1) to provide long retention, since only they can provide sustained (zero-order) release (see Chap. 4). Alternatively, matrix type of encapsulates (see Fig. 2.1) with a multiple order release close to zero-order might provide a release profile similar to the desired one. After release out of encapsulates, aroma may then diffuse further into the open atmosphere, depending on the affinity between the food product and its environment (see for this latter Druaux and Voilley 1997, and Van Ruth and Roozen 2002; De Roos 2003). Upon cooking under liquid conditions (boiling, frying, etc.) aroma release out of encapsulates should probably be based on diffusion or slow dissolution, since other release mechanisms (see Chap. 4) provide either too slow release (erosion), too fast release (burst release by fast dissolution, melting, or breakage), or are not applicable (enzymatic release).

To meet the first requirement mentioned above, encapsulates should be composed of water-insoluble carrier materials when cooking in water (boiling), and they should be composed of oil-insoluble carrier materials when cooking in oil (frying). The oil-insoluble carrier material might be the ones that are water-soluble. Please see the previous section for more information on such compositions. Melting of lipid coatings upon cooking might be used as a trigger for aroma release.

Open literature provides some examples on improved aroma retention upon cooking:

- Soper (1995) showed improved aroma retention when using complex coacervates in dough upon baking for 10 min at 177°C.
- Increased aroma retention upon baking was also found when using alginate beads or complex coacervates during baking of crackers (De Roos 2003). During preparation of hard bakery products, a lot of water is evaporated. Especially hydrophilic flavor will evaporate too, which can be prevented by encapsulation.
- Daragh and Stone (1975) prepared a water-soluble coating (gum arabic or dextrin) onto flavor in lipid particles. This double-coating was made by spray-chilling of lipid particles containing flavor, followed by fluid bed coating (see Sects. 5.3.3 and 5.3.4, respectively). These encapsulates were added to butter or margarine, and the inventors claimed superior flavor when one of these two products was used to bake cookies or cake. A disadvantage of this technology might be that these encapsulates need to be added after the pasteurization step of the butter or margarine (post-addition, which means extra costs) and will probably be noticeable by consumers due to their particle size.
- Redd and Sell (1997) from M-Cap Technologies filed a patent which disclosed a wax and fat coating of diacetyl butter applied on synthetic amorphous silica. The product was intended for use with microwaved popcorn and was designed to effect release of the aroma at a temperature above 66°C, but below the popping temperature of the corn. The product made in this manner is capable of

retaining the diacetyl for up to 4 h at 60°C and releases the flavoring material at a temperature above 66°C when it is heated in a microwave. Shelf life of the encapsulated product in contact with unpopped popcorn containing 5% moisture was claimed to be at least 3 months.

- Fluid bed coated encapsulates with a coating that was solubilising in water at 100°C in about 7 min provided slow release of aroma upon boiling and thus more retention than a spray-dried mixture (Doorn and Campanile, 2006). However, this will not be the case when fluid bed coating is performed with coatings that dissolve rapidly in water or with hard fat (since hard fat will melt during boiling).
- The use of cyclodextrin in boiling water had only a small effect on aroma retention compared to that one at room temperature (Yoshii et al. 2006). This agrees with the finding that the presence of β-cyclodextrin in spray-dried flavor did not affect the flavor release upon boiling in water or during rice cooking (Shiga et al. 2003).
- Coyne et al. (2005) entrapped pizza aroma in microspheres which then were lipid-coated by spray-chilling (see Sects. 5.3.4 and 5.3.8, respectively). Frozen pizzas with these encapsulates had a much stronger pizza aroma upon 3 min heating in a microwave. The inventors claimed that the aroma release could be triggered by absorbance of the microwave energy by the aqueous droplets in the inner part of the microcapsules (in the microspheres), which would then expand and fracture the fat barrier. Melting of the lipid material upon heating may also contribute to the release.
- A similar approach in which cookie aroma was released from fat particles on the surfaces of cookies during microwave heating was disclosed by Gaonkar and Ludwig (2005). The fat particles were applied on top of cookies by spraying a molten mixture of 5–40% aroma in ethanol and/or polypropylene glycol, medium chain fatty acid triglyceride (like glycerol tridecanoate fat) and high melting fat (like hydrogenated soybean oil) at about 70°C. The medium chain fatty acid triglyceride (35–75% of total fat) solubilised the aroma and the high melting fat (25–65%) provided structure. The fat particles protected the aroma from degradation during storage, and released aroma upon heating for 10 s in a microwave oven at 1,100 W. Those cookies had a pleasant flavor upon eating.

5.4.3 Aroma Release During Eating

Aroma that remains fully encapsulated will not be perceived in the mouth. Only aroma that has already released prior to eating or releases during eating will be perceived. For example, Reineccius et al. (2004) found that improved flavor retention by the use of β-cyclodextrin upon preparation of hard candy, fruit leathers, or angel food cake did not always result in improved sensory perception. They attributed this to poor aroma release from the inclusion complex during consumption.

Aroma release out of encapsulates during eating might be triggered by chewing, dissolution or dilution by water from the saliva, or by enzymes in the saliva.

In seasonings, a spray-dried aroma might be released when the food product (e.g. a snack) with the seasoning is eaten (Porzio 2004).

In chewing gum, the aroma can be encapsulated by spray-drying to allow faster (De Roos 2003) or slower aroma release upon hydration and chewing in the mouth than when the aroma is in the gum matrix. A flavor burst might occur when aroma and sweetener are entrapped within water-soluble encapsulates (such as spray-dried ones) or when the encapsulates burst upon chewing. For example, small gelatin capsules provide an aroma burst upon chewing (Uhlemann et al. 2002), although care must be taken that the capsules will not be broken during production. Aroma retention or differentiation in chewing gums might be obtained with aroma entrapped in relatively water-insoluble encapsulates, which might be spray-dried or fluid bed coated ones (Campanile 2004; Angelich 2005). Please see especially Angelich (2005) for more information on encapsulates in chewing gum and other confectioneries.

Complex coacervates or microspheres have been used by Unilever researchers to control the release of lipophilic volatiles in the mouth during eating (Malone et al. 2003; Malone and Appelqvist 2003). The complex coacervates and microspheres entrapped oil droplets containing aroma, which were termed ‘microstructured emulsions’ by them. It should be noted that the authors stated that it is the oil and not the aroma that is encapsulated. The oil droplets in the microstructures are enveloped in a gel phase, creating a static diffusion layer around the oil droplets. This increases the path-length through which the aroma must diffuse before coming under the influence of the advective conditions existing within the bulk of the product during eating. The result of these structures was to inhibit the rate at which the lipophilic aroma components replenish the continuous phase and reduce the rate of aroma release into the headspace. In this way, one could demonstrate a reduced aroma release from low fat systems similar to aroma release from high fat systems. A model for describing the aroma release from gel particles was developed by Lian et al. (2004). The model relates release rates to the composition (oil/water phase volume), particle size and the oil–water partition coefficient of the aroma, and takes into account the resistance to mass transfer in both the particle and the bulk liquid phase. It was possible to make microstructured emulsions from a range of biopolymer gels including calcium-alginate, gellan, gelatin, agar and starch and complex coacervates composed of gelatin/gum arabic. This provides the option to design particles that demonstrate controlled breakdown under physiological conditions during eating (Malone et al. 2003). Unfortunately, the particle size of the complex coacervates or microspheres used have to be relatively large (70–3,000 μm; the larger the better) to give an effect on aroma release rates, which limits their application.

5.5 Future Developments

The need for food products that are tasty, healthy (less sugar, less fat, and incorporation of nutrients which may often have an off-taste) and convenient will continue to demand improved aroma delivery systems. Furthermore, improved delivery systems may better overcome possible negatives already mentioned in Sect. 2.1, i.e. decreased costs, applicable in current processing and supply chain design, not noticed by consumers (visually or via mouthfeel) and more stable upon storage.

Improved aroma delivery systems might be achieved by modifying existing encapsulation processes (by choosing new carrier materials and new processing conditions), by developing new ones or by combining existing ones (as was done before by addition of cyclodextrin in the spray-drying or extrusion process, or by spray-chilling of microspheres). Furthermore, with the improvement of aroma encapsulates flavorists may need to adjust the aroma composition to make complete use of their potential (see also Porzio 2007a), in combination with other learnings from multisensory studies (Sijtsema et al. 2002). Last but not least, more studies about the incorporation, storage stability and release of aroma encapsulates in food products are needed to facilitate their success in the market.

References

Ajisaka N, Hara K, Mikuni K, Hara K (2000) Effects of branched cyclodextrins on the solubility and stability of terpenes. Biosci Biotechnol Biochem 64(4):731–734

Angelich A (2005) Encapsulation technology. Manuf Confectioner 85(10):51–55

Bakker MAE, Galema SA, Visser A (1999) Microcapsules of gelatin and carboxy methyl cellulose. European Patent EP937496

Baranauskienė R, Venskutonis PR, Galdikas A, Senulien D, Šetkus A (2005) Testing of microencapsulated flavor by electronic nose and SPME-GC. Food Chem 92:45–54

Baranauskienė R, Bylaitė E, Žukauskaitė J, Venskutonis RP (2007) Flavor retention of peppermint (*Mentha piperita* L.) essential oil spray-dried in modified starches during encapsulation and storage. J Agric Food Chem 55:3027–3036

Barnekow R, Fexer W (2007) Pressed agglomerates suitable for consumption having retarded aroma release. WO2007/098809

Begum SN (2005) Microencapsulation of lemon oil by precipitation method using sodium caseinate. Thesis, Food Science and Technology, School of Land and Food Sciences, The University of Queensland, St. Lucia

Benczédi D (2002) Flavor encapsulation using polymer-based delivery systems. In: Taylor AJ (ed) Food aroma technology. Sheffield Academic Press, Sheffield, UK

Bhandari BR, D'Arcy BR, Padukka I (1999) Encapsulation of lemon oil by paste method using β-cyclodextgrin: encapsulation efficiency and profile of oil volatiles. J Agric Food Chem 47(12):5194–5197

Bishop JRP, Nelson G, Lamb J (1998) Microencapsulation in yeast cells. J Microencapsul 15(6):761–773

Blake A, Cantergiani E, Harvey B (2003) Direct monitoring of flavor release from encapsulated flavor. In: Le Quéré JL, Étiévant PX (eds) Flavor research at the dawn of the twenty-first century. Proceedings of the 10th Weurman Flavor Research Symposium, London, Intercept Ltd, pp 154–158

Bohn DM, Cadwallader KR, Schmidt SJ (2005) Use of DSC, DVS-DSC, and DVS-fast GC-FID to evaluate the physicochemical changes that occur in artificial cherry Durarome upon humidification. J Food Sci 70(2):E109–E116

Bonnet EFJV (2006) Granular cooking aid comprising microbial encapsulates, and cubes and tablets comprising such granular cooking aids. WO2006045404

Bouquerand PE, Maio S, Normand V, Singleton S, Atkins D (2004) Swelling and erosion affecting aroma release from glassy particles in water. AIChE J 50(12):3257–3270
Bouwmeesters JFG, De Roos KB (1998) Process for preparation beads as food additive. WO1998015192
Brückner M, Bade M, Kunz B (2007) Investigations into the stabilization of a volatile aroma compound using a combined emulsification and spray drying process. Eur Food Res Technol 226:137–146
Buffo RA, Probst K, Zehentbauer G, Luo Z, Reineccius GA (2002) Effects of agglomeration on the properties of spray-dried encapsulated aromas. Flavor Frag J 17:292–299
Burdock GA (1994) Fenaroli's handbook of flavor ingredients, 3rd edn. CRC Press, Boca Raton
Bush Boake Allen Ltd (1973) Encapsulated compositions. GB1327761
Campanile F (2004) The flavor saviour? Food Processing 73(9):14
Carolina BC, Carolina S, Zamora MC, Jorge C (2007) Glass transition temperatures and some physical and sensory changes in stored spray-dried encapsulated flavors. Food Sci Technol 40:1792–1797
Chang CP, Dobashi T (2003) Preparation of alginate complex capsules containing eucalyptus essential oil and its controlled release. Colloids Surf B Biointerfaces 32:257–262
Charve J, Reineccius RA (2009) Encapsulation performance of proteins and traditional materials for spray dried flavors. J Agric Food Chem 57(6):2486–2492
Cho YH, Park J (2002) Characteristics of double-encapsulated flavor powder prepared by secondary fat coating process. J Foods Sci 67:968–972
Cho YH, Park J (2009) Encapsulation of flavors by molecular inclusion using β-cyclodextrin: comparison with spray-drying process using carbohydrate-based wall materials. Food Sci Biotechnol 18(1):185–189
Coyne B, Faragher J, Gouin S, Hansen CB, Ingram R, Isak T, Thomas LV, Tse KL (2005) Microcapsules. WO2005018794
Crothers M, Nelson G (2006) Targeted delivery. US2006/0127489
Daragh RT, Stone JL (1975) Fats with encapsulated flavors. US3867556
Dardelle G, Normand V, Steenhoudt M, Bouquerand PE, Chevalier M, Baumgartner P (2007) Flavor-encapsulation and flavor-release performances of a commercial yeast-based delivery system. Food Hydrocolloids 21:953–960
De Roos KB (2003) Effect of texture and microstructure on flavor retention and release. Int Dairy J 13:593–605
De Roos KB, Perren M, Sherman GA (2001) Encapsulation of active ingredients. EP1064856
Desai KGH, Park HJ (2005) Recent developments in microencapsulation of food ingredients. Drying Technol 23:1361–1394
Doorn L, Campanile F (2006) Encapsulation of food ingredients: principles and applications for flavor. Special Publications of the Royal Society of Chemistry 303:268–274
Druaux C, Voilley A (1997) Effect of food composition and microstructure on volatile flavor release. Trends Food Sci Technol 8:364–368
Firestein S (2001) How the olfactory system makes sense of scents. Nature 413:211–218
Fulger C, Popplewell M (1994) Flavor encapsulation. WO9406308, US5792505, or US5958502
Gaonkar AG, Ludwig CJ (2005) Aroma-producing compositions for foods. US2005287286AA, EP1611798 and US7442399BB
Goubet I, Le Quere J-L, Voilley AJ (1998) Retention of aroma compounds by carbohydrates: influence of their physicochemical characteristics and of their physical state. A review. J Agric Food Chem 46:1981–1990
Gouin S (2004) Microencapsulation: industrial appraisal of existing technologies and trends. Trends Food Sci Technol 15:330–347
Heinrich E (2006) Aroma release properties from yeast cell encapsulates in watery applications. Poster at 14th international workshop on bioencapsulation, Lausanne, Switzerland
Heinrich E (2007) Flavor encapsulation into yeast cells and cooking stability in aqueous food products. Podium presentation at the 34th Controlled Release Society meeting, Long Beach, CA

Henson L, Popplewell LM, Qi-Zheng J, Toth A, Bryant CM, Hans K, Pringgosusanto F (2006) Preparation and use of hydrogels. EP1716846, US2006239956, and US2006240076

Hong MM, Oh JM, Choy JH (2008) Encapsulation of flavor molecules, 4-hydroxy-3-methoxy benzoic acid, into layered inorganic nanoparticles for controlled release of flavor. J Nanosci Nanotechnol 8:5018–5021

Hsieh WC, Chang CP, Gao YL (2006) Controlled release properties of chitosan encapsulated volatile citronella oil microcapsules by thermal treatments. Colloids Surf B Biointerfaces 53:209–214

Jafari SM, He Y, Bhandari B (2007a) Encapsulation of nanoparticles of D-limonene by spray-drying: role of emulsifiers and emulsifying techniques. Drying Technol 25:1079–1089

Jafari SM, He Y, Bhandari B (2007b) Role of powder particle size on the encapsulation efficiency of oils during spray drying. Drying Technol 25:1091–1099

Jafari SM, Assadpoor E, He Y, Bhandari B (2008) Encapsulation efficiency of food flavor and oils during drying. Drying Technol 26:816–835

Kamaguchi Y, Mizutani K, Nishikawa M, Nagae K, Tanabe T (2001) Sugar-coated seamless capsule containing water-soluble material as content. JP2001238611

Kanakdande D, Bhosale R, Singhal RS (2007) Stability of cumin oleoresin microencapsulated in different combinations of gum arabic, maltodextrin, and modified starch. Carbohydr Polym 67:536–541

Kant A, Linforth RST, Hort J, Taylor AJ (2004) Effect of β-cyclodextrin on aroma release and flavor perception. J Agric Food Chem 52:2028–2035

Krishnan S, Bhosale R, Singhal RS (2005) Microencapsulation of cardamom oleoresin: evaluation of blends of gum arabic, maltodextrin and a modified starch as wall materials. Carbohydr Polym 61:95–102

Lemetter CYG, Meeuse FM, Zuidam NJ (2009) Control of the morphology and size of complex coacervate microcapsules during scale up. AIChE J 55(6):1487–1496

Lian G, Malone M, Homan JE, Norton IT (2004) A mathematical model of volatile release in mouth from the dispersion of gelled emulsion particles. J Control Release 98:139–155

Liu XD, Furuta T, Yoshii H, Linko P (2000) Retention of emulsified flavor in a single droplet during drying. Food Sci Technol Res 6(4):335–339

Liu XD, Atarashi T, Furuta T, Yoshii H, Aishima S, Ohkawara M, Linko P (2001) Microencapsulation of emulsified hydrophobic flavors by spray drying. Drying Technol 19(7):1361–1374

López-Carballo G, Cava D, Lagarón JM, Catalá R, Gavara R (2005) Characterisation of the interaction between two food aroma components, α-pinene and ethyl butyrate, and ethylene-vinyl alcohol copolymer (EVOH) packaging films as a function of environmental humidity. J Agric Food Chem 53:7212–7216

Madene A, Jacquot M, Scher J, Desobry S (2006) Flavor encapsulation and controlled release – a review. Int J Food Sci Technol 41:1–21

Malone ME, Appelqvist IAM (2003) Gelled emulsion particles for the controlled release of lipophilic volatiles during eating. J Control Release 90:227–241

Malone ME, Appelqvist IAM, Norton IT (2003) Oral behaviour of food hydrocolloids and emulsions. Part 2. Taste and aroma release. Food Hydrocolloids 17:775–784

Mangos T, Fischer N, Chauchadis P, Fexer W, Schütte C (2007) Mononuclearly filled microcapsules. EP1775017

McIver RC, Vlad F, Golding RA Jr, Leichssenring TD, Benczédi D (2002) Encapsulated flavor and/or fragrance composition. WO2002065858

Meyer A (1992) Perfume microencapsulation by complex coacervation. Chimia 46:101–102

Morishita T, Sunohara H, Tanoue S (1984) Method and apparatus for production of microcapsules. US4481157

Normand V, Dardelle G, Bouquerand PE, Nicolas L, Johnston DJ (2005) Flavor encapsulation in yeasts: limonene used as a model system for characterization of the release mechanism. J Agric Food Chem 53:7532–7543

Parris N, Cooke PH, Hicks KB (2005) Encapsulation of essential oils in zein nanospherical particles. J Agric Food Chem 53:4788–4792

Popplewell LM, Porzio MA (1998) Fat-coated encapsulation compositions and method for preparing the same. WO1998018338
Popplewell LM, Henson L, Toth A (2003) Keeping a full flavor. Food Ingredients Anal Int August: 4–7
Porzio M (2004) Flavor encapsulation: a convergence of science and art. Food Technol 58(7):40–47
Porzio MA (2007a) Flavor delivery and product development. Food Technol 01(07):22–29
Porzio M (2007b) Spray drying. An in-dept look at the steps in spray drying and the different options available to flavorists. Perfumer Flavorist 32(11):34–39
Porzio M (2008) Melt extrusion and melt injection. Perfumer Flavorist 33(6):48–53
Porzio MA, Madsen MG (1997) Double encapsulation process and flavorant compositions prepared thereby. WO1997013416
Rasp C, Brod H, Hinderer J, Renz KH (2000) Encapsulating medium and method for encapsulating perfumes and flavor. EP1034705
Ré MI (1998) Microencapsulation by spray drying. Drying Technol 16(6):1195–1236
Redd RV, Sell JL (1997) Method for the encapsulation of liquids. WO1997045195
Reineccius GA (2004) Spray-drying of food flavors. Drying Technol 22(6):1289–1324
Reineccius TA, Reineccius GA, Peppard TL (2004) Utilisation of cyclodextrin for improved flavor retention in thermally processed foods. J Food Sci 69(1):58–62
Reineccius TA, Reineccius GA, Peppard TL (2005) The effect of solvent interactions on cyclodextrin/flavor molecules inclusion complexes. J Agric Food Chem 53:388–392
Schleifenbaum B, Uhelemann J, Boeck R, Hinderer J (2000) Encapsulated flavor and/or fragrance preparations. WO2000036931
Schleifenbaum B, Voigt I, Aickele F (2006) Seamless filled capsules. EP1426045
Shefer A, Shefer SD (2003) A controlled release system for pharmaceutical and nutraceutical use. Patents US20030152629 and WO2004/082660
Shiga H, Yoshii H, Nishiyama T, Furuta T, Forssele P, Poutanen K, Linko P (2001) Flavor encapsulated and release characteristics of spray-dried powder by the blended encapsulatant of cyclodextrin and gum arabic. Drying Technol 19(7):1385–1395
Shiga H, Yoshii H, Taguchi R, Nishiyama T, Furuta T, Linko P (2003) Release characteristics of flavor from spray-dried powder in boiling water and during rice cooking. Biosci Biotechnol Biochem 67(2):426–428
Shiga H, Yoshii H, Ohe H, Yasuda M, Furuta T, Kuwahara H, Ohkawara M, Linko P (2004) Encapsulation of Shiitake (*Lenthinus Edodes*) flavors by spray drying. Biosci Biotechnol Biochem 68(1):66–71
Sijtsema S, Linnemann A, Van Gaasbeek T, Dagevos H, Jongen W (2002) Variables influencing food perception reviewed for consumer-oriented product development. Crit Rev Food Sci Nutr 42(6):565–581
Soottitantawat A, Yoshii H, Furuta T, Ohkawara M, Linko P (2003) Microencapsulation by spray drying: influence of emulsion size on the retention of volatile compounds. J Food Sci 68(7):2256–2262
Soottitantawat A, Yoshii H, Furuta T, Ohkawara M, Forssell P, Partanen R, Poutanen K, Linko P (2004) Effect of water activity on the release characteristics and oxidative stability of D-limonene encapsulated by spray drying. J Agric Food Chem 52:1269–1276
Soottitantawat A, Bigeard F, Yoshii H, Furuta T, Ohkawara M, Linko P (2005a) Influence of emulsion and powder size on the stability of encapsulated D-limonene by spray drying. Innovative Food Sci Emerg Technol 6:106–114
Soottitantawat A, Takayama K, Okamura K, Muranaka D, Yoshii H, Furuta T, Ohkawara M, Linko P (2005b) Microencapsulation of L-menthol by spray drying and its release characteristics. Innovative Food Sci Emerg Technol 6:163–170
Soper JC (1995) Utilization of coacervated flavors. In: Risch SJ, Reineccius GA (eds) Encapsulation and controlled release of food ingredients. ACS Symposium Series 590. American Chemical Society, Washington, DC, chapter 10, pp 104–112
Soper JC, Thomas MT (1998) Protein-encapsulated oil particles. EP856355
Soper JC, Molnar J, Vale JM (2004) Alginate matrix particles. WO2004098318

Subramaniam A, Reilly A (2004) Preparation of microcapsules. WO2004022221
Subramaniam A, McIver RC, Van Sleeuwen RMT (2006) Process for the incorporation of a flavor or fragrance ingredient or composition into a carbohydrate matrix. WO2006038067
Szente L, Szejtli J (2004) Cyclodextrins as food ingredients. Trends Food Sci Technol 15:137–142
Thies C (2007) Microencapsulation of flavors by complex coacervation. In: Lakkis JM (ed) Encapsulation and controlled release technologies in food systems. Blackwell, Ames, pp 149–170
Trophardy G (2007a) Flavor and/or fragrance capsules. WO2007/054583
Trophardy G (2007b) One step spray drying process. WO2007/135583
Ubbink J, Schoonman A (2003) Flavor delivery systems. In: Kirk-Othmer encyclopedia of chemical technology, on-line and 5th edn, vol 11. Wiley-Interscience, New York
Uhlemann J, Schleifenbaum B, Bertram HJ (2002) Flavor encapsulation technologies: an overview including recent developments. Perfumer Flavorist 27:52–61
Vaidya S, Bhosale R, Singhal RS (2006) Microencapsulation of cinnamon oleoresin by spray drying using different wall materials. Drying Technol 24:983–992
Valentinotti S, Armanet L, Porret J (2006) Encapsulated polyunsaturated fatty acids. WO2006067647
Van Ruth S, Roozen JP (2002) Delivery of flavor from food matrices. In: Taylor AJ (ed) Food flavor technology. Sheffield Academic Press, Sheffield, UK
Veith SR, Hughes E, Pratsinis SE (2004a) Restricted diffusion and release of aroma molecules from sol-gel-made porous silica particles. J Control Release 99:315–327
Veith SR, Pratsinis SE, Perren M (2004b) Aroma retention in sol-gel made silica particles. J Agric Food Chem 52:5964–5971
Veith SR, Perren M, Pratsinis SE (2005) Encapsulation and retention of decanoic acid in sol-gel-made silicas. J Colloid Interf Sci 283:495–502
Voilley A, Etiévant P (2006) Flavor in food. Woodhead Publishing Limited, Cambridge
Weinbreck FCJ, De Kruiff CG, Schrooyen P (2003) Complex coacervates containing whey proteins. WO03106014 and EP1371410
Wonschik J, Machinek A, Koschorreck C (2005) Coated spherical seamless filled capsules. WO2005077521
Yilmaz G (2003) Thermoplastic starch matrices for encapsulation and controlled release of volatile compounds. PhD Thesis, Utrecht University, Utrecht, The Netherlands
Yilmaz G, Jongboom ROJ, van Soest JJG, Feil H (1999) Effect of glycerol on the morphology of starch-sunflower oil composites. Carbohydr Polym 38:33–39
Yilmaz G, Jongboom ROJ, Feil H, Hennink WE (2001) Encapsulation of sunflower oil in starch matrices via extrusion: effect of the interfacial properties and processing conditions on the formation of dispersed phase morphologies. Carbohydr Polym 45:403–410
Yilmaz G, Jongboom ROJ, Van Dijk C (2005) Thermoplastic starch as a biodegradable matrix for encapsulation and controlled release. In: Mallapragada SK, Narasimhan B (eds) Handbook of biodegradable polymeric materials and their applications. Iowa State University, Ames, USA
Yoshii H, Yasuda M, Furuta T, Kuwahara H, Ohkawara M, Linko P (2005) Retention of cyclodextrin complexed shiitake (Lentinus edodes) flavors with spray drying. Drying Technol 23:1205–1215
Yoshii H, Neoh TL, Beak SH, Furuta T (2006) Release behavior of flavor encapsulated CD in slurry solution under boiling conditions. J Incl Phenom Macrocycl Chem 56(1–2):113–116
Yuliani S, Bhandari B, Rutgers R, D'Arcy B (2004) Application of microencapsulated flavor to extrusion product. Food Rev Int 20(2):163–185
Yuliani S, Torley PJ, D'Arcy B, Nicholson T, Bhandari B (2006a) Extrusion of mixtures of starch and D-limonene encapsulated with β-cyclodextrin: flavor retention and physical properties. Food Res Int 39:318–331
Yuliani S, Torley PJ, D'Arcy B, Nicholson T, Bhandari B (2006b) Effect of extrusion parameters on flavor retention, functional and physical properties of mixtures of starch and D-limonene encapsulated in milk proteins. Int J Food Sci Technol 41(S.2):83–94
Zasypkin D, Porzio M (2004) Glass encapsulation of flavor with chemically modified starch blends. J Microencapsul 21(4):385–397

Chapter 6
Microencapsulation of Fish Oil

Christiaan M. Beindorff and Nicolaas Jan Zuidam

6.1 Introduction

6.1.1 Marine Oils

For those fortunate to live near rivers, lakes and the sea, fish has been part of their diet for many centuries, and trade in dried fish has a long history. The important fishing industry developed when fishermen started to fish over wider areas of the seas and when improvements in freezing facilities allowed storage at sea, and subsequent distribution to urban consumers. For many, fresh fish and fried fish are now a part of their standard diet.

The original medicinal use of fish oils began with cod liver oil in the 1780s in England for arthritis and rheumatism. By the 1800s, it was used to prevent rickets. The prevention of rickets also depended on the vitamin D content. Fish oil contains several special types of fatty acids, the so called long-chain polyunsaturated fatty acids (LCPUFA, with more than 20 carbon atoms). The chemical structures of selected long-chain and other PUFAs are shown in Fig. 6.1.

Omega-3 fatty acids are those that have the first double bond at the third carbon-carbon bond from the terminal methyl group of the carbon chain, the omega (chemists normally start counting from the other side, the alpha). In Fig. 6.1, the omega-3 fatty acids are α-linolenic acid (ALA), eicosapentaenoic acid (EPA), and docosahexaenoic acid (DHA). Omega-6 fatty acids are those which have the first double bond at the sixth place from the end, here linoleic, γ-linoleic, and arachidonic acids. The characteristics of the omega-3 fatty acids are described in more detail below, since most people have a shortage of these (Garg et al. 2006).

C.M. Beindorff (✉) and N.J. Zuidam
Unilever R&D Vlaardingen, Olivier van Noortlaan 120, Vlaardingen, 3133 AT, The Netherlands
e-mail: christiaan.beindorff@unilever.com

N.J. Zuidam and V.A. Nedović (eds.), *Encapsulation Technologies for Active Food Ingredients and Food Processing*,
DOI 10.1007/978-1-4419-1008-0_6, © Springer Science+Business Media, LLC 2010

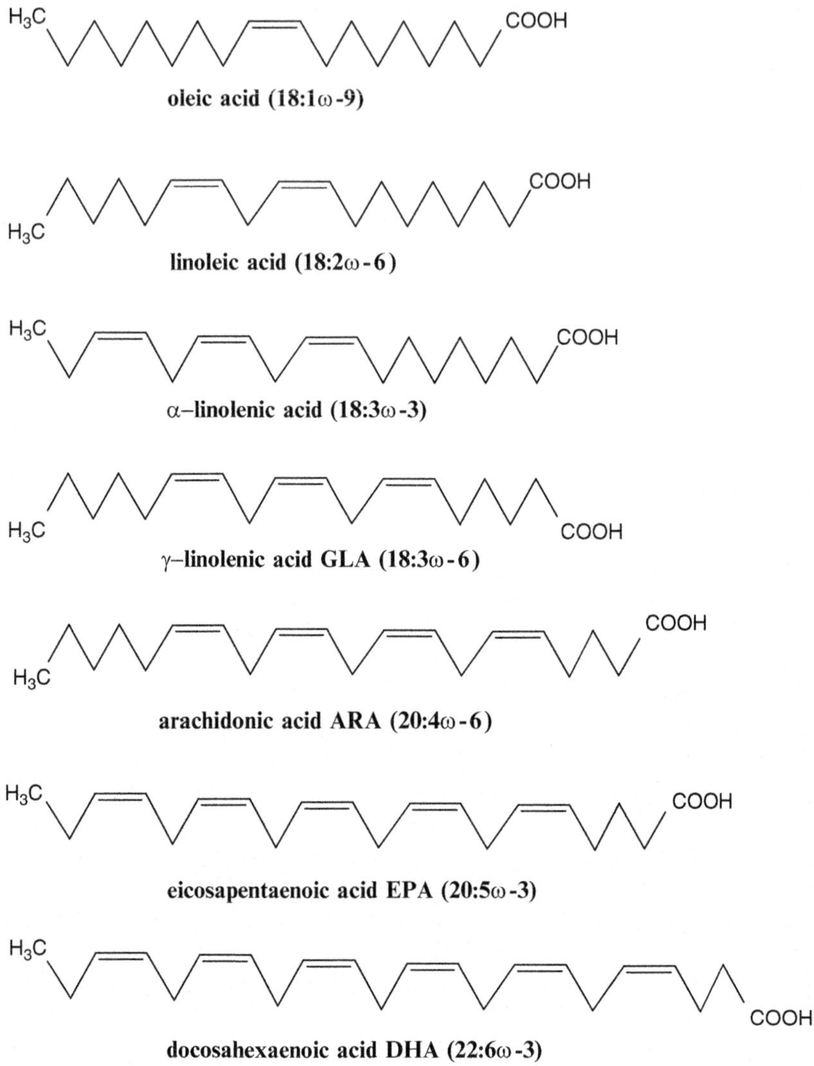

Fig. 6.1 Chemical structures of oleic acid and selected PUFAs

6.1.2 Omega-3 Fatty Acids

ALA is an essential fatty acid, which means that it is necessary to human health but cannot be manufactured by the body (Van Seeventer and Systermans 2006). Hence, ALA must be obtained from food. ALA is highly concentrated in certain plant oils such as flaxseed oil and to a lesser extent in canola, soy, perilla, and walnut oils. ALA is also found in wild plants, such as purslane (*Claytonia perfoliata*). Once ingested, the body converts ALA into EPA and DHA, the two types of omega-3 fatty acids more readily used by the body. ALA is the only omega-3 fatty acid found in vegetable products (Garg et al. 2006). The unique biochemical structure of ALA is important and helps to make it a key player in immunity, vision, cell membranes, and the production of hormone-like compounds.

EPA and DHA are polyunsaturated fatty acids that can be found in marine products (20–40% of the total fat contents) and algae (40% DHA and EPA). Both are important fatty acids that enter the body through consumption of marine products, fortification, or as ALA. A number of studies have been conducted to determine the role these fatty acids play in maintaining a healthy mind and body (Garg et al. 2006). Currently, more and more studies suggest that omega-3 fatty acids lower the risk of cardiovascular disease, some forms of cancer, and play a key role in brain development (Garg et al. 2006; Kolanowski and Laufenberg 2006; Eilander et al. 2007). However, due to their highly unsaturated nature they are very sensitive to lipid oxidation. This leads to an entire scale of off-smells and off-tastes, varying from grass- and bean-like, to cardboard- and fish-like flavors.

6.1.3 Composition of Different Sources of Fish Oil

The fatty acids of fish oil vary according to fish species. Table 6.1 presents the fatty acid composition of different fish oils. Fish oil is a by-product of fish meal production, and is obtained by filtration of cooked fish, followed by several purification steps, such as de-acidification to remove free fatty acids, bleaching to remove color and oxidation products, and deodorization to remove fish off-flavors. Optionally, the LCPUFA content can be concentrated, and anti-oxidants such as vitamin E are added subsequently (Kolanowski and Laufenberg 2006).

Of the fatty acids mentioned in Table 6.1, the saturated group (SAFA, mainly myristic and palmitic acids) is obtained from the fish diet, but it can also be biosynthesized by marine animals, if and when required. The mono-ethylenic fatty acids (MUFA, mainly palmitoleic, oleic, and eicosenoic acids) can also be biosynthesized by them. Many species of phytoplankton produce the highly unsaturated EPA (C20:5) and DHA (C22:6). Accordingly, these are passed up the animal food chain, and conserved as desirable fatty acids at each step. The leading countries producing

Table 6.1 Fatty acid profile (in %) of various fish species

Species	Herring	Tuna	Sardine	Cod liver	Menhaden	Salmon	Mackerel
SAFA	26	33	29	19	34	21	25
MUFA	54	18	31	50	24	29	47
EPA (C20:5)	10	6	22	11	15	13	7
DHA (C22:6)	6	26	9	12	10	18	8
LC omega-3	16	32	33	23	25	31	15

Data are taken from Firestone 1999. Values are averages, and may, for example, depend on the time of the year and region fish oil are Peru and Chile, followed by Denmark, the United States, Iceland, and Norway. The main fish species in Peru and Chile are anchovy, jack mackerel, Pacific mackerel, and sardine, while those in north-western Europe are capelin, Atlantic horse mackerel, sandeel, herring, and cod (Klootwijk 2006). New types of fish oils include salmon oil from Norway and tuna oil from Thailand and Australia. They are generally high quality oils that provide a useful source of omega-3 fatty acids (Gunstone 2004).

6.1.4 Food Fortification with LCPUFA

Growth in humans requires new cells and new membranes that are built up by phospholipids, for which EPA and DHA are necessary, the latter also reflecting development of vision and muscle control. The dietary recommendation for LCPUFA is still under debate; it varies with the various health organizations and also depends on disease prevention. The recommended daily doses of EPA and DHA for adults may vary from 180 to 500 mg for prevention of heart diseases, to 1,000 mg for decrease in mental illness (Garg et al. 2006; Kolanowski and Laufenberg 2006). According to the FDA in 2000, the daily intake of EPA and DHA should not exceed 3.0 g. Recently, the European Food Safety Authority (EFSA) recommended as the daily dose 250 mg of EPA and DHA and 2 g of ALA. However, the average amount of omega-3 fatty acids in a Western style diet is about 0.15 g per day, which is far below the recommended level. Food fortification with LCPUFA may fill this gap, since it is not likely that all consumers will start eating two to three fish servings a week to meet the recommended minimum doses.

6.2 Lipid Oxidation

As mentioned above, unsaturated fatty acids are sensitive to oxidation. The oxidation of lipid proceeds by a free radical mechanism, which can be described in terms of initiation, propagation, and termination processes (Chan 1987; Angelo 1992; Frankel 1996).

6.2.1 Initiation Reaction

The oxidation process can be initiated by a number of factors, including heat, UV radiation, ionizing radiation, or catalyzed-breakdown of peroxides (lipohydroperoxide or hydrogen peroxide). Metal-catalyzed oxidation (also called autoxidation) is the most relevant for foods, since trace elements are always present in raw ingredients.

Hydrogen peroxide might be formed in a metal ion-catalyzed reaction with oxygen:

$$O_2 + Fe^{2+} \Rightarrow \left[Fe(O_2)\right]^{2+}$$

$$\left[Fe(O_2)\right]^{2+} + H^+ \Rightarrow Fe^{3+} + HOO\bullet$$

$$2HOO\bullet \Rightarrow H_2O_2 + O_2$$

The hydrogen peroxide thus formed can be broken into hydroxyl radicals. These radicals initiate the oxidative chain.

$$H_2O_2 + Fe^{2+} \Rightarrow HO\bullet + HO^- + Fe^{3+}$$

$$H_2O_2 + Fe^{3+} \Rightarrow HOO\bullet + H^+ + Fe^{2+}$$

$$HO\bullet + LH \Rightarrow L\bullet + H_2O$$

Followed by:

In the presence of metal ions, fatty acid hydro-peroxides generate the corresponding alkoxyl radicals:

$$LOOH + Fe^{2+} \Rightarrow LO\bullet + HO^- + Fe^{3+}$$

$$LOOH + Fe^{3+} \Rightarrow LOO\bullet + H^+ + Fe^{2+}$$

Followed by the initiation of a new chain (chain branching):

$$LO\bullet + H_2L \Rightarrow LOH + HL\bullet$$

LH represents the fatty acid chain residue with an allylic proton. In this reaction, the role of Fe^{2+}/Fe^{3+} can also be fulfilled by copper ions or any other transition metal that can give a one-electron redox reaction. Iron and copper are the most common metals involved in lipid oxidation in foodstuff.

The initiation reaction when oil or fat is subjected to high temperature is completely different.

Under these circumstances, cleavage of peroxides takes place, leading to an alkoxy or hydroxyl radical. These can both initiate radical chains.

$$ROOH \Rightarrow RO\bullet + \bullet OH \ (\text{with } R = L \text{ or } H)$$

6.2.2 *Propagation*

The propagation reaction is predominantly a radical chain reaction. The peroxide radicals react with the weak allylic and bisallylic C–H bonds, which react further as:

$$L\bullet + O_2 \Rightarrow LOO\bullet$$

$$LOO\bullet + LN \Rightarrow LOOH + L\bullet$$

The first reaction takes place very fast, and the oxygen level should be very low to prevent this reaction from happening.

6.2.3 Termination

The propagation reaction will continue until one of the following termination reactions occur:

$$LOO\bullet + LOO\bullet \Rightarrow LOOL + O_2$$

$$LOO\bullet + L\bullet \Rightarrow LOOL$$

$$L\bullet + L\bullet \Rightarrow L - L$$

$$LOO\bullet + LOO\bullet \Rightarrow L = L + L - OH + O_2$$

These reactions lead to formation of dimers, polymers, ketones, and alcohols. At the beginning of the oxidation process, when oxidation proceeds at a low pace before becoming exponential, these reactions are of little importance since the concentration of radicals is low.

6.2.4 Formation of Off-Flavor

The primary products of the oxidation reaction described above are lipohydroperoxides and lipoperoxides. These compounds do not contribute to any off-flavor. The off-flavor is due to the homolytic or heterolytic cleavage of these primary products into the so-called secondary oxidation products. These secondary oxidation products consist of volatiles (aldehydes, alkanes, and ketones) which are responsible for (fishy) off-flavor (Jacobsen 1999; Van Ruth and Roozen 2000; Jónsdóttir et al. 2005; Velasco et al. 2006; Kolanowski et al. 2007; Shen et al. 2007).

6.2.5 Prevention of Lipid Oxidation

To prevent or slow-down the lipid oxidation, the following can be done (McClements and Decker 2000; Appelqvist et al. 2007):

- Prevent the initiation reaction from occurring (see Sect. 6.2.1)
- Reduce the oxygen concentration (see Sects. 6.2.1 and 6.2.2)

- Add anti-oxidants (to scavenge radicals, and thus terminate the oxidation reaction similar to those shown in Sect. 6.2.3)
- Lower the storage temperature
- Scavenge metal ions (iron or copper) using sequestrants in the water phase, if present (see Sect. 6.2.1)
- Exclude light (see Sect. 6.2.1)

The speed of the above mentioned reactions also depends on the initial quality and subsequent handling of the raw materials, and on the food structure and composition. For example, oxidation in a water-in-oil emulsion might differ from that in an oil-in-water emulsion, the use of positively charged emulsifiers and thickeners of the interfacial layer of emulsions may repel metal ions, etc. (McClements and Decker 2000; Appelqvist et al. 2007). Alternatively, microencapsulation can reduce oxidation (discussed in the subsequent sections).

6.3 Microencapsulation Technologies of Fish Oil

6.3.1 Requirements

Fish oil can be encapsulated to prevent off-flavor (Garg et al. 2006) by:

- preventing contact between oxygen and fish oil,
- preventing contact between metal ions and fish oil,
- preventing direct exposure to light, and
- trapping off-flavor.

These conditions can be achieved by the entrapment of the fish oil in a glassy state. Below the glassy state, the molecules in amorphous materials have little relative mobility. Within glassy state microencapsulates, it is possible to entrap volatiles like aroma (see Chap. 5), and maltodextrin powder with internal voids can be loaded with pressurized gas above its glass transition temperature and is then cooled to trap the gas (Zeller et al. 2006). Thus, this indicates that the exclusion of oxygen and entrapment of volatile off-flavors might be possible. Of course, the inclusion of oxygen from air during the microencapsulation process should be avoided as much as possible. Another advantage of microencapsulation might be the conversion of a liquid into a powder, which may ease the handling during supply chain or incorporation into food powder products. However, a common feature of many fish oil microencapsulates is that they disintegrate or dissolve in aqueous food products, and will therefore not be storage stable in many food products. Insolubility of microencapsulates in aqueous products may therefore be an important requirement, although this may lead to sandiness and consumer notice. Another important prerequisite of microencapsulation is that the fish oil will be bio-available upon consumption, which might be a concern if the microencapsulates do not dissolve or disintegrate in gastro-intestinal fluids. Without being bio-available, addition of it to food products is of no use, as claims about the physiological functionality of the fish oil cannot be made. These two aforementioned requirements, that is, stability upon storage and high bio-availability in the human gastro-intestinal tract, are somewhat contradictory in nature.

The next sections will first discuss the water-soluble microencapsulates and then the water-insoluble ones.

6.3.2 *Water-Soluble Microencapsulates*

6.3.2.1 Spray-Drying

Spray-drying is one of the oldest methods of microencapsulation in the food industry, and it is used most frequently to prepare fish oil microcapsules. In principle, an aqueous dispersion of an oil-in-water emulsion containing carrier material dissolved in the water phase is converted into a dry powder by spraying the feed into hot dry air resulting in moisture evaporation (see also Chap. 2). It is a one-step, continuous particle processing operation involving drying. Spray-drying normally contains the following stages: atomization, air contact, evaporation, and product recovery. The drying of the spray proceeds until the desired moisture content in the dried particles is obtained and the product is then recovered from the air. The resulting dried product conforms to powders, granules, or agglomerates, the form of which depends upon the physical and chemical properties of the feed, the dryer design, and the operation. Relatively low spray-drying temperatures are advantageous to minimize the lipid oxidation (Baik et al. 2004; Drusch and Schwarz 2006). The use of nitrogen instead of air to dry the dispersion can also be applied, but it increases the costs. Other options are two step processes where the particles are not dried completely during the first step but thereafter. Examples are further drying of spray-dried material on a moving belt with a Filtermat spray dryer or in a fluidized bed, or spray-drying of powder onto another carrier powder (like maltodextrin).

For example, maltodextrin, glucose syrup, proteins, sugars, gums, pectin, modified cellulose (e.g., hydroxypropyl methylcellulose or methylcellulose), and/or modified starch (e.g., octenylsuccinate-derivatized starch) can be used as a carrier material (Keogh et al. 2001; Kagami et al. 2003; Jónsdóttir et al. 2005; Tan et al. 2005; Drusch et al. 2006a, b; Drusch and Schwarz 2006; Kolanowski et al. 2004; Kolanowski et al. 2006; Drusch 2007; Drusch et al. 2007). Silica derivatives and tricalciumphosphate can be added to improve flowing properties of the powders (Drusch et al. 2006a). The fish oil loading of microencapsulates obtained via spray-drying varies from 1 to approximately 60 wt%. The particle size of these microcapsules varies from 10 to 400 μm. The higher the powder particle size, the higher the retention of fish oil (Jafari et al. 2007).

Although the spray-drying technology is commonly used, it has certain disadvantages. The spray-drying process may induce oxidation and the fish oil may have a limited shelf life due to the porous structure of the obtained spray-dried powder, and thus high access of air to the oil (Hogan et al. 2003; Kolanowski et al. 2006; Kolanowski et al. 2007). Kolanowski et al. (2006, 2007) even concluded that their spray-dried fish oil was less stable upon storage against oxidation than fish oil. However, this is not a common finding and might be due to the process conditions applied and/or materials used.

Encapsulates with a larger particle size might be more storage stable than those with a smaller particle size (Fang et al. 2005). Not only the temperature, but also the high moisture content affects the shelf life of spray-dried fish oil (Baik et al. 2004; Fang et al. 2005; Drusch et al. 2006a). This is probably because of a decrease in the glass transition temperature to values below the storage temperature, leading to relatively higher mobilities of molecules and reaction rates. Further improvements might be obtained by cross-linking of the carrier material prior to spray-drying, for example, by Maillard reaction of proteins (sodium caseinate, whey protein isolate, soy protein or skim milk powder) with carbohydrates (glucose, dried glucose syrup, or oligosaccharides) at 60–100°C for 30–60 min (Sanguansri and Augustin 2001; Augustin et al. 2006; Luff 2007). It is not clear if these improvements in storage stability were because of changes in powder morphology and/or anti-oxidative action of Maillard products. This product is commercialized under the name Driphorm by Nu-Mega. Another improvement might be obtained by the induction of simple coacervation (see also Sect. 6.3.3.1), addition of maltodextrin (with or without gum Arabic) into an emulsion of fish oil and hydroxypropyl methylcellulose (HPMC) prior to spray-drying (Ke-Gang et al. 2005). The addition of anti-oxidants, such as tocopherol or Trolox C, to the oil prior to spray-drying might also increase the storage stability of spray-dried fish oil (Hogan et al. 2003; Baik et al. 2004), although it may act as a pro-oxidant at high concentrations (Kolanowski et al. 2006). Serfert et al. (2009) found that a combination of anti-oxidants, that is, tocopherol, ascorbyl palmitate, and rosemary extract retarded the autoxidation significantly upon storage. However, the use of trace metal chelation like citrem did not affect this, although it improved the storage stability of the emulsion prior to spray-drying.

The shelf life of spray-dried fish oil can be enhanced by packing the fish micro-encapsulates alone or in food powders under nitrogen or vacuum in metalized packing material (Kolanowski et al. 2007).

6.3.2.2 Melt Injection

The melt injection process is based on the mixing of fish or vegetable oil in a starch matrix with anti-oxidants, sugars, emulsifiers, and water at a temperature above 100°C. The mixture is then extruded through a filter and collected in a bath filled with cold organic solvent (e.g., with iso-propanol or liquid nitrogen; Valentinotti et al. 2006; Subramaniam et al. 2006; see also Sects. 2.2.4 and 5.3.5). The cold organic solvent solidifies the sugar matrix and transforms it into a glassy state material which is then washed with a terpene (e.g., with limonene) to remove the surface oil present. This gives maximum protection to the microencapsulated oil against oxidation, resulting in no unpleasant odour or taste development during dry storage. The storage conditions are important as at a high water activity (A_w >0.3) the glassy sugar structures may start to disintegrate.

The fish oil load within microencapsulates prepared by melt injection is relatively low (10–20 wt%) and the particle size varies from 200 to 2,000 μm.

Figure 6.2 shows two confocal scanning laser microscopic (CSLM) images of a sugar extrudate with fish oil included. The images clearly show that the oil droplets, represented by light spheres, are well distributed through the whole matrix of the microencapsulate. This type of fish oil microencapsulate is protein free (and thus no gelatin), a feature that distinguishes it from other microencapsulate types.

6.3.2.3 Extrusion

Fish oil can also be encapsulated into a formable mixture or dough by using an extruder with one or more screws in a continuous process (Van Lengerich et al. 2004; Van Lengerich et al. 2007a, b; see also Sect. 2.2.4 of this book). The extrusion process is done at relatively low temperatures (below 30°C) and low pressures (500–5,000 kPa), in contrast to the melt extrusion process to encapsulate aroma (see Chap. 5). This can be achieved by using relatively high amounts of plasticizer (15–35 wt% water and 10–20 wt% of glycerol). First, emulsions using protein (such as sodium caseinate, wheat protein, or whey protein isolate), gum, or modified

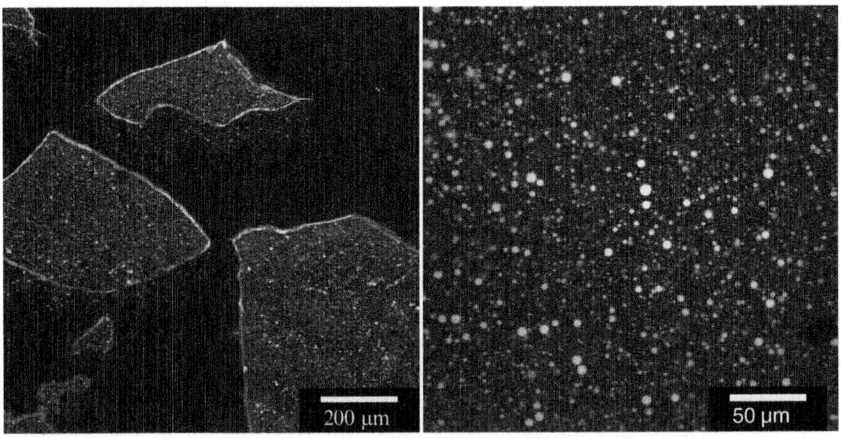

Fig. 6.2 CSLM images of Duralife® from Firmenich. The fish oil microencapsulate was embedded in low acryl embedding material and colored with a Nile blue solution. Sectioning of the embedded material was done using a microtome. Two laser lines from the argon/krypton laser were used to excite the fluorochrom Nile blue. The light spheres depict the fish oil droplets. The pictures were kindly provided by Ellen Drost (Unilever Research & Development Vlaardingen) and the Duralife® microencapsulates were obtained from Firmenich, Switzerland

starch as an emulsifier were prepared, and then mixed into 45–75 wt% of matrix materials (flours, starch, proteins, gums, etc.) with relatively high amounts of plasticizer (water and glycerol) and 0.5–4 wt% of an acidic anti-oxidant (ascorbic acid or erythorbic acid) using a twin screw extruder with a barrel temperature between 5°C and 10°C. The anti-oxidant is claimed to scavenge any oxygen before it can reach the fish oil, and hence, is not added to the emulsion phase in an attempt to avoid deleterious interactions between the protein and the acidic anti-oxidant. Flavor might be added to mask any off-flavor of fish oil, titanium dioxide or zinc oxide might be used as a whitener, and a flow agent such as starch might also be added. The extrudate is shaped by the die geometry (operating at 15–30°C) and can be cut immediately after extrusion or granulated by grinding or milling afterwards. The particle size of the extrudate will in general be around 0.5–2 mm. The water level of the extrudate (15–35 wt%) can be reduced by rotary or fluid bed drying to a water activity of less than or equal to 0.7 to increase the shelf life stability. The fish oil load is relatively low (typically 10–25%). Optionally, the particle can be coated with protein, starch or fat via, for example, pan coating. Alternatively, the fish oil can be added in an encapsulated (e.g., spray-dried) form when preparing food products by extrusion (Diguet et al. 2005).

6.3.2.4 Submerged Co-extrusion

Another type of extrusion technology is the so-called submerged co-extrusion, based on the simultaneous dropping of an oil droplet and shell material through a vibrating, concentric nozzle in a stream of cooling oil (see Sect. 2.2.9). The shell material may consist of gelatin or other polysaccharides and glycerol (as a plasticizer), and the fat content of the obtained particle is approximately 90 wt% (Sunohara et al. 2000). This patent of Sunohara et al. also discloses the use of an intermediate layer gelatin layer with citric acid to prevent insolubility with time due to interactions of unsaturated fatty acid with gelatin. The obtained capsules may have a relatively high fish oil load of about 70–95 wt% and a particle size between 0.8 and 8 mm. This technology is currently used by Morishita Jintan, Japan.

6.3.2.5 Cyclodextrins

Cyclodextrins are cyclic oligosaccharides of six to eight D-glucose molecules, which are enzymatically joined through alpha 1–4 linkages in such a way that they to form a ring (see Sect. 2.2.10 and Fig. 3.7). Cyclodextrins containing six, seven, or eight glucose molecules are referred to as α-, ß-, and γ-cyclodextrin, respectively. They may reduce the off-flavor perception by either forming a complex with LCPUFA or absorption of off-flavors. Schmid et al. (2001) disclose in their patent that the fish oil could form a complex with γ-cyclodextrin in water during stirring under a nitrogen atmosphere, and excluded from light at 45°C for 24 h. The resulting paste was filtered to remove water and then dried under vacuum at 50°C for

48 h. The fish oil did not oxidize for more than 24 h upon storage at 100°C under air when using γ-cyclodextrin, in contrast to the use of α- or ß-cyclodextrins. The fish oil loading was 15–40 w/w% of γ-cyclodextrin, and higher loads resulted in more oxidation. This product is now marketed as OmegaDry® by Wacker Chemical Corporation, using refined menhaden oil stabilized with mixed natural tocopherols and/or tert-butylhydroquinone. The powder particle size is 100 μm max.

6.3.3 Water-Insoluble Microencapsulates

6.3.3.1 Complex Coacervation

Coacervation is a method in which a solution is separated into a phase with a lot of polymers and a phase with hardly any polymers. In the case of one polymer, one speaks about "simple" coacervation (Ke-Gang et al. 2005; see also Sect. 6.3.2.1). Complex coacervation involves two or more polymers, and is most commonly used. One of the colloids is usually gelatin or whey proteins, while the other is an oppositely charged colloid, like gum arabic, sodium polyphosphate, or carboxy methyl cellulose. Coacervation is started by making an emulsion of oil droplets in the aqueous colloidal solution. By decreasing the pH, the phase separation of the solution in a polymer-rich and a polymer-poor phase is obtained, and the polymers precipitate on the interface of the oil droplets. Because the reaction is reversible and the polymers dissolve again if the pH is increased, hardening of the polymer layers might be necessary. This can be done by heat treatment and consequently de-naturation of the protein gelatin, or by cross-linking of the protein chains by glutaraldehyde or transglutaminase (see also Sect. 2.2.7). The release of ingredients may be controlled by:

- varying the amount or type of cross-linking material,
- varying the amount or ratio of the colloids,
- varying the wall-to-content ratio,
- using alternatives for the colloids with somewhat different properties,
- process conditions during preparation.

Complex coacervates may have a fish oil load of about 40–90 wt% and a particle size between 10 and 800 μm, depending on the materials and process conditions used.

Lumsdon et al. (2005) found that complex coacervates composed of gelatin or ß-lactoglobulin, gum arabic, starch, and cross-linked with glutaraldehyde could decrease the oxidation of fish oil. Complex coacervates containing fish oil can be present in two different morphologies: mono- and poly-nucleated format (see Fig. 6.3). The first type consists of a single oil droplet core surrounded by a hydrocolloid shell, whereas the second type consists of a multi-oil droplet core surrounded by a common hydrocolloid shell (Yan 2003), here composed of gelatin and polyphosphate. The complex coacervates might be double coated by repetition of the complex coacervation process (Yan et al. 2004) or by entrapment in another glassy

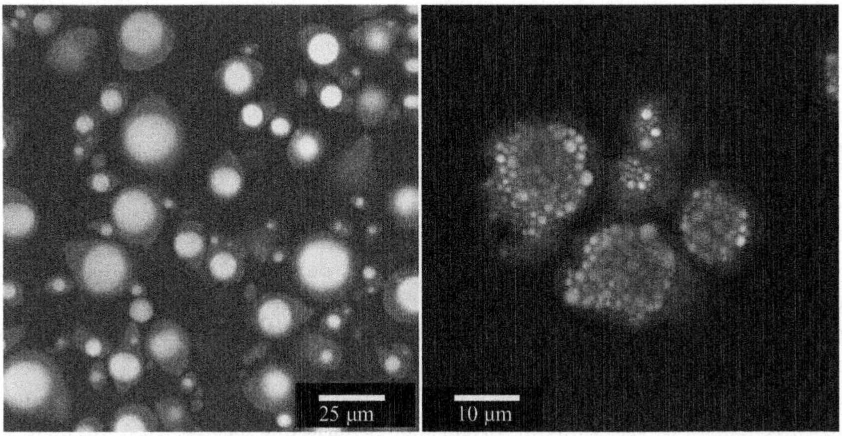

Fig. 6.3 CSLM images of two types of complex coacervates. The left side shows a mono-nucleated type with sunflower oil and on the right a poly-nucleated type of complex coacervate with fish oil is shown. In the CSLM picture, the light sphere depicts the oil core of complex coacervates and the dark circle around is the protein wall material. The images were kindly provided by Ellen Drost (Unilever Research & Development Vlaardingen) and the microencapsulates were obtained from the International Special Products, United Kingdom (*left picture*) and Ocean Nutrition Canada, Canada (*right picture*)

matrix (e.g., composed of maltodextrin and modified starch) by spray-drying, spray-granulation, or melt extrusion (Bouquerand et al. 2007) to enhance the storage stability and to prevent the dissolution in the mouth. Interestingly, complex coacervation has also been used to prepare ~100 μm nanoparticles composed of DHA, ß–lactoglobulin, and pectin (Zimit and Livney 2009). First, 0.22 mM DHA molecules interacted with 0.05% ß–lactoglobulin (1:2 molar ratio), and then 0–0.6 wt% pectin was added and the pH adjusted to pH 4.5 with hydrochloric acid. These nanoparticles were colloidal stable and transparent in an aqueous solution, and protected DHA against oxidation. However, their load is quite low compared to other microencapsulates.

A modified coacervation process was recently published by Gan et al. (2008). Soy protein isolate (SPI) was preheated at 95°C for 30 min and cooled to room temperature to promote de-naturation of SPI, followed by emulsification with fish oil and the emulsifier SPAN 80. The emulsion was subsequently treated at pH 5 and 40°C with transglutaminase. The obtained coacervates were filtered and freeze-dried. After freeze-drying, the powder were dry-mixed with sucrose or ribose and heated at 80°C for 1 h to induce cross-linking by Maillard reactions. The freeze-dried samples were less sensitive towards oxidation than fish oil at 50°C. No or small effect was obtained by the transglutaminase treatment. The use of ribose, however, delayed the oxidation of fish oil, most likely due to release of anti-oxidative Maillard reaction products during heating and a slower rate of gas permeability through the capsules.

Protein cross-linking of complex coacervates may have profound consequences for the structure of food proteins, and the subsequent properties of the final product. Cross-linking involves the chemical reaction of a bifunctional reagent, such as glutaric dialdehyde (1,5-pentanedial) with two amino acid residues (arginine or lysine) in the same protein, in adjacent subunits, or in two proteins that associate with each other. Although this reaction looks quite simple, a large variety of subsequent reaction pathways may be involved in this cross-linking. It is generally assumed that glutaric dialdehyde cross-linking of proteins takes place through reaction of the aldehyde groups of glutaric dialdehyde (II) with ε-amino groups of arginine or lysine residues (Tabor et al. 1992). According to these authors, the effect of cross-linking on the oxidative stability is rather limited. The amino acids involved in the cross-linking reaction (arginine and lysine) are only a small percentage (<10%) of the total amount present. Studies performed to investigate the complex mechanism of cross-linking reveal that glutaric dialdehyde cross-linking is a process resulting in the formation of a large variety of cross-linked entities. Although isolation and characterization of these products would seem to be the most direct approach to understanding the mechanism of glutaric dialdehyde cross-linking, the complex nature of the reactions involved suggests the difficulty of this approach. Furthermore, reaction time, glutaric dialdehyde concentration, temperature, and pH are very important parameters influencing the subsequent reaction pathways. Unfortunately, the chemical cross-linking with glutaric dialdehyde is legally not allowed to be used in foods in the European community.

Another way of cross-linking is through the use of enzymes. Transglutaminase is an (microbial) enzyme that connects glutamine and lysine residues with the release of NH_3. It is safe to be used in food products. Transglutaminase can be used for cross-linking several proteins, like gelatin, casein, whey protein, and certain soy proteins.

Not only is the need of protein cross-linking for use in complex coacervates a disadvantage of this technology but also the common use of gelatin in this microencapsulation product can give rise to difficulties with either Kosher or Halal status of the food matrix. Due to the fact that most gelatin used for complex coacervation is obtained from pig skin, these status problems can easily arise. Of course, an alternative source of gelatin is available, for example, from beef, poultry and fish, but in most cases they are more expensive than the gelatin from the pig.

6.3.3.2 Microspheres

An example of fish oil encapsulation in microspheres is the use of calcium alginate as an encapsulation material. Alginate is a polysaccharide, which is obtained by extraction from brown algae. The use of alginate as an encapsulation material is based on the capacity of alginate to bind di- and trivalent ions (except with Mg^{2+}) resulting in gel formation. The most common way to form alginate beads is to drop an alginate solution (1.5–4 wt%) into a calcium chloride solution (50 mM–1.5 M). The beads are formed instantaneously. Blatt et al. (2003) used this microencapsulation procedure followed by an additional coating of hydropropyl cellulose on the dried beads via fluid bed coating.

The Norwegian company Denomega Nutritional Oils sells fish oil encapsulates in microspheres with the Austrian encapsulation company GAT. They are prepared by first making water-in-oil emulsions, with calcium in the water phase. The addition of an aqueous alginate solution during stirring produces a phase inversion, and calcium alginate begins to deposit on the newly formed drops (Casana Giner et al. 2006). Another colloid is then added that deposits on the surface of the microspheres (e.g., xanthan gum) and then a primary surfactant is added to reduce the size of residual water in the oil drops. Agglomeration or deagglomeration may occur (depending on the process conditions), and finally the microspheres are hardened at elevated temperature (75°C for 120 min). These microspheres might be relatively small (1–30 μm) and may have a load between 20 and 50 wt%. The GAT encapsulates are claimed to protect the fish oil against oxidation and degradation during food processes and release their contents below a pH of 3.5 (e.g., in the stomach) (Thueringer 2009). The encapsulates can also be used to entrap, for example, water-soluble active agents like vitamin C, by maximising the residual water phase in the oil droplets during the preparation process.

The use of cross-linked, isolated soy proteins is another example of the use of microspheres for fish oil entrapment (Cho et al. 2003). These microspheres were prepared by making a duplex emulsion of fish oil-in-water-in-corn oil, in which the protein is cross-linked with transglutaminase to entrap the fish oil. A reduction in oxidation of the microencapsulated fish oil was found compared to non-encapsulated fish oil upon storage for up to 14 days at 50°C. Microspheres with fish oil can also be prepared by dropping emulsions composed of 20 wt% fish oil, 1 % chitosan, and 10 wt% maltodextrin or 20 wt% fish oil, 1 wt% chitosan, and 1 wt% whey protein isolate via ultrasonic atomization into a sodium hydroxide solution (Klaypradit and Huang 2008). The basic solution solidifies the chitosan and after centrifugation, filtering, washing with distilled water, and freeze-drying, microspheres with entrapped fish oil droplets and with a particle size of about 8–10 μm were obtained.

6.3.3.3 Calcium Carbonate Capsules

The production of calcium carbonate capsules is based on the electrostatic adsorption of calcium carbonate particles on the negatively charged oil surface of a fish oil-in-water emulsion during stirring (Nakahara et al. 2006). Salts, proteins, thickening, and/or stabilising agents might then be added, followed by freeze- or spray-drying. An anti-oxidant in the oil is also used. The advantage of calcium carbonate shells is that it only dissolves at low pH. This low pH condition can be achieved in the human body only when the food matrix, containing the calcium carbonate encapsulates, is present in the stomach. A scanning electronic microscopic image of a calcium carbonate microcapsule containing fish oil is presented in Fig. 6.4. These encapsulates have been commercialized by the KITII Corporation under the

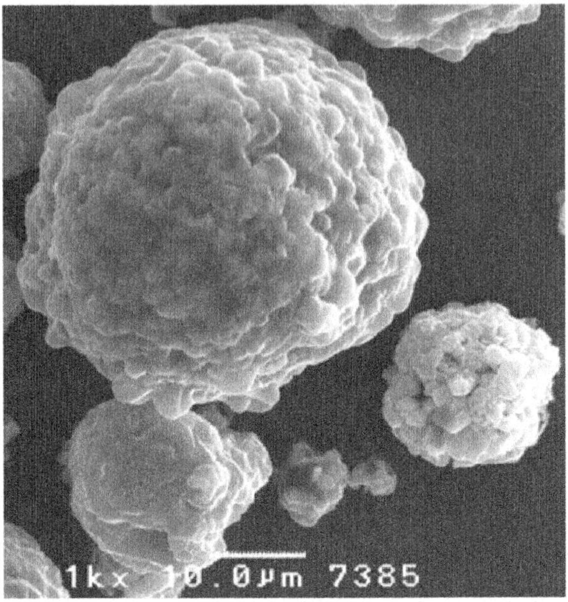

Fig. 6.4 A scanning electron microscope image of a Calshell microcapsule containing fish oil. Courtesy of KIITI Corporation

name Calshell. The particle size of Calshell is approximately 20 μm which is rather small compared to other types of encapsulates. This can be an advantage because these small particles will not be noticed. Optionally, the particle size can be enlarged. Also, the fish oil loading of Calshell, which is approximately 30–40 wt%, is favorable when compared with other types of fish oil encapsulates. According to the supplier information, less fish off-flavor is produced when Calshell encapsulates are used in a food matrix (compared to the free oil), and the capsules are high temperature and pressure resistant. Unfortunately, this technology is currently rather expensive.

6.3.3.4 Emulsions with Protein or Polymer Shell

The oxidation of emulsion depends on the food structure and composition, for example, water-in-oil emulsion, oil-in-water emulsion, emulsifiers, and composition of the aqueous phase (McClements and Decker 2000; Appelqvist et al. 2007). Also, the type and charge of emulsifiers may influence oxidation. Anionic emulsifiers attract positively charged metal ions such as iron, which brings them closer to the lipid phase and enhances oxidation. Positively charged emulsifiers (such as proteins) may repel metals and reduce oxidation, unless they have a metal-binding capability (such as sodium caseinate). In addition, a protein layer may protect oil by steric hindrance and/or free radical scavenging properties, and excess of proteins in the bulk may scavenge metals.

Further reduction might be achieved by the formation of layers around emulsions. The use of a single layer composed of sodium caseinate cross-linked with transglutaminase could not inhibit lipid oxidation (Kellerby et al. 2006). The use of two layers, which were composed of lecithin and chitosan, resulted in a slight inhibition of oxidation (Ogawa et al. 2003; Klinkerson et al. 2005a). These emulsions were prepared by first making a so-called primary emulsion with lecithin, and then mixing this emulsion with a solution of a positively charged chitosan (Klinkerson et al. 2005a, b). These two-layered emulsions could be freeze- or spray-dried after addition of corn syrup solids to improve the oxidative stability (Klinkerson et al. 2005b; Klinkesorn et al. 2005c; Klinkerson et al. 2006; McClements and Decker 2005, 2007; Shaw et al. 2007). Corn syrup solids also improve the physical stability of the emulsions against other food processing conditions such as salt, thermal processing, chilling, and freezing. The addition of the chelator ethylenediaminetetraacetic acid (EDTA) and mixed tocopherols further improved the oxidative stability of dried emulsions. Surprisingly, the oxidation was more rapid at 11% and 33% relative humidity than at 52%, which might be due to the sugar crystallization or formation of Maillard products.

Further inhibition of lipid oxidation may be achieved with multi-layered interfaces around oil droplets in oil-in-water emulsions, which may also enable their application as a controlled release system or triggered release system (Guzey and McClements 2006; McClements and Decker 2005). For example tuna oil-in-water emulsion droplets (particle size 0.2–1 μm) can be prepared by first emulsification with a food-grade anionic surfactant such as lecithin or fatty acid salts (so-called primary emulsion), followed by incubations with a cationic bio-polymer such as chitosan (so-called secondary emulsion), and then with an anionic bio-polymer such as pectin (so-called tertiary emulsions). One may repeat these alternative incubations to adsorb another cation/anion bilayer on the emulsion, although care must be taken to prevent the agglomeration due to the presence of excess free polyelectrolyte.

An increase in the oxidative stability has been reported by Guzey and McClements (2006) when using emulsions with a SDS-chitosan-pectin wall. This can be related due to the repulsion of iron ions by the chitosan layer and/or physically shielding of the iron ions by the multi-layers.

The lipids in the multi-layered emulsions might be bio-available (Park et al. 2007). However, applications of this technology in food matrixes remain difficult because chitosan is not a food grade ingredient in the USA and Europe and is only allowed in food supplements and pharmaceutical applications. Furthermore, chitosan has no Kosher or Halal status because it is obtained from crustaceans.

6.3.3.5 Coated Microencapsulates

Powder with fish oil (e.g., obtained by spray-drying) might be coated with a lipid layer to prevent oxidation upon storage. Ponginebbi and Publisi (2008) disclosed in their patent that fish oil powder in the presence of 0.5% silica flow aid could be coated in a fluid bed by spraying with molten 30% (w/w) hydrogenated palm wax. Presence of the lipophilic anti-oxidant L-ascorbic acid 6-palmitate in the lipid coating improved the oxidation stability of these microencapsulates upon storage better than the presence of the hydrophilic anti-oxidant Gravinol-T. A similar lipid coating might also be used by spray-chilling of 25–33 wt% spray-dried fish oil in molten lipid (Subramanian et al. 2008).

6.4 Analytical Characterization of Fish Oil Microencapsulates

6.4.1 Chemical Composition

Fish oil microencapsulates can be analyzed with several routine analytical techniques to measure the amount of protein, fat, moisture, and inorganic matter (ash content). Please see "method collections" from American Oil Chemists' Society (AOCS), Association of Analytical Communities (AOAC) or pharmacopeia for more details. In short, protein levels in foods are calculated from the total nitrogen content determined using the Dumas or Kjehldahl method. Fat is determined gravimetrically after acid hydrolysis of microencapsulates and extraction using petroleum ether. Moisture is determined from the weight loss of the sample during heating in a microwave oven. The ash content is determined gravimetrically by heating the sample in an open vessel to 550°C under air to constant weight. The remaining carbohydrate content of microencapsulates can then be calculated by subtraction of protein, fat, inorganic matter, and water contents from 100%. Not only is the total amount of fat in microencapsulates (the loading) important, but also the knowledge on the presence and level of the so-called surface oil is informative. Surface fish oil is easily oxidized, because it is located on or just in the outer shell of microencapsulates. Indeed, Velasco et al. (2006) concluded that free fish oil and microencapsulated fish oil oxidized differently, probably due to differences in the oxygen availability between encapsulated oil globules. In general, the level of surface oil should be as low as possible. However, the amount of surface oil alone cannot be used to predict shelf-lives of microencapsulated fish oil (Drusch and Berg 2008). The amount of surface oil can be determined by mild extraction with hexane or petroleum ether if they do not dissolve microencapsulates (as is the case with, e.g., spray-dried ones; Velasco et al. 2006; Drusch and Berg 2008).

Not only is the fat content of microencapsulates important, but also the fatty acid composition of the microencapsulated fat. Also, the percentage of DHA and EPA in the fat phase and the ratio between these two ingredients are important from the nutritional perspective. Especially, when a nutrient or health claim is made on the label of the product, the amounts of DHA and EPA must be specified. Furthermore, it is important to measure the percentage of other polyunsaturated fatty acids, mono-unsaturated fatty acids (mainly oleic acid), and saturated fatty acids (predominantly, myristic, palmitic, and stearic acid). Ideally, the level of saturated fatty acids should not exceed the acceptable value, and should be in accordance with the current nutritional recommendations. Standardized gas chromatographic methods can be used for the fatty acid analysis.

Very important with regard to the initiation of fish oil oxidation are the amounts of both iron and copper in microencapsulates. These metals can be present in the fish oil or in the carrier material. For food safety reasons, the levels of the trace elements mercury, lead, arsenic, and cadmium are also important. Fish oil may contain high amounts of heavy metals due to the pollution of oceans and seas, for example, gelatin can be high in trace metals both from the source (bone, skin) and the process chemicals (acids, lye).

6.4.2 *Physico-Chemical Characteristics*

Particle size of micoencapsulates can be measured by standard techniques like laser scattering or particle image analyzer (O'Hagan et al. 2005 and references therein). Microscopic observations with either conventional light or confocal scanning laser techniques can obtain very useful information about the size, distribution of sizes, and structure of the microcapsule. CSLM can be combined with staining techniques that allow staining of either the hydrophobic core or the hydrophilic shell of the microencapsulate (see Figs. 6.2 and 6.3). Microencapsulates can be visualized either from a dispersion in water if the microencapsulate does not dissolve in water (e.g., cross-linked complex coacervates), or from a dry powder. Scanning electron microscopy might be used to study the details of the morphology (see Fig. 6.4).

6.4.3 *Stability Testing*

A more instrumental technique is the measurement of the oxygen head space concentration and the amount of secondary oxidation products formed. The decrease of the head space oxygen concentration in a sealed bottle containing a certain amount of fish oil powder dissolved or dispersed in an aqueous solution is a good indication of the oxidative stability of the fish oil. Oxygen consumption reflects the propagation step in the lipid oxidation cascade. Head space oxygen concentration in sealed bottles and their variation over time can easily be measured with an oxygen analyzer, in combination with an air pressure measurement.

The measurement of secondary oxidation products (e.g., aldehydes, ketones, and alcohols) requires more sophisticated analytical equipment. The formation of these volatile products during time can be followed by the static headspace gas chromatography (Van Ruth and Roozen 2000; Jónsdóttir et al. 2005). Key off-flavor compounds are hexanal, acetaldehyde, trans-2-pentenal, pentanal, propanal, propenal, etc. There might be a good correlation between the oxygen consumption over time and the formation of secondary oxidation products (but not always), because both might be part of the same lipid oxidation cascade.

Standardized application tests where fish oil powder is dispersed in, for example, warm milk and directly consumed can give good indications of taste and fishy off-flavor. Of course, these organoleptical observations are from a consumer point of view the most important ones to make choices between different fish oils and their microencapsulates.

In most cases instrumental and organoleptical analyses of the fish oil microencapsulate give the same conclusion regarding the oxidative stability. However, contradictory results can sometimes be obtained.

6.5 Selection of Fish Oil Microencapsulates

Currently, both fish oil and fish oil microencapsulates are widely used in food applications (bread, orange juice, cereals, bars, etc.). Microencapsulation broadens the scope of application of sensitive fish oil in diverse food applications by enhancing the stability. A large set of different fish oil microencapsulates is currently available in the market. Roughly, these microencapsulates can be categorized in different groups according to the percentage of oil loading, species of fish oil which reflects in the amount of EPA and DHA, and the type of used microencapsulation technology. The combination of these three parameters results at the end in a price indication of the microcapsule. A selected list of fish oil microencapsulate suppliers, their commonly used microencapsulation technique, and some characteristics are presented in Table 6.2. Most of the listed suppliers use the well-known spray-drying technology to produce fish oil microencapsulates.

To select the type of fish oil microencapsulate suitable for a certain food application type from the list of commercial microencapsulates or own developments is a challenge. The following criteria might be relevant:

- quality and safety of the fish oil (including the absence of off-flavors, pesticides, EPA/DHA ratio, and high EPA and DHA content),
- physico-chemical characteristics of the microencapsulates,
- stability during supply chain, processing, and/or storage in a food product (which may depend on the composition of the food matrix and packaging),
- bio-availability of the fish oil,
- commercial available quantity,
- costs,
- legal status (food grade and, e.g., the use of cross-linking agents), and
- availability in Halal or Kosher quality.

In general, one has to consider the fact that the microencapsulated fish oil is more expensive than native fish oil. As a golden standard, it is therefore advised to test always as a first option the use of native fish oil and only if this does not give an acceptable end product, consider the use of a microencapsulated version. For

Table 6.2 Commercially available fish oil microencapsulates

Supplier	Trade name	Technology	Water soluble?	Expected price range
Arjuna, India	Zepufa	Spray-dried powder	Yes	Low
BASF, Germany	Dry n-3	Spray-dried gelatin/caseinate/sugar matrix with a starch coating	Yes	Low
DSM, The Netherlands	Ropufa	Spray-dried cornstarch-coated matrix of gelatin and sucrose	Yes	Low
The Wright Group, USA	Supercoat Omega-3	Spray-dried starch/protein matrix	Yes	Low
National Starch, USA	Novomega	Spray-dried powder based on modified starch and soy protein	Yes	Low
Kievit, The Netherlands	Vana-Sana	Spray-dried powder produced by spraying on a moving belt	Yes	Low
Nu-Mega, Australia	Driphorm	Spray-dried powder with Maillard products	Yes	Low
Wacker, Germany	OmegaDry	Complexation with γ-cyclodextrin in water, followed by vacuum drying	Yes	Middle
Firmenich, Suisse	Duralife	Melt injection with matrix based on maltodextrin and sugar	Yes	Middle
Ocean Nutrition Canada, Canada	Meg-3	Dried complex coacervates of gelatin and polyphosphate	No	Middle
KITII Corporation, Japan	Calshell	Calcium carbonate precipitation	Only at low pH	High
Morishita Jintan, Japan		Submerged co-extrusion	Yes	High
Denomega, Norway	Denomega GAT 100	Calcium alginate microspheres prepared via duplex emulsions and phase inversion	No	Middle

example, as an alternative to microencapsulation one may also use an edible coating of zein and fish oil on cookies, cereals, or fruit piece (Bello et al. 2007).

Much depends on what the structure of the food matrix is. For example, is it a dry product or in a liquid form? Water-soluble microencapsulates will dissolve rapidly in aqueous food products and can therefore not give further protection. Presence of different macro- and micro-nutrients may also influence the oxidative stability of fish oil. For example, the presence of vitamin C (ascorbic acid) may act as an anti-oxidant or as a pro-oxidant (Gunstone 2004). Sometimes, only for supply chain and logistic reasons, companies decide to select microencapsulated fish oil for application in their products instead of the use of native fish oil.

Spray-drying is generally the cheapest option available; submerged co-extrusion and calcium carbonate capsules are available at relatively higher costs. The price range of the other microencapsulation technologies in Table 6.2 might be more in the middle of these two extremes. The prices of the several commercial microencapsulates vary widely, and depends on many factors (including fish oil and material costs, cost structure of the company, processing costs, and volume of sales). Hence, one should consider these price estimates as only indicative.

6.6 Future Developments

Currently, there are a lot of different types of fish oil microencapsulates that are commercially available. When looking at different food applications the following requirements of fish oil microencapsulate are currently missing:

- better resistance to oxygen burden and more proof for prevention of oxidation and off-flavor formation in food products,
- water-insoluble microencapsulates that are storage stable in an aqueous food product and do not sediment,
- evidence for bio-availability of fish oil from especially water-insoluble microencapsulates.

Future research might be aimed at filling these gaps and improving existing technologies further by choosing new carrier materials and new processing conditions. One may make use of similar technologies as used for the microencapsulation of aroma (see Chap. 5), and/or combine existing ones. For example, one may spray-chill microspheres.

References

Angelo AJ (1992) Lipid oxidation in food, 500. American Chemical Society, Division of agriculture and food, Washington, DC

Appelqvist IAM, Golding M, Vreeker R, Zuidam NJ (2007) Emulsions as delivery systems in foods. In: Lakkis JM (ed) Encapsulation and controlled release technologies in food systems. Blackwell Publishing, Ames, IA, pp 41–81

Augustin MA, Sanguansri L, Bode O (2006) Maillard reaction products as encapsulants for fish oil powders. J Food Sci 71(2):E25–E32

Baik M-Y, Suhendro EL, Nawar WW, McClements DJ, Decker EA, Chinachoti P (2004) Effects of antioxidants and humidity on the oxidative stability of microencapsulated fish oil. J Am Oil Chem Soc 81(4):355–360

Bello A, Gimmler N, Roy P (2007) Omega-3 fatty acids encapsulated in zein coatings and food products incorporating the same, WO2007/030718

Blatt Y, Pinto R, Safronchik O, Sedlov T, Zelkha M (2003) Stable coated microcapsules. WO2003018186

Bouquerand PE, Dardelle G, Gouin S, Schleifenbaum B, Trophardy G (2007) Encapsulated active ingredients, methods of preparation and their use, WO2007026307

Casana Giner V, Gimeno Sierra M, Gimeno Sierra B, Moser M (2006) Continuous multi-micro-encapsulation process for improving the stability and storage life of biologically active ingredients, EP1702675

Chan HW-S (1987) Autoxidation of unsaturated lipids. Academic, London

Cho Y-H, Shim HK, Park J (2003) Encapsulation of fish oil by an enzymatic gelation process using transglutaminase cross-linked proteins. J Food Sci 68(9):2717–2723

Diguet S, Feltes K, Kleemann N, Leuenberger B, Ulm J (2005) Extrusion-stable poly-unsaturated fatty-acid compositions for food products, WO2005/089569

Drusch S (2007) Sugar beet pectin: a novel emulsifying wall component for microencapsulation of lipophilic food ingredients by spray-drying. Food Hydrocoll 21:1223–1228

Drusch S, Berg S (2008) Extractable oil in microcapsules prepared by spray-drying: Localisation, determination and impact on oxidative stability. Food Chem 109:17–24

Drusch S, Schwarz K (2006) Microencapsulation properties of two different types of n-octenyl-succinate-derivatised starch. Eur Food Res Technol 222:155–164

Drusch S, Serfert Y, Schwarz K (2006a) Microencapsulation of fish oil with n-octenylsuccinate-derivatised starch: flow properties and oxidative stability. Eur J Lipid Sci Technol 108:501–512

Drusch S, Serfert Y, Van den Heuvel A, Schwarz K (2006b) Physicochemical characterization and oxidative stability of fish oil encapsulated in an amorphous matrix containing trehalose. Food Res Intern 39:807–815

Drusch S, Serfert Y, Scampicchio M, Schmidt-Hansberg B, Schwarz K (2007) Impact of physico-chemical characteristics on the oxidative stability of fish oil microencapsulated by spray-drying. J Agric Food Chem 55:11044–11051

Eilander A, Hundscheid DC, Osendarp SJ, Transler C, Zock PL (2007) Effects of n-3 long chain polyunsaturated fatty acid supplementation on visual and cognitive development throughout childhood: a review of human studies. Prostaglandins Leukot Essent Fatty Acids 76:189–203

Fang X, Shima M, Adachi S (2005) Effects of drying conditions on the oxidation of lineoleic acid encapsulated with gum arabic by spray-drying. Food Sci Technol Res 11(4):380–384

Firestone D (1999) Physical and chemical characteristics of oils, fats, and waxes. AOCS, Washington, DC

Frankel EN (1996) Antioxidants on lipid foods and their impact on food quality. Food Chem 57(1):51–55

Gan C-Y, Cheng L-H, Easa AM (2008) Evaluation of microbial transglutaminase and ribose cross-linked soy protein isolate-based microcapsules containing fish oil. Innovative Food Sci Emerg Technol 9:563–569

Garg ML, Wood LG, Singh H, Moughan PJ (2006) Means of delivering recommended levels of long chain n-3 polyunsaturated fatty acids in human diets. J Food Sci 71(5):66–71

Gunstone FD (2004) The chemistry of oils and fats. Blackwell, Oxford, UK

Guzey D, McClements DJ (2006) Formation, stability and properties of multilayer emulsions for application in the food chemistry. Adv Colloid Interface Sci 128–130:227–248

Hogan SA, O'Riordan ED, O'Sullivan M (2003) Microencapsulation and oxidative stability of spray-dried fish oil emulsions. J Microencapsul 20(5):675–688

Jacobsen C (1999) Sensory impact of lipid oxidation in complex food systems. Fett/Lipid 101(12):S484–S492

Jafari SM, He Y, Bhandari B (2007) Role of powder particle size on the encapsulation efficiency of oils during spray drying. Drying Technol 25:1091–1099

Jónsdóttir R, Bragadóttir M, Arnarson GO (2005) Oxidatively derived volatile compounds in micro-encapsulated fish oil monitored by solid-phase micro-extraction. J Food Sci 70(7):433–440

Kagami Y, Sugimura S, Fujishima N, Matsuda K, Kometani T, Matsumura Y (2003) Oxidative stability, structure, and physical characteristics of microcapsules formed by spray drying of fish oil with protein and dextrin wall materials. J Food Sci 68(7):2248–2255

Ke-Gang W, Chai X-H, Chen Y (2005) Microencapsulation of fish oil by simple coacervation of hydroxypropyl methylcellulose. Chin J Chem 23:1569–1572

Kellerby SS, Gu YS, McClements DJ, Decker EA (2006) Lipid oxidation in a menhaden oil-in-water emulsion stabilised by sodium caseinate cross-linked with transglutaminase. J Agric Food Chem 54:10222–10227

Keogh MK, O'Kennedy BT, Kelly J, Auty MA, Kelly PM, Fureby A, Haahr A-M (2001) Stability of oxidation of spray-dried fish oil powder microencapsulated using milk ingredients. J Food Sci 66(2):217–224

Klaypradit W, Huang YW (2008) Fish oil encapsulation with chitosan using ultrasonic atomizer. Lebenson Wiss Technol 41:1144–1139

Klinkerson U, Sophanodora P, Chinachoti P, Decker EA, McClements DJ (2005a) Encapsulation of emulsified tuna oil in two-layered interfacial membranes prepared using electrostatic layer-by-layer deposition. Food Hydrocoll 19:1044–1053

Klinkerson U, Sophanodora P, Chinachoti P, McClements DJ, Decker EA (2005b) Stability of spray-dried tuna oil emulsions encapsulated with two-layered interfacial membranes. J Agric Food Chem 53:8365–8371

Klinkerson U, Sophanodora P, Chinachoti P, Decker EA, McClements DJ (2006) Characterization of spray-dried tuna oil emulsified in two-layered interfacial membranes prepared using electrostatic layer-by-layer deposition. Food Res Intern 39:449–457

Klinkesorn U, Sophanodora P, Chinachoti P, McClements DJ, Decker EA (2005) Increasing the oxidative stability of liquid and dried tuna oil-in-water emulsions with electrostatic layer-by-layer deposition technology. J Agric Food Chem 53:4561–4566

Klootwijk W (2006) De goede visgids. Fontaine Uitgevers, Kortehoef, The Netherlands

Kolanowski W, Laufenberg G (2006) Enrichment of food products with polyunsaturated fatty acids by fish oil addition. Eur Food Res Technol 222:472–477

Kolanowski W, Laufenberg G, Kunz B (2004) Fish oil stabilisation by microencapsulation with modified cellulose. Int J Food Sci Nutr 55(4):333–343

Kolanowski W, Ziolkowski M, Weissbrodt J, Kunz B, Laufenberg G (2006) Micro encapsulation of fish oil by spray drying-impact on oxidative stability. Part I. Eur Food Res Technol 222:336–342

Kolanowski W, Jaworska D, Weissbrodt J, Kunz B (2007) Sensory assessment of microencapsulated fish oil powder. J Am Oil Chem Soc 84(1):37–45

Luff F (2007) Omega-3 and micro-encapsulation technology – making functional foods taste better for longer. Food Sci Technol 21(1):30–31

Lumsdon SO, Friedmann TE, Green JH (2005) Encapsulation of oils by coacervation, WO2005105290

McClements DJ, Decker EA (2000) Lipid oxidation in oil-in-water emulsions: impact of molecular environment on chemical reactions in heterogeneous food systems. J Food Sci 65:1270–1282

McClements DJ, Decker EA (2005) Biopolymer encapsulation and stabilization of lipid systems and methods for utilisation thereof, WO2005/086976

McClements DJ, Decker EA (2007) Encapsulated emulsions and methods of preparations, WO2007038616

Nakahara M, Kurosaki A, Otake H, Kanai H, Yokokawa H, Kumabe K, Yanaka H (2006) Method for producing calcium component powder containing oil-soluble substance, EP1736532A1

O'Hagan P, Hasapidis K, Coder A, Helsing H, Pokrajac G (2005) Particle size analysis of food powders. In: Onwulata C (ed) Encapsulated and powdered foods. CRC, Boca Raton, FL, pp 215–245

Ogawa S, Decker EA, McClements DJ (2003) Influence of environmental conditions on the stability of oil in water emulsions containing droplets stabilized by lecithin-chitosan membranes. J Agric Food Chem 51:5522–5527

Park GY, Mun S, Park Y, Rhee S, Decker EA, Weis J, McClements DJ, Park Y (2007) Influence of encapsulation of emulsified lipids with chitosan on their in vivo digestibility. Food Chem 104:761–767

Ponginebbi L, Publisi C (2008) Coating for oxygen sensitive materials, EP1920633 & US2008112987

Sanguansri L, Augustin MA (2001) Encapsulation of food ingredients, WO2001074175

Schmid G, Harrison M, Polchinski P (2001) Process for the stabilisation of acylglycerols comprising high amounts of w-3 polyunsaturated fatty acids by means of gamma-cyclodextrin, EP1041136, US20030087879 and US6878696

Serfert Y, Drusch S, Schwarz K (2009) Chemical stabilisation of oils rich in long-chain polyunsaturated fats during homogenisation, microencapsulation and storage. Food Chem 113:1106–1112

Shaw LA, McClements DJ, Decker EA (2007) Spray-dried multilayered emulsions as a delivery method for ω-3 fatty acids into food systems. J Agric Food Chem 55:3112–3119

Shen Z, Rusli JK, Sanguansri L, Augustin MA (2007) Retention of propanal in protein-stabilised tuna oil-in-water emulsions. Food Chem 101:746–752

Subramaniam A, McIver RC, Van Sleeuwen RMT (2006) Process for the incorporation of a flavour or fragrance ingredient or composition into a carbohydrate matrix, WO2006/038067

Subramanian S, Connolly BJ, Hendrickson WA (2008) Encapsulated labile compound compositions and methods of making the same, US2008026108

Sunohara H, Kamaguchi R, Kozaki T, Yoshikado M, Higuchi J (2000) Double-layered capsule of unsaturated fatty acid or derivative thereof and process for producing the same, EP1025842

Tabor BE, Owers R, Janus JW (1992) The crosslinking of gelatin by a range of hardening agents. J Photographic Sci 40(5–6):205–211

Tan LH, Chan LW, Heng PWS (2005) Effect of oil loading on microspheres produced by spray drying. J Microencapsul 22(3):253–259

Thueringer S (2009) How to stabilise your omegas. Functional Ingredients, April issue, 28–29 (see also http://www.functionalingredientsmag.com/article/Formulations/how-to-stabilise-your-omegas.aspx)

Valentinotti S, Armanet L, Porret J (2006) Encapsulated polyunsaturated fatty acids, WO2006/067647

Van Lengerich BH, Leung L, Robie SC, Kang Y, Lakkis J, Jarl TM (2004) Encapsulation of sensitive components using pre-emulsification, US2004/0017017

Van Lengerich BH, Walther G, Van Auken B (2007a) Encapsulation of readily oxidizable components, US2007/0098853

Van Lengerich BH, Walther G, Van Auken B (2007b) Encapsulation of readily oxidizable components, US2007/0098854

Van Ruth SM, Roozen JP (2000) Aroma compounds of oxidised sunflower oil and its oil-in-water emulsion: volatility and release under mouth conditions. Eur Food Res Technol 210:258–262

Van Seeventer PB, Systermans FWG (2006) Micro encapsulation of nutritional oils: human milk lipids as a model for infant formula. AgroFood Industry Hi-tech 17(5):9–12

Velasco J, Marmesat S, Dobarganes C, Marquez-Ruiz G (2006) Heterogeneous aspects of lipid oxidation in dried microencapsulated oils. J Agric Food Chem 54(5):1722–1729

Yan N (2003) Encapsulated agglomeration of microcapsules and method for the preparation thereof, WO2003086104

Yan N, Jin Y, Moulton S, Perry TC (2004) Microcapsules having multiple shells and method for the preparation thereof, WO2004041251

Zeller BL, Van Seeventer PB, Poortinga AT (2006) Non-protein foaming compositions and methods of making the same, WO2006023564

Zimit P, Livney YD (2009) Beta-lactoglobulin and its nanocomplexes with pectin as vehicles for w-3 polyunsaturated fatty acids. Food Hydrocoll 23:1120–1126

Chapter 7
Encapsulation of Iron and Other Micronutrients for Food Fortification

Michael B. Zimmermann and Erich J. Windhab

7.1 Micronutrient Deficiencies

Iodine, vitamin A and iron deficiencies are important global public health problems, particularly for preschool children and pregnant women in low-income countries (World Health Organization 2000). These deficiencies are mainly due to monotonous, poor-quality diets that do not meet nutrient requirements. In countries where existing food supplies and/or limited access fail to provide adequate levels of these nutrients in the diet, food fortification is a promising approach. Co-fortification of foods with iron, iodine and vitamin A may be advantageous due to beneficial interactions of these micronutrients in metabolism. Studies in animals and humans have shown that iron deficiency anemia (IDA) impairs thyroid metabolism (Zimmermann et al. 2000a, 2000b; Hess et al. 2002a, 2002b). Vitamin A deficiency may exacerbate anemia through impairment of iron metabolism (Semba and Bloem 2002). Vitamin A, together with iodine, may reduce thyroid hyperstimulation and risk for goiter (Zimmermann et al. 2007). These micronutrient interactions strongly argue for multiple micronutrient fortification. However, food fortification with iron is not straightforward.

7.2 Encapsulation of Iron

The main advantage of Fe encapsulation is that it may allow addition of Fe compounds of high bioavailability to difficult-to-fortify food vehicles, such as cereal flours, milk products and low-grade salt (Zimmermann 2004). Fe encapsulation may

M.B. Zimmermann (✉)
Laboratory for Human Nutrition, Institute for Food Science and Nutrition, Swiss Federal Institute of Technology (ETH), Schmelzbergstrasse 7, LFV E19, Zürich, Switzerland
e-mail: michael.zimmermann@ilw.agrl.ethz.ch

E.J. Windhab
Laboratory of Food Process Engineering, Institute of Food Science and Nutrition, Swiss Federal Institute of Technology (ETH), Zürich, Switzerland

N.J. Zuidam and V.A. Nedović (eds.), *Encapsulation Technologies for Active Food Ingredients and Food Processing*,
DOI 10.1007/978-1-4419-1008-0_7, © Springer Science+Business Media, LLC 2010

decrease Fe-catalyzed oxidation of fatty acids, amino acids, and other micronutrients that can cause adverse sensory changes and decrease the nutritional value of these foods (Schrooyen et al. 2001). Also, it may reduce interactions of Fe with food components that cause color changes and lower Fe bioavailability, such as tannins, polyphenols and phytates (Hurrell 2002).

A number of encapsulated Fe compounds are in development or commercially available. These include forms of ferrous sulfate, ferrous fumarate, ferric pyrophosphate, and elemental Fe (see next section). Water-soluble coating materials, such as maltodextrin and cellulose, typically do not provide adequate protection against iron oxidation in moist environments. Therefore, most encapsulated Fe compounds are coated with hydrogenated oils that provide an effective water barrier at relatively low cost. These hard fat encapsulates can be prepared by fluid bed coating, or spray chilling/spray cooling (see Chap. 2 of this book).

A new concept goes beyond regarding microcapsules as a container from which functional components can be released in a controlled manner within the gastro intestinal tract. The new principle focuses on microcapsules as microreactors containing subcapsules with two or more functional components, which chemically or physically interact in a synergistic manner concerning the objective to be achieved. With respect to micronutrient encapsulation, typical synergistic reactions between the micronutrient component and the reaction partner component are the suppression of oxidation and complexation, but it could as well be a synthesis reaction generated by the functional molecules. The major objective of this approach is the improvement of bioavailability and the reduction of functional component losses during storage, preparation and perception.

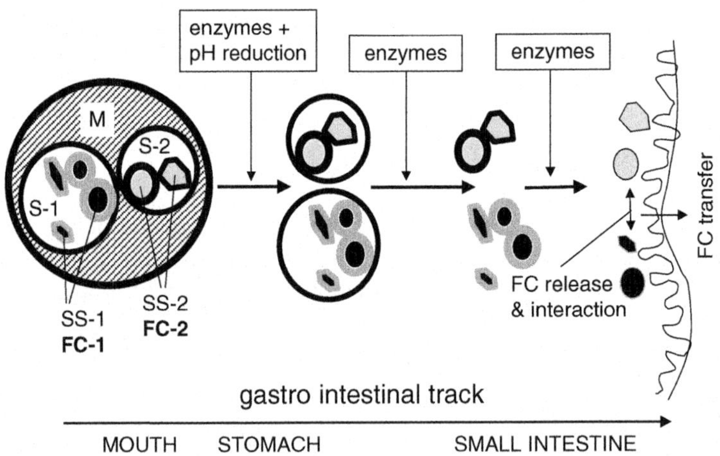

Fig. 7.1 The Multicapsule (MCAPS) concept showing schematically (i) the des-integration of MCAPS (M) including sub-capsules (Si, SSi) and encapsulated functional components (FCs) on the gastro intestinal track and (ii) controlled release and interaction of the FCs in a microenvironment at the intestinal wall

Figure 7.1 gives a schematic view of a multi-microcapsule morphology and its stepwise structure degradation in the digestive process (Windhab 1999, 2004).

The processing of multi-microcapsules comprises the operations of dispersing (1), mixing (2) and spray chilling or spray drying (3). Details are given in the following stepwise description :

1. Dispersing: The dispersing step can be divided into two sub-steps : (1a) Dispersing of functional components within a watery or lipid fluid phase and (1b) Dispersing/emulsifying the liquid functional component dispersion/solution (received from 1a within another non-miscible fluid phase, thereby forming a duplex emulsion).
   The dispersing steps include formation of interfacial structures to stabilize the interface. Subsequent to each dispersing step, partial solidification of the disperse fluid phase(s) is desirable to support structure stability for the next processing step (Windhab et al. 2005).
2. Mixing – I Two or more emulsion systems processed according to 1a/b and containing at least two different functional component dispersions/solutions are gently mixed such that the disperse phase structures are mostly kept.
3. Spray Chilling/Spray Drying The multiple emulsion/dispersion system resulting after mixing step 2 will be spray chilled (3a) or spray dried (3b) in order to receive a powder end product. In case of an o/w/o multiple emulsion/dispersion, multiple phase structure spray chilling will be applied and lead to lipid coated hydrophilic capsules with lipid sub-capsules (o/w/o-MCAPS). In the case of reverse phase arrangement, the respective water soluble phase- coated w/o/w-MCAPS are received from spray drying (Windhab and Wagner 1994).
   To incorporate the multi component capsules generated within processing steps 1–3 into a food system, another mixing step has to be added.
4. Mixing – II: Mixing of MCAPS into food matrices can be done with powder mixers (4a) if powdery component mixtures are provided for application, such as e.g., in the case of triple-fortified salt (example given below).
   In case of embedding the MCAPS into a more complex food matrix, another fluid mixing device can be applied, such as an extruder system (4b) as has been demonstrated for highly viscous wheat or rice flour-based dough systems.

Figure 7.2 gives a schematic view of an arrangement of the four processing operations (1–4) considered. In Fig. 7.3 powder particles with embedded MCAPS are shown.

For each of the processing operations (1–4) the processing parameters have to be adjusted such that specific structure criteria (A–D) are fulfilled, allowing the required encapsulation quality to be achieved. These are:

A. Disperse structures with small size distribution.
B. Homogeneous distribution of the disperse phases within the respective bulk phase entities.
C. Negligible structure degradation during subsequent processing steps.
D. Structure stability on shelf life time scale under typical environmental conditions.

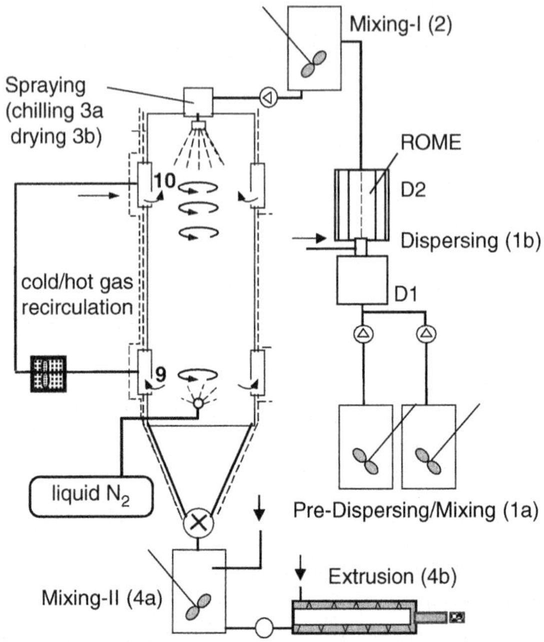

Fig. 7.2 Schematic drawing of the Multicapsule (MCAPS) processing steps: pre-mixing (i), dispersing (ii), spray processing (chilling or drying) (iii) and MCAPS-food matrix mixing/shaping (extrusion) (iv)

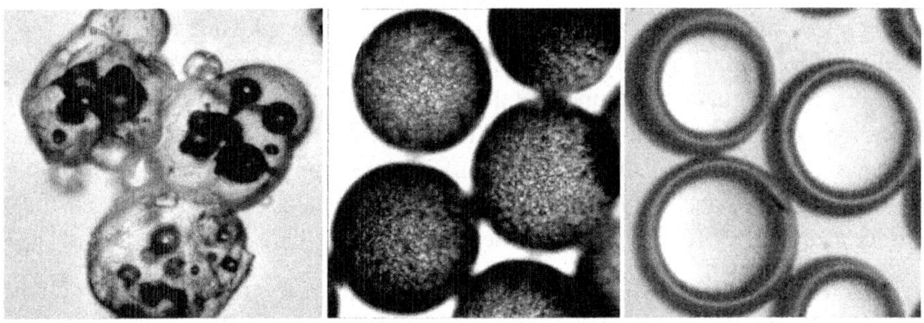

3.1 **3.2a** **3.2b**

Fig. 7.3 Powder MCAPS with embedded multiple capsule structures (**a**) and different morphologies of the disperse sub-capsule phase (**b**, **c**) in the liquid state containing functional components

To improve processing conditions with respect to these criteria, new process design tools (μPRO: Micro-Fluidics + CFD based Process Design) have been introduced and innovative processing devices (e.g., ROtating MEmbrane dispersing device (ROME)) derived (Schadler and Windhab 2006). For MCAPS processing, the Rotating Membrane dispersing device (ROME) has proved to be a crucial tool to achieve adjustable disperse drop/capsule structure size distributions under gentle mechanical processing conditions (Müller-Fischer et al. 2007).

Figure 7.4 gives an insight into a micro-fluidics crossflow channel as used for membrane emulsification investigations. Figure 7.5 demonstrates a small pilot scale Rotating Membrane device (ROME) for continuous multiple capsule emulsion processing.

The crossflow field is adjusted by the rotational velocity of the membrane cylinder. Disperse drops are generated at the rotating membrane surface by feeding the disperse fluid phase through a hollow shaft into the membrane cylinder and pressing it through the membrane pores with trans-membrane pressures ΔP_M in the range of about 0.5–5 bars, depending on mass flow rate d*m*/d*t* and dynamic viscosity function $\eta(dv/dr)$. The continuous (bulk) fluid phase is pumped axially through the annular gap between the rotating membrane cylinder and the concentric cylinder housing. The interplay of interfacial tension between disperse fluid phase, continuous (bulk) fluid phase and membrane surface with the flow shear stresses generated by the concentric cylinder rotational flow field, determines the critical drop/capsule size which is detached from the membrane surface.

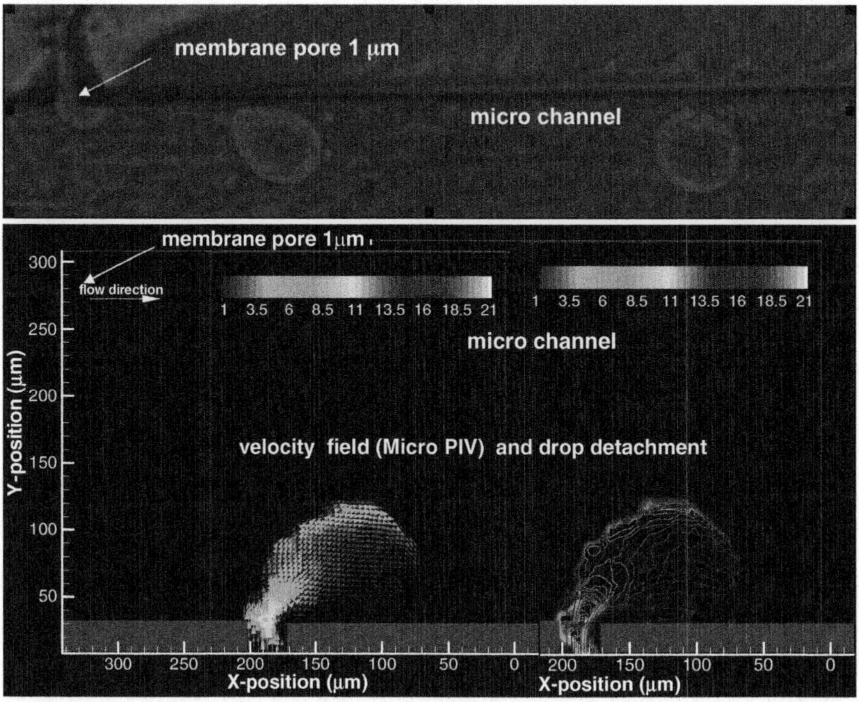

Fig. 7.4 Drop/capsule generation at membrane pore in a Micro-fluidics cross-flow channel (**a**) and velocity field within drop during forming at the pore measured by micro PIV (Particle Imaging Velocimetry) (**b**)

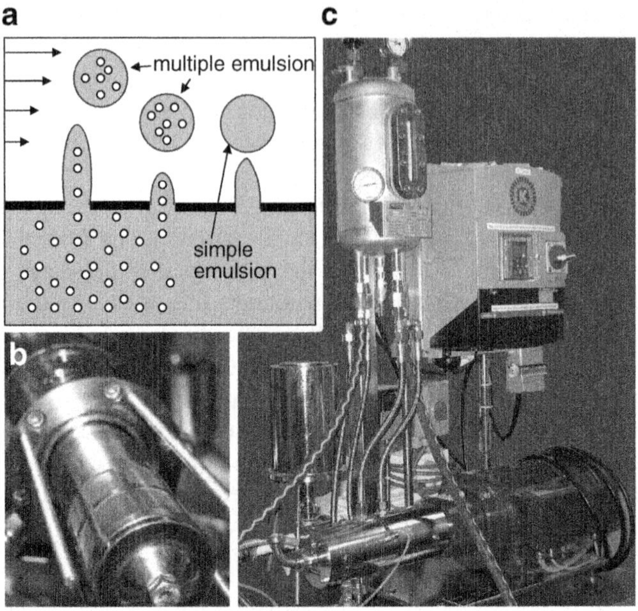

Fig. 7.5 Rotating membrane dispersing device (ROME): schematic view of multiple emulsion dispersing process (**a**), membrane cylinder (**b**) and patented lab/pilot-scale commercialized ROME device (by Kinematica AG, Littau-Luzern, Switzerland) (**c**)

For multiple emulsification/encapsulation processing ROME devices can be arranged in series. However, a first dispersing step for simple emulsification can also be carried out within alternative dispersing devices like Micro-Fluidizer/high pressure homogenizer, rotor/stator disperser or ultrasound disperser. In the second multiple emulsion dispersing step, the ROME device proved clearly advantageous compared to conventional dispersing devices. Due to the uniform and gentle mechanical dispersing treatment in the ROME device, disperse multiple emulsion structures were not destroyed (e.g., by re-coalescence), even up to very high disperse phase fractions of 50–60 Vol%.

7.3 Bioavailability

However, the effect of encapsulation per se on Fe bioavailability is a concern. Several factors may influence Fe bioavailability from encapsulated products, including capsule material, ratio of capsule material to Fe, and the technology and process used for encapsulation. The relative bioavailability (RBV) of a series of encapsulated Fe compounds has been tested in rat Hb repletion tests (Hurrell 1985; Hurrell et al. 1989). The Hb repletion test measures the bioavailability of Fe compounds based on their ability to increase hemoglobin (Hb) in anemic rats, relative to ferrous sulfate. The compounds were obtained from Durkee (Cleveland, Ohio, USA) and had a 40:60 ratio of capsule material: $FeSO_4 \cdot H_2O$. As shown in Table 7.1, the RBV of ferrous sulfate encapsulated with hydrogenated soybean oil, hydrogenated palm oil, or mono- and diglycerides was comparable to that of ferrous sulfate. Ferrous sulfate encapsulated with maltodextrin showed a nonsignificant reduction in RBV compared to ferrous sulfate. Ethyl cellulose encapsulated ferrous sulfate (Eurand International, Milan, Italy) was also tested (20:80 capsule:substrate ratio), and had 33% higher RBV. The RBV of ferrous sulfate encapsulated in hydrogenated soybean oil (Balchem, Slate Hill, NY, USA) varies between 78 and 99, depending on the capsule material and the capsule:substrate ratio (Table 7.1) (D. Barclay, Nestec Ltd., personal communication, 2003). Ferrous sulfate encapsulated in hydrogenated soybean oil at a 50:50 capsule:substrate ratio showed an RBV equivalent to nonencapsulated ferrous sulfate. However, there was a 20% reduction in RBV when the capsule:substrate ratio was increased to 67:33. Also, there was a 20% reduction in RBV when the capsule was modified to include glycerol-monostearate and tricalcium phosphate.

The relative bioavailability (RBV) of encapsulated forms of ferrous fumarate was also tested in rat Hb repletion tests (D. Barclay, Nestec Ltd., personal communication, 2003) (Table 7.2). Ferrous fumarate encapsulated in hydrogenated castor oil (Biodar, Yavne, Israel) at 70:30 and 60:40 ratios of capsule: substrate, and encapsulated in hydrogenated soybean oil (Balchem) at a 40:60 ratio had RBVs equal to non-encapsulated ferrous sulfate. However, ferrous fumarate encapsulated in carnauba wax (Westreco, Marysville, Ohio, USA) had a RBV of only 66% compared to ferrous sulfate. In a separate study, the RBV of ferric ammonium citrate was equivalent to ferrous sulfate, but when encapsulated with hydrogenated palm oil in a 60:40 ratio, its RBV was decreased by 25%.

Table 7.1 Relative bioavailability (RBV) of encapsulated forms of ferrous sulfate as measured by the rat hemoglobin repletion test

Form	Fe content[a]	RBV[b]
Ferrous sulfate·H_2O (nonencapsulated)	33	102 ± 5
Ferrous sulfate encapsulated		
In ethyl cellulose (20:80 ratio of capsule:Fe $SO_4 \cdot H_2O$)	27	133 ± 10
In 40:60 ratio with:		
Part. hydrogenated soybean oil	20	114 ± 7
Part. hydrogenated palm oil	20	95
Mono- and diglycerides	20	116 ± 7
Part. hydrogenated soybean and cottonseed oil	20	79
Maltodextrin	20	87
In 50:50 ratio with part. hydrogenated soybean oil	16.7	99
In 50:50 ratio with part. hydrogenated soybean oil (44% hydro. soy oil, 5% glyceromonostearate, 1% tricalcium phosphate)	16.7	78
In 67:33 ratio with part. hydrogenated soybean oil	10.9	81

[a]As percent
[b]As mean or mean ± SEM

Table 7.2 Relative bioavailability (RBV) of encapsulated forms of ferrous fumarate and ferric ammonium citrate as measured by the rat hemoglobin repletion test

Compound	RBV[a]
Ferrous fumarate	101 ± 7
Encapsulated ferrous fumarate	
In hydrogenated castor oil (30% ferrous fumarate)	103 ± 9
In hydrogenated castor oil (40% ferrous fumarate)	111 ± 10
In hydrogenated soybean oil (60% ferrous fumarate)	98 ± 10
In carnauba wax	66 ± 6
Fe ammonium citrate	107
Encapsulated ferric ammonium citrate in hydrogenated palm oil (60% capsule: 40% ferric ammonium citrate)	82

[a]As mean or as mean ± SEM

The bioavailability of encapsulated ferric pyrophosphate (FePP) was evaluated by Wegmuller et al. (2003) The hemoglobin (Hb) repletion method in weanling rats (Forbes et al. 1989) was used to compare the relative bioavailability (RBV) of FePP with a mean particle size (MPS) ≈ 2.5 μm to the same compound encapsulated in hydrogenated palm oil at a capsule:substrate ratio of 60:40. The encapsulated FePP was produced by spray cooling. The capsule was made of hydrogenated palm oil (Nutriswiss) containing 1% lecithin (Loders Croklaan). The palm oil was heated to 85°C and a suspension containing the FePP was made. This was then passed through a screw pump (Scheerle) into a stainless steel spraying tower, atomized using air as the second medium, and cooled with liquid nitrogen. The RBV compared with ferrous sulfate was calculated by the slope-ratio technique. The RBV was 43% for encapsulated MPS ≈ 2.5 μm, significantly lower than the nonencapsulated FePP compounds (69%).

The variable RBV reported in these studies demonstrates the importance of testing the bioavailability of new encapsulated Fe compounds in rat Hb repletion studies before they can be recommended for use in food fortification (Zimmermann 2004).

7.4 Food Fortification with Encapsulated Iron

7.4.1 Sensory Testing in Cereals

Encapsulated Fe was dry mixed into precooked wheat flour (10 mg/100 g) and the samples were stored at 37°C for 3 months (Hurrell 1985). The compounds tested were ferrous sulfate encapsulated with hydrogenated soybean oil, hydrogenated palm oil, mono- and diglycerides, maltodextrin, or ethyl cellulose. All the compounds performed as well as electrolytic Fe and an unfortified control. They did not cause fat oxidation as measured by pentane generation, nor did they generate off-flavors as measured by a taste panel. However, when prepared at temperatures above the melting points of the capsules (52–66°C), the encapsulated Fe compounds produced green-gray color changes in porridges made with hot water and wheat flour. Ferrous sulfate encapsulated in either hydrogenated soybean oil or mono- and diglycerides was dry mixed into a wheat cereal at 10 mg and 50 mg/100 g, and the samples were stored at 37°C for 3 months (Hurrell et al. 1989). Both compounds performed as well as electrolytic Fe and an unfortified control. Ferrous sulfate encapsulated with ethyl cellulose, dry mixed into a rice cereal at 18.5 mg Fe/100 g and stored at 37°C for 3 months, also performed as well as carbonyl Fe and the control (Hurrell et al. 1989).

7.4.2 Ferrous Sulfate and Ascorbic Acid Encapsulated in Soy Lecithin Liposomes

If liquid food products are to be fortified with Fe, liposome technology can be used (Schrooyen et al. 2001). A series of studies in Argentina reported on the bioavailability of ferrous sulfate (6.6% Fe) with ascorbic acid (2%) encapsulated in soy lecithin liposomes in a fluid suspension (SFE-171, Biofer™) (Boccio 1997). The bioavailability of SFE-171 was comparable to that of nonencapsulated ferrous sulfate in animals and humans. Absorption of SFE-171 given with milk was not significantly different from ferrous sulfate or ascorbate given with water, but was greater than ferrous sulfate given with milk ($p<0.01$). The authors suggested that encapsulation of the Fe improved its bioavailability by reducing its interaction with protein, calcium, and other milk components (Boccio 1997).

In a series of further absorption studies in rats, the absorption of SFE-171 was comparable to or greater than ferrous sulfate when given with different meals. In humans ($n=29$) without IDA (mean Hb 16.8±1.4; serum ferritin 113±74 μg/L), mean absorption±SD of SFE-171 given with 250 ml of milk (6.6% Fe) was 10.4±4.9%. In an uncontrolled study, preschool children with Fe-deficiency anemia ($n=10$, mean age 23.1±8.9 months) received a daily milk portion (250 ml) fortified with SFE-171. All the subjects receiving the fortified milk attained normal Hb concentrations; the mean number of days required was 50.7±19.2 (Boccio 1997).

An alternative approach to fortify milk with encapsulated Fe is the use of ferric ammonium sulfate coated with polyglycerol monostearate (Kwak et al. 2003). The microencapsulates were made by spraying a heated (50°C) solution of polyglycerol monostearate with ferric ammonium sulfate (3:1 up to 20:1 w/w) into a solution of 0.05% Tween 60 at 5°C. The organoleptic properties of milk fortified with this encapsulated compound were superior to milk fortified with other Fe compounds; lipid peroxidation in the milk was reduced by using the encapsulated Fe. Microcapsules containing $FeCl_3$ or $FeSO_4$ plus ascorbic acid coated with hydrogenated milk fat or stearine have been used to fortify cheese (Jackson and Lee 1992); the stearine-coated Fe caused less oxidation of fat in cheese compared to those fortified with nonencapsulated Fe (Jackson and Lee 1992).

7.4.3 *Emulsifications of Ferric Pyrophosphate*

SunActive Fe™ (Taiyo Kagaku Co., Ltd., Yokkaichi, Mie, Japan) is a commercially available (but relatively expensive), patented form of ferric pyrophosphate (FePP). It contains micronized FePP (mean particle size 0.3–0.5 μm) stabilized using emulsifiers including polyglycerol esters and hydrolyzed lecithin. It is available in liquid form (12% Fe) and as a powder (8% Fe). It remains in suspension in liquid products and can be used to fortify dairy-based products, such as milk, drinkable yogurts and soymilk. In rats, the RBV of dispersible SunActive Fe™ is similar to ferrous sulfate (Sakaguchi et al. 2004). Several stable isotope studies in humans have reported Fe absorption rates of 15–25% for rice, 63–83% for infant cereals and 94% for a milk-based drink (Fidler et al. 2004; Moretti et al. 2006).

7.4.4 *Encapsulated Ferrous Fumarate Plus Ascorbic Acid As "Sprinkles" in Ghana*

Encapsulated ferrous fumarate has been used successfully in complementary food fortification. In a feeding trial, anemic infants in Ghana received a daily sachet of "sprinkles" containing ferrous fumarate encapsulated in hydrogenated soybean oil (80 mg elemental Fe) (Descote ferrous fumarate 60 Ultra, Particle Dynamics, St. Louis, USA) (Zlotkin et al. 2001). The sachets also contained ascorbic acid (50 mg) and were mixed into complementary foods at homes just before feeding. A second group received ferrous sulfate drops 3 times/d for 2 months (total daily dose: 40 mg elemental Fe). In both groups, there was a significant increase in serum ferritin and Hb concentrations from baseline to the end of treatment, and the change in Hb concentrations was not significantly different between groups. Despite the presence of ascorbic acid in the sprinkles, the bioavailability of Fe appeared to be lower from the sprinkles than from the drops. This may have been due to the fact that the sprinkles were mixed with food, which may have inhibited Fe absorption relative to the drops that were taken without food. The authors suggested that improved safety and ease of use may favor the use of sprinkles to deliver Fe (Zlotkin et al. 2001). Subsequent studies have shown the efficacy of sprinkles containing a lower dose of Fe (12 mg) as encapsulated ferrous fumarate (Hirve et al. 2007).

A stable isotope study compared the Fe absorption from sprinkles containing encapsulated ferrous fumarate or non encapsulated ferrous fumarate, mixed into rice-based and wheat-based complementary foods in non-anemic infants (Liyanage and Zlotkin 2002). The absorption of the ferrous fumarate was significantly reduced by encapsulation. A subsequent isotope absorption study tested sprinkles containing encapsulated ferrous fumarate plus ascorbic acid added to a maize-based porridge in infants (Tondeur et al. 2004). Mean Fe absorption was 4.5% in iron-sufficient infants and 8.3% in those with iron-deficiency anemia.

7.4.5 Dual Fortified Salt with Encapsulated Ferrous Sulfate and Iodine

Although dual fortification of salt with iodine and Fe could be an effective fortification strategy, ensuring iodine stability and Fe bioavailability in dual fortified salt (DFS) is difficult. Water-soluble Fe compounds, which are the most bioavailable, react with moisture and impurities in salt and produce unacceptable color changes. Moreover, in the presence of ferrous ions and oxygen, the iodine in DFS is unstable due to catalytic oxidation of iodate or iodide to I_2. Insoluble Fe compounds, such as elemental Fe powders or Fe phosphate compounds, cause fewer sensory changes but may be so poorly absorbed as to be of little nutritional value. Encapsulated Fe has excellent potential for overcoming unwanted sensory changes and iodine losses in salt, while maintaining high bioavailability (Zimmermann et al. 2003).

A DFS containing encapsulated ferrous sulfate was tested in rural villages in the Rif Mountains of northern Morocco (Zimmermann et al. 2003). Because the local diet was high in phytic acid and low in ascorbic acid, estimated non-heme Fe bioavailability was only 0.4–4.3%. Local salt was dual fortified with 25 μg iodine/g salt as (non-encapsulated potassium iodide) KI and 1 mg Fe/g salt as ferrous sulfate hydrate encapsulated with partially hydrogenated vegetable oil (Cap-Shure FS-165E-50, Balchem, Slate Hill, NY, USA), or fortified with iodine only (Iodized Salt, IS). The encapsulated Fe compound contained 50% $FeSO_4{\cdot}H_2O$. After storage for 20 weeks, there were no significant differences in iodine content between the DFS and IS.

The efficacy of the DFS was evaluated in a randomized, double-blind trial in iodine-deficient 6–15 year-old children ($n=377$). Each participating family shared a 2 kg monthly salt portion for 9 months. Mean Hb and all indices of Fe status significantly increased in the DFS group compared to the IS group at 40 weeks. The prevalence of IDA and Fe-deficiency without anemia was sharply lower in the DFS group at 40 weeks. Although the DFS was well accepted, Fe encapsulation did not entirely prevent color changes. During the dry season, when the moisture content of the local salt was low (<1%), there was no significant color difference between IS and DFS. However, during the damp winter season, when the moisture content of the local salt was high (~3%), the DFS developed a yellow color during storage. Although the partially hydrogenated vegetable oil capsule is water resistant, some loss of capsule likely occurred from abrasion during mixing. When the salt moisture content was high, the ferrous Fe was oxidized in the presence of water and oxygen and precipitation of yellow-brown ferric hydroxide occurred.

The DFS demonstrated good bioavailability without an enhancer, despite the high phytic acid content of the diet. To fully liberate Fe encapsulated in partially hydrogenated vegetable oil requires lipase digestion in the proximal duodenum. This brief delay in Fe appearance may reduce Fe binding to inhibitory substances (such as phytic acid) in the stomach and upper duodenum, and thereby potentially increase Fe absorption, although this has not been directly tested. Although a stable and efficacious DFS could provide new opportunities for controlling iron deficiency, the cost-effectiveness of this approach remains to be determined. Current cost estimates for encapsulated ferrous sulfate are 4–7× that of non-encapsulated ferrous sulfate, but the price could fall if production is on a large scale. Because performance of encapsulated Fe may vary depending on climate, salt quality and diet, the authors suggested that more studies were needed in other countries under local conditions (Zimmermann et al. 2003).

7.4.6 Fortification of Salt with Encapsulated Fe: Stability and Acceptability Testing

In the Morocco intervention trial described above (Zimmermann et al. 2003), salt containing ferrous sulfate encapsulated with hydrogenated soybean oil was effective in reducing the prevalence of anemia, but it caused a yellow color in the salt during the damp winter season. Therefore, Wegmuller et al. (2003) tested the potential of various encapsulated Fe compounds as salt fortificants. In temperate Morocco and tropical Côte d'Ivoire, native salt was dual fortified with iodine and 15 different encapsulated Fe compounds were commercially available or in development (Table 7.3). In Morocco, the fortified salts were prepared using unground, unwashed salt from a local cooperative. In Côte d'Ivoire, the fortified salts were prepared using unwashed imported Senegalese salt. Tests were done comparing the following:

- Ferrous sulfate to encapsulated ferrous sulfate (seven different capsule materials)
- Ferrous fumarate to encapsulated ferrous fumarate (six different capsules)
- Ferric pyrophosphate to encapsulated micronized ferric pyrophosphate
- Electrolytic Fe to encapsulated electrolytic Fe

The salt fortification level chosen for the different Fe compounds was equivalent to 1 mg of Fe as ferrous sulfate/g salt, considering the estimated RBV of the Fe compounds compared to ferrous sulfate (Hurrell 2002). All DFS and an IS control were stored in loosely woven high-density polyethylene bags, and in transparent low-density polyethylene bags. The color and iodine content were measured at baseline and after storage for 1, 2, 4 and 6 months. Color changes were evaluated using reflectance colorimetry. Acceptability of the salt color was judged at 4 months by standardized interviews comparing 100 g DFS and IS samples side-by-side on white backgrounds.

For almost all compounds, encapsulation did not prevent adverse sensory changes. There was a rapid and marked color change in many of the DFS, particularly the ones fortified with forms of elemental Fe and ferrous sulfate. In general, the color change in Morocco was greater than in Côte d'Ivoire. Color instability was likely the result of the formation of ferric oxides of Fe with a rusty orange color, and the formation of I_2 which has a dark brown color. Two of the salts forti-

Table 7.3 The color and acceptability of local salt fortified with encapsulated iron compounds in Morocco and Côte d'Ivoire

Fe compound	Capsule material/additives	Fe [%]	Producer	Color in stored salt after 2 months
Electrolytic				
Electrolytic Fe	None	10	OMG Americas	Unacceptable dark brown
Electrolytic Fe	Partially hydrogenated soybean oil	70	Balchem	Unacceptable dark brown
Pyrophosphate				
Fe pyrophosphate	None	21	Lohmann	Acceptable pale beige
Fe pyrophosphate	Dextrin, glycerol esters of fatty acids, hydrolyzed lecithin	8	Taiyo Kagaku	Acceptable pale beige
Fumarate				
Fe fumarate	None	33	Lohmann	Unacceptable orange pink brown
Fe fumarate	Partially hydrogenated palm oil	16	Lohmann	Unacceptable orange pink brown
Fe fumarate	Hydrogenated soybean oil	20	Watson Foods	Unacceptable orange pink brown
Fe fumarate	Cellulose derivative	28	Bio Dar	Unacceptable orange pink brown
Fe fumarate	Granulated with dextrin, cellulose, SHMP, coated with soy stearine	12	MI	Acceptable pale gray + dark flecks
Fe fumarate	Wax matrix	18	Particle Dynamics	Acceptable pale gray + dark flecks
Fe fumarate	Mono- and diglycerides	21	Particle Dynamics	Unacceptable orange pink brown
Sulfate				
Fe sulfate	None	29	Lohmann	Unacceptable yellow brown
Fe sulfate	Partially hydrogenated soybean oil	17	Balchem	Unacceptable yellow brown
Fe sulfate	Mono and diglycerides	15	Bio Dar	Unacceptable yellow brown
Fe sulfate	Cellulose derivative	27	Bio Dar	Unacceptable yellow brown
Fe sulfate	Partially hydrogenated palm oil	23	Lohmann	Unacceptable yellow brown
Fe sulfate	Partially hydrogenated oil	20	Watson Foods	Unacceptable yellow brown
Fe sulfate	Soy stearine	20	MI	Unacceptable yellow brown
Fe sulfate	Stearic acid	20	Particle Dynamics	Unacceptable yellow brown

fied with encapsulated ferrous fumarate were judged acceptable at 4 months. One was encapsulated in soy stearine and granulated with dextrin, cellulose and sodium hexametaphosphate (SHMP). Its color stability was likely to be at least partly due to the presence of a stabilizing agent, SHMP (Narasinga Rao 1985). The other ferrous fumarate that performed well and showed only a slight color change was encapsulated in an edible wax matrix, but the bioavailability of the Fe in this edible wax capsule has not been tested.

Two forms of encapsulated ferric pyrophosphate with small particle size showed promise as a salt fortificant; their color in the DFS was acceptable, and at 4 months, the DFS retained 92 and 56% of their original iodine content in Morocco and Côte d'Ivoire, respectively, similar to iodized salt. Their color and iodine stability may be due to the fact that Fe phosphate compounds are insoluble in water and only poorly soluble in dilute acid, which reduces their reactivity, even in salt with a high moisture content.

The findings of this study suggest currently available encapsulation techniques for Fe do not sufficiently reduce moisture penetration and/or Fe solubility when the encapsulated Fe is mixed into salt. Mechanical abrasion of the capsules during mixing may also contribute by damaging capsules and exposing and/or freeing Fe. Also, the Fe in these compounds may not be fully coated by the encapsulation process. Spray cooling is typically referred to as "matrix" encapsulation, because the particles are aggregates of the iron compound buried in the fat matrix, while "true" encapsulation is usually reserved for processes leading to core/shell type microcapsules. A matrix encapsulation process leaves a significant proportion of the active ingredients lying on the surface of the microcapsules or sticking out of the fat matrix, thus having direct access to the environment and generally releasing their content easily (Gouin 2004). The strong binding of the ingredient to the fat matrix can prevent the release of the ingredient even though the fat matrix is melted and/or damaged during processing (Gouin 2004).

7.4.7 Fortification of Salt with Encapsulated Ferrous Fumarate

A new formulation of encapsulated, agglomerated ferrous fumarate (EFF) has been developed by the Micronutrient Initiative (Ottawa, Canada). The encapsulation provides a physical barrier between iodine and the salt, its impurities and the ferrous fumarate. The EFF premix includes encapsulated ferrous fumarate with a soy stearine coating made in a fluidized bed, which was then agglomerated with titanium dioxide, hydroxypropyl methylcellulose (HPMC), and sodium hexametaphosphate (SHMP). The acceptability and efficacy of this premix, when added to iodized salt, was recently tested in school children in rural southern India with low iron stores. Local salt fortified with the EFF caused color changes in local foods when added during cooking, and the EFF segregated in the salt when the moisture content was low. However, the iodine content of the salt was stable and the EFF was efficacious in a household-based feeding trial over 10 months, where body iron stores were significantly increased in iron-deficient children (Andersson et al. 2008).

7.4.8 Triple Fortification of Salt with Iron, Iodine and Vitamin A

In remote, rural areas of subsistence farming in Africa and Indonesia, salt is one of the few regularly purchased food items and therefore is often the only suitable food vehicle for food fortification (Hess et al. 1999; Melse-Boonstra et al. 2000). Because salt is consumed daily at fairly steady levels even by low socio-economic population groups, it could be a promising food vehicle for triple fortification with iodine, Fe, and vitamin A. Adding vitamin A to salt is particularly challenging, due to its fat solubility and its instability in the presence of oxygen and other oxidizing agents, particularly when exposed to light and heat. Encapsulation of vitamin A and iodine could protect them from interactions with salt impurities, light, air, and moisture. Fortification of salt with a single dry mix containing these three micronutrients may be simpler and less expensive compared to separate addition into different food vehicles.

In a series of studies (Zimmermann et al. 2004; Wegmueller et al. 2006), triple fortified MCAPS (see Sect. 7.2) containing iron, iodine and vitamin A for salt fortification were produced and tested. Spray cooling was chosen for the production of the microcapsules because of the desirable coating characteristics produced by the technique, and because spray cooling is currently the least expensive encapsulation technology (Gouin 2004). Fully hydrogenated palm fat was chosen as coating material due to its high melting point (63°C) for several reasons: (a) the sprayed capsules need to resist high temperatures encountered during storage and transport of salt in tropical climates; (b) its hydrophobic properties prevent the entrance of water and reduce reactions between different core ingredients; and (c) fat is an ideal matrix for fat-soluble vitamin A due to the stabilization and the delay of oxidation of the vitamin. Lecithin was added to the molten palm fat together with the Fe to reduce the viscosity of the suspension and enable a capsule:substrate ratio of 60:40.

The particle size of the microcapsules was set according to the average particle size of the salt to be fortified (approximately 1.5 mm). A small mean particle size of ≈100 μm was chosen for the capsules to increase van der Waals forces with the larger salt crystals and thereby reduce segregation in the salt. An additional consideration was the need to maintain an adequate volume–surface ratio in order to protect the ingredients, but not interfere with the release kinetics of the nutrients from the capsules to maintain bioavailability.

For the production of the microcapsules, ferric pyrophosphate (FePP) (≈21% Fe) with a mean particle size (MPS) of 2.5 μm (Dr. Paul Lohmann, Emmerthal, Germany) was used. KIO3 was chosen over KI because it is less reactive with ferric forms of iron. Reagent-grade potassium iodate (KIO3) (Riedel-de Haen, Hannover, Germany) was chosen as the iodine source. The KIO3 was ground to a smaller particle size (<10 μm) before spraying, using a laboratory bead mill K8 (Bühler, Uzwil, Switzerland). An oily form of vitamin A was used due to its solubility in the fat matrix, its higher stability and lower costs compared to dried forms (Dary and Mora 2002). The vitamin A fortificant used was liquid retinyl palmitate 1.7 MIU/g, stabilized with 10 mg of BHT/million IU (BASF ChemTrade GmbH, Burgbernheim, Germany).

The ferric pyrophosphate (40% w/w) and lecithin (1% w/w) were stirred into molten palm fat (80–90°C) and than filled into a heated tank above a rotary pump. This tank was continuously stirred to avoid sedimentation of the Fe particles and then, just before spraying, the iodine and retinyl palmitate were added. Before spraying, the mixture contained a ratio of Fe:iodine:retinol of 100 mg:2 mg:5 mg per g mixture. To minimize losses of iodine and retinol due to heat, light and/or oxidation, the final suspension was immediately sprayed into a pre-cooled tower through a two-fluid nozzle with heated air of ≈90°C as the second medium. Liquid nitrogen was sprayed into the middle zone of the tower to cool and rapidly solidify the atomized particles. The capsule:substrate ratio of the produced microcapsules was 60:40. The mean particle size was ≈135 μm.

Figure 7.6 shows a light microscope image of the microcapsules (100× magnification). The spherical particles have a smooth surface characteristic of spray cooled or spray chilled products because there is no mass transfer by evaporation as in spray drying (Shahidi and Han 1993). There is a relatively wide range of particle size. The presence of agglomerates is likely due to the particles being insufficiently solidified during settling in the spraying tower. Figure 7.7 shows a SEM micrograph of the surface of microcapsules at two different magnifications. Crystals of the palm fat can be seen on the capsule surface. Small capsules have fewer fat crystals on the surface than larger ones, most probably due to the shorter solidification time for small capsules. To visualize the internal structure, microcapsules were broken (Fig. 7.8). The large crystals of the palm fat can be recognized on the surface of the cross section, but the individual nutrient components are not visible on the surface or in the cross section.

The microcapsules were tested as a salt fortificant in rural northern Morocco (Zimmermann et al. 2004), where deficiencies of iron, iodine and vitamin A are common and salt is regularly consumed at a level of 5–15 g/d. Using local salt, iodized salt (IS) was prepared by adding KIO_3 at a concentration of 25 μg I/g salt, and triple-fortified salt (TFS) was prepared using microcapsules at a concentration of 2 mg Fe, 25 μg I, 60 μg vitamin A/g salt, and stored for 6 months. Although there was no significant color change in the IS and TFS over 6 months of storage, there was a slight difference in color between the TFS (light beige) and the IS (milky white) after mixing. This resulted in an absolute color difference measured by reflectance colorimetry of (ΔE) of ≈9. Although Wegmuller et al. (2003) reported that fortified salts with $\Delta E < 10$ compared to IS were considered acceptable in interviews at local African markets, an improvement in salt color should be a future goal in the refinement of the encapsulation process. Because fats crystallize in various forms that exhibit very different crystal sizes, hydrophobicities and densities (Gouin 2004), refinement of the crystallization during production and storage could help prevent discoloration due to reactions of the Fe compound with the salt matrix.

The spraying process resulted in iodine and vitamin A losses of approximately 40% and 30%, respectively. Additional losses of iodine after storage for 6 months were ≈20% in both the TFS and IS. The retinyl palmitate used in the experiments was surprisingly stable during storage, with losses of only 12% after 6 months (Table 7.4). This was likely due to the hydrogenated fat acting as an excellent

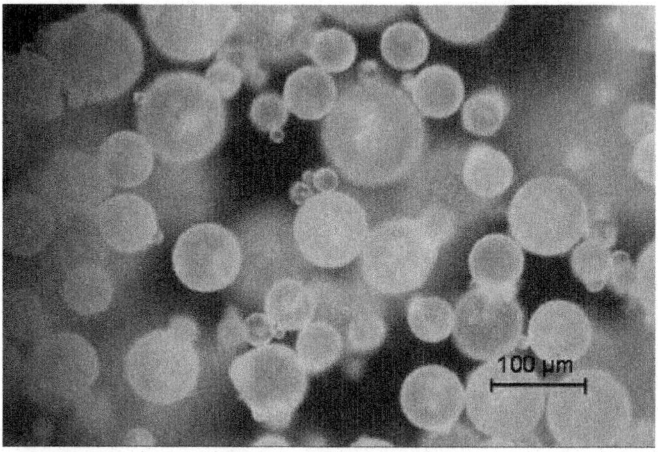

Fig. 7.6 Sprayed microcapsules containing iron, iodine and vitamin A designed for salt fortification analyzed by light microscopy (100× magnified)

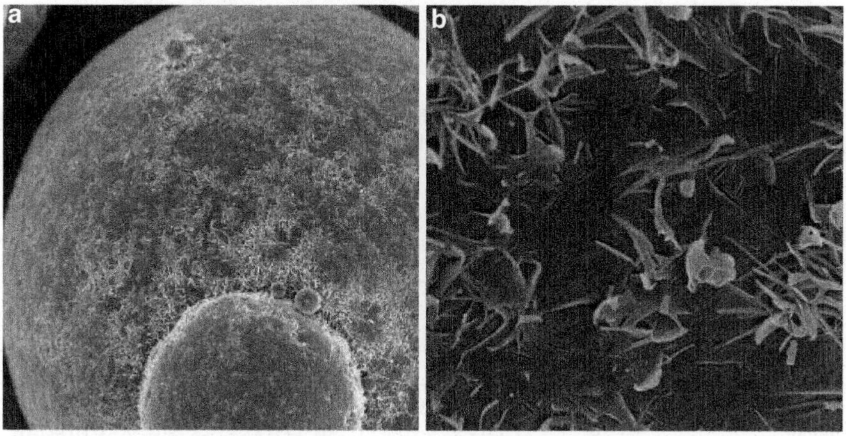

Fig. 7.7 (**a**, **b**) Surface of microcapsules containing iron, iodine and vitamin A designed for salt fortification analyzed by SEM micrographs: (**a**) 700× magnified, picture width: 154 µm and (**b**) 5,000× magnified, picture width: 21.6 µm

barrier to oxygen. However, vitamin A and iodine losses during production of the capsules were high, and the encapsulation process should be improved to reduce these losses, particularly for vitamin A, which is expensive. Producing the microcapsules in an oxygen-free area and without direct light could be the next step. In this study, the losses are anticipated by overage before spraying at a ratio of iron 1 mg: iodine 15 µg:vitamin A 50 µg, to achieve the final target ratio for fortification, that is: iron 1 mg:iodine ≈ 12 µg:vitamin A ≈ 30 µg.

Fig. 7.8 Inner structure of a broken microcapsule containing iron, iodine and vitamin A designed for salt fortification analyzed by SEM micrographs (700× magnified, picture with: 154 μm)

Table 7.4 Color, iodine and vitamin A concentration in iodized salt (IS) and triple fortified salt (TFS) containing iron, iodine and vitamin A after storage for 0, 2, 4 and 6 months and fortification level before storage[a]

	Vitamin A	Iodine[b]		Color		
	μg/g salt	μg/g salt		Lightness[c]		ΔE_{ab}[d,e]
	TFS	IS	TFS	IS	TFS	TFS/IS
Fortification level[f]	100	25	40			
0 month	69.7±7.1	27.4±3.8	25.7±3.0	80.9±0.88	79.82±0.26	9.01±1.05
2 months	74.5±3.2	18.6±2.1	22.2±6.1	80.3±0.45	79.87±0.31	9.17±0.07
4 months	69.4±2.7	21.0±3.1	26.5±3.7	80.7±1.01	79.73±0.38	9.02±0.28
6 months	61.2±4.5	19.4±1.4	20.6±0.9	80.3±0.75	79.59±0.11	9.02±0.52

[a]Values are means ± SD ($n=3$)
[b]Significant main effect of time, $P<0.001$ (ANOVA)
[c]Lightness scale: 0 = black, 100 = white
[d]ΔE_{ab} absolute color difference between TFS and IS
[e]Significant main effect of fortification, $P<0.001$ (ANOVA)
[f]Iodine and vitamin A added before spraying for TFS and I added at mixing for IS

Once these stability and acceptability tests were complete, we compared the efficacy of the TFS and IS in a randomized, double-blind trial. The subjects were 6–14-year-old children. Each participating family was provided with 2 kg of salt at the beginning of each month for 10 months to supply all household needs. There were significant improvements in iron, iodine and vitamin A status in the TFS group, and a reduction in the prevalence of anemia and vitamin A deficiency. Although this trial showed the feasibility of triple fortifying salt using encapsulation, the price increment for a triple-fortified salt would be high and a significant barrier to its implementation.

7.4.9 Dual Fortification of Wheat Flour with Encapsulated Ferrous Sulfate and Iodine

For iron fortification of low extraction wheat flour, the WHO recommends both nonencapsulated and encapsulated iron compounds (World Health Organization 2006). If ferrous sulfate is used, it may need to be encapsulated to avoid sensory problems during long-term storage (Hurrell 2002). However, commercially available products cannot be used by most flour mills because their particle size is too large to pass the sieves used during milling. Salt iodization is the preferred strategy for iodine prophylaxis. But it is ineffective where salt is not the major condiment (e.g., in Southeast Asia, where it is replaced by soy and fish sauce) or where the presence of multiple small salt producers makes installation of fortification infrastructure difficult. In these situations, alternate vehicles for iodine should be explored. For example, in rural areas of North Africa, India and Pakistan, wheat flour is a staple food and is centrally processed, while salt is provided by many small local producers. For these reasons, Biebinger et al. (2008) recently developed and tested the efficacy of microcapsules containing ferrous sulfate and iodine specifically designed for wheat flour fortification. A spray-cooling process was used to package micronized ferrous sulfate and potassium iodate into small dense microcapsules of hydrogenated palm oil (mean particle size ≈40 μm) (Fig. 7.9). The capsule to substrate ratio was approximately 50:50; the mean ± SD iron and iodine content of the capsule was $15.7 \pm 0.4\%$ and $0.45 \pm 0.02\%$, respectively. During spraying, mean iodine losses were ≈25%; the baking process produced no measurable iodine losses.

The microcapsules were used to fortify wheat-based biscuits with 10 mg iron and 150 μg iodine per serving, and their efficacy was compared to a non-fortified control in a randomized feeding trial in young Kuwaiti women. The fortified biscuits were indistinguishable in taste and color from the unfortified biscuits. In the group receiving the microcapsule-fortified biscuits, body iron stores and urinary iodine concentration increased, and the prevalence of iron deficiency was reduced by nearly half. This is the first report showing the efficacy in humans of wheat flour fortification with encapsulated ferrous sulfate. The authors estimated that 11% of the iron from the encapsulated ferrous sulfate was absorbed. This value is similar to that from a study in Thailand using unencapsulated ferrous sulfate, where ≈11% of the administered iron dose was absorbed after a comparable time frame of 20 weeks (Zimmermann et al. 2005). This suggests the encapsulation of the ferrous sulfate used in the study did not adversely affect its bioavailability. However,

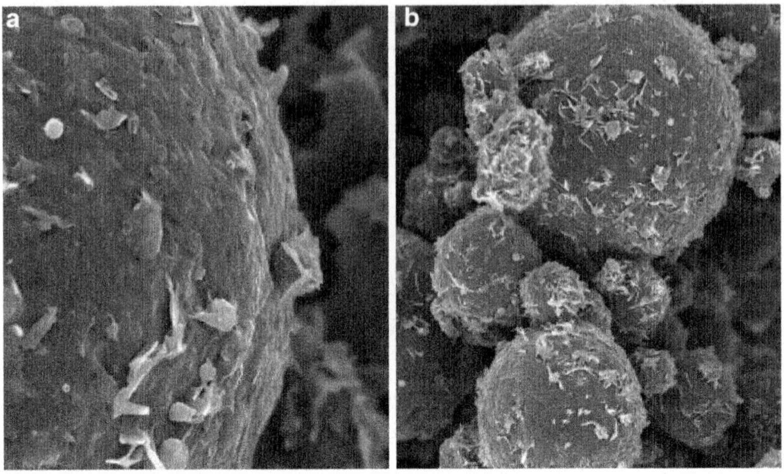

Fig. 7.9 Structural appearance of the fat-based microcapsules containing iron and iodine designed for wheat flour fortification, analyzed by scanning electron microscopy (SEM): (**a**) 1,000× magnified, picture width: 107 μm and (**b**) 4,000× magnified, picture width: 27 μm

because the vehicle was a baked snack, heating during cooking likely melted most if not all of the capsule and released its contents before consumption.

An innovation of this study was the production of dense microcapsules small enough to pass the sieve sizes of 325 mesh, widely used in the milling industry, without compromising iron bioavailability or sensory qualities. Commercially available encapsulated ferrous sulfates are typically too large (>100 μm) to pass through these sieves. The use of micronized dried ferrous sulfate and an optimized spray-cooling method made it possible to reduce microcapsule particle size and yet ensure an adequate surrounding physical barrier. Future refinements in the encapsulation process for wheat flour fortification should be directed toward: (a) further reduction in the amount of iron exposed on the capsule surface to reduce potential adverse sensory changes in stored flours or fortified products without compromising desirable particle size and iron bioavailability; (b) reduction of iodine losses during processing of capsules; and (c) potential addition of an enhancer of iron absorption, such as vitamin C or EDTA, to the capsule. The feasibility of adding encapsulated ferrous sulfate to wheat flour on a national scale, including issues of production, cost, distribution and storage, require further investigation.

7.5 Final Considerations

Iron (Fe) encapsulation has the potential to help overcome several major challenges in Fe fortification of foods. It may decrease unwanted sensory changes and reduce interactions of Fe with food components that lower Fe bioavailability. However, the effect of encapsulation per se on Fe bioavailability is a concern. Studies comparing encapsulated ferrous sulfate, ferric ammonium citrate and ferrous fumarate in most often hard fat to non-encapsulated compounds indicate that a ratio of capsule:substrate ≤50:50 is unlikely to decrease the relative bioavailability (RBV) of the Fe. Encapsulated ferrous fumarate given with ascorbic acid as a complementary food supplement has been shown to be efficacious in anemic children. Salt has been successfully fortified with microcapsules containing ferric pyrophosphate, iodine and vitamin A. In order to fortify salt with soluble Fe compounds, further refinements in capsule design are needed to increase resistance to moisture and abrasion. Encapsulated ferrous sulfate specifically designed for wheat flour fortification has been shown to be efficacious in improving iron status in young women. More studies evaluating the potential efficacy of encapsulated Fe in staple cereals (wheat and maize flours) are needed. A potential barrier to the use of encapsulated forms of Fe in staple food fortification is the relatively low melting point of the hard fat capsules, which may cause unwanted sensory changes during food preparation. Process costs for encapsulation can be high, and unless they can be reduced, may limit applications. Further research is needed to determine which encapsulation technologies are most effective in ensuring iron bioavailability from encapsulated compounds.

References

Andersson M, Thankachan P, Muthayya S, Goud RB, Kurpad AV, Hurrell RF, Zimmermann MB (2008) Dual fortification of salt with iodine and iron: a randomized, double-blind, controlled trial of micronized ferric pyrophosphate and encapsulated ferrous fumarate in southern India. Am J Clin Nutr 88(5):1378–1387

Biebinger R, Zimmermann MB, Al-Hooti SN, Al-Hamed N, Al-Salem E, Zafar T, Kabir Y, Al-Obaid I, Petry N, Hurrell RF (2009) Efficacy of wheat-based biscuits fortified with microcapsules containing ferrous sulphate and potassium iodate or a new H-reduced elemental iron: a randomised, double-blind, controlled trial in Kuwaiti women. Br J Nutr (in press)

Boccio JR (1997) A new procedure to fortify milk and dairy products with high-bioavailable ferrous sulfate. Nutr Rev 55:240–246

Dary O, Mora JO (2002) Food fortification to reduce vitamin A deficiency: International Vitamin A Consultative Group recommendations. J Nutr 132(9 Suppl):2927–2933

Fidler MC, Walczyk T, Davidsson L, Zeder C, Sakaguchi N, Juneja LR, Hurrell RF (2004) A micronised, dispersible ferric pyrophosphate with high relative bioavailability in man. Br J Nutr 91(1):107–112

Forbes AL, Adams CE, Arnaud MJ, Chichester CO, Cook JD, Harrison BN, Hurrell RF, Kahn SG, Morris ER (1989) Comparison of in vitro, animal and clinical determinations of iron bioavailability: International Nutrition Anemia Consultative Group Task Force report on iron bioavailability. Am J Clin Nutr 49:225–238

Gouin S (2004) Microencapsulation: industrial appraisal of existing technologies and trends. Trends Food Sci Technol 15(7–8):330–347

Hess SY, Zimmermann MB, Adou P, Torresani T, Hurrell RF (2002a) Treatment of iron deficiency in goitrous children improves the efficacy of iodized salt in Cote d'Ivoire. Am J Clin Nutr 75(4):743–748

Hess SY, Zimmermann MB, Arnold M, Langhans W, Hurrell RF (2002b) Iron deficiency anemia reduces thyroid peroxidase activity in rats. J Nutr 132(7):1951–1955

Hess SY, Zimmermann MB, Staubli-Asobayire F, Tebi A, Hurrell RF (1999) An evaluation of salt intake and iodine nutrition in a rural and urban area of the Cote d'Ivoire. Eur J Clin Nutr 53(9):680–686

Hirve S, Bhave S, Bavdekar A, Naik S, Pandit A, Schauer C, Christofides A, Hyder Z, Zlotkin S (2007) Low dose 'Sprinkles' – an innovative approach to treat iron deficiency anemia in infants and young children. Indian J Pediatr 44(2):91–100

Hurrell RF, Furniss DE, Burri J, Whittaker P, Lynch SR, Cook JD (1989) Iron fortification of infant cereals: a proposal for the use of ferrous fumarate or ferrous succinate. Am J Clin Nutr 49(6):1274–1282

Hurrell RF (1985) Nonelemental sources. In: Clydesdale FM, Wiemer KL (eds) Iron fortification of foods. Academic Press, Orlando, pp 39–53

Hurrell RF (2002) Fortification: overcoming technical and practical barriers. J Nutr 132(4 Suppl):806S–812S

Jackson LS, Lee K (1992) Fortification of cheese with microencapsulated iron. Cult Dairy Prod J 27:4–7

Kwak HS, Yang KM, Ahn J (2003) Microencapsulated iron for milk fortification. J Agric Food Chem 51:7770–7774

Liyanage C, Zlotkin S (2002) Bioavailability of iron from micro-encapsulated iron sprinkle supplement. Food Nutr Bull 23:133–137

Melse-Boonstra A, Pee S, Martini E, Halati S, Sari M, Kosen S, Muhilal, Bloem MW (2000) The potential of various foods to serve as a carrier for micronutrient fortification, data from remote areas in Indonesia. Eur J Clin Nutr 54(11):822–827

Moretti D, Zimmermann MB, Wegmüller R, Walczyk T, Zeder C, Hurrell RF (2006) Iron status and food matrix strongly affect the relative bioavailability of ferric pyrophosphate in humans. Am J Clin Nutr 83(3):632–638

Narasinga Rao BS (1985) Salt. In: Clydesdale FM, Wiemer KL (eds) Iron fortification of foods. Academic Press, Orlando, pp 155–164

Müller-Fischer NF, Bleuler H, Windhab EJ (2007) Dynamically enhanced membrane foaming. Chem Eng Sci 62:4409–4419

Sakaguchi N, Rao TP, Nakata K, Nanbu H, Juneja LR (2004) Iron absorption and bioavailability in rats of micronized dispersible ferric pyrophosphate. Int J Vitam Nutr Res 74:3–9

Schadler V, Windhab EJ (2006) Kontinuierliches Membran-Emulgieren mittels rotierender Mikromembranen mit definierten Porenabständen. Desalination 189(1–3):130–135

Schrooyen PM, van der Meer R, De Kruif CG (2001) Microencapsulation: its application in nutrition. Proc Nutr Soc 60:475–479

Semba RD, Bloem MW (2002) The anemia of vitamin A deficiency: epidemiology and pathogenesis. Eur J Clin Nutr 56(4):271–281

Shahidi F, Han XQ (1993) Encapsulation of food ingredients. Crit Rev Food Sci 33(6):501–547

Tondeur MC, Schauer CS, Christofides AL, Asante KP, Newtonm S, Serfass RE, Zlotkin SH (2004) Determination of iron absorption from intrinsically labeled microencapsulated ferrous fumarate (sprinkles) in infants with different iron and hematologic status by using a dual-stable-isotope method. Am J Clin Nutr 80:1436–1444

Wegmuller R, Zimmermann MB, Hurrell RF (2003) Dual fortification of salt with iodine and encapsulated iron compounds: stability and acceptability testing in Morocco and Cote d'Ivoire. J Food Sci 68(6):2129–2135

Wegmueller R, Zimmermann MB, Buehr V, Windhab E, Hurrell RF (2006) Development, stability, and sensory testing of microcapsules containing iron, iodine, and vitamin A for use in food fortification. J Food Sci 71(2):S181

Windhab E, Wagner T (1994) Apparatus for the preparation of ice cream, frozen dessert and fruit masses. European Patent EP 0659351 B1, published 1999

Windhab E (1999) Method for the manufacture of cold sprayed, solidified, storage stable and free flowing micro-capsule systems and their application. Patent DE 197 50 679 B4

Windhab E (2004) Method and device for the production of storage stable Multi-Micro-Capsules comprising adjustable synergistically active functional components. European Patent EP1722885 A1

Windhab EJ, Dressler M, Feigl K, Fischer P, Megias-Alguacil D (2005) Emulsion processing – from single-drop deformation to design of complex processes and products. Chem Eng Sci 60(8–9):2101–2113

World Health Organization (2000). Turning the tide of malnutrition: responding to the challenge of the 21st century. WHO/NHD/00.7. WHO, Geneva

World Health Organization (2006) Guidelines on food fortification with micronutrients. WHO, Geneva

Zimmermann MB (2004) The potential of encapsulated iron compounds in food fortification: a review. Int J Vitam Nutr Res 74(6):453–461

Zimmermann MB, Adou P, Torresani T, Zeder C, Hurrell RF (2000a) Persistence of goiter despite oral iodine supplementation in goitrous children with iron deficiency anemia in Cote d'Ivoire. Am J Clin Nutr 71(1):88–93

Zimmermann MB, Adou P, Torresani T, Zeder C, Hurrell RF (2000b) Iron supplementation in goitrous, iron-deficient children improves their response to oral iodized oil. Eur J Endocrinol 142(3):217–223

Zimmermann MB, Jooste PL, Mabapa NS, Schoeman S, Biebinger R, Mushaphi LF, Mbhenyane X (2007) Vitamin A supplementation in iodine-deficient African children decreases thyrotropin stimulation of the thyroid and reduces the goiter rate. Am J Clin Nutr 86(4):1040–1044

Zimmermann MB, Wegmueller R, Zeder C, Chaouki N, Biebinger R, Hurrell RF, Windhab E (2004) Triple fortification of salt with microcapsules of iodine, iron, and vitamin A. Am J Clin Nutr 80(5):1283–1290

Zimmermann MB, Zeder C, Chaouki N, Saad A, Torresani T, Hurrell RF (2003) Dual fortification of salt with iodine and microencapsulated iron: a randomized, double-blind, controlled trial in Moroccan schoolchildren. Am J Clin Nutr 77(2):425–432

Zimmermann MB, Winichagoon P, Gowachirapant S, Hess SY, Harrington M, Chavasit V, Lynch SR, Hurrell RF (2005) Comparison of the efficacy of wheat-based snacks fortified with ferrous sulfate, electrolytic iron, or hydrogen-reduced elemental iron: randomized, double-blind, controlled trial in Thai women. Am J Clin Nutr 82(6):1276–1282

Zlotkin S, Arthur P, Yeboah Antwi K, Yeung G (2001) Treatment of anemia with microencapsulated ferrous fumarate plus ascorbic acid supplied as sprinkles to complementary (weaning) foods. Am J Clin Nutr 74:791–795

Chapter 8
Encapsulation of Carotenoids

Henelyta S. Ribeiro, Heike P. Schuchmann, Robert Engel, Elke Walz, and Karlis Briviba

8.1 Introduction

Carotenoids are natural pigments, which are synthesized by microorganisms and plants. More than 600 naturally occurring carotenoids have been found in the nature. The main sources of carotenoids are fruits, vegetables, leaves, peppers, and certain types of fishes, sea foods, and birds. Carotenoids may protect cells against photosensitization and work as light-absorbing pigments during photosynthesis. Some carotenoids may inhibit the destructive effect of reactive oxygen species. Due to the antioxidative properties of carotenoids, many investigations regarding their protective effects against cardiovascular diseases and certain types of cancers, as well as other degenerative illnesses, have been carried out in the last years (Briviba et al. 2004; Krinsky et al. 2004; Kirsh et al. 2006). A diet rich in carotenoids may also contribute to photoprotection against UV radiation (Stahl et al. 2006). In vitro studies have shown that carotenoids such as β-cryptoxanthin and lycopene stimulate bone formation and mineralization. The results may be related to prevention of osteoporosis (Kim et al. 2003; Yamaguchi and Uchiyama 2003; 2004; Yamaguchi et al. 2005).

Carotenoids are very sensitive to heat, oxidation, and light, due to their unsaturated chemical structures (Fig. 8.1). They are almost insoluble in water and only slightly oil soluble at room temperature (about 0.2 g/L_{oil}), but their solubility in oil

H.S. Ribeiro (✉)
Unilever Discover, Colworth Science Park, Sharnbrook, Bedfordshire, UK
e-mail: henelyta.ribeiro@unilever.com

H.P. Schuchmann
Institute of Process Engineering in Life Sciences, University of Karlsruhe, 76131 Karlsruhe, Germany

R. Engel
BASF SE, 67056 Ludwigshafen, Germany

E. Walz and K. Briviba
Max Rubner-Institute (MRI), Federal Research Institute of Nutrition and Food, Department of Food and Bio Process Engineering, 76131 Karlsruhe, Germany

N.J. Zuidam and V.A. Nedović (eds.), *Encapsulation Technologies for Active Food Ingredients and Food Processing*,
DOI 10.1007/978-1-4419-1008-0_8, © Springer Science+Business Media, LLC 2010

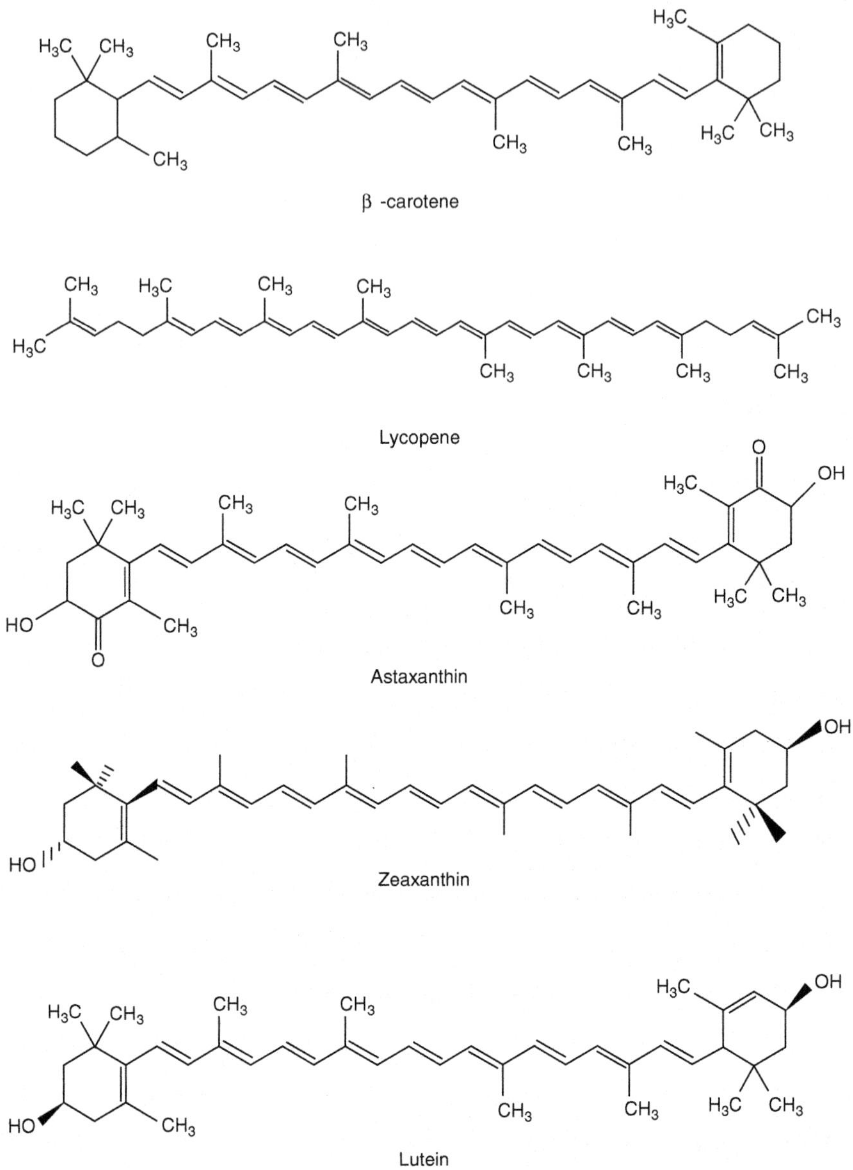

Fig. 8.1 Chemical structures of some carotenoids

increases greatly with increasing temperature (Ax et al. 2001). It has been found that only a minor part of the carotenoids in raw fruits or vegetables is absorbed in the intestines, probably due to the fact that carotenoids in the nature exist as crystals or are bound in protein complexes. In contrast, carotenoids dissolved in vegetable oils show a higher bioavailability (Parker 1997).

Incorporation of carotenoids into micro and nano structures may influence their solubility and crystallinity. After formulating carotenoids into such particulate systems, they may be easily delivered into cellular compartments, improving their bioavailability. The carotenoids considered in this chapter are β-carotene, lycopene, astaxanthin, lutein, and zeaxanthin (Fig. 8.1) which have been applied for food coloration, and human nutrition due to their antioxidant biological property.

β-carotene is the most widespread of all carotenoids. It is responsible for the orange–yellow color of carrot, palm fruit, pumpkin, mango, acerola, yellow cassava, etc. β-carotene has provitamin A activity. It is an apolar, bicyclic carotenoid that contains 11 conjugated double bonds. The chemical structure of β-carotene is accountable for its high antioxidative activity. β-carotene supplementation may reduce risks of heart disease deaths and stimulate proliferation and differentiation of osteoblast cells (Buijsse et al. 2008; Sahni et al. 2009). Köpcke and Krutmann (2008) have recently reported that dietary supplementation of humans with β-carotene may protect against sunburn.

Lycopene also belongs to the group of carotenoids. It shows a very high quenching rate of reactive singlet oxygen, due to its apolar and acyclic molecular structure. Lycopene is a natural red pigment mainly found in tomato, tomato products, and lycopene-carrots. Epidemiological evidence has suggested that lycopene can protect individuals from colorectal cancer (Vrieling et al. 2007), and men from prostate cancer (Schwarz et al. 2008).

Astaxanthin exist in salmon, lobster, fish eggs, crabs, trout, fish eggs, and flamingo. Its molecule has two cyclic end groups with polar groups, similar to lutein and zeaxanthin.

Lutein and zeaxanthin are isomers found in various foods including green leafy vegetables, fruits, and marigold flower. They show identical chemical formulas with different position of a double bond in one of the end rings. They are present in the human retina. Christen et al. (2008) have shown to possibly reduce the risk of developing cataracts by a diet rich in these carotenoids. Zeaxanthin might be effective to prevent age-related macular degeneration (AMD) (Sajilata et al. 2008).

With modern methods of encapsulation technology, solubility, stability, and bioavailability of carotenoids can be considerably improved (Hoppe et al. 1986; Horn and Rieger 2001; Ax 2003; Ribeiro and Schubert 2004; Engel et al. 2005; Ribeiro et al. 2005a, 2006a, 2008; Schuchmann et al. 2005; Auweter et al. 2006; Flanagan and Singh 2006; End et al. 2007; Feldthusen et al. 2007; Garti and Aserin 2007; Garti et al. 2007; Leuenberger 2007; McClements et al. 2007). In order to achieve optimized physical and chemical stability, and bioavailability of carotenoids in these advantageous formulations, innovative processes for their production have been developed and investigated, which will be discussed in this chapter.

8.2 Processes to Encapsulate Carotenoids

Since carotenoids are very expensive, some cost-effective processes are desirable to produce carotenoid formulations, in order to reduce the capital cost of the final products. The challenge for manufactures is to formulate easy-to-disperse powders, colloidal dispersions, emulsions, or suspensions, which are suitable for various applications.

Optimization of processing conditions may prevent carotenoid oxidation and isomerization, and improve its solubility to achieve satisfactory bioavailability. Choice of optimal process, emulsifier(s), and other ingredients are the most important variables to achieve droplets size and desired color for each application. As differences in particle size may also give rise to variations in color shade, a good process control is essential during the manufacture of coloring-formulations (Müller and Tamm 1966; Runge et al. 1998; Lüddecke et al. 2004a).

8.2.1 O/W Emulsions

Oil-in-water (O/W) emulsions containing carotenoids dissolved in finely dispersed oil droplets (Fig. 8.2) can be produced using various emulsification processes. This type of formulations may increase absorption of these substances in vitro and in vivo. For preparing carotenoid-loaded O/W emulsions, the carotenoid is dissolved in a vegetable oil or in an apolar solvent at elevated temperatures, and subsequently emulsified with an aqueous phase containing an emulsifier to stabilize the droplets. Carotenoids in the forms of fine droplets show a better water dispersibility compared to those in bulk form.

Submicron-sized oil droplets allow for a supersaturation of carotenoid molecules in O/W emulsions (Schweikert and Kolter 1997; Ax 2003). There are two reasons for this as follows:

(1) There is a lack in crystallization germs or seed crystals after emulsification (Bunnell et al. 1958)
(2) The concentration of saturation is significantly increased within small droplets (Mersmann et al. 2005) as described by (8.1):

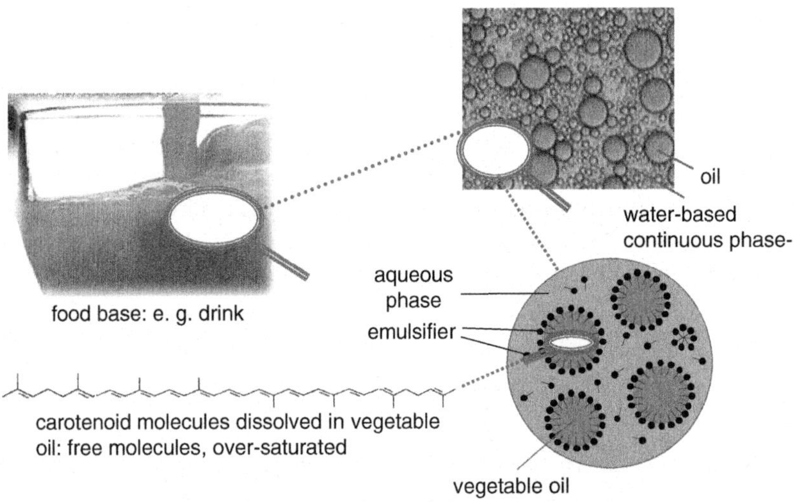

Fig. 8.2 Microstructure of carotenoid-loaded O/W emulsion-based products

$$\ln\left(\frac{c}{c^*}\right) = \frac{4\gamma_{CL}M}{RT\rho_c L^*_{crit}}. \tag{8.1}$$

Herein c is the saturation concentration as a function of droplet radius r, c^* the saturation concentration for droplets of infinite radius ($r \to \infty$ or planar flat interfaces), γ_{CL} is the interfacial tension between the crystals and the liquid phase, R the Avogadro constant, M the molar mass, T the temperature, ρ_c the crystal density, and $L_{crit}{}^*$ is the critical germ forming diameter.

Submicron-sized, vegetable oil-based emulsion droplets are stable against sedimentation and coalescence over long storage time and thus account for stable, homogeneous, and water-dispersable products. In their production, however, some hurdles have to be taken. Fine carotenoid O/W emulsions have been prepared by various methods, such as high-pressure homogenizers, as well as membrane and microchannel emulsification processes, and phase inversion temperature (PIT) method. These will be described here.

8.2.1.1 High-Pressure Homogenization

Challenges in the production of high-pressure homogenized submicron-sized carotenoid-loaded emulsions involve the adaptation of temperature profiles and residence times in each operational unit to the kinetics of carotenoid molecule solution, their degradation and crystallization as well as to the kinetics of droplet breakup, stabilization by interfacial active material and coalescence.

The process line as developed by Ax (2003) and patented in 2005 (Feldthusen et al. 2005) involves the unit operations of (1) carotenoid molecule solution, (2) emulsion premix preparation, (3) high-pressure-homogenization, and (4) emulsion droplet stabilization (Fig. 8.3).

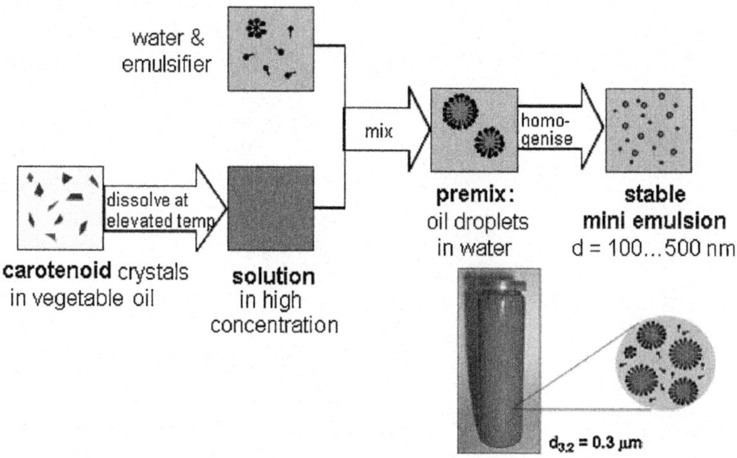

Fig. 8.3 Principle of carotenoid-loaded O/W emulsion preparation with unit operations involved (Ax, 2003) (Courtesy of Shaker Verlag)

For the dissolution of carotenoid molecules in vegetable or middle-chained triglycerides (MCT) temperatures above 100°C are required (Fig. 8.4). At these temperatures dissolution time is limited to some seconds. High temperatures result in structural changes, as depicted by Fig. 8.5. With increasing dissolution temperature a higher amount of diverse *cis*-isomers are found after dissolving *all-trans* isomers of carotenoid molecules in oil and emulsifying them. Therefore, temperatures should be decreased as fast as possible. Most easily this is realized by quenching the carotenoid loaded oil with a cold emulsifier solution, being the continuous phase of the emulsions. In these emulsion premix droplets the carotenoid molecules are oversaturated as their radius is too high for stabilizing them in the dissolved state [see (8.1)]. Droplet sizes have to be reduced without retention time.

High-pressure homogenization is the ideal mechanical process for producing droplet sizes below 1 μm. High-pressure homogenizers have been traditionally used in the dairy industry. These machines are operated continuously at throughputs up to several thousands L/h. They consist essentially of a high-pressure pump and a homogenizing nozzle. The pump creates the pressure, which is then transferred within the nozzle to kinetic energy being responsible for droplet disintegration. Droplet elongation, deformation, and break-up in high-pressure homogenization nozzles are mainly due to elongational and shear stresses as well as inertia turbulent and cavitational stresses acting at the interfaces. Recent developments in high-pressure homogenizing concentrate on nozzle design (Schuchmann 2005, 2007; Schubert 2005). Examples of new homogenizing nozzles (Fig. 8.6) are opposing jets, for example, Microfluidizer® (Cook and Lagace 1985, 1990), jet dispersers (Hovestad et al. 2000), and a simple orifice valve (Stang 1998) as well as developments

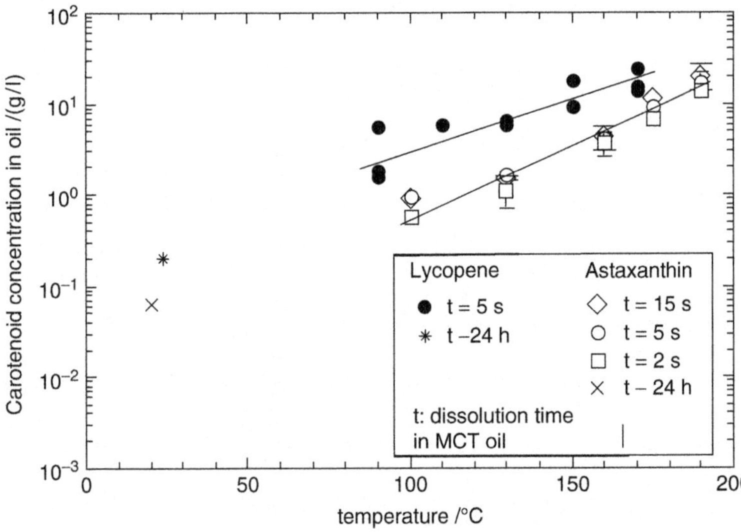

Fig. 8.4 Dissolution of carotenoid molecules in MCT oil: carotenoid concentration as a function of dissolution temperature and time *t* (Ax 2003) (Courtesy of Shaker Verlag)

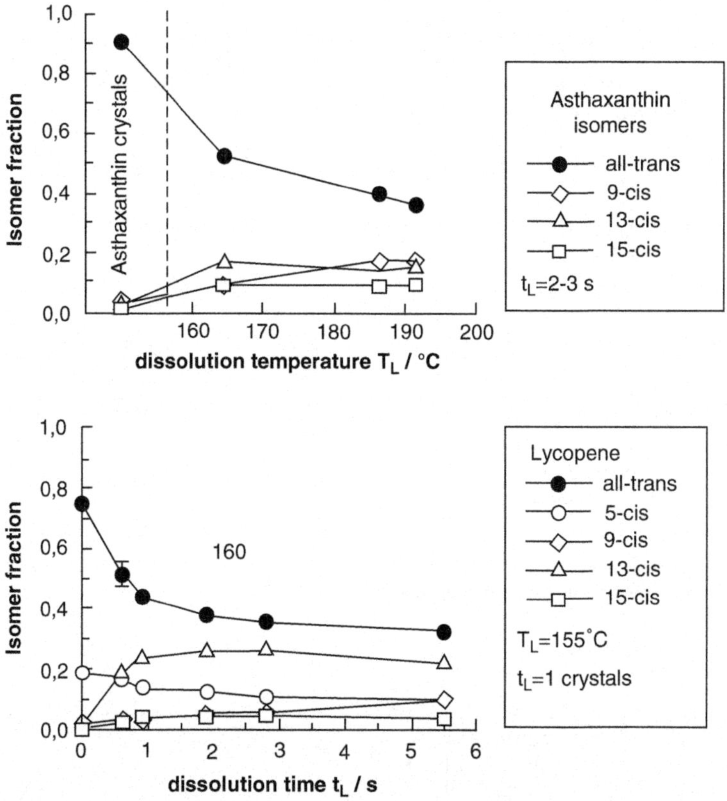

Fig. 8.5 Isomer fractions of astaxanthin and lycopene after dissolution at different temperatures T_L and time t in oil and emulsification (Ax 2003) (Courtesy of Shaker Verlag)

based on it (Kolb 2001; Freudig 2004; Aguilar et al. 2004, 2008; Scheid and Buchholz 2005; Köhler et al. 2007, 2008).

The design of the homogenizing nozzle influences the flow pattern of the emulsion in the nozzle and hence droplet disruption (Stang et al. 2001). In Fig. 8.7, the elongational rate and the kinetic turbulent energy are given for a relatively simple but efficient homogenization nozzle consisting mainly of a simple round valve (Köhler et al. 2008).

Droplet elongation, deformation, and break-up are realized within milliseconds. A huge new interfacial area is created which has to be covered and stabilized by newly adsorbing emulsifier molecules. The adsorption kinetics has to follow break-up kinetics as droplet collision rates are quite high in emulsions of reasonable disperse phase contents (as, for example above 30 vol% in carotenoid emulsions). Therefore at least one emulsifier of small molecular mass and easy molecular structure has therefore to be added, for example, Tween® 20. This emulsifier is responsible for droplet stabilization against re-coalescence within the homogenization nozzle. For O/W emulsions containing carotenoids additional emulsifiers are usually added being required for long-term shelf-life stability and bioavailability (see Sect. 8.3).

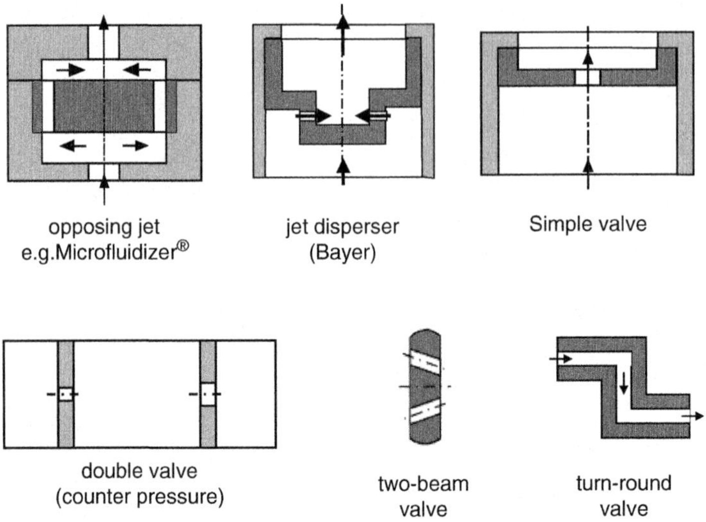

Fig. 8.6 Recent developments in high-pressure homogenizing: New homogenizing nozzles for efficient droplet disruption (Schuchmann 2007)

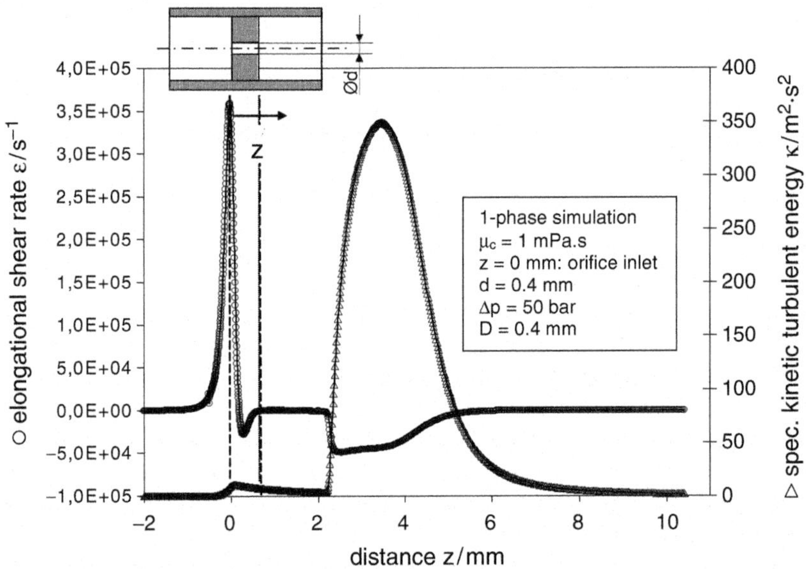

Fig. 8.7 Elongational shear rate and specific kinetic turbulent energy in a simple orifice valve, as required for efficient droplet break-up to the submicron-sized range in homogenization of carotenoid-loaded emulsions (CFD simulation: Fluent®) (Aguilar et al. 2008; Köhler et al. 2007)

Droplet size distribution depend on oil type (viscosity), emulsifiers applied (dynamic interfacial tension), homogenizing nozzle geometry, and homogenization pressure. An example is given in Fig. 8.8.

The carotenoid molecules themselves do not show any relevant interfacial activity and do not change droplet viscosity to a relevant amount. Thus droplet disruption is not influenced by dissolved carotenoid molecules (Fig. 8.9). Absorption spectra depict that carotenoid molecules are mainly found in their molecular dissolution status within droplet of submicron size range (Fig. 8.10) (Ax 2003; Velikov and Pelan 2008).

Solid lipid nanodispersions can also be produced by high-pressure homogenization by melting the lipid at least 10°C above its melting point. Since these kind of particles show a lipid core in solid state, they may provide controlled release of encapsulated lipophilic molecules (Farhang 2007). Hentschel et al. (2008) have successfully applied high-pressure melt-emulsification to encapsulated β-carotene into solid lipid nanodispersions, which particle sizes were smaller than 1 μm.

8.2.1.2 Dead-End Membrane Emulsification

The membrane emulsification process offers the opportunity to produce emulsions with narrow droplet size distribution without high mechanical stress at low energy input (10^4–10^6 J/m^3) compared to conventional mechanical methods (10^6–10^8 J/m^3) (Schubert and Engel 2004). Membranes usually applied in separation processes have been successfully used for emulsification as well. However, this new field of application has just been recently explored (Vladisavljević and Williams 2005, 2006).

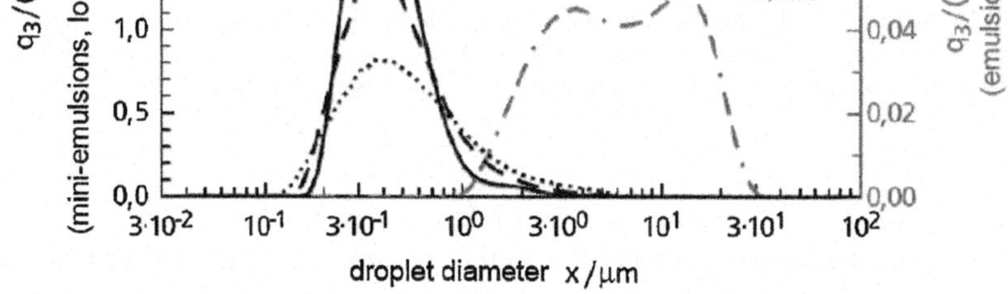

Fig. 8.8 Droplet size distributions (volume density distribution) of lycopene emulsions (premix and high-pressure homogenized at different pressure p_H) (Ax, 2003) (Courtesy of Shaker Verlag)

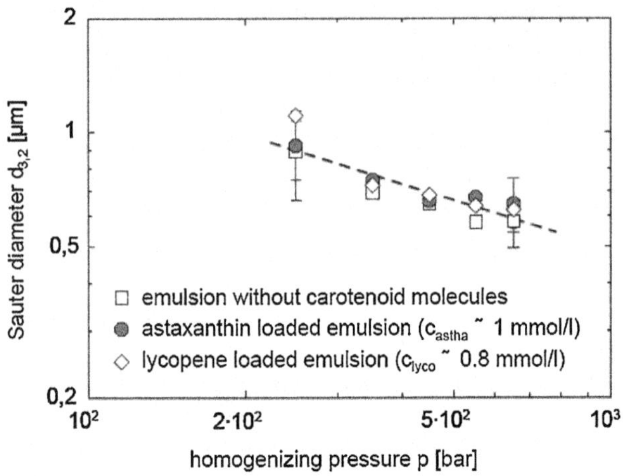

Fig. 8.9 Mean droplet diameter (sauter diameter) of O/W emulsions without carotenoid molecules or loaded with astaxanthin or lycopene, respectively. Varied parameter: Homogenization pressure (Ax 2003)

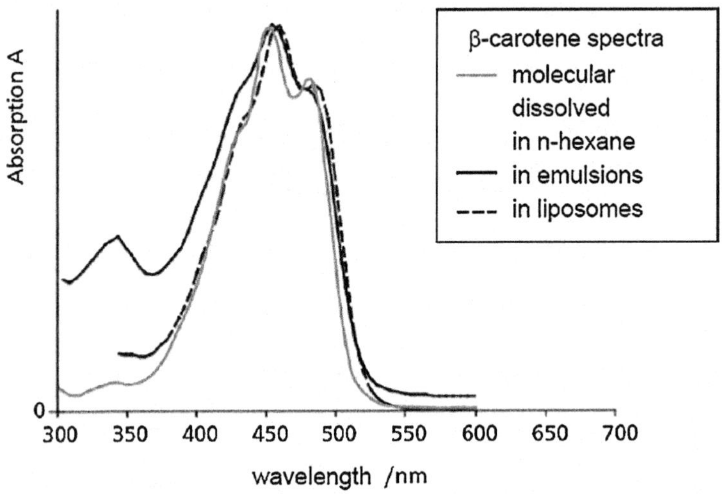

Fig. 8.10 Absorption spectra of carotenoid molecules different formulations: dissolved in *n*-hexane, formulated in submicron-sized O/W emulsion droplets at a concentration well above saturation concentration compared to pure crystals and formulated in liposomes (Ax 2003) (Courtesy of Shaker Verlag)

Premix membrane emulsification is a relative new emulsification process that was developed by Suzuki et al. (1998). In this process, emulsions are produced by pressing a pre-emulsion through a membrane. Defined characteristics of different membranes render this process very flexible. Depending on their pore size and the membranes being hydrophobic or hydrophilic, water-in-oil (W/O) or oil-in-water

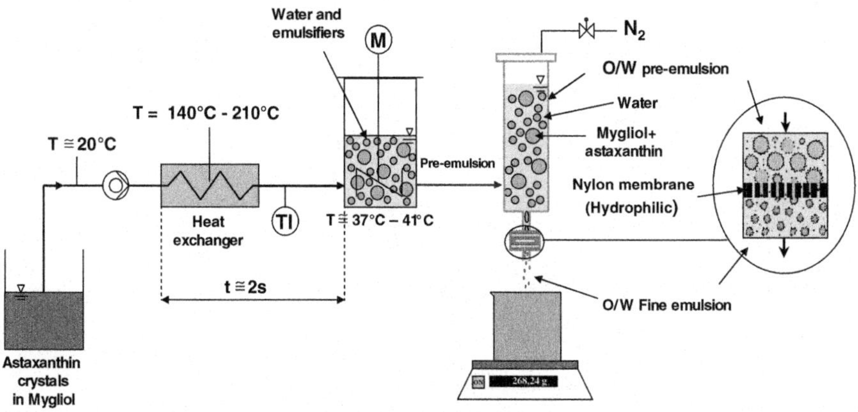

Fig. 8.11 Schematic process for producing O/W emulsions containing carotenoids by premix membrane emulsification (Courtesy of Wiley-Blackwell Publishers)

(O/W) emulsions with different droplet sizes can be produced. The characteristics of membrane emulsification, like low energy input and narrow droplet size distribution, add to other advantages of the premix emulsification compared to direct membrane emulsification, like the facility to clean and disinfect, and high flux. This makes the premix membrane emulsification process very promising and well suited for life science applications.

Ribeiro et al. (2005b) have investigated the effect of repeated premix membrane processing on the microstructure of astaxanthin-loaded O/W emulsions at pressures between 5 and 15 bar and dispersed phase fractions between 10% and 40% (Fig. 8.11). For studying the effect of repeated premix membrane emulsification on emulsion microstructure, the premixes were emulsified three times using the same membrane made of polyamide 6,6. Increased pressure and number of repetitions of the emulsification process resulted in narrower droplet size distributions (Fig. 8.12) and smaller Sauter mean diameters ($d_{3,2}$). Astaxanthin degradation and particle-size distributions were investigated after certain storage times. No coalescence was observed.

8.2.1.3 Microchannel Emulsification

Several studies on microfluidic channels used as an emulsification process have been carried out in the last 10 years. Microchannel (MC) emulsification is a relatively new technique to produce emulsions with a very narrow droplet size distribution (Kawakatsu et al. 1997; Kikuchi et al. 2000; Sugiura et al. 2001; Kobayashi and Nakajima 2006). The fact that narrow droplet size distributions is achieved, without applying strong mechanical forces, makes MC and membrane emulsification interesting methods for shear sensitive components. MC emulsification can be used for producing monodispersed emulsions with a coefficient of variation of less than 5%.

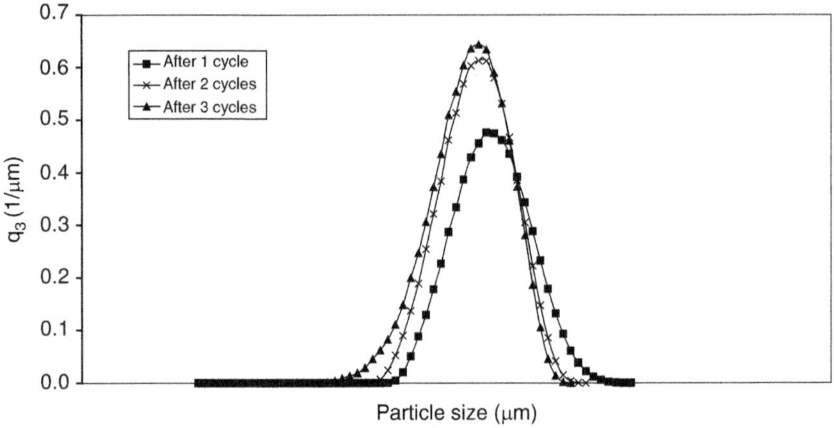

Fig. 8.12 Influence of the repeated premix membrane emulsification on droplet size distributions ($\phi = 0.4$; $\Delta P = 10$ bar)

In this way, microfluidic channel is the most appropriate emulsification process to control the droplet sizes and droplet size distribution of emulsions.

Monodisperse O/W emulsions containing β-carotene has been produced by MC emulsification (Ribeiro et al. 2006a; Neves et al. 2008a; 2008b). The scheme of the process for producing O/W emulsions is shown in Fig. 8.13. The microfluidic channels setup consists of a uniformly sized MC arrays manufactured on silicone microchip, a module, accessory for supplying the continuous and disperse phases, and a microscope video system, which enables the observation of emulsification behavior. A grooved, and an asymmetric through-holes with a slit and circular channels of surface-oxidized, silicon-based microfabricated devices were used for preparing the emulsions. Because of oxidation treatment, MC plates possess negative charge on their surfaces. The grooved type consists of an array of microfluidic channels and a slit-like terrace, which produces droplets in a relatively low throughput between 0.01 and 0.1 mL/h. The asymmetric straight-through microchannel plate used consists of channels with a diameter of 10 μm and slits with a length of 50 μm in longer line and 10 μm in shorter line. The flow rate of the continuous and disperse phases were 20 mL/h and 0.5 mL/h, respectively. Laminar flow rates of disperse and continuous phases are very important, independent variables for controlling the droplet size (Vladisavljević et al. 2006). Sucrose laurate was used as the emulsifier.

Figure 8.14 shows images of β-carotene droplets generation during and after MC emulsification using grooved (a) and straight-through (b) types, respectively. As can be seen, MC emulsification with well-defined channels provides formation of monosized O/W emulsions containing lipophilic compounds, such as β-carotene. Monodispersed droplets with an average diameter of approximately 9 μm and 27 μm, and coefficient of variation of about 5.0% and 2.6%, could be produced

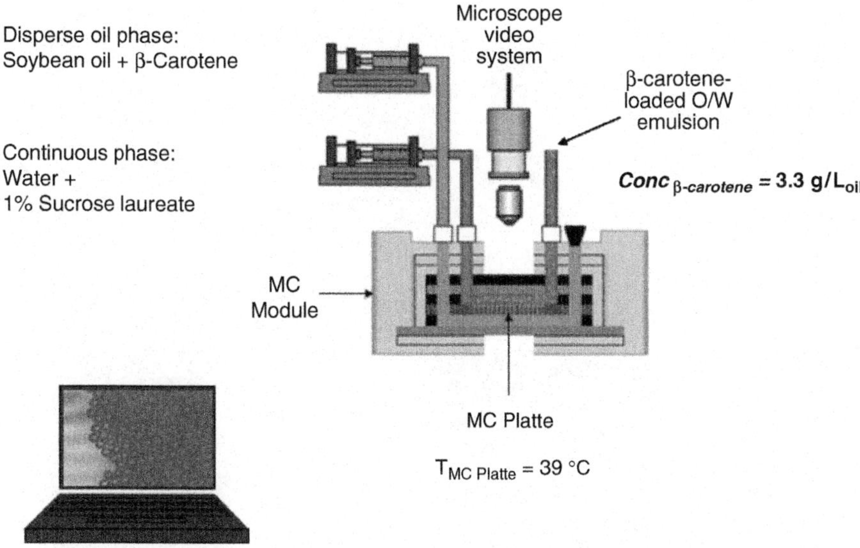

Fig. 8.13 Scheme of the microfluidic channels for monosized droplet formation

using grooved and asymmetric straight-through MC plates, respectively. Due to the hydrophilic surface of the silicon-oxidized microchannel plate, β-carotene-loaded O/W emulsions could be successfully prepared. Total β-carotene concentration in the oil phase was 3.3 g/L_{oil}. Because of added α-tocopherol, β-carotene concentration remained unchanged for more than 3 weeks. Emulsions were chemically and physically stable for at least 5 months at 4°C.

8.2.1.4 Phase Inversion Temperature Method

Oil-in-water (O/W) emulsions containing ethoxylated non-ionic emulsifiers may undergo a phase inversion to water-in-oil (W/O) emulsions upon heating. At the phase inversion the hydrophilic and lipophilic properties of the emulsifier are balanced, resulting in minimum interfacial tensions between oil and water phases. This temperature-induced phase inversion is utilized in the so-called PIT method. This technique is a thermodynamic phenomenon that produces nanodispersed bluish transparent O/W emulsions with narrow droplet size distributions (Engels et al. 1995).

Non-food-grade, nanosized carotenoid-loaded O/W emulsions were successfully produced using cetostearyl poly(oxyethylene (12) glycol) and glyceryl monostearate as the emulsifying agents by PIT method (Ribeiro and Schubert 2004). Dycapryl ether was used as the oil phase. The PIT emulsion was prepared conventionally by heating up to 74°C and subsequently rapid cooling down to 25°C. The total astaxanthin and lycopene concentration was around 0.3 and 3.5 $g/L_{emulsion}$. No detectable degradation of lycopene and astaxanthin was observed

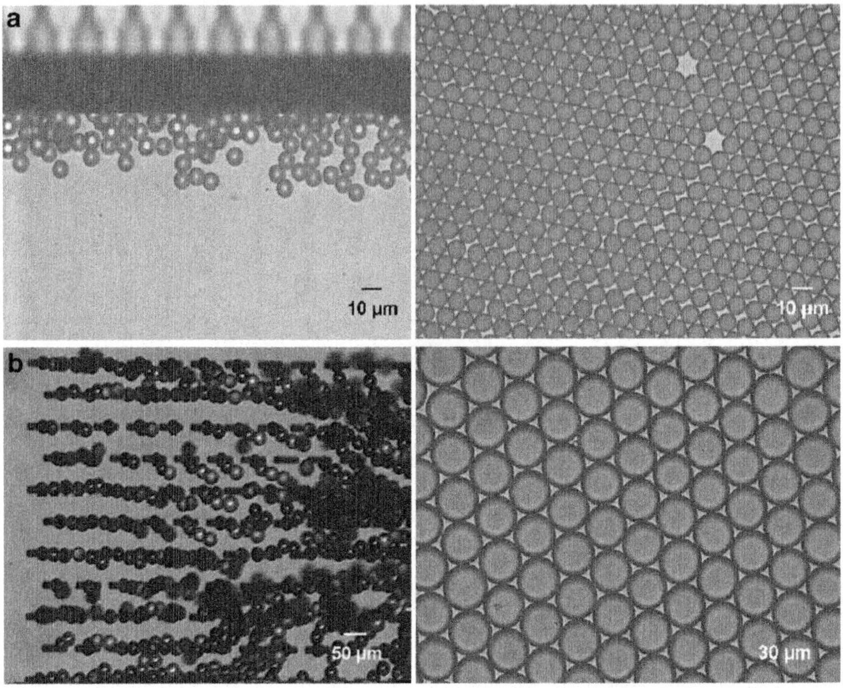

Fig. 8.14 Images of the droplets formation during and after emulsification process using two different types of MC plates: (**a**) grooved and (**b**) asymmetric straight-through

during a month of storage. The Sauter mean diameter of the finely dispersed droplets was around 80 nm and droplet size distributions were between 40 and 200 nm, as determined by means of laser diffraction combined with polarization intensity differential scattering (PIDS) technology (Fig. 8.15). These data suggest that PIT emulsions could be an excellent vehicle for delivering carotenoids.

8.2.1.5 High Internal Phase Emulsions (HIPE)

The continuous phase of an O/W emulsion can be extracted through centrifugation or evaporation under vacuum. This yields a transparent, gel-like cream which can be stored and later redispersed in water. The oil droplet deforms into polyhedral structures, separated by extremely thin films of emulsifier, which can be redispersed in water. Thus, the cream is called highly concentrate emulsion or "biliquid foam".

Ribeiro et al. (2005a) have produced biliquid foams containing carotenoid in submicron size. Fine emulsions were prepared by mixing vegetable oil with aqueous solutions of Tween® 20 and Xanthan by high-pressure homogenization (Fig. 8.16). The average droplet diameter was controlled by changing the

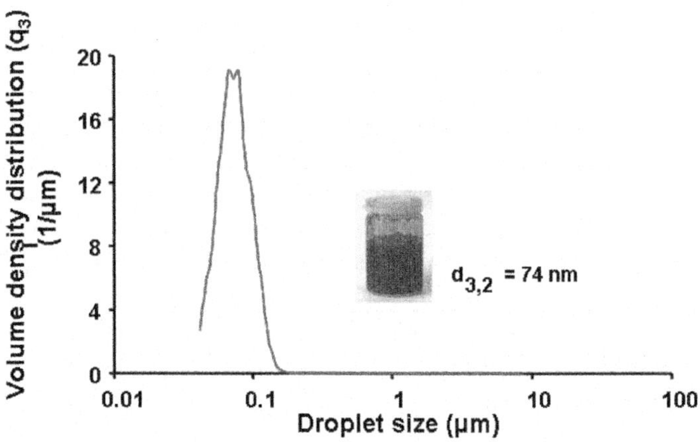

Fig. 8.15 Droplet size distribution of lycopene O/W emulsion and sample appearance

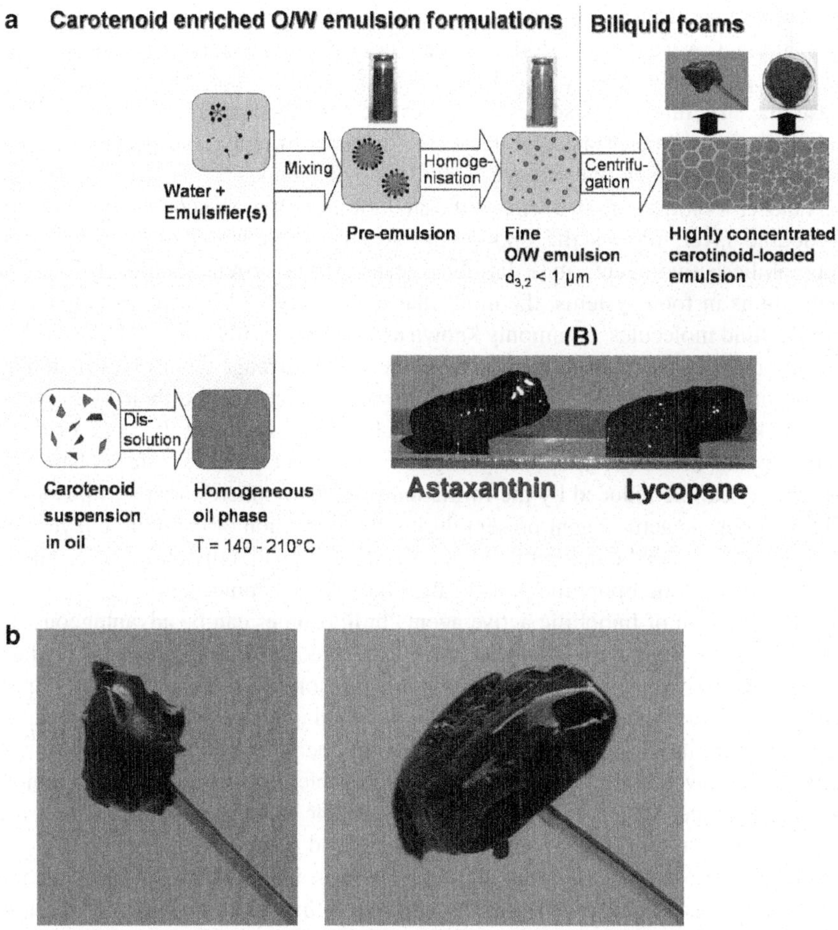

Fig. 8.16 (**a**) Schematic procedure for producing biliquid foams, (**b**) appearance of the samples

emulsifier-to-oil ratio. It could be set in the range from 100 to 800 nm by choosing the concentration of the disperse phase between 10% and 50%. The aqueous phase of the emulsions was removed by centrifugation. After centrifugation, the cream turns into a biliquid foam with an oil volume fraction up to 0.99, which shows a transparent and elastic (gel-like) structure. After centrifugation the droplet still retained their integrity and did not coalesce. The most important factors that have been identified are the physical properties of the continuous and dispersed phase as viscosity and the droplet size distribution before and after centrifugation.

8.2.2 *Liposomes*

As described in Chap. 2, liposomes consist of a molecular double layer that separates an inner, watery phase from the outer, continuous watery phase. Except from the inner part of the spherical particle not being "empty" but filled with a watery phase, the structure of liposomes could best be described as hollow spheres. With both the inner and outer phase of liposomes being hydrophilic, the system seems to be suited for encapsulation of water soluble active agents.

A closer look into the structure of the shell of these hollow spheres, the molecular double layer, reveals the opportunity for the formulation of lipophilic and amphiphilic active agents in a dissolved state in this system (Senior 1987). For applications in food systems, the molecular double layer of liposomes is made of phospholipid molecules, commonly known as lecithins, of high purity. Upon hydration and input of mechanical energy, these molecules arrange themselves in double layers with their lipophilic parts oriented towards the center and their hydrophilic parts making up the outer surface of the double layer. Since, at the edge of these double layers, the inner, lipophilic part gets into contact with the watery phase, the free energy can be reduced by the formation of hollow spheres, that is, liposomes. With a lipophilic active agent present during the formation of the double layer and the liposomes, it can be incorporated into the double layer's lipophilic core. There, it is dissolved in the lipophilic parts of the phospholipid molecules.

A formulation of lipophilic active agents in liposomes can be advantageous for several reasons. First of all, the active agent is formulated as a solution, thus exhibiting a high bioavailability. Furthermore, since liposomes form spontaneously upon hydration of the phospholipid, they have a good physical stability over time. Depending on the process chosen for production, their size can be adjusted even to diameters below 100 nm so that the water-dispersible formulations have an almost clear appearance (Fig. 8.17). Several processes for production of liposomes are described in literature (Lasic 1993; Winterhalter and Lasic 1993).

For the formulation of lipophilic active agents in liposomes in general, the active agent has to be dissolved in the substance that will later, upon hydration, build the double layer. The solubility of carotenoids in any phospholipid fraction at ambient conditions is by far too low in order to achieve reasonable carotenoid

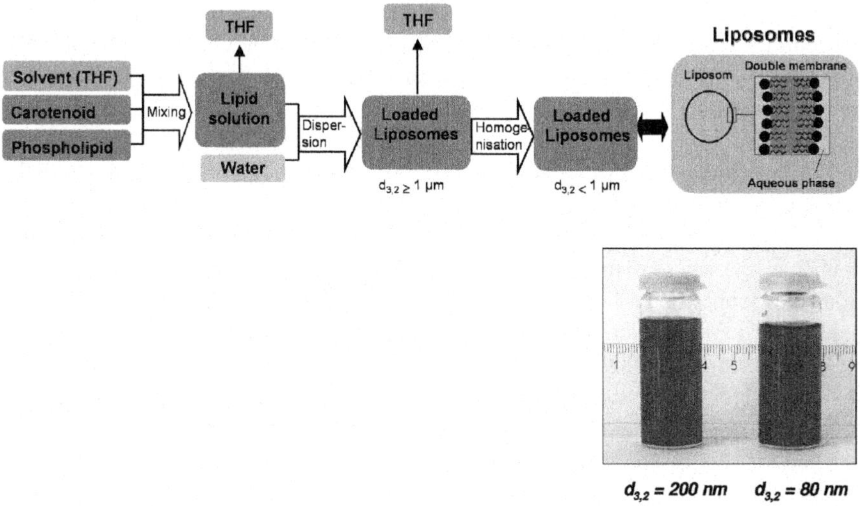

Fig. 8.17 Carotenoid-loaded liposomes formulation and appearance of liposomes containing lycopene (Ax 2003) (Courtesy of Shaker Verlag)

concentrations in the final product. Although solubility can usually be increased at elevated temperatures, this method is unsuitable for reaching sufficient carotenoid concentrations. Phospholipids decompose at approximately 180°C before a satisfactory concentration of carotenoids can be achieved. Therefore, a solvent which is capable of solving reasonable concentrations of both the phospholipid and the active agent, that is, the carotenoid, must be found. Above that, as will be described in a later step, the solvent's vapor pressure in the temperature range up to approximately 100°C should be high enough for allowing its removal from the solution under reduced pressure.

The processes for production of liposomes known from literature could be adapted for the production of carotenoid-loaded liposomes as described by Ax (2003) and in the following. Having selected an appropriate solvent, a solution of phospholipids, the active agent, and, if necessary, an antioxidant, is prepared. From this solution, the solvent is extracted, for example, in a rotary evaporator, until a semi-solid concentrated solution is obtained, which is usually the case at a solvent concentration of approximately 60%. By means of introducing mechanical energy and high shear, this solution is dispersed in water. In this step, liposomes form spontaneously. In order to obtain the desired liposome size distribution, this dispersion is subjected to high-pressure homogenization, for example, in a microfluidizer. Any excess solvent is then extracted from the formulation by nitrogen-stripping. The resulting liposome dispersion has a clear appearance if the size of the liposomes is below 100 nm, which can be achieved by applying a homogenization pressure of 1,000 bar or higher, depending on the homogenization valve (Ax et al. 2000).

Solvents identified as suitable for producing carotenoid-loaded liposomes are chloroform or tetrahydrofuran (THF). Carotenoid concentrations in the final liposome-dispersion are reported to have reached 0.3 mmol/L at a phospholipid concentration of 10%. The remaining solvent concentration was below 0.05%. By comparison of the absorption spectra of carotenoids dissolved in hexane and formulated in liposomes, it could be proven that the carotenoids existed in a dissolved state in the double layer of liposomes (see Fig. 8.10; Ax 2003).

The maximum achievable concentration of active agents in liposomes greatly depends on their polarity that determines their position in the double layer. Less polar carotenoids like β-carotene and lycopene tend to accumulate in the core of the double layer, whereas more polar structures will orient their polar groups away from the double layer's center. Therefore, higher concentrations of polar carotenoids than of less polar ones can be incorporated into liposomes (Grolier et al. 1992; Socaciu et al. 2000; Barenholz et al. 2006).

The need for organic solvents with rather well-specified properties renders production of liposomes for food applications difficult if not impossible. Above that, the highly purified lecithin essential for the formation of double layers and liposomes is very expensive compared to most food ingredients used in concentrations around 10% in formulations for active agents in food systems. Until now, only cell culture studies have been carried out in order to measure the bioavailability of carotenoids from liposomes. In these trials, the bioavailability from liposomes was superior to most other formulations like O/W emulsions. Due to the high specific surface of the disperse liposome-system, this is not surprising. However, levels of bioavailability of carotenoids in O/W emulsions stabilized with certain emulsifier systems by far exceeded those reported for carotenoids in liposomes. Bioavailability and stability issues will be further dealt with in Sect. 8.3.

8.2.3 Production of Nanoparticles by Precipitation

Nanoparticles containing carotenoids can be produced by precipitation and condensation processes with remarkable droplet size from 100 to 300 nm (Horn 1989; Paust 1991, 1994; Horn and Lüddecke 1996; Auweter et al. 1999; Köpsel 1999; Horn and Rieger 2001; Lüddecke & Schweikert, 1999; Lüddecke et al. 2004b). There are three different methods applied for producing carotenoid-loaded organic nanoparticles: Emulsification evaporation, emulsification diffusion, and solvent-displacement methods. Emulsification evaporation technique is applied when an apolar organic solvent is used. Solvent-displacement method is appropriated when a polar solvent (acetone, ethanol) is used as the dispersed phase to produce O/W nanoemulsions containing carotenoids. The nanodroplets formation process is related to the decreasing droplet size due to the rapid diffusion of solvent to the aqueous phase. Emulsification diffusion method, which uses amphiphilic solvent (benzyl alcohol, ethyl acetate) as the dispersed phase, could also be successfully applied to produce nanoemulsions containing carotenoids with remarkable narrow droplet size distributions and small droplets size. Solvent-displacement and emulsification diffusion methods are very low energy-input processes.

Ribeiro et al. (2008) have produced non-food grade nanoparticles containing β-carotene by interfacial deposition of the biodegradable polymers, poly(D,L-lactic acid) (PLA) and poly(D,L-lactic-co-glycolic acid) (PLGA), following displacement of acetone from the organic phase towards the aqueous phase (Fig. 8.18). In this kind of formulation, the polymer assumes the function of protective colloid, providing physico-chemical stabilization of the nanodispersed particles. The total β-carotene concentration in the dispersed phase was about 150 mg/L. The results showed that polymer-encapsulated β-carotene nanodispersions, with approximately 100% entrapment efficiency, could be prepared by the solvent-displacement method. Sauter mean diameters of the nanodroplets were found to hardly vary using gelatin or Tween® 20 as emulsifiers, with a $d_{3,2}$ of about 70 nm and narrow particle size distributions in both cases (Fig. 8.19). After freeze-drying, the particles were resuspended in water, showing no changes in the droplet sizes and their distribution.

Chu et al. (2007) have prepared milk protein-stabilized β-carotene nanodispersions using acetone. Particles stabilized by sodium caseinate have shown a bimodal size distribution from 13 to 171 nm. Morphological experiments confirmed the presence of particles stabilized by casein micelles and submicelles, which 88% of the β-carotene precipitated was stabilized by submicelles. Similar research work has been patented by Auweter et al. (2006), where carotenoids were suspended in an azeotropic isopropanol/water mixture. Temperature and pressure were applied for allowing the complete dissolution of carotenoid crystals. The homogeneous organic solution was mixed with sodium caseinate solution. Nanoparticles around 100 nm were produced, then filtered and freeze-dried.

Bortlik et al. (2005) patented a highly bioavailable whey protein-based formulation which may be applied to encapsulate carotenoids. Polar organic solvent were used for dissolving the active molecule. When lycopene was encapsulated into the

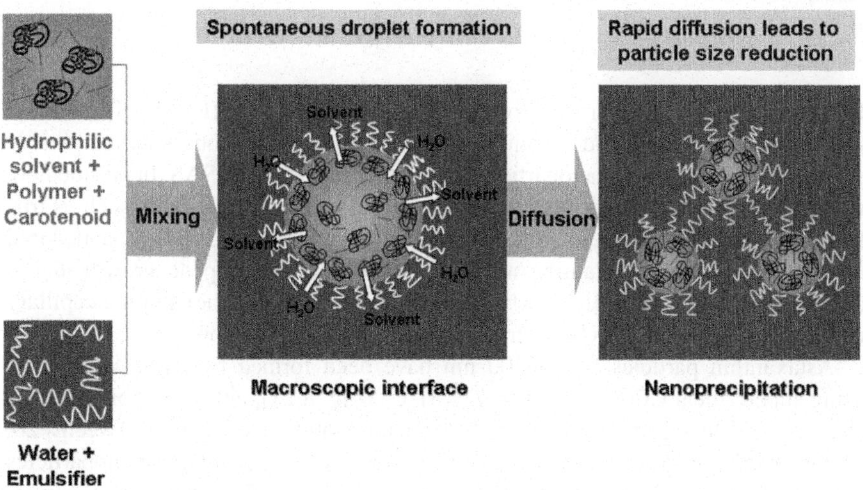

Fig. 8.18 Schematic representation of the spontaneous formation of β-carotene nanoparticles by solvent displacement method

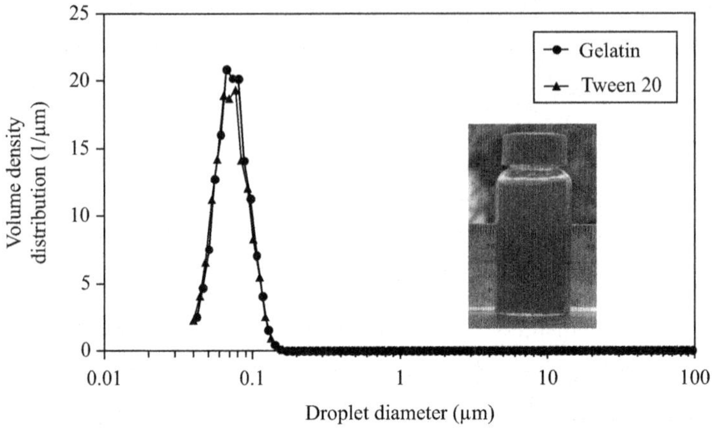

Fig. 8.19 Droplet size distribution of β-carotene nanoparticles stabilized with gelatin or Tween® 20, and appearance of a sample with gelatin (Ribeiro et al. 2008)

whey protein matrix, called "lactolycopene", it exhibited similar bioavailability in plasma and buccal mucosa cells in humans compared to tomato paste (Richelle et al. 2002).

Solvent evaporation technique has been used to microencapsulate astaxathin in a chitosan matrix cross-linked with glutaraldehyde (Higuera-Ciapara et al. 2004). Microcapsules were chemically stable over a storage time of 8 weeks.

These methods show some advantages, such as spontaneous particle formation, consequently low energy input, high entrapment efficiency, and high reproducibility.

8.2.4 Nanoparticles Produced by RESSAS Method

Rapid expansion of supercritical solution into aqueous solution (RESSAS) is a novel process for producing and stabilizing nanoparticles in aqueous solution. Stable suspension of carotenoid nanoparticles can be produced by RESSAS. In this method, a mixture of a poorly water soluble substance and a supercritical fluid, such as trifluormethane (CHF_3), carbon dioxide (CO_2), or nitrous oxide (N_2O), is expanded into an aqueous solution containing a surfactant, which may impede growth and/or agglomeration of nanoparticles which can happen due to collisions in the capillary nozzle into the expansion chamber (Fig. 8.20) (Türk and Lietzow 2004).

Astaxanthin particles below 300 nm have been formed by RESSAS process using supercritical CO_2 ($p_c = 74$ bar, $T_c = 31°C$) (Fig. 8.21), although their particles show a very low concentration in the aqueous solution containing Tween® 20. Similar results have been observed after production of astaxanthin nanoparticles using supercritical CHF_3, and N_2O. According to Horn and Rieger (2001), carotenoids show poor solubility in supercritical fluids, and they may also promote carotenoid oxidation. Sakaki (1992) investigated the solubility of β-carotene in

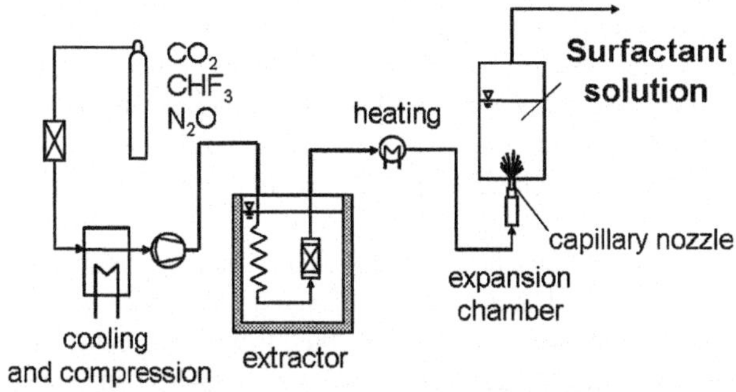

Fig. 8.20 Schematic RESSAS process (diameter of the capillary nozzle = 50 μm)

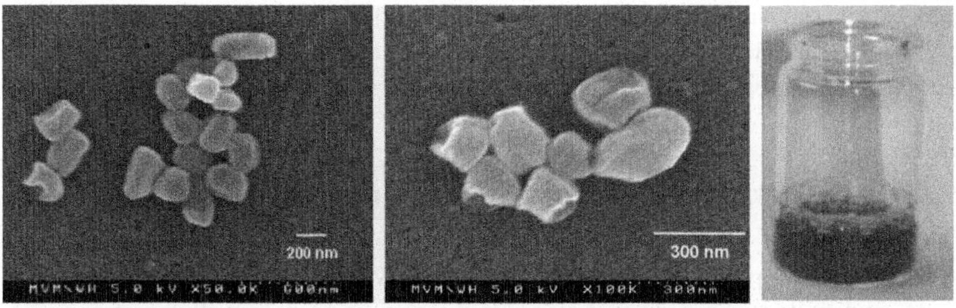

Fig. 8.21 SEM images of astaxanthin nanoparticles and appearance of the sample

supercritical CO_2 and N_2O at different pressure (9.64–29.5 MPa) and temperature (308–323 K) ranges. The maximal solubility of β-carotene in supercritical CO_2 and N_2O was about 5.3 and 66 g/m^3 at 29.5 MPa and 323 K, respectively.

8.2.5 Carotenoid Formulation Using Cyclodextrins

Cyclodextrins are able to serve as host molecules to encapsulate many food ingredients (Shahidi and Han 1993; Hedges et al. 1995), including carotenoids as well (Matioli and Rodriguez-Amaya 2003; Reuscher et al. 2004; Blanch et al. 2007). Complex inclusion of carotenoid with (the non-food grade) hydroxypropyl-β-cyclodextrin can also be successfully produced (Yuan et al. 2008). Cyclodextrins can be used to increase solubility of active molecules and improve their absorption, protect against light and oxygen, and may prevent off-flavor.

Supercritical fluid extraction and a conventional method using an amphiphilic organic solvent were used for entrapping lycopene into cyclodextrin molecules. Better encapsulation yield was observed by the conventional method using solvent (94% vs. 68%). β-Cyclodextrin–lycopene complex was chemically and physically stable over 6 months (Blanch et al. 2007). Reuscher et al. (2004) coated γ-cyclodextrin–lutein complexes with a vegetable oil or a biopolymer (e.g., hydroxypropyl methylcellulose) to enhance their stability against oxidation. γ-Cyclodextrin complexes contained approximately 5–25% carotenoid. They also claim that the cellular uptake of lutein was much higher from a cyclodextrin complex than when using micelles or liposome formulations.

Anantachoke et al. (2006) have established a method using a spinning disk device to produce controlled size nanoparticles containing β-carotene with maximal mean diameter of 82 nm. Either α- or β-cyclodextrin or sulfonato-calixarenes were used as stabilizer. Nanoparticles stabilized by cyclodextrins can be easily applied in food products.

An innovative carotenoid self-assembled rigid-rod β-barrel has been developed by Baumeister and Matile (2000). Carotenoids were entrapped in a hydrophobic interior of rigid-rod lipocalin, composed by oligo(*p*-phenylene) peptides. The encapsulation stoichiometry of peptides:carotenoid was 4:1 or 1:1 for β-carotene or astaxanthin, respectively.

8.2.6 Encapsulation of Carotenoids in Carbon Nanotubes

A hypothesis supported by a recent study has shown that degradation of β-carotene can be delayed after its encapsulation into the cavity of single-walled carbon nanotubes (SWCNT) (Yanagi et al. 2006). Although SWCNT cannot be applied in foods, the technique may contribute to other types of application of carbon nanotubes as carrier system of active molecules.

8.2.7 Encapsulation of Carotenoids in Microspheres

The entrapment of carotenoids in microspheres has not been frequently reported in the literature, most likely since it might not be the best choice to improve carotenoid stability and bioavailability.

Furcellaran beads have been reported as an efficient matrix to encapsule β-carotene (Laos et al. 2007); however, a low concentration of β-carotene has been achieved in this system. Beads were produced by furcellaran gelation. First, hot furcellaran solution (4–8%) with sea buckthorn juice was dropped into a 1.3 M salt solution containing either $CaCl_2$ or KCl, allowing its complete gelation. Afterwards, the spherical gel beads (d=4.6 mm) were filtered and washed. KCl solution provided firmer gel beads than $CaCl_2$ solution.

Dunaliella salina, a pink micro-algae rich in β-carotene, was encapsulated in calcium alginate beads and rapidly dried by fluidized bed (Leach et al. 1998). The dry process caused isomerization of *all-trans* carotenoids, which is expected due to temperature sensitivity showed by carotenoids in general. Alginate concentration played an important role on carotenoid stability over storage time.

Kittikaiwan et al. (2007) have developed a novel encapsulation method using chitosan (80 mol% deacethylation) to encapsulate *Haematococcus pluvialis*, a rich source of astaxanthin. *H. pluvialis* beads were coated five times with chitosan solution forming multilayers of chitosan film. Chitosan capsules containing the algal have shown uniform shape with diameters from 0.35 to 0.5 cm and 100 μm film thickness. Storage studies on chemical stability have proven that chitosan as wall material can prevent degradation of astaxanthin.

8.2.8 Production of Microcapsules by Drying Processes

Drying can also be considered as an important unit operation to transform moist or fluid carotenoid formulations into dry solids. It allows the production of carotenoid-microcapsules of a certain structure, size, and shape (Horn et al. 1985; 1988; Cathrein et al. 1991; Vilstrup et al. 1997). Colombo and Gerber (1991) have reported several drying processes to encapsulate carotenoids and fat-soluble vitamins, such as drum-cooling and spraying processes. In this section, the use of spray-drying and freeze-drying to obtain dry carotenoid formulations are discussed in more detail.

8.2.8.1 Spray-drying

Rodríguez-Huezo et al. (2004) have successfully produced dried O/W and W/O/W multiple emulsions containing oleoresin from red chilies and marigold petals by spray drying. Gellan and mesquite gum, maltodextrin, and gum arabic were used as wall materials. Morphological studies have shown microcapsules with large central voids when W/O/W emulsions were prepared with high solid contents (35%); they have shown inner and outer particles sizes ($d_{4,3}$) of 1.5 and 34 μm, respectively.

Reuscher et al. (2004) spray-dried complexes of lutein and γ-cyclodextrin, followed by spraying with a hydroxypropyl methylcellulose coating or dispersion in oil. They claimed enhanced stability of the carotenoid against oxidation by these oil or biopolymer coating.

Spray-dried lycopene-loaded O/W emulsion stabilized with a gum arabic and sucrose mixture (8:2) has been studied as photostabilizer of vitamins A and D3 in skimmed milk (Montenegro et al. 2007). Vitamins degradation has slowed down by 45% in presence of lycopene formulation.

Loksuwan (2007) has encapsulated β-carotene into three different wall materials: Acid-modified and native tapioca starch, and maltodextrin. Use of modified tapioca starch provided higher retention of β-carotene (82%) during spray drying process and resulted in a wider particle size distribution (75–150 μm) but smaller particle size than native tapioca starch and maltodextrin. Chen et al. (2004) have patented a spray-dried formulation in which carotenoids and other fat-soluble molecules have been successfully entrapped in a carbohydrate matrix. Another patent

(Leuenberger et al. 2006) describes the process for preparing beadlets containing lipophilic compounds, for example carotenoids. Emulsions stabilized by gelatin were produced and converted to dry powder using spray drying. Afterwards, the gelatin matrix in the coated particles was cross-linked by exposure to radiation. In another process, after preparing emulsions, a cross-linking enzyme was added, then emulsions were spray-dried and the gelatin matrix was cross-linked by incubating.

Lycopene macrocapsules with a combination of gelatin and sucrose as wall materials have been produced by Shu et al. (2006). Emulsions were previously prepared by high-pressure homogenization and then spray-dried. Encapsulation yield and efficiency varied with the ratio of core and wall (M_{core}/M_{wall}), ratio of gelatin and sucrose (M_g/M_s), homogenization pressure, temperatures during drying process, and lycopene purity. The highest values of encapsulation yield and efficiency found were about 91% and 82%, respectively. Lycopene slightly degraded after spray drying.

Santos et al. (2005) have encapsulated paprika oleoresin in either a combination of rice starch/gelatin or arabic gum microcapsules with particle sizes of 20 μm or 16 μm, respectively. Rice starch/gelatin as wall materials resulted into porous microparticles, which provided low effective barrier against carotenoid oxidation.

New combinations of carbohydrates and proteins have also been investigated to encapsulate carotenoids. For example, Feldthusen et al. (2005) firstly dissolved astaxanthin in organic solvent and then mixed it with an aqueous phase containing one of the following combinations: trehalose/sodium caseinate, lactose/sodium caseinate, lactose/soy bean protein, or glucose syrup/sodium caseinate. Afterwards the dispersions were spray-dried and the dried product contained 10–25% astaxanthin. Polysaccharides, such as acacia gum, pectin, cellulose, cellulose derivates, and/or modified polysaccharides, are other examples of wall materials of microcapsules containing carotenoids (Leuenberger et al. 2008). Milling and spray-drying processes were used for the preparation of the encapsulated systems.

8.2.8.2 Freeze-Drying

β-Carotene encapsulated in a trehalose-gelatin matrix with and without magnesium chloride ($MgCl_2$) by freeze-drying was studied by Elizalde et al. (2002). First, a coarse β-carotene-loaded O/W emulsion was prepared using a rotor-stator system, and then it was frozen in liquid nitrogen (−190°C) before freeze-drying. Although $MgCl_2$ was added to delay trehalose crystallization and improve microcapsule stability against enzymes, no positive effect was observed. Amorphous phase played an important role on carotenoid stability, and a noncrystallized sample showed 80% retention of β-carotene for at least 6 months. Crystallization of the sugar matrix was intimately related to the loss of β-carotene.

Chiu et al. (2007) reported an interesting encapsulation study done with 4.8% lycopene extract from tomato pulp waste. 4.5% Gelatin and 10% poly(γ-glutamic acid) (γ-PGA) were applied as carrier matrix. Results showed that microcapsules around 39 μm were obtained, and freeze-drying caused a loss of 23.5% lycopene. As expected, different pHs affected the total lycopene release and isomers profile.

In another example, using a combination of drying processes, O/W emulsion containing β-carotenoid stabilized by low methoxylated pectin was atomized into aqueous solution of calcium salts. The microgels were then centrifuged and freeze-dried to yield the final product containing about 2% of β-carotene (Carle et al. 2004). Desobry et al. (1997) compared three drying processes to encapsulate β-carotene with maltodextrin. Before drying, β-carotene was dispersed in 40% maltodextrin solution and high-pressure homogenized three times at 253 MPa. Spray, drum, and freeze-drying provided particle sizes of approximately 30 μm, 105 μm, and 80 μm, respectively. These processes operate at different residence times and temperatures, consequently distinct dried structures were obtained. The highest β-carotene degradation (14%) was observed with drum-drying. Freeze-drying provided less loss of the carotenoid, about 8%.

8.3 Carotenoid Product Formulations

A wide range of processes to encapsulate carotenoids have been previously discussed in Sect. 8.2. Nutritional enhancement of food products with carotenoids will only be useful if they are able to be taken up by the body.

The bioavailability of carotenoids might depend on the following:

(a) Chemical and physical stability of the carotenoids upon storage within the food product
(b) Stability of the carotenoids during the gastro-intestinal tract passage
(c) Uptake by intestinal cells

The next sections will review these three aspects, using emulsion and liposome formulations as examples.

8.3.1 Chemical and Physical Stability of Carotenoids in Food Formulations

Experiments investigating both physical and chemical stability of carotenoid emulsions have been carried out. Droplet size and its distribution is an important characteristic of carotenoid dispersions, which is associated with many properties, such as shelf-life stability, homogeneity, creaminess, viscosity, carotenoid bioavailability, and color. Small oil droplets with supersaturated carotenoids may be kept stable over long storage time (Figs. 8.22 and 8.23), as well as liposomes (Matsushita et al. 2000).

Food products containing carotenoids and providing health benefits by improving antioxidant levels in individuals, so-called "functional foods", are, for example, beverages, breads, candies, cookies, dairy products (ice-cream, cheddar cheese, etc.), dry cake mix, salad dressings, macaroni products, mayonnaise, savory, shortenings, soups, and spreads (Bauernfeind et al. 1958; Bunnell et al. 1958; Garnett et al. 2003;

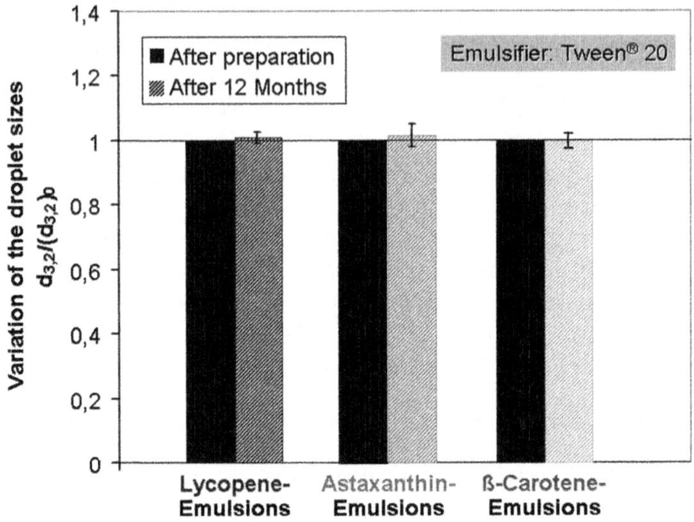

Fig. 8.22 Emulsion stability: Variation of emulsion droplet size (sauter diameter) over 1 year storage time. $d_{3,2o}$ = initial sauter diameter

Leuenberger 2007; Manz 1967; Santipanichwong and Suphantharika 2007; Upritchard et al. 2003). When carotenoid emulsion-based formulations are mixed with products as foods, they have to remain stable even under chemical and physical conditions changed.

The kinetics of carotenoid degradation follows a first-order kinetic model (Henry et al. 1998; Ax et al. 2003). Chemical stability of lycopene-loaded O/W emulsions is given in Fig. 8.24, as an example. The carotenoid formulations were diluted in three different products: skimmed milk, orange juice, and water (control). Three different emulsifiers were used in these investigations. It was found that lycopene stability strongly depends on the food system (Fig. 8.25). In orange juice ($pH < 4.5$), the lycopene was particularly stable, whereas in milk ($pH \cong 6.6$) and water (pH 7), respectively, lycopene concentration decreased quite fast. Stability of lycopene did not depend on the emulsifier type. Addition of α-tocopherol to the dispersed oil phase, however, allowed for prolonged lycopene stability in all three food systems (Fig. 8.26). Antioxidants can be used to protect O/W emulsions against degradation over a long period of time. As can be seen in Fig. 8.27, coalescence of oil droplets was not observed in any of the systems investigated over 3 weeks (Ribeiro et al. 2003).

O/W emulsions containing lycopene and vitamin C, and stabilized by Tween® 20 was also applied in full- and low-fat yogurt, 3.8% and 1.5% fat, respectively. The initial concentration of lycopene in each dairy product was 10 mg/L. It remains stable for 3 weeks in the full-fat yogurt, however, approximately 20% of lycopene degradation could be observed in the low fat one over the same storage time.

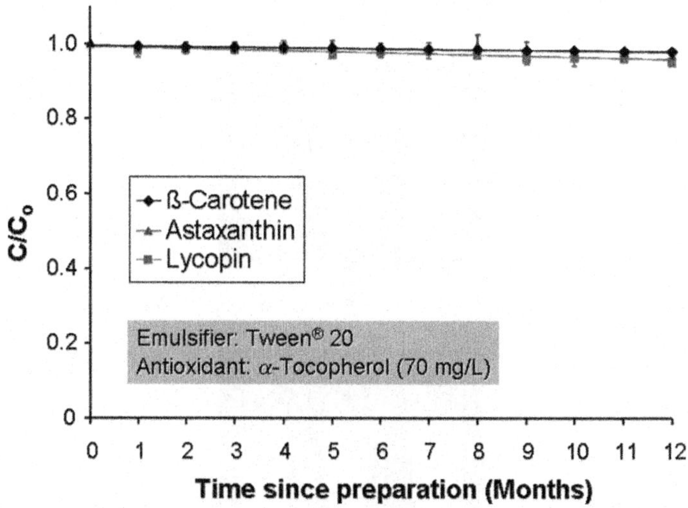

Fig. 8.23 Carotenoid stability in O/W nanoemulsions: Concentration of different carotenoids in nanoemulsions as a function of storage time and α-tocopherol concentration

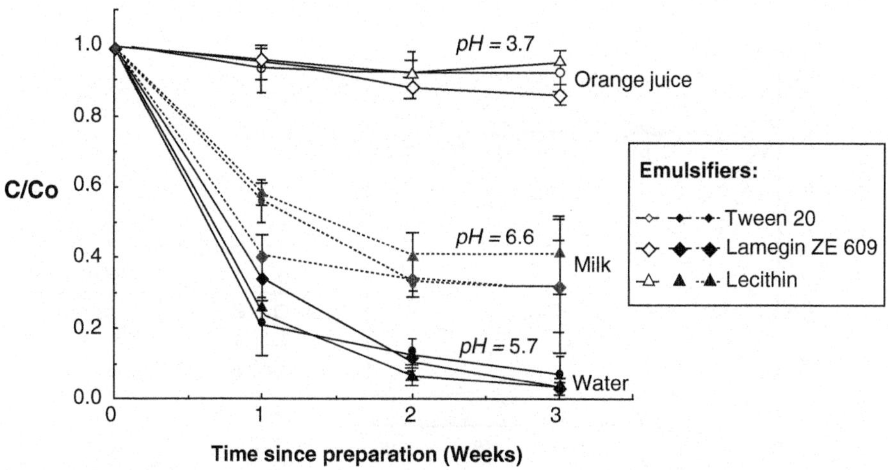

Fig. 8.24 Influence of food matrix on the chemical stability of lycopene without α-tocopherol. C = concentration and C_0 = initial concentration. The *bars* represent standard deviations (Ribeiro et al. 2003) (Courtesy of Wiley-Blackwell Publishers)

By combining emulsifiers a positive influence on carotenoid stability in nanoemulsion-based formulations was found. Ribeiro et al. (2006b) have shown decreased lycopene and astaxanthin degradation in O/W emulsions stabilized with a combination of emulsifiers, especially for a combination of a whey protein isolate and a sugar ester, compared to emulsions prepared with a single emulsifier.

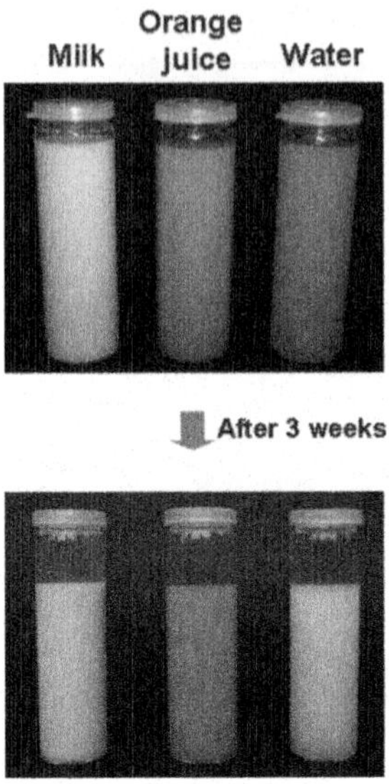

Fig. 8.25 Appearance of the three model systems after incorporation of O/W emulsion containing lycopene and after 3-weeks storage. Emulsifier: Tween® 20

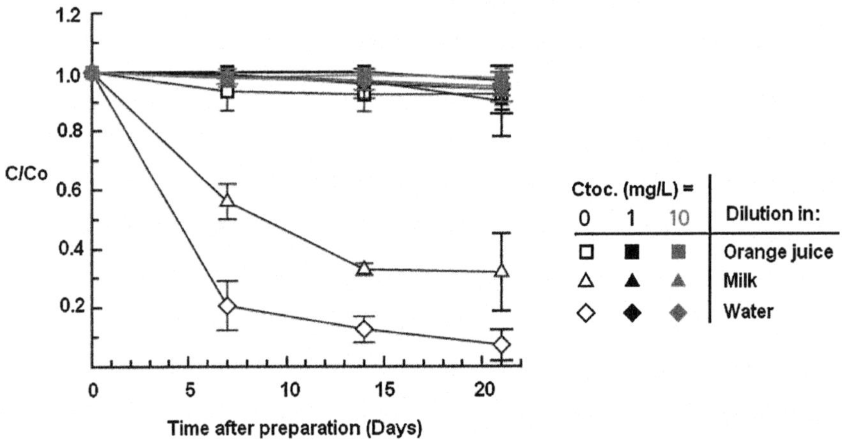

Fig. 8.26 Influence of tocopherol content and food matrix on lycopene stability in O/W emulsions during storage at 4°C. C=concentration and C_0 = initial concentration. The *bars* represent standard deviations (Ribeiro et al. 2003)

Physical stability of carotenoid loaded nanoemulsions includes stability against droplet coalescence and flocculation or aggregation. Droplets can be stabilized against these by steric hindrance of long emulsifier molecules adsorbed at the interfaces and/or by electrical charges. The droplets zeta-potential is a measure of stabilizing electrostatic interaction forces between the droplets. The droplet zeta-potential of O/W emulsions containing lycopene, prepared with combination of whey protein isolates (BiPro®, BioZate®), with either Tween® 20 or sucrose laurate (L-1695), two non-ionic emulsifiers, as a function of pH is shown in Fig. 8.28. At the lowest pH the droplets had a relatively high positive zeta-potential because the pH was below the isoelectric point of the whey proteins. When the pH was increased, the magnitude of the positive charge on the droplets decreased. Eventually the zeta-potential of the droplets became zero (isoelectric point at a pH between 4 and 5). A further increase in pH caused the droplets to gain a net negative charge.

Droplet aggregation of the O/W emulsions was observed to occur around the isoelectric point (pH 4–5). Figure 8.29 shows the lycopene-loaded O/W emulsions in different pH ranges (2–9) 4 and 24 h after changing the pH, respectively. Droplet agglomeration could be observed after 24 h in samples at pH 4. Droplet flocculation/aggregation caused a significant increase in the creaming of the droplets in the emulsions. Aggregation and creaming stability of whey protein stabilized emulsions has been previously found to depend strongly on pH (Phillips et al. 1994). Emulsion droplets were highly susceptible to aggregation near the isoelectric point of the whey proteins because of the relatively low electrostatic repulsion between the droplets. Samples prepared with astaxanthin showed the same physical and chemical stability behavior. The droplet size of the emulsion containing lycopene and using the emulsifiers L-1695 and BioZate® is shown in the Fig. 8.30 in a pH range between 2 and 9. The measurements were carried out 4 h after changing the pH. It can be seen that droplet sizes increased around the isoelectric point, which means that the emulsions were unstable at this pH due to droplet agglomeration.

8.3.2 *Stability of Carotenoid Formulations During Passage Through the In Vitro Gastro-Intestinal Tract*

The in vivo efficacy of formulations containing functional ingredients depends on the one hand, on the stability of the emulsions during processing and storage, and on the other hand, on their behavior while passing through the gastro-intestinal tract. After oral ingestion of the emulsions, the chemical and enzymatic conditions in the stomach and small intestine can lead to extensive chemical and physical changes in the emulsion system and thereby influence the absorption of the active substance at the area of resorption. To simulate the influence of the physiological conditions during digestion, a model system (Fig. 8.31) was used. This model consists of a stomach section and a small-intestine compartment. Simulated gastric, biliary, and pancreatic secretions are introduced into the corresponding compartment. The pH is monitored and controlled to the specific values for the stomach and small intestine. The temperature is kept at 37°C. As a basis for the preparation of the digestive secretions the systems described by Minekus (1998) and the German Institute for Standardization (DIN 19738 2004) (see Table 8.1)

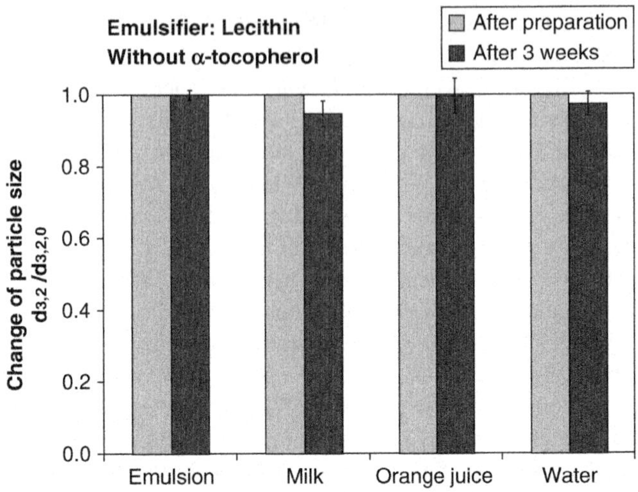

Fig. 8.27 Changes in the sauter diameter ($d_{3,2}$) over 3 weeks. $d_{3,2_0}$ = initial sauter diameter. The *bars* represent standard deviations

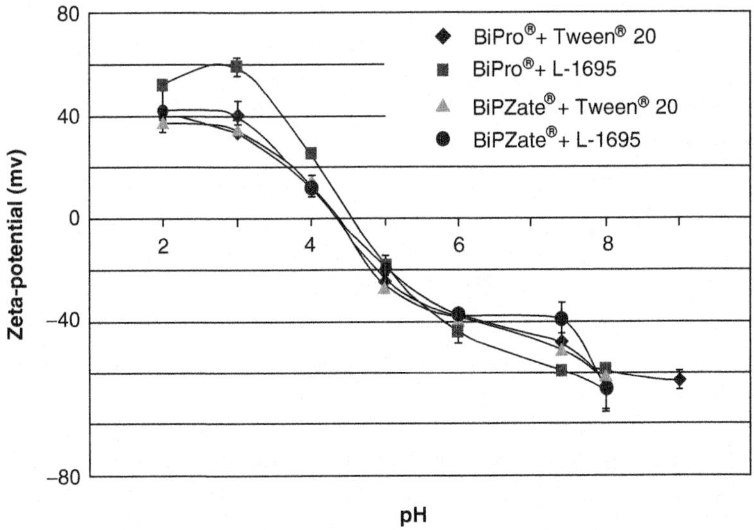

Fig. 8.28 Droplet zeta-potential as function of the pH for O/W emulsions containing lycopene

were adopted. To study the effect of digestion on the stability of different formulations, two test samples were investigated: an emulsion stabilized by Tween® 20 and a sample of liposomes composed of Phospholipon 80®, both containing lycopene (Ax 2003).

As an indicator of the stability of the formulations against aggregation or coalescence, particle size distributions and zeta potential were measured. For carotenoid emulsions, the zeta potential changed from −35 mV at the beginning to about −20 mV after stomach incubation. The zeta potential of liposomes increased from −60 mV to −20 mV. In Fig. 8.32, particle size distributions of untreated carotenoid emulsions and liposomes as well as formulations after incubation are shown. No significant change in droplet size distribution after processing in the stomach was observed in any of the investigated carrier systems.

Under the conditions in the small intestine, dispersed lipid systems were disintegrated under the influence of pancreatic lipase enzymes. Addition of pancreatic lipase to the emulsion led, as a result of lipolysis, to a prompt decline in pH. The appearance of the emulsions changed from opaque to translucent, they darkened in color, and viscosity changes occurred. All these observations indicate that

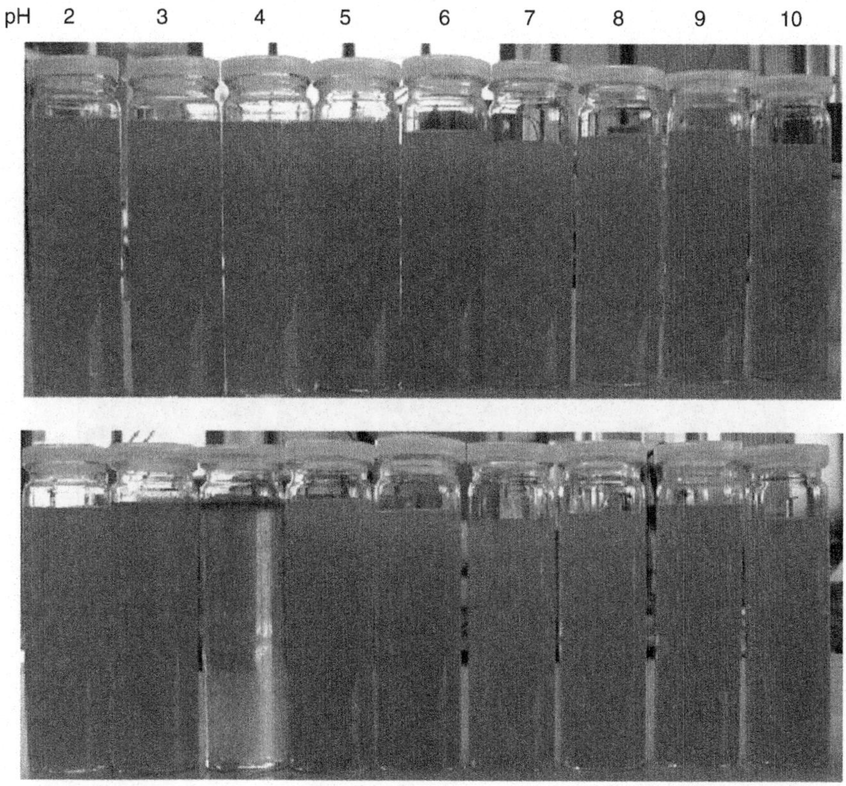

Fig. 8.29 Appearance of diluted astaxanthin-loaded O/W emulsions stabilized by sucrose laurate (L-1695) + BioZate® after changing the pH and over 24 h. In both figures, the emulsion on the left is the emulsion with pH 2.0. The pH increases successively from left to right, with the emulsion with pH 9.0 on the right

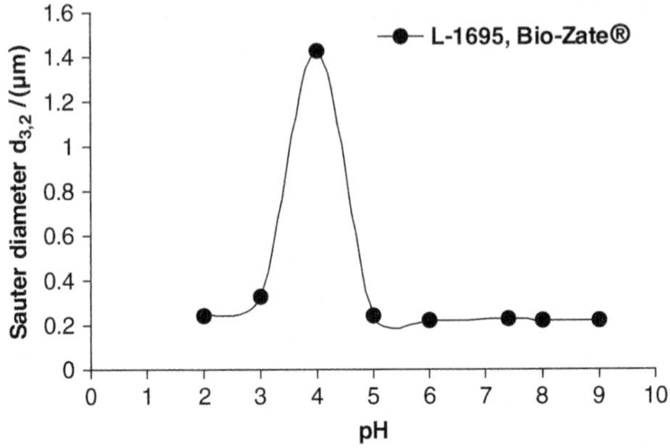

Fig. 8.30 Sauter diameter of the emulsion containing lycopene stabilized by sucrose laurate and BioZate® in different pH

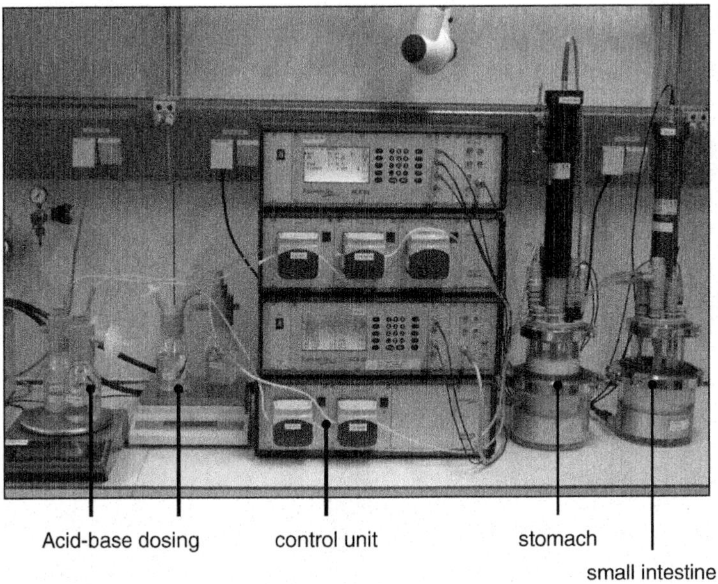

Fig. 8.31 In vitro gastrointestinal tract model

the dispersed oil droplets were destroyed. Liposome incubation under the conditions of the upper intestine resulted in a comparatively slow reaction. The liposome formulation became more transparent and acquired a more viscous, gel-like consistency.

Table 8.1 Modeling of the gastrointestinal conditions

	Stomach	Intestine
Temperature (°C)	37	37
pH	1–5	7–9
Dilution (dispersion:secretion)	3:5	1:4
Electrolyte	NaCl; KCl; KH_2PO4	NaCl; KCl; $CaCl_2 \cdot 2H_2O$; Bile acid
Enzyme	Pepsin	Pancreatin
Incubation time (h)	3	5

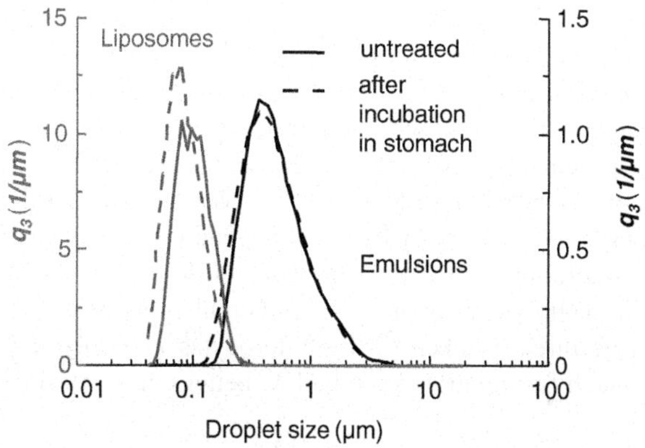

Fig. 8.32 Influence of stomach digestion on carotenoid emulsion and liposome (Ax 2003) (Courtesy of Shaker Verlag)

8.3.3 Bioavailability and Cellular Uptake of Carotenoids

Intestinal carotenoid absorption in human depends on a number of factors such as the carotenoid species, molecular linkage, the amount of carotenoids consumed in a meal, the matrix in which the carotenoid is incorporated, effectors of absorption, the nutrient status of the host, genetic factors, host-related factors, and the interaction of these factors (Castenmiller and West 1998; Parada and Aguilera 2007).

This complexity is responsible for the large inter-individual variability in bioavailability and plasma concentration of carotenoids observed in many intervention studies.

Fruit and vegetables and their products are the major dietary source of carotenoids. In plant tissue, carotenoids are localized in cellular plastids, where carotenoids are associated with light-harvesting complexes or crystalline structures. Carotenoids are generally insoluble in water. In the small intestine carotenoids are extracted from the food matrix, solubilized in bile acid micelles and absorbed by the enterocytes. The intracellular localization, form of carotenoid (free, protein-bound, crystalline), and intactness of the plant cellular matrix are important determinants of carotenoid bioavailability. There are pronounced differences in the relative bioavailability of carotenoids from different fruit and vegetables and their products, for example, the consumption of fruits resulted in plasma concentrations of β-carotene which were six times higher than those found after consumption of green leafy vegetables (de Pee et al. 1998). Disruption of the plant tissue matrix by mechanical homogenization or thermal treatment enhances the bioavailability of carotenoids (Gärtner et al. 1997; Porrini et al. 1998). Physico-chemical properties of carotenoid species are also important for bioavailability. The bioavailability of lutein, which is more polar than β-carotene, is fivefold to ninefold higher than that of β-carotene from vegetables such as spinach. Dietary fat (3–5 g per meal) and endogenous emulsifiers (bile acids) are necessary to incorporate carotenoids released from the food matrix into mixed micelles. It appears that 3–5 g of fat is sufficient for carotenoid absorption (van het Hof et al. 2000). All these factors which depend on food, food treatment and composition, secretion of digestive enzymes and bile acids into the small intestine make it difficult to predict the plasma concentration (relative bioavailability) for an individual person on the basis of his/her fruit and vegetable consumption. The same amount of fruit and vegetables can result in plasma concentrations with a difference of one order of magnitude. Carotenoid formulations, such as emulsions, supply carotenoids, fat, and emulsifiers in a matrix from which carotenoids can be directly taken up by enterocytes or easily incorporated into mixed micelles. Such formulations could allow better control of the bioavailability of carotenoids.

In vitro cell culture models make it possible to investigate the uptake of carotenoids by intestinal cells and to avoid the effects of some factors which are difficult to control: efficacy of food matrix disruption, solubilization of carotenoids, and incorporation into mixed micelles (Briviba et al. 2001).

There are several ways to solubilize carotenoids in water in order to investigate the cellular uptake or effects of carotenoids in cell culture model systems. Carotenoids dissolve readily in apolar solvents such as chloroform, acetone, or tetrathydrofuran (THF). THF is often used to solubilize carotenoids in water. A final concentration of 0.5% (v/v) of tetrahydrofuran in water is high enough to solubilize most of carotenoids at concentrations of up to 10 μM (Bertram et al. 1991). This THF concentration usually did not affect viability in most cell lines. Cells incubated with astaxanthin delivered in THF at the same concentration as β-carotene (10 μM) showed about a 50-fold higher amount cellular amount of astaxanthin, as detected by HPLC. These observations were also confirmed by confocal resonance Raman microspectroscopy image analysis – a strong Raman signal of astaxanthin but no detectable signal of β-carotene was found in cells (Briviba et al. 2006).

Incubation of HT 29 cells with O/W emulsions prepared using Tween® 20 as an emulsifier and loaded with 10 μM β-carotene or astaxanthin resulted in an increase in the cellular concentration of carotenoids. A slight dependence of the cellular carotenoid concentration on droplet size was observed (Fig. 8.33). The cellular uptake of

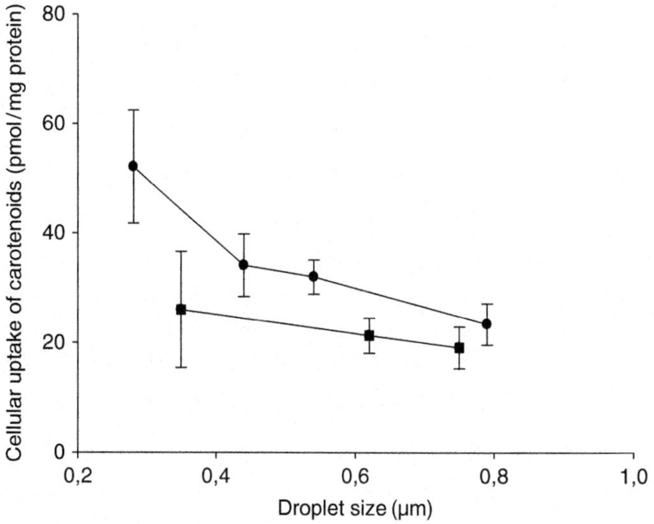

Fig. 8.33 Effect of droplet size of O/W emulsions loaded with carotenoids on cellular uptake of astaxanthin (*squares*) or β-carotene (*circles*)

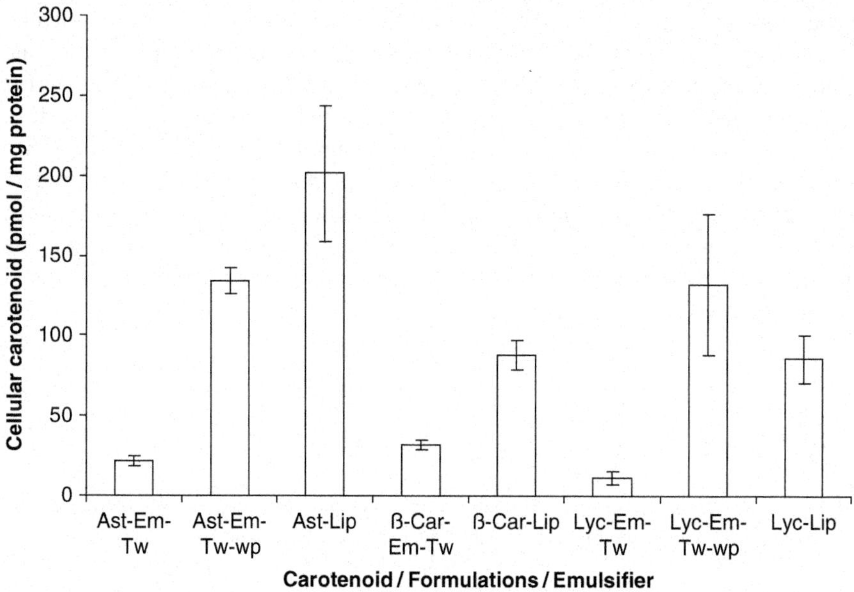

Fig. 8.34 Effect of carrier type and carotenoid species on cellular uptake

astaxanthin and β-carotene from O/W emulsions was comparable. The cellular uptake of carotenoids from emulsions depends on the emulsifiers used. Whey proteins combined with Tween® 20 or sucrose laurate significantly increased carotenoid (astaxanthin, lycopene) uptake from emulsions (Ribeiro et al. 2004; Ribeiro et al. 2006b). Effective cellular uptake of carotenoids (astaxanthin, β-carotene, lycopene) from carotenoid-enriched liposomes prepared from phosphatidylcholine was observed. Carotenoid uptake from liposomes was more effective than from Tween® 20 emulsions, but comparable to or lower than that derived from whey protein/Tween® 20 emulsions (Fig. 8.34). Thus, the uptake of carotenoids by the intestinal cells depends on carotenoid, species, carrier type (emulsions, liposomes), and the chemical ingredients of the formulation (e.g., whey proteins). HT 29 cells line were treated with cell culture medium supplemented with either carotenoid-enriched O/W emulsions with emulsifier Tween® 20 (Em-Tw) or emulsions with a combination of two emulsifiers Tween® 20 and whey protein isolate (Em-Tw-wp) or liposomes (Lip). The whey protein isolate used was BioZate®. Cell culture medium was replaced daily. Cellular levels of carotenoids were estimated after incubation for 72 h. Carotenoids were extracted and determined by reversed-phase HPLC.

8.4 Final Considerations

This chapter has shown a wide range of process to encapsulate carotenoids. Optimal powder or liquid carotenoid formulations allow the production of a physically/chemically stable and bioavailable structure after their preparation, storage, and application in food systems. These requirements are very important for delivering the carotenoid into the gastro-intestinal tract, and later for being absorbed. Encapsulation of carotenoids may meet these requirements. Other criteria to select the right type of encapsulates might be food grade processes, easiness of production, application in dry or liquid products, clarity/transparence of liquid product, large volume production, costs, IPR situation, etc. The ideal encapsulation process and type of formulation will depend on specific application, market requirement, and current regulations.

References

Aguilar FAP, Freudig B, Schuchmann HP (2004) Herstellen von Emulsionen in Hochdruckhomogenisatoren mit modifizierten Lochblenden. Chemie Ingenieur Technik 4:396–399

Aguilar FAP, Köhler K, Schubert H, Schuchmann HP (2008) Herstellen von Emulsionen in einfachen und modifizierten Lochblenden: Einfluss der Geometrie auf die Effizienz der Zerkleinerung und Folgen für die Maßstabsvergrößerung. Chemie Ingenieur Technik 80:607–613

Anantachoke N, Makha M, Raston CL, Reutrakul V, Smith NC, Saunders M (2006) Fine tuning the production of nanosized β-carotene particles using spinning disk processing. J Am Chem Soc 128:13847–13853

Auweter H, Haberkorn H, Hechmann W, Horn D, Lüddecke E, Rieger J, Weiss H (1999) Die supramolekulare Struktur ausgefällter, nanometergroβer β-Carotinpartikel. Angewandte Chemie 111:2325–2328

Auweter H, Bohn H, Lüddecke E, Hinz W, Runge F, Pfeiffer A-M (2006) Production of solid preparations of water-soluble, sparingly water-soluble or water-insoluble active compounds. US Patent 7,105,176 B2

Ax K (2003) Emulsionen und Liposomen als Trägersysteme für Carotinoide. Doctoral dissertation, University of Karlsruhe. Shaker Verlag, Germany

Ax K, Lambrich U, Schubert H (2000) Zerkleinerung von Liposomen in mechanischen Homogenisierapparaten. Lebensmittelverfahrenstechnik 45:277–280

Ax K, Schubert H, Briviba K, Rechkemmer G, Tevini M (2001) Oil-in-water emulsions as carriers of bioavailable carotenoids. In: PARTEC 2001, Nuremberg, 27–29 March 2001

Ax K, Mayer-Miebach E, Link B, Schuchmann H, Schubert H (2003) Stability of lycopene in oil-in-water emulsions. Eng Life Sci 3:199–201

Barenholz Y, Diminsky D, Cohen R (2006) Carotenoid-loaded liposomes. US Patent 7,048,943 B2

Bauernfeind JC, Smith EG, Bunnell RH (1958) Coloring fat-base foods with β-carotene. Food Technol 12:527–535

Baumeister B, Matile S (2000) Rigid-rod β-barrels as lipocalin models: Probing confined space by carotenoid encapsulation. Chem Eur J 6:1739–1749

Bertram JS, Pung A, Churley M, Kappock TJ, Wilkins LR, Cooney RV (1991) Diverse carotenoids protect against chemically induced neoplastic transformation. Carcinogenesis 12:671–678

Blanch GP, Castillo MLR, Caja MM, Pérez-Méndez M, Sánchez-Cortés S (2007) Stabilization of *all-trans*-lycopene from tomato by encapsulation using cyclodextrins. Food Chem 105:1335–1341

Bortlik K, Saucy F, Duruz E, Richelle M, Lambelet P, Baur M, Pfeifer AMA (2005) Primary composition comprising a lipophilic bioactive compound. US 2005/0106219 A1

Briviba K, Schnäbele K, Schwertle E, Blockhaus M, Rechkemmer G (2001) β-Carotene inhibits growth of human colon carcinoma cells in vitro by induction of apoptosis. Biol Chem 382:1663–1668

Briviba K, Kulling SE, Möseneder J, Watzl B, Rechkemmer G, Bub A (2004) Effects of supplementing a low-carotenoid diet with a tomato extract for 2 weeks on endogenous levels of DNA single strand breaks and immune functions in healthy non-smokers and smokers. Carcinogenesis 25:2373–2378

Briviba K, Bornemann R, Lemmer U (2006) Visualization of astaxanthin localization in HT29 human colon adenocarcinoma cells by combined confocal resonance Raman and fluorescence microspectroscopy. Mol Nutr Food Res 50:991–995

Buijsse B, Feskens EJM, Kwape L, Kok FJ, Kromhout D (2008) Both α- and β-carotene, but not tocopherols and vitamin C, are inversely related to 15-year cardiovascular mortality in dutch elderly men. J Nutr 138:344–350

Bunnell RH, Driscoll W, Bauernfeind JC (1958) Coloring water-base foods with β-carotene. Food Technol 12:536–541

Carle R, Schieber A, Mutter S (2004) Novel composition comprising carotenoids. WO 2004/066750 A1

Castenmiller JJ, West CE (1998) Bioavailability and bioconversion of carotenoids. Annu Rev Nutr 18:19–38

Cathrein E, Stein H, Stoller H, Viardot K (1991) Verfahren zur Herstellung von Carotinoidpräparaten. EP 0410236

Chen, C.-C., Leuenberger, B., Zedi, E. (2004). Compositions containing fat-soluble substances in a carbohydrate matrix. EP 1 066 761 B1

Chiu YT, Chiu CP, Chien JT, Ho GH, Yang J, Chen BH (2007) Encapsulation of lycopene extract from tomato pulp waste with gelatin and poly(γ-glutamic acid) as carrier. J Agric Food Chem 55:5123–5130

Christen WG, Liu S, Glynn RJ, Gaziano JM, Buring JE (2008) Dietary carotenoids, vitamins C and E, and risk of cararact in women – A prospective study. Arch Ophthalmol 126:102–109

Chu B-S, Ichikawa S, Kanafusa S, Nakajima M (2007) Preparation and characterization of β-carotene nanodispersions prepared by solvent displacement technique. J Agric Food Chem 55:6754–6760

Colombo VE, Gerber F (1991) Structures and properties of stabilized vitamin and carotenoid dry powders. Food Struct 10:161–170

Cook EJ, Lagace AP (1985) Apparatus for forming emulsions. US Patent 4,533,254

Cook EJ, Lagace AP (1990). Method of forming a microemulsion. US Patent 4,908,154

de Pee S, West CE, Permaesih D, Martuti S, Hautvast JG (1998) Orange fruit is more effective than are dark-green, leafy vegetables in increasing serum concentrations of retinol and beta-carotene in schoolchildren in Indonesia. Am J Clin Nutr 68:1058–1067

Desobry SA, Netto FM, Labuza TP (1997) Comparison of spray-drying, drum-drying and freeze-drying for β-carotene encapsulation and preservation. J Food Sci 62:1158–1162

DIN 19738 (2004) Bodenbeschaffenheit – Resorptionsverfügbarkeit von organischen und anorganischen Schadstoffen aus kontaminiertem Bodenmaterial. Normenausschuss Wasserwesen (NAW) im DIN

Elizalde BE, Herrera MI, Buera MP (2002) Retention of β-carotene encapsulated in a trehalose-based matrix as affected by water content and sugar crystallization. J Food Sci 67:3039–3045

End L, Haber B, Schulte S, Obermüller-Jevic U (2007) Nanoformulations of health ingredients. In: PARTEC 2007, Nuremberg

Engel R, Ribeiro HS, Briviba K, Rechkemmer G, Schuchmann HP, Schubert H (2005) Untersuchungen zur Formulierung von Carotinoiden und Phytosterolen in Mizellen zur Verbesserung der ernährungsphysiologischen Wirksamkeit. German Nutr Soc 7:82

Engels T, Förster T, von Rybinski W (1995) The influence of coemulsifier type on the stability of oil-in-water emulsions. Colloids Surf A 99:141–149

Farhang B (2007) Nanotechnology and lipids. Lipid Technol 19:132–135

Feldthusen JJ, Auweter H, Habich A, Lüddecke E, Pfeiffer A-M (2005) Method for producing dry powders of one or several carotenoids. WO 2005/075385 A2

Feldthusen JJ, Köpsel C, Schuchmann HP, Engel R, Ribeiro HS, Schubert H, Ax K (2007) Emulsifier system, emulsion and use thereof. BASF AG, Ludwigshafen, Germany. WO 2007/003599

Flanagan J, Singh H (2006) Recent advances in the delivery of food-derived bioactives and drugs using microemulsions. In: Mozafari MR (ed) Nanocarrier technologies: Frontiers of nanotherapy. Springer, The Netherlands

Freudig, B. (2004). Herstellen von Emulsionen und Emulgieren von Milch in modifizierten Lochblenden. Doctoral dissertarion, University of Karlsruhe. Shaker Verlag, Germany

Garnett KM, Guerra-Santos LH, Gierhart DL (2003) Zeaxanthin formulatin for human ingestion. US Patent US2003108598A1

Garti N, Aserin A (2007) Nanoscale liquid self-assembled dispersions in foods and the delivery of functional ingredients. In: McClements DJ (ed) Understanding and controlling the microstructure of complex foods. Woodhead Publishing, England

Garti N, Pinthus E, Aserin A, Spernath A (2007) Improved solubilisation and bioavailability of nutraceuticals in nanosized self-assembled liquid vehicles. In: Lakkis JM (ed) Encapsulation and controlled release technologies in food systems. Blackwell, USA, pp 13–40

Gärtner C, Stahl W, Sies H (1997) Lycopene is more bioavailable from tomato paste than from fresh tomatoes. Am J Clin Nutr 66:116–122

Grolier P, Azais-Braesco V, Zelmire L, Fessi H (1992) Incorporation of carotenoids in aqueous systems: uptake by cultured rat hepatocytes. Biophys Acta 1111:135–138

Hedges AR, Shieh WJ, Sikorski T (1995) Use of cyclodextrin for encapsulation in the use and treatment of food products. In: Risch SJ, Reineccius GA (eds) Encapsulation and controlled release of food ingredients. ACS Symposium Series 590, USA

Henry LK, Catignani GL, Schwartz SJ (1998) Oxidative degradation kinetics of lycopene, lutein, and 9-*cis* and all-*trans* β-carotene. J Am Oil Chem Soc 75:823–829

Hentschel A, Gramdorf S, Müller RH, Kurz T (2008) β-Carotene-loaded nanostructured lipid carriers. J Food Sci 73:N1–N6

Higuera-Ciapara I, Felix-Valenzuela L, Goycoolea FM, Argüelles-Monal W (2004) Microencapsulation of astaxanthin in a chitosan matrix. Carbohydr Polym 56:41–45

Hoppe PP, Scheneider JU, Schulz B, Tiefenbacher H (1986) Stabile injizierbare beta-Carotin-Solubilisate und Verfahren zu ihrer Herstellung. EP 0055817

Horn D (1989) Preparation and characterization of microdisperse bioavailable carotenoid hydrosols. Die Angewandte Makromolekulare Chemie 166/167:139–153

Horn D, Lüddecke E (1996) Preparation and characterization of nano-sized carotenoid hydrosols. In: Pelizzetti E (ed) Fine particles science and technology. Kluwer, The Netherlands

Horn D, Rieger J (2001) Organische Nanopartikel in wässriger Phase – Theorie, Experiment und Anwendung. Angewandte Chemie 113:4460–4492

Horn D, Schmidt H-W, Ditter W, Hartmann H, Lüddecke E (1985) Verfahren zur Herstellung von feinverteilten, pulverförmigen Carotinoid- bzw. Retinoidpräparaten. EP 0065193

Horn D, Lüddecke E, Schäfer P (1988) Pulverförmige, wasserdispergiebare Carotinoid-Zubereitungen und Verfahren zu ihrer Herstellung. DE 3702030

Hovestad W, Klinksiek B, Melchiors M (2000) Homogenenes Vermischen von wässrigen 2 K-Polyurethanlacken. Farbe Lack 7:40–44

Kawakatsu T, Kikuchi Y, Nakajima M (1997) Regular-sized cell creation in microchannel emulsification by visual microprocessing method. J Am Oil Chem Soc 74:317–321

Kikuchi Y, Kikuchi HE, Kuboi Y, Nakajima M (2000) Micro- and nanotechnology for biomedical and environmental applications. Proc SPIE 3912:95–104

Kim L, Rao AV, Rao LG (2003) Lycopene II – Effect on osteoblasts: The carotenoid lycopene stimulates cell proliferation and alkaline phosphatise activity of SaOS-2 cells. J Med Food 6:79–86

Kirsh VA, Hayes RB, Mayne ST, Chatterjee N, Subar AF, Dixon LB, Albanes D, Andriole GL, Urban DA, Peters U (2006) Supplemental and dietary vitamin E, β-carotene, and vitamin C intakes and prostate cancer risk. J Natl Cancer Inst 98:245–254

Kittikaiwan P, Powthongsook S, Pavasant P, Shotipruk A (2007) Encapsulation of *Haematococcus pluvialis* using chitosan for astaxanthin stability enhancement. Carbohydr Polym 70:378–385

Kobayashi I, Nakajima M (2006) Generation and multiphase flow of emulsions in microchannels. In: Kockmann N (ed) Micro process engineering. Wiley-VCH Verlag, Weinheim

Köhler K, Aguilar FA, Hensel A, Schubert K, Schuchmann HP (2007) Advanced optimisation of orifice-type high pressure emulsification valves. In PARTEC 2007, Nuremberg

Köhler K, Kulozik U, Schuchmann HP, Karasch S (2008) Energiesparende Homogenisierung von Milch mit etablierten sowie neuartigen Verfahren. Chemie Ingenieur Technik 80:1107–1116

Kolb G (2001) Zur Emulsionsherstellung in Blendensystemen. Doctoral dissertation, Halle-Wittenberg University

Köpcke W, Krutmann J (2008) Protection from sunburn with β-carotene – A meta-analysis. Photochem Photobiol 84:284–288

Köpsel C (1999) Struktur von Carotinoidaggregaten. Doctoral dissertation, Heinrich-Heine-University Düsseldorf

Krinsky NI, Mayne ST, Sies H (2004) Carotenoids in health and disease. Marcel Dekker, New York

Laos K, Lõugas T, Mändmets A, Vokk R (2007) Encapsulation of β-carotene from sea buckthorn (*Hippophaë rhamnoides* L.) juice in furcellaran beads. Innovat Food Sci Emerg Technol 8:395–398

Lasic DD (1993) Liposomes: From physics to application. Elsevier, Amsterdam

Leach G, Oliveira G, Morais R (1998) Production of a carotenoid-rich product by alginate entrapment and fluid-bed drying of *Dunnaliella salina*. J Sci Food Agric 76:298–302

Leuenberger BH (2007) Conceptual design of carotenoid product forms. In: Bröckel U, Meier W, Wagner G (eds) Product design and engineering, Vol. 2: Rawmaterials, additives and applications. Wiley-VCH Verlag, Weinheim, pp 539–567

Leuenberger BH, Tritsch J-C, Ulm J (2006) Process for preparing beadlets containing fat-soluble substances. EP 1 074 592 B1

Leuenberger BH, Schlegel B, Voelker KM (2008) Process for the manufacture of a powder containing carotenoids. WO 2008/098694 A1

Loksuwan J (2007) Characteristics of microencapsulated β-carotene formed by spray drying with modified tapioca starch, native tapioca starch and maltodextrin. Food Hydrocolloid 21:928–935
Lüddecke E, Schweikert L (1999) Finely dispersed carotenoid and retinoid suspension and their preparation. US 5,895,659
Lüddecke E, Auweter H, Schweikert L (2004a) Use of carotenoid aggregates as colorants. US 6,827,941 B1
Lüddecke E, Runge F, Auweter H (2004b) Herstellung und Charakterisierung von Carotinoid-Nanoteilchen. In: Product design and process engineering course. BASF AG, Ludwigshafen
Manz VU (1967) Die Anwendung und Bedeutung von synthetischen Carotinoiden in der Lebens- und Futtermittel- sowie in der pharmaceutischen Industrie. Chimie 21:329–335
Matioli G, Rodriguez-Amaya DB (2003) Microencapsulation of lycopene with cyclodextrins. Ciência e Technologia de Alimentos 23:102–105
Matsushita Y, Suzuki R, Nara E, Yokoyama A, Miyashita K (2000) Antioxidant activity of polar carotenoids including astaxanthin-β-glucoside from marine bacterium on PC liposomes. Fish Sci 66:980–985
McClements DJ, Decker EA, Weiss J (2007) Emulsion-based delivery systems for lipophilic bioactive components. J Food Sci 72:R109–R124
Mersmann A, Kind M, Stichlmair J (2005) Thermische Verfahrenstechnik Grundlagen und Methoden, Verein Deutscher Ingenieure-Buch.
Minekus M (1998) Development and validation of a dynamic model of the gastrointestinal tract. Doctoral dissertation, University of Utrecht
Montenegro MA, Nunes IL, Mercadante AZ, Borsarelli CD (2007) Photoprotection of vitamins in skimmed milk by an aqueous soluble lycopene-gum arabic microcapsule. J Agric Food Chem 55:323–329
Müller P, Tamm R (1966) Verfahren zur Herstellung eines insbesondere zum Färben von Lebens- und Futtermitteln geeigneten Carotinoidpräparates. DE1211911
Neves MA, Ribeiro HS, Fujiu K, Nakajima M (2008a) Formulation of controlled size PUFA-loaded oil-in-water emulsions by microchannel emulsification using β-carotene rich palm oil. Ind Eng Chem Res 47:6405–6411
Neves MA, Ribeiro HS, Kobayashi I, Nakajima M (2008b) Encapsulation of lipophilic bioactive molecules by microchannel emulsification. Food Biophys 32:126–131
Parada J, Aguilera JM (2007) Food microstructure affects the bioavailability of several nutrients. J Food Sci 72:R21–R32
Parker RS (1997) Bioavailability of carotenoids. Eur J Clin Nutr 51:86–90
Paust J (1991) Recent progress in commercial retinoids and carotenoids. Pure Appl Chem 63:45–58
Paust J (1994) Herstellung und Anwendung von Carotinoiden. Chimia 48:494–498
Phillips LG, Whitehead DM, Kinsella J (1994) Structure–function properties of food proteins. Academic, New York
Porrini M, Riso P, Testolin G (1998) Absorption of lycopene from single or daily portions of raw and processed tomato. Br J Nutr 80:353–361
Reuscher H, Kagan DI, Madhavi DL (2004) Coated carotenoid cyclodextrin complexes. US Patent 2004/0109920 A1
Ribeiro HS, Schubert H (2004) Lycopene enriched PIT emulsion. In: 3rd Euro fed lipid congress, Scotland
Ribeiro HS, Ax K, Schubert H (2003) Stability of lycopene emulsions in food systems. J Food Sci 68:2730–2734
Ribeiro HS, Briviba K, Rechkemmer G, Schubert H (2004) Zelluläre Aufnahme von flüssigen Lycopin-Formulierungen. German Nutr Soc 6:43
Ribeiro HS, Cruz RCD, Schubert H (2005a) Biliquid foams containing carotenoids. Eng Life Sci 5:84–88
Ribeiro HS, Rico LG, Badolato GG, Schubert H (2005b) Production of O/W emulsions containing astaxanthin by repeated premix membrane emulsification. J Food Sci 70:E117–E123
Ribeiro HS, Chu BS, Nakajima M (2006a) Production of β-carotene-loaded O/W emulsions by microchannel emulsification. In: Congres Mondial de L'Emulsion (CME), France

Ribeiro HS, Guerrero JMM, Briviba K, Rechkemmer G, Schuchmann H, Schubert H (2006b) Cellular uptake of carotenoid-loaded O/W emulsions in colon carcinoma cells in vitro. J Agric Food Chem 54:9366–9369

Ribeiro HS, Chu BS, Ichikawa S, Nakajima M (2008) Preparation of nanoemulsions containing β-carotene. Food Hydrocolloid 22:12–17

Richelle M, Bortlik K, Liardet S, Hager C, Lambelet P, Baur M, Applegate LA, Offord EAA (2002) Food-based formulation provides lycopene with the same bioavailability to humans as that from tomato paste. J Nutr 132:404–408

Rodríguez-Huezo ME, Pedroza-Islas R, Prado-Barragán LA, Beristain CI, Vernon-Carter EJ (2004) Microencapsulation by spray drying of multiple emulsions containing carotenoids. J Food Sci 69:E351–E359

Runge F, Zwissler GK, End L, Schweikert L, Dieter H (1998) Verwendung von Carotinoid-Solubilisaten zum Färben von Lebensmitteln und pharmazeutischen Zubereitungen. DE 19653410 A1

Sahni S, Hannan MT, Blumberg J, Cupples LA, Kiel DP, Tucker KL (2009) Inverse association of carotenoid intakes with 4-y change in bone mineral density in elderly men and women: The Framingham Osteoporosis Study. Am J Clin Nutr 89:416–424

Sajilata MG, Singhal RS, Kamat MY (2008) The carotenoid pigment zeaxanthin – A review. Compr Rev Food Sci Food Saf 7:29–49

Sakaki K (1992) Solubility of β-carotene in dense carbon dioxide and nitrous oxide from 308 to 323 K and from 6 to 30 MPa. J Chem Eng Data 37:249–251

Santipanichwong R, Suphantharika M (2007) Carotenoids as colorants in reduced-fat mayonnaise containing spent brewer's yeast β-glucan as a fat replacer. Food Hydrocolloid 21:565–574

Santos AB, Favaro-Trindade CS, Grosso CRF (2005) Preparation and characterization of paprika oleoresin microcapsules obtained by spray drying. Ciência e Tecnologia de Alimentos 25:322–326

Scheid S, Buchholz M (2005) Nanomix – New developments in high-pressure homogenisation technology for cosmetics and pharmaceutics. In: 1st European congress on life science process technology, Nuremberg

Schubert H (2005) Emulgiertechnik. Behrs Verlag, Hamburg

Schubert H, Engel R (2004) Product and formulation engineering of emulsions. Chem Eng Res Des 82:1137–1143

Schuchmann HP (2005) Emulsionen und Schäume. In: Schuchmann HP, Schuchmann H (eds) Lebensmittelverfahrenstechnik: Rohstoffe – Prozesse – Produkte. Wiley, Weinheim, pp 219–252

Schuchmann HP (2007) Emulsification techniques for the formulation of emulsions and suspensions. In: Bröckel U, Meier W, Wagner G (eds) Product design and engineering, Vol. 1: Basics and technologies. Wiley, Weinheim, pp 63–93

Schuchmann HP, Bub A, Schubert H, Mayer-Miebach E, Ribeiro HS (2005) Pflanzeninhaltsstoffe im Kampf gegen Krebs- und Herz-Kreislaufkrankheiten – Die Rolle der Verfahrenstechnik zur Verbesserung der Bioverfügbarkeit von Carotinoiden. Chemie Ingenieur Technik 77:1179–1180

Schwarz S, Obermüller-Jevic UC, Hellmis E, Koch W, Jacobi G, Biesalski H-K (2008) Lycopene inhibits disease progression in patients with benign prostate hyperplasia. J Nutr 138:49–53

Schweikert L, Kolter K (1997) Stabile zur parenteralen Verabreichung geeignete Carotinoid-Emulsionen. EP 0 800 824 B1

Senior JH (1987) Fate and behaviour of liposomes *in vivo*: A review of controlling factors. Crit Rev Ther Drug Carrier Syst 3:123–193

Shahidi F, Han X-Q (1993) Encapsulation of food ingredients. Crit Rev Food Sci Nutr 33:501–547

Shu B, Yu W, Zhao Y, Liu X (2006) Study on microencapsulation of lycopene by spray-drying. J Food Eng 76:664–669

Socaciu C, Jessel R, Diehl HA (2000) Competitive carotenoid and cholesterol incorporation into liposomes: effects on membrane phase transition, fluidity, poarity and anisotropy. Chem Phys Lipids 106:79–88

Stahl W, Heinrich U, Aust O, Tronnier H, Sies H (2006) Lycopene-rich products and dietary photoprotection. Photochem Photobiol Sci 5:238–242

Stang M (1998) Zerkleinern und Stabilisieren von Tropfen beim mechanischen Emulgieren. VDI Fortschrittsberichte, series 3, No. 527. VDI Verlag, Düsseldorf

Stang M, Schuchmann HP, Schubert H (2001) Emulsification in high-pressure homogenizers. Eng Life Sci 4:151–157

Sugiura S, Nakajima M, Iwamoto S, Seki M (2001) Interfacial tension driven monodispersed droplet formation from microfabricated channel array. Langmuir 17:5562–5566

Suzuki K, Shuto I, Hagura Y (1998) Preparation of corn oil/water and water/corn oil emulsions using PTFE membranes. Food Sci Technol Int 42:164–167

Türk M, Lietzow, R. (2004). Stabilized nanoparticles of phytosterol by rapid expansion from supercritical solution into aqueous solution. AAPS PharmSciTech 5:1–10

Upritchard JE, Schuurman CRWC, Wiersma A, Tijburg LBM, Coolen SAJ, Rijken PJ, Wiseman SA (2003) Spread supplemented with moderate doses of vitamin E and carotenoids reduces lipid peroxidation in healthy, nonsmoking adults. Am J Clin Nutr 78:985–992

van het Hof KH, West CE, Weststrate JA, Hautvast JGAJ (2000) Dietary factors that affect the bioavailability of carotenoids. J Nutr 130:503–506

Velikov K, Pelan E (2008) Colloidal delivery systems for micronutrients and nutraceuticals. Soft Matter 4:1964–1980

Vilstrup P, Jensen NM, Krag-Andersen S (1997) Process for the preparation of a water-dispersible carotenoid preparation in powder form. EP 0684973

Vladisavljević GT, Williams RA (2005) Recent developments in manufacturing emulsions and particulate products using membranes. Adv Colloid Interface Sci 113:1–20

Vladisavljević GT, Williams RA (2006) Manufacture of large uniform droplets using rotating membrane emulsification. J Colloid Interf Sci 299:396–402

Vladisavljević G, Kobayashi I, Nakajima M (2006) Production of monodisperse O/W emulsions at high production scales using asymmetric microchannels. In: 7th Annual Meeting of the Japan Society for Food Engineering, Tsukuba, Japan

Vrieling A, Voskuil DW, Bonfrer JM, Korse CM, van Doorn J, Cats A, Depla AC, Timmer R, Witteman BJ, van Leeuwen FE, van't Veer LJ, Rookus MA, Kampman E (2007) Lycopene supplementation elevates circulating insulin-like growth factor-binding protein-1 and -2 concentrations in persons at greater risk of colorectal cancer. Am J Clin Nutr 86:1456–1462

Winterhalter M, Lasic DD (1993) Liposome stability and formation: Experimental parameters and theories on the size distribution. Chem Phys Lipids 64:35–43

Yamaguchi M, Uchiyama S (2003) Effect of carotenoid on calcium content and alkaline phosphatase activity in rat femoral tissues in vitro: The unique anabolic effect of β-cryptoxanthin. Biol Pharm Bull 43:281–296

Yamaguchi M, Uchiyama S (2004) β-Cryptoxanthin stimulates bone formation and inhibits bone resorption in tissue culture in vitro. Mol Cell Biochem 258:137–144

Yamaguchi M, Igarashi A, Morita S, Sumida T, Sugawara K (2005) Relationship between serum β-cryptoxanthin and circulating bone metabolic markers in healthy individuals with the intake of juice (*Citrus unshiu*) containing β-cryptoxanthin. J Health Sci 51:738–743

Yanagi K, Miyata Y, Kataura H (2006) Highly stabilized β-carotene in carbon nanotubes. Adv Mater 18:437–441

Yuan C, Jin Z, Xu X, Zhuang H, Shen W (2008) Preparation and stability of the inclusion complex of astaxanthin with hydroxypropyl-β-cyclodextrin. Food Chem 109:264–268

Chapter 9
Encapsulation of Enzymes and Peptides

Gabrie M.H. Meesters

9.1 General Introduction

A large part of formulated peptides and proteins, e.g., enzymes used as food ingredients, are formulated in a liquid form. Often, they are dissolved in water to which glycerol or sorbitol is added to reduce the water activity of the liquid, thus reducing the change of microbial growth. Still, there are reasons to formulate them in a solid form. Often, these reasons are stability, since a dry formulation is often much better than liquid formulations, and less transportation cost, since less mass is transported if one gets rid of the liquid; however, most of the times, the reason is that the product is mixed with a solid powder. Here, a liquid addition would lead to lump formation.

Additional issues that play a role when formulating these products in a solid form are, for example, allergenicity, dust reduction, dosing accuracy, and dissolution.

Stability in a solid formulation is often much better than in a liquid formulation, since the water activity is low. Often, stabilizers are added to the solid forms to protect the proteins against denaturation.

The problem of using powdered proteins is that people can get allergic to these proteins when exposed to them for a long period of time or to high concentrations of them. Most of the time, the places of high risk are the factories where these proteins are produced or handled. By encapsulation or granulation, the tendency to form dust loaded with proteins is greatly reduced, and safe products can be obtained. Peptides often show little to no allergic reactions, since the molecules are too small to induce an allergic reaction.

When dosing these solid formulations, the dosing accuracy may be an issue. This means that the content of the active ingredients needs to be such that a homogeneous mixture can be obtained which gives the right application behavior. This can be achieved in many ways. Simple granulates may be mixed with powders in blenders, but products can also be made into tablets which should contain the right

G.M.H. Meesters
DSM Food Specialties, P.O. Box 1, 2600 MA Delft, The Netherlands
e-mail: gabrie.meesters@dsm.com

N.J. Zuidam and V.A. Nedović (eds.), *Encapsulation Technologies for Active Food Ingredients and Food Processing*,
DOI 10.1007/978-1-4419-1008-0_9,

amount of product to perform a certain behavior, reaction or allow for an accurate release profile in time.

Also, the concentration in a solid formulation may need to have the required homogeneity in a powder mixture. For example, when an enzyme formulation with a high concentration is mixed with a chicken feed mixture, it needs to be of a concentration that each chicken will get the right amount of product each day and well spread over the day. This means that each feed pellet needs to have the activity present and not one out of every 50 pellets that will be eaten by a chicken. This can be achieved by adding inert material to the solid formulation.

When the peptides or proteins need to dissolve quickly in the application, these solid products can be made instant in their dissolution behavior. This means that they wet and dissolve quickly in water by simple stirring. In many cases, the wettability of these powders is poor, and by granulation in the right way, the dissolution can be enhanced enormously.

Instant properties of a product are associated with the four phases that occur after pouring a powder on a liquid surface. These phases are the following (Schubert 1993):

(a) Penetration of the liquid into the pore structure owing to capillarity. The ease of penetration is called *wettability*.
(b) Sinking of the particles or portions of the powder into the liquid, *sinkability*.
(c) Dispersion of the powder in the liquid, *dispersibility*.
(d) If the substance is soluble, dissolution of the particles in the liquid, *solubility*.

The four phases take place in the sequence indicated, with occasional overlapping of two phases. For a powder with good instant properties the four phases are completed within a few seconds (for a height of the powder material of 10 mm) (Schubert 1981, 1993). The phases are governed by different physical laws and can be influenced in different ways.

9.2 Types of Techniques Used to Encapsulate or Granulate

There are several techniques that are used to granulate and encapsulate proteins and peptides. The simplest way is by spray-drying combined with agglomeration. Other main techniques are fluid bed techniques, high and low shear granulation, and extrusion. Also techniques of encapsulating enzymes in liquid matrices like liposomes or gels like alginate and carrageenans are used. These techniques will be discussed below.

9.2.1 Spray-Drying and Combined Granulation

The technique of spray-drying and combined granulations is shown in Fig. 9.1. It is also called multi-stage drying or integrated spray-drying. The technique is mostly used for the production of solid protein and peptide formulations. Material is sprayed

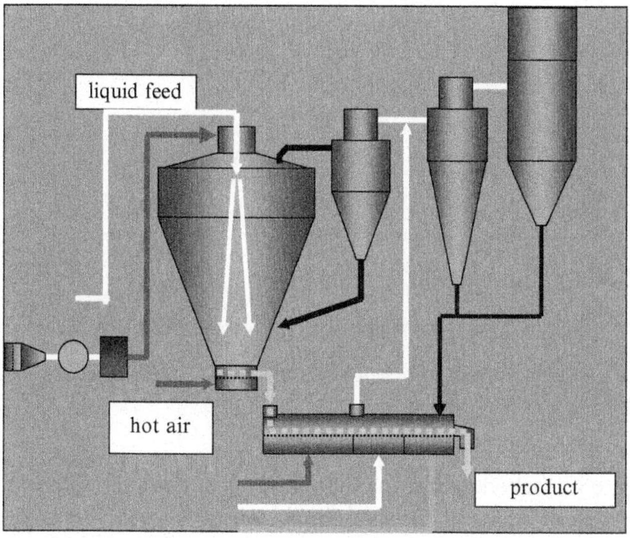

Fig. 9.1 Typical multistage dryer. Liquid is sprayed in a tower and agglomerated and dried in several places in the system (courtesy of GEA/Niro, Denmark, www.niro.com)

into the spray-dryer, where it is dried to single particles. These single particles collide with particles which are recycled from the integrated or connected fluid bed. Since these drying particles get sticky during drying, they form agglomerates with already dried ones. Often they recycle several times to reach the desired particle size. The advantage of this technique over simple single stage drying is that the product is free flowing, which makes dosing easy and it has very good instant properties. These instant properties are important since many proteins and peptides have very poor wettability. This also means that water does not wet the particles easily. Since these particles are quite porous, the capillaries formed in these agglomerates enhance the uptake of water by the capillary sorption. This makes the wetting surface much larger, the particle fall apart when wetted from the inside, and do not form lumps. With the selection of a spray-dryer there are three main design variables that influence final product properties. They are selection of the atomizer, most suitable airflow pattern and the drying chamber design. Several operating conditions parameters are also important. Design rules that establish the final product properties qualitatively are not yet available. Figure 9.2 shows schematically how these granules are formed. Due to the larger size of these agglomerates, the formation of dust is greatly reduced as well. This make the handling of these powders less dangerous, but care still needs to be taken, since these agglomerates are fragile and may break during handling though, thus giving rise to formation of some dust particles.

To get the right amount of protein in the final formulation, so one can dose the right amount in the final application with the right distribution of activity, diluents can be introduced in the spray-drying section. This is often referred to as co-drying. Here

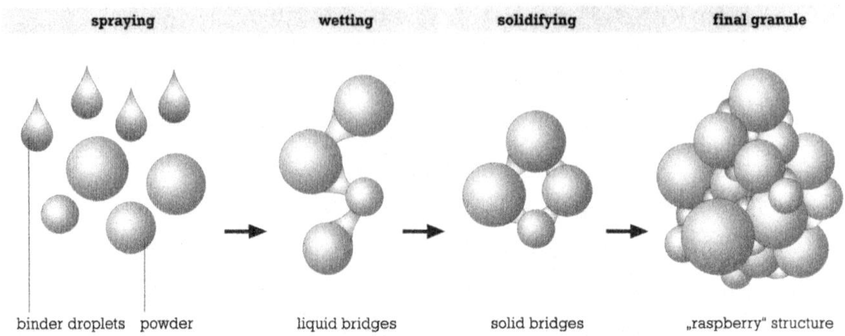

Fig. 9.2 Schematic description of particle formation (courtesy of Glatt GmbH, Binzen, Germany, www.glatt.de)

a dry powder, e.g., maltodextrins, (modified) starches, flour and salts are often used. This diluents can be added to the spray-liquid which is sprayed into the tower (but this may increase viscosity, or make the proteins precipitate), but most of the time these diluents are blown into the drying chamber as solids. The protein-loaded droplets will stick to these diluents and form mixed products. Of course blending the protein solids with inert material can be done in a mixer after the granulation step, but here the product and the diluents may segregate, since no physical bond is formed.

9.2.2 High Shear Granulation

In high shear granulation one mixes a powder with a liquid, and often a binder is used as well. This is schematically given in Fig. 9.3. By using high shear granulation, much denser particles are formed than by using spray-drying. They have the advantage of being much stronger, which makes the handling much safer, since less dust will be formed. These granules usually dissolve not as quickly as the powders from the spray-dried and agglomerated units. Dissolution times are still one to a few minutes though. Figure 9.4 gives a typical example of a high shear granule.

High shear granulation is used by the company Novozymes in Denmark as their most important formulation technology to granulate their enzymes. These enzymes are used, e.g., in detergents, as food and feed additives, and as paper and pulp processing aids. Since the enzyme is dispersed through the whole particle, dust formed during handling will contain enzymes, which may induce allergic reactions. To prevent this, the high shear granules are subsequently coated to protect the granules against attrition during product handling. Coating may take place in a high shear granulator, but most of the time the coating is applied by fluid bed coating. The coating reduces dust formation and enhances the storage stability of the enzymes as well.

When using the peptides and proteins as the solid in this process, one first needs to get the powder in a dry form, after which this powder is wetted again in a high

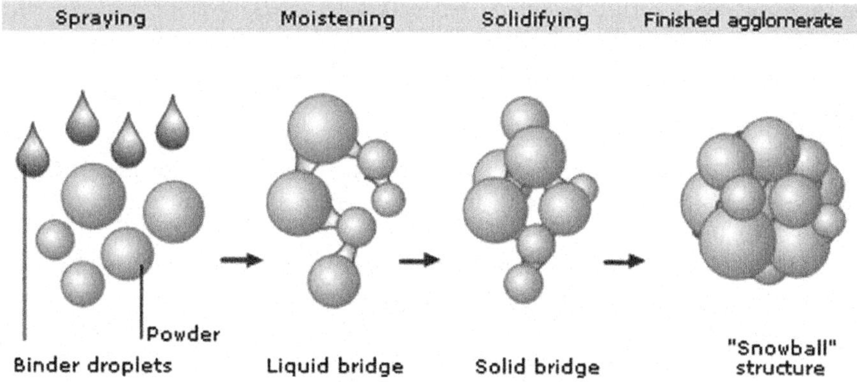

Fig. 9.3 Schematic drawing of high shear granulation process. Due to the high shear a compacted "snowball" structure is obtained (courtesy of Glatt GmbH, Binzen, Germany, www.glatt.de)

Fig. 9.4 Typical example of a high shear granulate (courtesy of Glatt GmbH, Binzen, Germany, www.glatt.de)

shear mixer and dried for a second time. Water needs to be evaporated twice, which is not very efficient. The other option is to administer the proteins and peptides as the liquid used in the high shear process. The powder used in this case is then an inert diluent. The drawback here is that only limited amount of liquid can be added to the powder mixture for high shear granulation. When too little liquid is added, there will be no agglomeration taking place. When on the other hand too much liquid is added, the powder will be over-wetted, resulting in a mud-like mixture. This means that there is a minimum and maximum amount of liquid that can be added to the powder. This operating window depends strongly on the powders used and may vary from less than 10% (w/w) to more than 50% (w/w). A typical enzyme formulation is given in Table 9.1.

When the primary reason for using the high shear techniques is just to make granules with high protein content, usually the solid is the protein. On the other hand, when high potent proteins like enzymes are granulated, the amount added to the granules is important for dosing accuracy; then the protein is often in the liquid form used in high shear granulation.

Figure 9.5 shows two different types of batch high shear granulator, and Fig. 9.6 shows a continuous high shear granulator. Both systems are used in the food industry. The pharmaceutical industries typically use the batch systems, often the types with a vertical shaft. In the food but also in the non-food and non-pharmaceutical industries like the fertilizer industries a lot of large continuous high shear granulators are used.

Table 9.1 Novo-Nordisk high shear granulation formulation (Markussen and Schmidt 1977)

Compound	Wt%
Enzyme powder	5–30
Binder (e.g., PVP, sugars)	1–10
Water-soluble salts	40–60
Water-insoluble salts	5–20
Coconut mono-ethanol-amide	4–7
TiO_2	0–4
water	

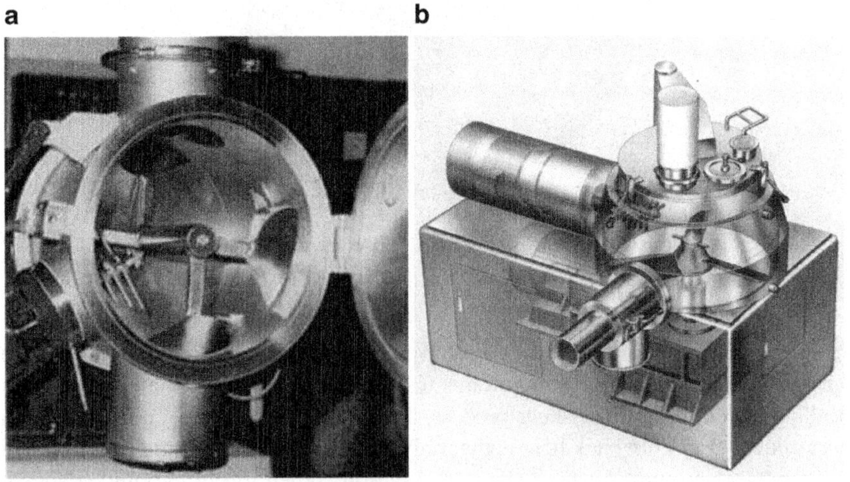

Fig. 9.5 Batch high shear granulator horizontal axis (**a**), and vertical axis (**b**) (courtesy of Lödige GmbH, Paderborn, Germany (**a**), www.loedige.com and Glatt GmbH, Binzen, Germany (**b**) www.glatt.de)

Fig. 9.6 Continuous high shear granulator (Lödige, Paderborn, Germany www.loedige.com)

9.2.3 Low Shear Granulation

Low shear granulation is a technique where a solid powder (mixture) is mixed with a liquid where there is short contact time and the mixing is not so intense. Here rather open granules are formed with good instant properties (see Fig. 9.7). Contrary to spray-drying where the protein is often present in the liquid form, in a low shear granulator the solid contains the protein.

A typical example of a low shear granulator is given in Fig. 9.8, the so-called Flex-o-mix of Hosokawa. The Flex-o-mix is a continuous system. The powder falls down a tube where a mixing device is present. The liquid is sprayed in between the mixer blades and mixed with the powder while falling down the tube. This tube is made of flexible material, which is moved by rollers. This prevents buildup of material at the walls of the granulator. Typical residence times in the mixer are 1–5 s. The resulting product is a wet agglomerated product; with a relative low bulk density compared to e.g., high shear granulates, but comparable to or slightly higher compared to multi stage drying. The bulk density can be changed by adjusting the amount of liquid that is added per weight of powder proceeding through the device. The wet product is dried in a fluid bed dryer, which is fed directly from the Flex-o-mix which is placed most of the time on top of the drier. As is shown in Table 9.2, mixtures of powders can be agglomerated into mixed solid systems which do not segregate, and in the formulation shown in Table 9.2, here the cocoa powder is easily dispersed into milk or water, resulting in a ready to drink cocoa milk.

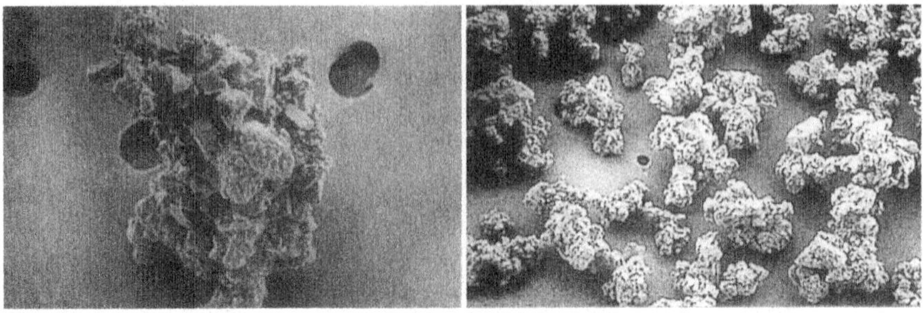

Fig. 9.7 Typical product from a low shear granulation process (courtesy of Hosokawa, Doetinchem, The Netherlands, www.hosokawa.com)

a b

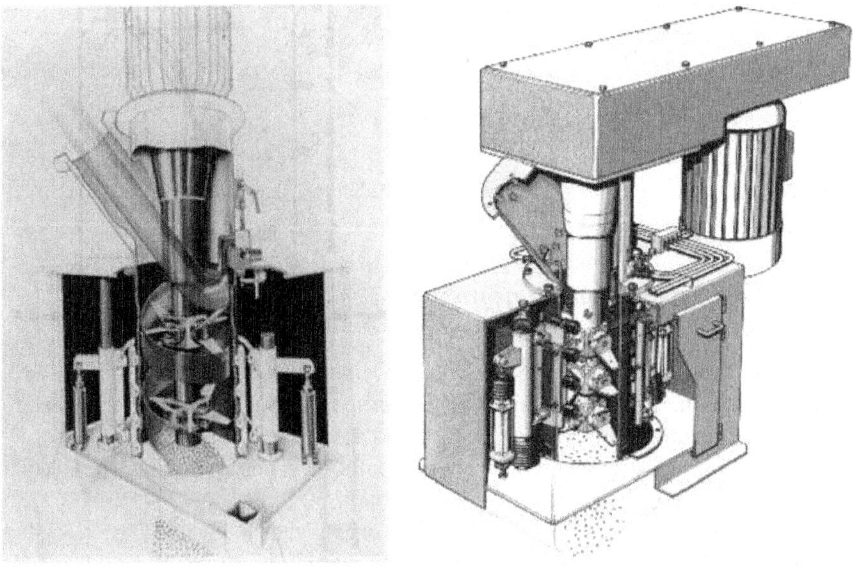

Fig. 9.8 The Hosokawa Flex-o-mix (courtesy of Hosokawa, Doetinchem, The Netherlands)

Table 9.2 Typical formulation of instant Cocoa powder as made in a Flex-o-mix, where mainly sugar, cocoa powder and milk powder are premixed and agglomerated with water (courtesy of Hosokawa, Doetinchem, The Netherlands, www.hosokawa.com)

Component	For use in milk [w/w%]	For use in water [w/w%]
Sugar	78	39
Cocoa powder	19	28
Milk powder	0	31
Flavorings	<1	<1
Thickeners	<1	<1
Emulsifiers	<1	<1
Vitamins	3	2

The products from the Flex-o-mix also possess good dosing properties which makes them very suitable for use in vending machines. In spray-dryers similar instant forms can be made, but usually not so many different components are granulated together as can be done in this low shear system.

9.2.4 Fluid Bed Agglomeration

In fluid bed agglomeration a powder is fluidized in a fluid bed into which a binder liquid is sprayed. This binder liquid wets the solid particles and forms bridges between the particles. Due to the drying of the air, these liquid bridges dry and solidify, thus gluing the particles together. This is schematically given in Fig. 9.9.

Like in spray-drying and agglomeration and low shear granulation as in the Flex-o-mix, the product is of relative low bulk density and shows good instant behavior. Fluid bed agglomeration can be done in a batch or continuous system, as is shown in Fig. 9.10. Figure 9.11 shows a typical fluid-bed agglomerate. The open structure can be clearly seen, giving the product very good instant properties.

9.2.5 Low Pressure Extrusion and Spheronization

Low pressure extrusion is a technology which was already used in the late 1960s and early 1970 by Novo Nordisk to encapsulate their enzyme granules (Nielsen and Markussen 1971). The technology was used in the late 1990s by Gist-Brocades, The Netherlands (now DSM) and others to encapsulate their feed enzymes (Barendse et al. 1998; Harz et al. 2000).

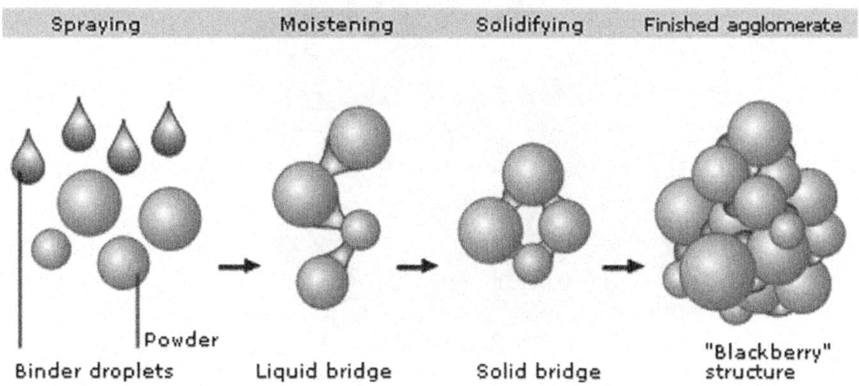

Fig. 9.9 Schematic drawing of the particle formation in a fluid bed agglomeration process (courtesy of Glatt GmbH, Binzen, Germany, www.glatt.de)

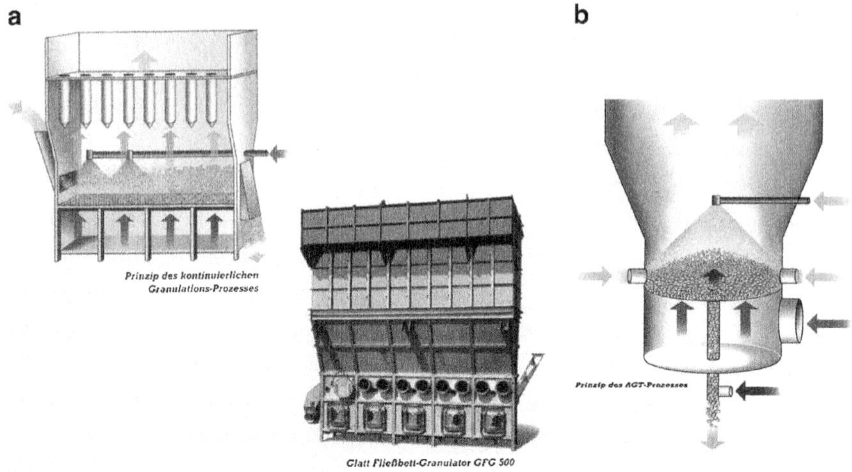

Fig. 9.10 Two types of continuous fluid bed agglomeration. *Left* a continuous fluid bed system with top spray nozzle systems of Glatt, on the *right* the Advanced Granulation Technology (AGT) system of Glatt which includes a classifier in the middle of the circular fluid bed system (courtesy of Glatt GmbH, Binzen, Germany, www.glatt.de)

Fig. 9.11 Typical example of fluid bed agglomerate (courtesy of Glatt GmbH, Binzen, Germany, www.glatt.de)

In a wet massing step, powder and liquid are mixed. The protein can either be part of the solids or can be dissolved in the liquid. The material is intensively mixed to obtain a wet mass. This wet mass is pressed through a dye plate in an extruder, forming strings of wet material (spaghetti-like), see Fig. 9.12.

These strings are emptied in a spheronizer, which is a rotating plate where these strings are broken into small particles. These particles are rounded off and compacted to form spherical to rounded-off ones as shown in Fig. 9.13.

These wet particles are further dried in a fluid bed dryer, and can be coated with a protective coating if desired. Figure 9.14 gives a schematic drawing of the mechanism.

A typical recipe of a formulation made by low pressure extrusion and spheronization is given in Table 9.3 and Fig. 9.15 shows the formed particles based on this recipe.

A simpler but similar technique is reported by Öngen et al. (2002). They report the encapsulation of an α-amylase into a starch matrix, the substrate of the enzyme. This would allow a slow release of the enzyme upon water take up. The starch was a pre-gelatinized potato starch. The dry encapsulated enzyme was obtained by kneading the starch in a kneader at about 80°C. The enzyme was dosed in to the kneader as a liquid (solution in water). The resulting mass was cooled down and grinded into smaller particles. The enzyme concentration in the final product was reported up to 8% (w/w) on dry base of starch. The activity was released when wetting the particles, due to the degradation of the starch by the enzyme. Typical times reported of full release of the enzyme activity were in the range of hour to days. This formulation was also used to modulate flavor release from starch matrixes via α-amylase controlled degradation (Yilmaz et al. 2002).

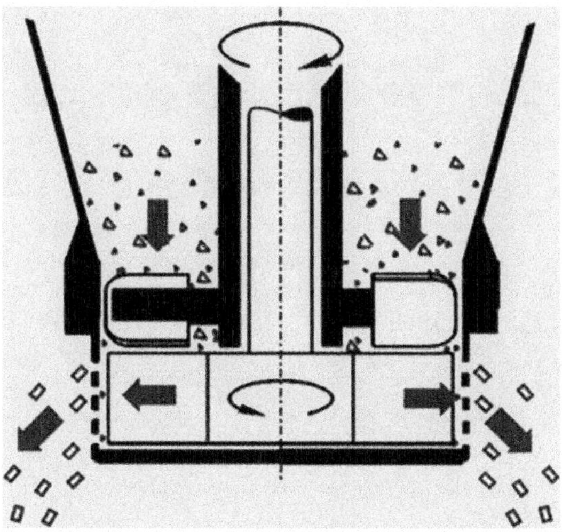

Fig. 9.12 Schematic drawing of a Bextruder BX of Hosokawa Micron, which can be used to extrude wet powders (courtesy of Hosokawa Micron, The Netherlands, www.hosokawa.com)

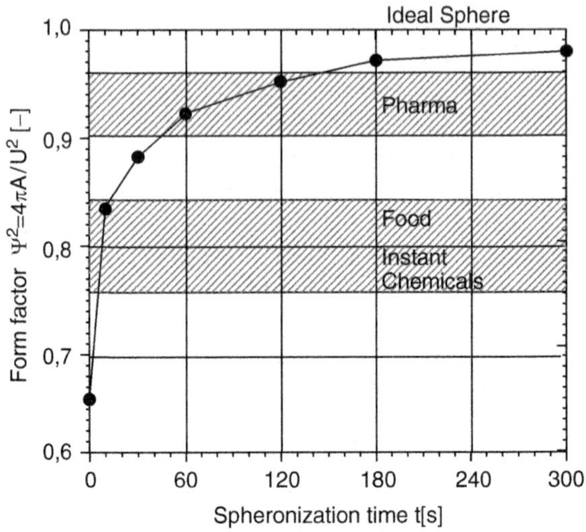

Fig. 9.13 Shape change as function of spheronization time showing the rounding off of the extrudates (courtesy of Hosokawa Micron BV, The Netherlands, www.hosokawa.com)

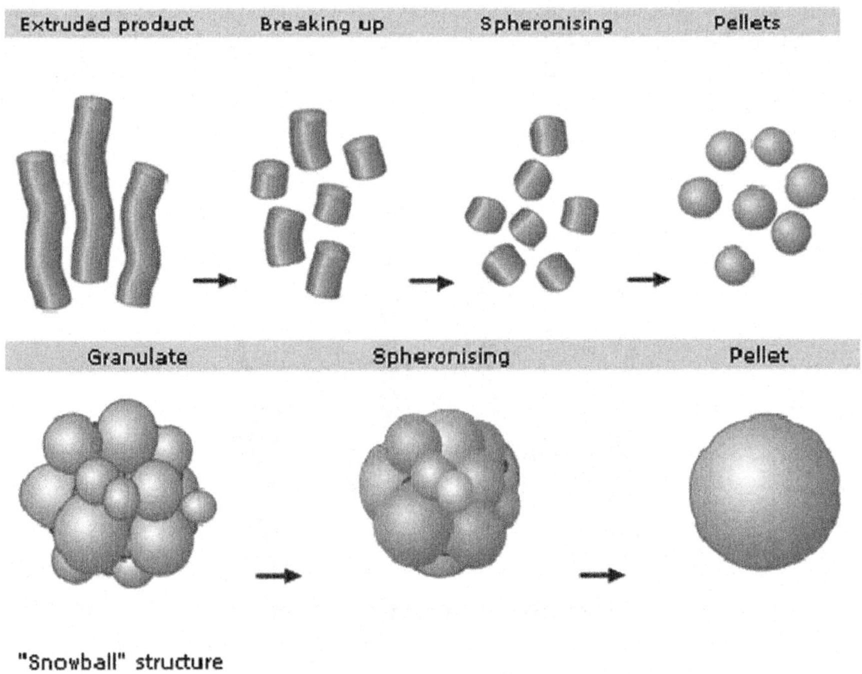

Fig. 9.14 Schematic drawing of the extrusion process (*top*). The lower drawing shows in detail the spheronizing of the particles (courtesy of Glatt GmbH, Binzen, Germany, www.glatt.de)

Table 9.3 The Nathuphos® formulation as developed by Gist-Brocades (now DSM) used in an extrusion and spheronization set up (Barendse et al. 1998; Harz et al. 2000)

Compound	Wt%
Enzyme liquid	15–30
Starch	65–80
Stabilizing salts	0.1–5
Outer PEG-6000 coating	1–10

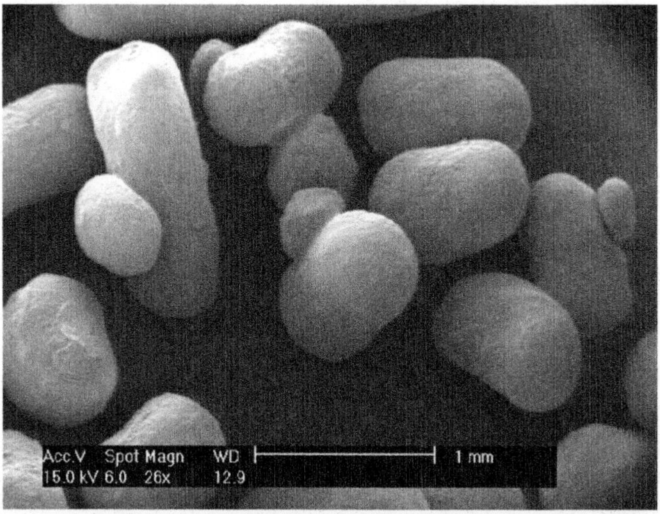

Fig. 9.15 Sample of the final coated Natuphos® enzyme granules (courtesy of DSM-Food Specialties BV, Delft, The Netherlands)

9.2.6 Encapsulation in Gel Systems

Anjani et al. (2007) describe a technique where a liquid in which a protease is dissolved, is mixed with an alginate, pectin or κ-carrageenan polymer solution. By dropping this mixture in a $CaCl_2$ solution, the drops of polymer solution containing the enzyme forms a gel bead, thus encapsulating the enzymes. They used a nozzle to form uniformly size droplets in the range of 0.5–2 mm (see also Meesters 1992 and Heinzen 1995). Figure 9.16 shows a typical setup which is used to make these gel beads. Scaled up versions are available as well.

The gel particles formed by using this technique can be further processed. They can be dried and additionally coated with a wax, a polylysine or polylactate coating. In the above mentioned paper of Anjani et al. (2007), applications are found in the area of accelerating of cheese ripening. Here the enzyme is slowly released in the curt and enhances the flavor formation.

Kailasapathy and Lam (2005) describe the formation of similar gel beads but here they use a water-in-oil emulsification process to prepare gellan or κ-carrageenan beads with protease. This process is already described by Magee and Olsen (1981).

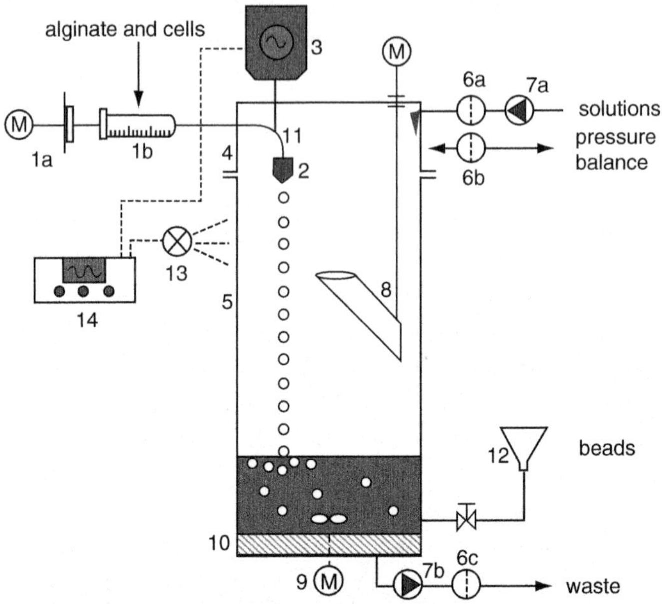

Fig. 9.16 System to produce uniformly sized droplets of alginate, which are gelated in a $CaCl_2$ containing bath (courtesy of Inotech Biosystems International Inc., Switzerland, www.inotechintl.com). 1 = syringe pump, 2 = nozzle, 3 = vibration unit, 4 = pulsation unit, 5 = reaction vessel, 6 = filters, 7 = pump, 8 = by-pass unit, 9 = stirrer, 10 = filtration unit, 11 = capillary, 12 = collection unit, 13 = stroboscope, 14 = frequency generator, M = motor

9.2.7 Liposome Technologies

Liposomes are used to encapsulate many types of molecules. They can also be applied to encapsulate protein such as enzymes. Walde and Ichikawa (2001) reported an extensive literature review on the use of liposome technology to encapsulate enzymes. They specify the types of enzymes encapsulated as such and the applications of use (mainly in the biomedical field or for acceleration of cheese ripening).

Liposomes are lipid vesicles formed from amphiphilic molecules which aggregate in water, forming a sphere. The liposome can be built out of a single or many layers of amphiphilic, polymolecular shells (lamellae). Sizes of the liposomes range from typically 10–100 nm for a single shell liposome to several micrometers for multi-layered liposomes. The enzyme is entrapped in the aqueous inter-lamellar spaces of the liposome during the preparation process. The encapsulation efficiency of enzymes depends on the preparation method and conditions used. Generally it is more difficult to entrap large enzymes in small liposomes than small enzymes. The enzyme may interact with the lipid vesicles, which may increase its encapsulation efficiency (but also may induce vesicle aggregation). The more the layers, the more stable the liposome might be. Liposomes are relatively stable in liquid systems, but entrapped enzymes may leak out upon storage. Also the long-term stability of the liposomes themselves can be an issue due aggregation of the vesicles or hydrolysis and/or oxidation of the lipid components. Scale up of liposomal technologies is not always easy. Membrane extrusion techniques are reported in the overview using pore of ~100 nm to produce large quantities of liposomes.

9.3 Conclusions

Most of the dry formulations of peptides, proteins and enzymes are granulated into a product fit for its use in application. Instant properties, strength, controlled release, size, etc. determine the technique which should be applied to get these required properties into the final product. A large variety of techniques are developed by several companies, which have grown into mature techniques. Since scale up is not always that easy and the product properties should be as required, piloting is required to fine tune the specifications. Most equipment suppliers have piloting facilities available and master the right scale-up rule.

Liquid formulations of proteins and enzymes are often simple mixtures of an enzyme concentrate with glycerol or sorbitol as main components. For certain application (like cheese ripening), more sophisticated formulations like the liposome technology or gel matrices might be used. These technologies were initially developed in the pharmaceutical world, but are now expanding in the other fields like food as well. Cost and scale up are still the main issues which are limiting the use of these types of technologies in the nonpharma fields.

References

Anjani K, Kailasapathy K, Philips M (2007) Microencapsulation of enzymes for potential application in acceleration of cheese ripening. Int Dairy J 17:79–86

Harz HP, Betz R, Schöner FJ, Meesters GM, Andela CSM (2000) Polymer-coated, granulated enzyme-containing feed additives and method for production thereof. Patents DE 19929257, EP 1189518 or WO 0100042

Barendse RCM, Meesters G, Harz HP (1998) Carbohydrate-based enzyme granulates. EP986313/WO9854980

Heinzen Ch. (1995) Herstellung von monodispersen Mikrokugeln durch Hydroprillen. Thesis, Swiss Federal Institute of Technology, Zurich

Kailasapathy K, Lam SH (2005) Application of encapsulated enzymes to accelerate cheese ripening. Int Dairy J 15:929–939

Magee EL, Olsen NF (1981) Microencapsulation of cheese ripening systems; stability of microcapsules. J Dairy Sci 64:611–615

Meesters GMH (1992) Mechanisms of droplet formation. Thesis, Delft University of Technology

Markussen EK, Schmidt AW (1977) Verfahren zur herstellung eines Enzymgranulats, das dabei erhaltene Produkt und seine Verwendung. DE2730481 or US4106991

Nielsen TK, Markussen EK (1971) Enzympräparat und Verfahren zu seiner Herstellung. DE2137043

Öngen G, Yilmaz G, Jongboom ROJ, Feil H (2002) Encapsulation of α-amylase in starch matrix. Carbohydr Polym 50:1–5

Schubert H (1993) Instantization of powdered food product. Int Chem Eng 33(1):28–45

Schubert H (1981) Instanteigenschaften von Agglomerierten stoffen. International symposium of agglomeration, Nuremberg, A68–A82

Walde P, Ichikawa S (2001) Enzymes inside lipid vesicles: preparation, reactivity and application. Biomol Eng 18:143–177

Yilmaz G, Öngen G, Jongboom ROJ, Feil H, van Dijk C, Hennink WE (2002) Modulated release of a volatile compound from starch matrixes via enzymatically controlled degradation. Biomacromolecules 3:305–311

Chapter 10
Encapsulation of Probiotics for use in Food Products

Verica Manojlović, Viktor A. Nedović, Kasipathy Kailasapathy, and Nicolaas Jan Zuidam

10.1 Probiotics From Research to Consumer

The history of the role of probiotics for human health is one century old and several definitions have been derived hitherto. One of them, launched by Huis in't Veld and Havenaar (1991) defines probiotics as being "mono or mixed cultures of live microorganisms which, when applied to a man or an animal (e.g., as dried cells or as a fermented product), beneficially affect the host by improving the properties of the indigenous microflora". Probiotics are living microorganisms which survive gastric, bile, and pancreatic secretions, attach to epithelial cells and colonize the human intestine (Del Piano et al. 2006). It is estimated that an adult human intestine contains more than 400 different bacterial species (Finegold et al. 1977) and approximately 10^{14} bacterial cells (which is approximately ten times the total number of eukaryotic cells in the human body). The bacterial cells can be classified into three categories, namely, beneficial, neutral or harmful, with respect to human health. Among the beneficial bacteria are *Bifidobacterium* and *Lactobacilli*. The proportion of bifidobacteria represents the third most common genus in the gastro-

V. Manojlović
Department of Chemical Engineering, Faculty of Technology and Metallurgy, University of Belgrade, Karnegijeva 4, 11000 Belgrade, Serbia

V.A. Nedović
Department of Food Technology and Biochemistry, Faculty of Agriculture, University of Belgrade, Nemanjina 6, PO Box 127, 11081 Belgrade-Zemun, Serbia

K. Kailasapathy
Probiotics and Encapsulated Functional Foods Unit Centre for Plant and Food Science, School of Natural Sciences, University of Western Sydney, Hawkesbury Campus, Locked Bag 1797, Penrith South DC, NSW 1797, Australia

N.J. Zuidam (✉)
Unilever R&D Vlaardingen, Structured Materials & Process Science, Olivier van Noortlaan 120, 3133 AT Vlaardingen, The Netherlands
e-mail: klaas-jan.zuidam@unilever.com

N.J. Zuidam and V.A. Nedović (eds.), *Encapsulation Technologies for Active Food Ingredients and Food Processing*,
DOI 10.1007/978-1-4419-1008-0_10,

intestinal tract, while *Bacteroides* predominates at 86% of the total flora in the adult gut, followed by *Eubacterium*. Infant-type bifidobacteria *B. bifidum* are replaced with adult-type bifidobacteria, *B. longum* and *B. adolescentis*. With weaning and aging, the intestinal flora profile changes. Bifidobacteria decrease, while certain kinds of harmful bacteria increase. Changes in the intestinal flora are affected not only by aging but also by extrinsic factors, for example, stress, diet, drugs, bacterial contamination and constipation. Therefore, daily consumption of probiotic products is recommended for good health and longevity. There are numerous claimed beneficial effects and therapeutic applications of probiotic bacteria in humans, such as maintenance of normal intestinal microflora, improvement of constipation, treatment of diarrhea, enhancement of the immune system, reduction of lactose-intolerance, reduction of serum cholesterol levels, anticarcinogenic activity, and improved nutritional value of foods (Kailasapathy and Chin 2000; Lourens-Hattingh and Viljoen 2001; Mattila-Sandholm et al. 2002). The mechanisms by which probiotics exert their effects are largely unknown, but may involve modifying gut pH, antagonizing pathogens through production of antimicrobial and antibacterial compounds, competing for pathogen binding, and receptor cites, as well as for available nutrients and growth factors, stimulating immunomodulatory cells, and producing lactase (Kopp-Hoolihan 2001).

Probiotics can be delivered commercially either as nutritional supplements, pharmaceuticals or foods. A large number of probiotic products are available in the market in the form of milk, drinking and frozen yoghurts, probiotic cheeses, ice-creams, dairy spreads and fermented soya products. Also, special freeze-dried pharmaceutical dietary preparations are available in the form of tablets, but the marketing as a pharmaceutical product requires long, complex and costly research, and a demonstration of a well-defined therapeutic target. Together with prebiotics, probiotics are often consumed as functional foods, demonstrated to be effective for the treatment or control of several diseases. Prebiotic substances, such as lactulose, lactitol, xylitol, inulin and certain non-digestive oligosaccharides, selectively stimulate the growth and activity of, for example, bifidobacteria in the colon (Zubillaga et al. 2001). Most widely and commercially used probiotic species are *Lactobacillus* (*L. acidophillus, L. casei, L. fermentum, L. gasseri, L. johnsnli, L. lactis, L. paracasei, L. plantarum, L. reuteri, L. rhamnosus, L. salivarius*), *Bifidobacterium* (*B. bifidum, B. breve, B. lactis, B. longum*), *Streptoccocus* (*S. thermophilus*) species, yeasts and molds (*Saccharomyces boulardii*). The presence of a specific enzyme, the fructose-6-phosphate-phospoketolase (F6PPK) in bifodobacteria, is the main criteria to distinguish them from Lactobacillus.

International standards (e.g., from the International Dairy Federation) require that products claimed to be 'probiotic products' contain a minimum of 10^7 viable probiotic bacteria per gram of product or 10^9 cells per serving size when sold, in order to provide 10^{6-8} cells/g feces. However, many products failed to meet these standards when they are consumed. This is due to death of probiotics cells in food products during storage, even at refrigerating temperatures. Consequently, industrial demand for technologies ensuring stability of bifidobacteria in foods remains strong, which leads to the development of immobilized cell technology to produce probiotics with increased cell resistance to environmental stress factors (Doleyres and Lacroix 2005).

This chapter will briefly review the isolation and selection of probiotic strains, and then focus on the use of several microencapsulation techniques to protect probiotics. The use of these microencapsulates in several food applications and their future developments are then discussed.

10.2 Isolation and Selection of Probiotic Strains

The isolation of potential probiotic strains proceeds from animal or human planktonic flora, or adhesive bacteria (adhered on the surface of the epithelial cells and interacting with the intestinal mucosa and surfactants). This can be achieved by preendoscopic biopsies or brushing. The procedure of brushing is more physiological, less invasive, requires optimal intestinal preparation and permits the withdrawn of almost complete planktonic flora. After a strain has been isolated, and purified to obtain a pure culture, it must be taxonomically classified. The identification of a strain is performed by comparison of rDNA gene sequences with those available in the GeneBank database. The minimum DNA genomic similarity of 70% and a 16 S rRNA sequence similarity of 97% are required to recognize a probiotic strain. After taxonomic classification, growth curves are developed and duplication parameters are determined for the specific strain. The presence of plasmid DNA is also assessed during the preliminary stage in order to obtain information on the genomic stability of the strain. As a general rule, the presence of plasmids is not a reason to discard the strain as a potential probiotic, but the role of this extra-chromosomal DNA in establishing phenotypes relevant for the technological and probiotic properties must be assessed (Del Piano et al. 2006).

A probiotic strain must be resistant to stomach and upper intestine microenvironment, to be able to reach the colon and be effective by conferring health benefits to the host. Therefore, in vitro studies are conducted to test the survival of a potential probiotic strain to gastric, bile or pancreatic juices. The survival of a strain depends on both strain characteristics and intestinal juice type (simulated gastric, bovine or pig bile and various types of animal pancreatic extracts).

Except the stability, the safety of a novel and existing starter, and probiotic cultures must be evaluated. There was a constant requirement for antibiotic resistant probiotics in the past few decades. This, on the other hand, led to the prevalence of multi-drug resistant strains that caused diseases in humans. The establishing of a safety profile implies determination of strain resistance to a wide variety of common classes of antibiotics and subsequent confirmation of non-transmission of drug resistance genes or virulence plasmids. Ideally, probiotic bacteria should exhibit tolerance to antimicrobial substances, but should not be able to transmit such resistance to other bacteria (Charteris et al. 2000; Kheadr et al. 2004; Moubareck et al. 2005). Although studies on safety of probiotics are necessary, in general most of *Lactobacillus* and *Bifidobacterium* strains are recognized as safe and have long history of safe use in foods or present in normal human intestinal microflora. Cases of infection pathologies or allergic reactions caused by probiotics or food substances employed for their processing are very rare.

Since probiotic bacteria are very sensitive to the environmental factors, stability tests are a prerequisite to define conditions under which they should be produced and stored. Stability of probiotics depends on many factors, including the genus, species, strain biotype and the formulation, as well as parameters such as temperature, water, pH, osmotic pressure, mechanical stresses and oxygen. Especially, the viability of lactic acid bacteria is jeopardized after freezing. Therefore, a strain must be tested on growth conditions during fermentation (alkali used to neutralize pH), harvesting conditions (cell washing, medium in which cells are re-suspended after concentration) and freezing conditions (cryoprotectants, freezing temperature, rate, duration).

Last but not least, the health benefits of potential probiotics strains should be assessed. Some potential health benefits, ranging from maintenance of normal intestinal flora to anti-cancer effects, have already been mentioned in the previous section. However, such benefits might be very strain-specific, are relatively small (compared to drugs) and may be affected by the food matrix. Long-term clinical studies with many people are therefore required to get fully proven health effects, especially when people are generally healthy.

10.3 Microencapsulation Technology for Probiotics

10.3.1 Protection Needs of Probiotics

During the time from processing to consumption of a food product, probiotics in that food product need to be protected against the following:

- Processing conditions, like high temperature and shear.
- Desiccation if applied to a dry food product.
- Storage conditions in the food product on shelf and in-home, like food matrix, packaging and environment (temperature, moisture, oxygen).
- Degradation in the gastrointestinal tract, especially the low pH in stomach (ranging from 2.5 to 3.5) and bile salts in the small intestine.

Microencapsulation technologies have been developed and successfully applied to protect the probiotic bacterial cells from damage caused by the external environment at the conditions mentioned above.

Encapsulation technology is widely used for various food applications such as stabilizing food compounds, controlling the oxidative reactions, sustained or controlled release of active ingredients (probiotics, minerals, vitamins, phytosterols, enzymes fatty acids and antioxidants), masking unpleasant flavors and odors, or to provide barriers between the sensitive bioactive materials and the environment (see other chapters of this book). Encapsulation technology is based on packing solid, fluid or gas compounds in milli-, micro- or nano-scaled particles which release their contents upon applying specific treatments or conditions (e.g., heating, salvation, diffusion and pressure). Sealed capsules are coated with semipermeable, spherical, thin, strong membrane around the solid or liquid core. A coating can be designed to open in specific areas of the human body and microcapsules can gradually release active ingredients. For engineering probiotic containing capsules, a coating is usually employed which can withstand acidic conditions in the stomach and bile salts from the pancreas after consumption. In this way, the protection of the biological integrity of probiotic products is achieved during gastro-duodenal transit, which is a prerequisite for delivery of a high concentration of viable cells to the jejunum and the ileum. Probiotics should ideally be released in segments of the gastrointestinal tract where Peyer's patches and other mucosa-associated lymphatic tissues are found that are said to play a critical role in immunostimulation (Rescigno et al. 2001). Since encapsulates should provide protection of sensitive microorganism against harsh conditions in the gut environment, the produced particles should be tested on swelling, erosion, disintegration in simulated gastric/intestinal fluids prior to industrial and real-life applications. Another purpose of microencapsulation of probiotic bacteria is to stabilize them, that is, to ensure prolonged viability during storage. The so-called stabilization of microorganisms means providing metabolic activity after storage and intake by a new host (Viernstein et al. 2005). An average rate loss found for sophisticated formulations under excellent storage conditions is one log unit of cell number reduction per year, which still means a loss of 90% per year.

There are two main problematic issues when considering microencapsulation of probiotics: (1) the size of probiotics (between 1 and 5 μm diameter) which immediately excludes nanotechnologies, and (2) difficulties to keep them alive. The most common techniques currently used for microencapsulation of probiotics will be presented in this section (Sect. 10.3) and their application in food products in Sect. 10.4.

10.3.2 Spray-Drying

Microencapsulation by spray-drying is a well-established technique suitable for large-scale, industrial applications: a liquid mixture is atomized in a vessel with a nozzle or spinning wheel and the solvent is then evaporated by contacting with hot air or gas. The resulting particles are collected after their fall to the bottom. Spray-drying is probably the most economic and effective drying method in industry. It can be used for dehydration of materials and/or encapsulation. However, to our best knowledge spray-drying has not been developed commercially for probiotics for food use yet, because of low survival rate during drying of the bacteria and low stability upon storage. The conventional procedure requires exposing of cells to severe temperature and osmotic stresses due to dehydration, which results in relatively high viability and activity losses immediately after spray-drying and most likely also affects storage stability. Main parameters that affect these include the following:

Type of strain: One strain survives spray-drying much better than the other. Preferably stationary phase cultures should be used (Corcoran et al. 2004).

Drying temperature: The logarithmic number of probiotics decreases linearly with outlet air temperature (in the range of 50–90° C) of the spray-dryer (Brian and Etzel 1997; Chavez and Ledeboer 2007), and to a lesser extent with the inlet air temperature (typically in the range of 150–170° C). An optimal outlet air temperature might be as low as possible (using a low feed rate, also allowing low inlet air temperatures like 80° C); however, one should take care that the powder obtained has been dried sufficiently at such low temperature conditions. Alternatively, a second drying step in, for example, a fluid bed (Meister et al. 1999) or vacuum oven (Diguet 2000; Chavez and Ledeboer 2007) might be applied.

Drying time: The shorter the heating time, the better the viability of probiotics. Optimal drying time, however, is affected by the droplet size of the atomized liquid, which is influenced by viscosity and flow rate of the feed solution.

Type of atomization: High shear must be avoided during atomization, and the air pressure applied might also influence the droplet size and thus the optimal drying time.

Carrier material: Typically a mixture of about 20% (w/v) (dairy) proteins and/or carbohydrates are used, which may be in the glassy state at storage temperatures to minimize molecular mobility and thus degradation. Examples include skim milk powder (SMP), non-fat dry milk solids (NFDM), soy protein isolates, gum arabic, pectin, (modified) starch, maltodextrin and sugars.

Osmotic, oxidative and mechanical stresses should be minimized, during both spray-drying and rehydration. Antioxidants and osmoprotectants might be included in the carrier material. Furthermore, the use of 'pre-stressed' bacteria may improve survival. Desmond et al. (2001) found that heat-adapted (52°C for 15 min) or salt-adapted (0.3 M NaCl for 30 min) *Lactobacillus paracase*i had a, respectively, 18-fold or 16-fold greater viability upon spray-drying than controls.

Storage conditions: Survival of probiotics is optimal at low water activity (<0.25) and low temperatures. Oxygen and light might be detrimental, so a nitrogen or vacuum-sealed package with a proper barrier function should be selected.

Unfortunately, the conditions need to be optimized for each different type of probiotic strain, and a good survival upon spray-drying may not indicate a good survival upon storage in a spray-dried form.

Picot and Lacroix (2004) spray-dried fresh and freeze-dried bifidobacteria in the presence of an o/w emulsion composed of anhydrous milk fat and an aqueous solution of 10% heat-denaturated whey protein isolate. This resulted in the production of water-insoluble microcapsules (<100 μm). However, the viability of the probiotics was low and slightly better results were obtained in the absence of the milk fat (26 or 1.4% survival for fresh *Bifidobacterium breve* R070 and *Bifidobacterium longum* R023, respectively, in the absence of fat; the experiments with freeze-dried ones resulted in survival rates <1%). The authors used a relatively high outlet temperature of 80° C, which might be the cause of the low survival found upon spray-drying. Another reason for the low survival might be the sensitivity of their probiotics towards the spray-drying process. The authors claim higher storage stability in yoghurt (+2.6 log cycles after 28 days at 4° C) and survival in gastrointestinal (GI) tract of the encapsulated probiotics (+2.7 log cycles) compared to free ones.

Crittenden et al. (2006) also spray-dried probiotics in the presence of an o/w emulsion, but combined this with Maillard reaction products between protein and carbohydrates to improve film-forming and oxygen-scavenging properties of the shell. First, emulsions were prepared of canola vegetable oil, caseinate, fructo-oligosaccharides (a prebiotic), dried glucose syrup or resistant starch, and heated to 98° C for 30 min to promote Maillard reactions. The emulsion was then cooled to 10° C, probiotics were added and finally this mixture was spray-dried. Final formulations of the dried powders were 8% (w/w) probiotics, 32% (w/w) oil, 20% (w/w) caseinate, 20% (w/w) fructo-oligosaccharides and either 20% (w/w) dried glucose syrup or 20% (w/w) resistant starch. The encapsulated probiotics were more stable upon storage at 25° C and 50% relative humidity than non-encapsulated ones. The encapsulated probiotics were also more stable in in vitro gastrointestinal tract conditions.

10.3.3 Freeze- or Vacuum-Drying

Freeze-drying is performed by freezing probiotics in the presence of carrier material at low temperatures, followed by sublimation of the water under vacuum. In this way, water phase transition and oxidation are avoided. The addition of cryoprotectants helps to retain probiotic activity upon freeze-drying and stabilize them during storage. Many investigators have used SMP as the major drying medium, but other compounds like fructose, lactose, mannose, monosodium glutamate, sorbitol (Champagne et al. 1991; Carvalho et al. 2002, 2003, 2004a, b), trehalose (Garcia De Castro et al. 2000), 30% maltodextrin (Brian and Etzel 1997) and a mixture of 20% soy protein isolate and 20% maltodextrin (Chavez and Ledeboer 2007) have also been used as protective additives in recent investigations. The obtained dried mixture can be grounded (Picot and Lacroix 2003) and the final particles are of a wide size distribution and with a low surface area.

Freeze-dried probiotics are well stable upon storage, especially at low temperatures and in an inert atmosphere (nitrogen or vacuum). In general, the choice of optimal water content (in the order of 3–8%) is a compromise between high survival rates immediately after drying (more survival at higher water contents) and low inactivation upon storage (more survival at low water contents although not necessarily at 0%). The decrease in survival of freeze-dried bacteria under vacuum may follow first-order kinetics and the rate constants can be described by an Arrhenius equation (King et al. 1998). Extrapolation from results obtained at higher temperatures allows one to predict the degradation at any selected temperature. Based on the study of King et al. (1998) one can calculate that at 70, 60, 50, 20 and 4° C, a 50% reduction in cell viability of freeze-dried *Lactobacillus acidophilus* in originally 4.15% glycerol, 10% NFDM and 0.53% $CaCO_3$ and with a final moisture content of 3% is obtained after 0.2 h, 50 h, 9.6 days, 5.2×10^6 days and 1.3×10^{10} days, respectively. The Arrhenius relationship might be affected by phase transition (if any) and atmosphere (oxidation by oxygen may not follow first-order kinetics). Maybe water content will play a role as well.

Not much is known about the rehydration medium. When probiotics suspended in water are freeze-dried, the rehydration medium has a considerable effect on viability (Champagne et al. 1991). The situation is more complex when a better drying medium has been used. The rehydration medium must be free of RNase and probably near a neutral pH. Different temperature effects have been reported, depending on the type of strain. It has been recommended to rehydrate the culture back to the volume it had prior freeze-drying. Long rehydration periods might be detrimental as the bacteria themselves might form inhibitory compounds.

Unfortunately, freeze-drying is a very expensive technology [about 4–7 more expensive than spray-drying (Chavez and Ledeboer 2007)]. However, freeze-drying is one of the least harmful drying methods of probiotics and is therefore probably most often used to dry probiotics, also as a standard to compare with other drying techniques. Most freeze-dried probiotics only provide stability upon storage and not or limited in the gastrointestinal tract. An exception might be freeze-dried probiotics from Cell Biotech in Korea, which are called Duolac™ (http://www.cellbiotech.com/sub06/img/duolac.pdf). Bacteria with a soy protein coating (most likely made by a precipitation process prior freeze-drying) get a further polysaccharide coating (= dual coating) and are then freeze-dried in the presence of cryoprotectant. After grinding, the particle sizes are around 125–250 µm. Cell Biotech claims that the coating material shrinks and coagulates together at stomach pH to protect lactic acid bacteria. The coating material dissolves in the small intestine due to its neutral pH conditions, but it should still protect the bacteria for bile salts.

Some years ago, a new starch-based technology for probiotic microencapsulation was developed by VTT Biotechnology (Myllarinen et al. 2000, using freeze-drying; O'Riordan et al. 2001, using spray-drying). Both steps, the bacterial production and their encapsulation were preformed in one batch process (Myllarinen et al. 2000). Starch is a dietary component, having an important role in the colonic physiology. Starch consists of two types of molecules, amylose and amylopection (see Chap. 3). There are different forms: starch entrapped within food matrix, granular starch structure and retrograded starch formed after food processing. In the VTT technology large potato starch granules (50–100 µm), enzymatically treated to obtain a porous structure, were used as a carrier. The enzyme attacked the inside of the granules, making them porous. Subsequently, amylose, the linear polymer of starch was solubilized, cooled and precipitated over the bacteria-filled starch granules. The strength of adhesion of bifidobacteria to starch granules varied for different starches (Crittenden et al. 2001). Finally, the whole product together with the growth media was freeze-dried to a powder form. Different amylases (bacterial, malt, fungal, pancreatic) have been tested using a range of conditions to establish the optimal method to produce internal hollows for the encapsulating bacteria inside starch granules. Several probiotic strains have been used in starch encapsulation studies. In addition, the viability of encapsulated bacteria stored at room temperature was at least 6 months and when frozen, at least 18 months. The capsule material appeared to be resistant to intestine milieu in in vitro and in vivo studies. A new, interesting approach is to use starch granules that naturally form aggregates, such as small barley starch granules (Mattila-Sandholm et al. 2002).

Vacuum drying is a similar process as freeze-drying, but it takes place at 0–40° C for 30 min to a few hours. The advantages are that the products are not frozen, which prevents freezing damage and energy consumption, and that the drying is fast. King and Su (1993, 1995), and King et al. (1998) used controlled low-temperature vacuum dehydration (CLTVD) to dry *Lactobacillus acidophilus* at about 0° C. This temperature was maintained by a controlled combination of shelf heating and vacuum adjustment in a freeze-dryer. Vacuum drying at about 40° C resulted in poor survival numbers. At about 0° C the bacterial survival upon CLTVD was just slightly lower than the one upon freeze-drying (King and Su 1995). A lower decrease in viability of *Lactobacillus acidophilus* with time was found when 0.57% $CaCO_3$ and 4.1% glycerol was added to the growth medium and 10% NFDM prior CLTVD (King and Su 1995; King et al. 1998). The freeze-dried or CLTVD-dried bacteria could be stored at 5 or –20° C for more than 120 days without much degradation, in contrast to storage at room temperature. Based on the study of King et al. (1998) one can calculate that at 70, 60, 50, 20 and 4° C, a 50% reduction in cell viability of vacuum-dried *Lactobacillus acidophilus* in originally 4.1% glycerol, 10% NFDM and 0.57% CaCO3 and a final moisture content of 3% is obtained after 0.5 h, 4 h, 22 h, 999 days and 7.5×10^4 days, respectively.

10.3.4 Fluid Bed Coating

In spray-coating techniques, the core material needs to be in a solid form and is kept in motion in a specially designed chamber, either by injection of air at the bottom (fluid bed coating, see also Sect. 2.2.2 of this book) or by rotary action (e.g., pan coating). Solid forms of probiotics can be obtained by spray-drying or freeze-drying (see previous sections). A liquid coating material is sprayed through a nozzle over the core material in a hot environment. The film formation then begins, followed by successive wetting and drying or solidification stages which result in a solid, homogenous layer on the surface of a core. The small droplets of the sprayed liquid contact the particle surface, spread on the surface and coalesce. The spray liquid, also referred as shell, wall or coat material can be a solution, a suspension, an emulsion or a melt. Any edible material with a stable molten phase can be sprayed at high deposition rates, allowing coatings with a thickness of 100 μm up to 10 mm. The coating material can be injected from many angles and this influences the properties of the coating. In Fig. 10.1 three fluid bed coating technologies are presented, principally differing in the type of air fluidization employed and the site in the vessel where the coating material is sprayed: the top spray-, the bottom spray- and the tangential spray-coating. The probiotic bacteria are presented in fine powder particles prepared by traditional methods (fermentation, concentration, freeze-drying and granulation). The coating material is introduced into the vessel under compressed air. In food applications the coating of probiotics is mostly lipid based (e.g., waxes, fatty acids and specialty oils), but proteins (e.g., gluten and casein) or carbohydrates (e.g., cellulose derivatives, carrageenan and alginate) can also be used (Champagne and Fustier 2007).

Spray-coating technique is suitable for particles with a diameter from 50 μm to 5 mm. Product quality characteristics depend on numerous variables, which affect different steps of the process. The film characteristics, through the evaporation or congealing rate, are function of fluidization air velocity, temperature and humidity (Jacquot and Pernetti 2004; Sect. 2.2.2 of this book). The coating homogeneity and success are influenced by the stickiness of the coating material, the wettability of particles by the coating liquid and the operating conditions. The thickness of the

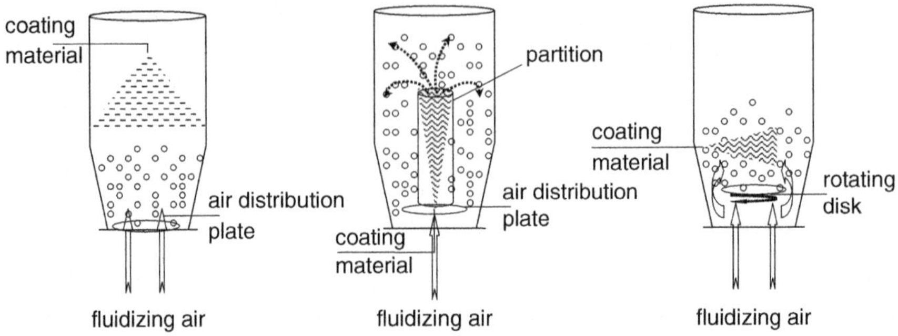

Fig. 10.1 Fluid bed coating methods for the microencapsulation of probiotics. (**a**) Fluid bed top spray-coating; (**b**) Fluid bed bottom spray-coating with the Würster device; (**c**) Fluid bed tangential spray-coating. Adopted from publications of Champagne and Fustier (2007) and Jacquot and Pernetti (2004)

final film coat is determined by the number of coating cycles (passages of the particles in the coating zone). An adequate choice of the coating material (with respect to viscosity and hygroscopicity), and control of the operating conditions, such as the particle velocity and bed moisture content, prevent collision between particles and agglomeration. During the spraying process bubbles might form due to shear and be trapped in the coating film, which affects porosity, permeability and mechanical properties of the shell layer.

In the top fluid bed coating mode, the spray liquid and the air flow are counter current (Fig. 10.1a) and the distance between powder particles and liquid droplets are relatively large. Therefore, there is a risk of spray liquid drying or solidifying before they coat the particles. Particles should travel fast to prevent agglomeration and the liquid droplets should be small enough to sediment on the core and create dense coating film. In practical applications, the motions of fluidized core particles are random, resulting in a nonuniform coating. In the bottom fluid bed coating, the spray liquid is introduced in the vessel through spray nozzles placed at the bottom, thus in the concurrent direction with the air flow (Fig. 10.1b). Würster (1950) improved the device by adding a cylindrical partition centrally placed and an air distribution plate. This improved device brings the powder particles in circulation and enables dense and homogenous coating. Collisions between spray liquid droplets and powder particles are increased, resulting in higher coating efficiency, lower droplets drying or solidifying before they coat the particles and minimal risk of agglomeration. In addition, the production capacity of the Würster coating device is increased compared to a conventional top spraying coating system. The tangential fluid bed coating device is also called rotary spray-coating system (Fig. 10.1c). A rotary disk, placed at the bottom of the chamber, maintains a complex fluidization pattern and the particles movement is influenced by centrifugal force, air stream and gravity (Jacquot and Pernetti 2004). The coating liquid is brought in tangentially, while air streams pass through the gap between the rotor disk and inside chamber wall, maintaining fluidization of the core particles. As with bottom spray device, the achieved coating is homogeneous. The main disadvantage of the technique is the high shear stress applied to the particles, thus it is limited to sturdy and resistant materials.

Fluid bed coating is among all, probably the most applicable technique for the coating of probiotics in industrial productions, since it is possible to achieve large batch volumes and high throughputs. As written above, most coatings used are lipid based. Commercially available encapsulates include Probiocap™ of Lallemand (Institut Rosell, see http://www.lallemand.com/HNAH/eng/probiotics.shtm; Makhal and Kanawjia 2003; Durand and Panes 2001). These particles are made by fluid bed coating of freeze-dried probiotics with low melting lipids, around 250 μm in size, and developed by Lallemand in collaboration with Balchem (see http://www.balchem.com/encapsulates and Wu et al. 2000).

10.3.5 Spray-Cooling

In spray-cooling, a molten matrix with low melting point containing the bioactive compound is atomized through a pneumatic nozzle into a vessel (see also Sect. 2.3.3 of this book). This process is similar to spray-drying with respect to the production of fine droplets. However, it is based on the injection of cold air into the vessel to enable solidification of the gel particle rather than on hot air which dries the droplet into a fine powder particle. The liquid droplet solidifies and entraps the bioactive product. Spray-cooling is considered as the least expensive encapsulation technology and offer few advantages over other encapsulation techniques. It may expand the range of matrices used. Further, it is possible to produce very small particles. However, so far it has been used rarely for probiotics (rather more suitable for encapsulation of other food ingredients, like water-soluble vitamins, fatty acids, antioxidants, fatty acids, yeasts, enzymes), since other technologies are easier to establish in laboratories. One example is the spray-cooling of a slurry of freeze-dried probiotics and molten lipids (e.g., 60–75% stearic acids at 60° C), which was atomized by a rotary disk in a cooling chamber to give 75–300 μm encapsulates (Rutherford et al. 1993). The contact time of the freeze-dried probiotics should remain very short, but no details about the survival rate of freeze-dried probiotics at 60° C were given by the inventors. Section 10.3.3 may indicate more about the storage stability of freeze-dried probiotics at this high temperature, assuming that molten lipids have no further detrimental effect on probiotics.

10.3.6 Encapsulation of Probiotics in Microspheres

10.3.6.1 Gel-Particle Techniques

Probiotics can be encapsulated in microspheres (gel beads or polymeric matrix beads), often coated with an outer layer which may be designed to dissolve under specific conditions allowing release of the encapsulated bacteria (Anal and Singh 2007).

Polymeric matrices are utilized mainly to protect probiotics against low pH and high bile concentrations, but they also ease handling and allow propagation of the probiotics in application.

Extrusion or emulsification techniques may be applied to produce spherical polymer beads ranging from 0.3 to 3 mm in diameter (Krasaekoopt et al. 2003). The first step in both techniques is mixing of bacterial culture with a polymer solution to create bacteria-polymer suspension (Fig. 10.2), which is then extruded through a needle to produce droplets collected in a bath where gelation occurs (ionotropic or thermal), or dispersed in a continuous phase applying mixing to create stable w/o emulsion.

Extrusion is the oldest and the most common approach to making capsules with hydrocolloids, and might be achieved by simply dropping an aqueous solution of probiotics into a gelling bath. Extrusion bead production techniques (like electrostatic, coaxial-air flow, vibration, atomization or jet-cutter) are based on applying the additional force to generate smaller spheres compare to those produce by simple dropping; the size of the particles can be adjusted by choosing needle diameter and manipulating the distance between the outlet and the coagulation solution and electric or piezzo parameters. Extrusion technology is more popular than emulsion technology due to its simplicity, easy to handle with the equipment, low cost at small scale and gentle formulation conditions ensuring high retention of cell viability. The main problem with respect to their applications on probiotics is the relatively large particle size, although it is possible to generate microspheres of very narrow size distribution.

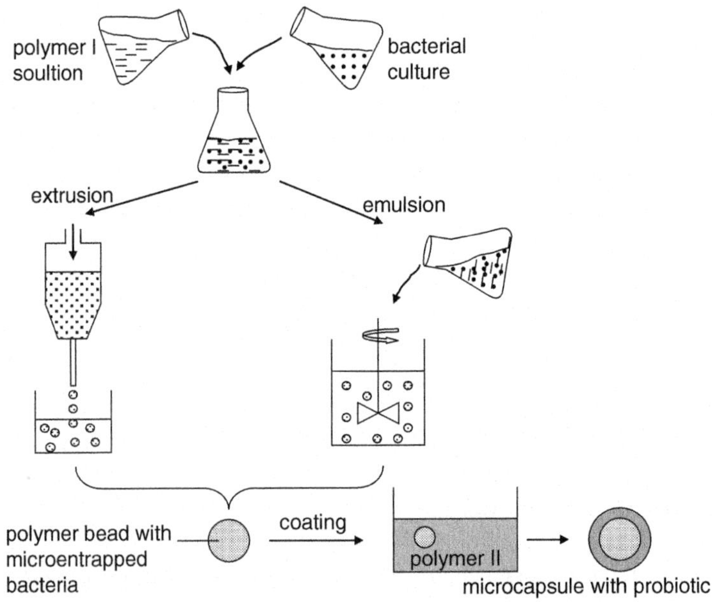

Fig. 10.2 Gel-particle technologies for the microencapsulation of probiotics

In the emulsion technique, a small volume of cell-polymer suspension (discontinuous phase) is added to a large volume of a vegetable oil (continuous phase) such as soybean oil, sunflower oil, canola oil or corn oil. In some studies, white light paraffin oil (Rao et al. 1989) and mineral oil (Groboillot et al. 1993) have been used. The mixture is homogenized to form a water-in-oil (w/o) emulsion. In some cases emulsifiers are added to form more stable emulsions, since these agents lower the surface tension of droplets leading to smaller spheres. The most common emulsifier used is Tween 80 at low concentrations. Once the emulsion is formed, solidification occurs after the addition of an adequate solidifying agent to the emulsion. In the emulsion procedure, adjustment of agitation speed and phase ratio enables production of the targeted bead size. The size of the beads can vary between 25 μm and 2 mm. The double emulsion technique (water-in-oil-in water, w/o/w), a modification of the basic technique in which an emulsion is made of an aqueous solution in a hydrophobic wall polymer can also be appropriate for incorporation of probiotics (Shima et al. 2006). The relative viability of the encapsulated microbial depends on operating parameters, such as inner phase volume ratio and the median diameter of the oil droplets.

The obtained polymer beads with entrapped microbials can be further introduced into a second polymer solution to create a coating layer which provides an extra protection to the cells and/or gives sensorial properties to the product. Another way to perform coating is to use co-extrusion devices, where beads formation and wrapping occur simultaneously (see Sect. 10.3.7). Coating can be performed with cationic polymers, such as polyethylenimine, polypropyleneimine, chitosan or combination of these. However, these polymers have no or limited food grade status. Formation of the membrane around the beads results in stronger microcapsules and minimized cell release.

Storage stability of obtained microspheres can also be enhanced by fluidization drying or by freeze-drying using cryoprotective additives like skimmed milk with or without 5% saccharose and/or 0.35% ascorbic acid (Goderska and Czarnecki 2008).

10.3.6.2 Encapsulation of Probiotics in Alginate

Alginate is the most widely used encapsulation matrix for various food-grade and non-food compounds. Alginate is used in the form of a salt of alginic acid. Alginates are naturally derived linear copolymers of 1,4-linked β-D-mannuronic acid (M) and α-L-guluronic acid (G) residues (Martinsen et al. 1989; Gombotz and Wee 1998; Sect. 3.2.1.4 of this book). The ratio and sequential distribution of uronic acid residues, along the length of the alginate chain, vary in alginates of different origins (brown seaweeds, certain bacteria) (Martinsen et al. 1989; Gombotz and Wee 1998). There is no regular repeat unit in alginate polymers, and the chains can be described as a varying sequence of regions which usually denotes as M blocks, G blocks and MG blocks. Aqueous solutions of polysaccharides form hydrogels in the presence of divalent Ca^{2+} ions via ionic interactions between the acid groups on G blocks and the gelating ions. As a result, calcium alginate gels are physically cross-linked polymers with mechanical and hosting properties dependant on the alginate composition. The mechanism of calcium-induced alginate gel formation occurs due to orderly alignment of the alginate polymers which interact with divalent cations such as calcium, where calcium ions occupy the space between two alginate polymers like an egg placed inside an egg box and is known as "egg box" gelling mechanism (Smidsrod and Skjak-Braek 1990; Skjak-Braek et al. 1986).

Both techniques (extrusion and emulsion) can be applied to generate calcium alginate microspheres. In the emulsion technique, the addition of an oil-soluble acid, such as acetic acid, reduces the alginate pH from 7.5 to approximately 6.5, enabling initiation of gelation with Ca^{2+} (Poncelet et al. 1993).

The survivability of probiotic cultures in calcium alginate beads in general depends on many factors, such as concentration of alginate and gelation solution ($CaCl_2$), the duration of gelation and cell concentration (Chandramouli et al. 2004; Lee et al. 2004; Lee and Heo 2000; Sheu et al. 1993). For example, the survival of *Lactobacillus casei* increased proportionately with increasing alginate concentrations from 2% to 4% (Mandal et al. 2006). The probiotic strain *L. acidophilus* encapsulated in Ca–alginate beads showed higher survival level under different conditions compared to the non-encapsulated cultures. The viability of encapsulated bacteria in simulated gastric fluid appeared to increase with the increase in bead size. Lee and Heo (2000) proposed a model to express the influences of gel concentration, bead size and initial cell numbers on the survival of bifidobacteria in calcium alginate beads in *in vitro* gastrointestinal conditions. While large alginate beads (>1 mm) get rough textural structure in the real microbial feed solution, small capsules (<100 μm) allow fast and easy diffusion of water and other molecules in and out of the matrix (Truelstrup Hansen et al. 2002). Out of nine different strains of *Bifidobacterium spp* encapsulated in calcium alginate spheres, only the strain *B. lactis* Bb-12 was found to be resistant to low pH and bile salts (Truelstrup Hansen et al. 2002). The loaded, 20–70 μm in diameter calcium alginate microspheres were produced by emulsification procedure using 2% alginate and showed good stability of Bb. 12 after storage up to 16 days in various surrounding media ($CaCl_2$, milk, yoghurt, sour cream) and for 1 h in simulating gastric fluid (37°C, pH 2.0). However, the small alginate spheres could not provide good protection to the other, more acid-sensitive bifidobacteria strains against low pH or upon storage in milk.

Selmer-Olsen et al. (1999) found that addition of protective solutes was very important when drying *Lactobacillus helveticus* CNRZ 303 in calcium alginate beads by a fluid bed. The bacteria fall in viability when the water content decreased below 100% (w/w). The best survival upon drying was found in the presence of adonitol (a sugar alcohol derived from ribose) and non-fat milk solids (respectively, 71% and 57% survival after drying to 20–30% water content). These were also the best for survival upon storage. Rehydration conditions also affected the survival rate; best results were obtained by Selmer-Olsen et al. (1999) when the cells were rehydrated in cheese whey permeate between 20 and 30°C and pH 6–7.

In vitro laboratory studies have shown that with alginate hydrogel microcapsules, the release of the probiotic bacteria can be accomplished by shaking the gel beads in 0.1 M phosphate or citrate buffer solutions in a laboratory stomacher blender.

The calcium ions holding the alginate polymers are pulled out of the beads due to their affinity for hydroxyl ions and hence the orderly gel structure disintegrates and releases the probiotic bacteria. This method of release of the bacteria is used to determine the encapsulation efficiency (Chandramouli et al. 2004).

Coated alginate microcapsules approve to have better protective characteristics compared to uncoated ones. Krasaekoopt et al. (2004) encapsulated three different probiotic strains in alginate particles coated with three types of materials (chitosan, sodium alginate and poly-L-lysine) and determined that chitosan-coated alginate beads provided the best protection for the strains *L. acidophilus* 547 and *L. casei* 01, while sensitive *B. bifidum* ATCC 1994 did not survive the acidic conditions of gastric juice. Chitosan, a positively charged polyamine, forms a semipermeable membrane around a negatively charged polymer such as alginate. This membrane does not dissolve in the presence of Ca^{2+} chelators or antigeling agents and thus enhances the stability of the gel (Smidsrod and Skjak-Braek 1990). As a consequence, the cell release is lowered down to 40% (Zhou et al. 1998). Low-molecular-weight chitosan diffuses more readily into the calcium alginate gel matrix resulting in a denser membrane than with high-molecular-weight chitosan (McKnight et al. 1988). A whey protein and pectin conjugation has also been used as a protective membrane around calcium alginate beads (Guérin et al. 2003). Protein–alginate composite beads were covalently bound by a transacylation reaction (Levy and Edwards-Levy 1996). The reaction involved the formation of amide bonds between protein and alginate, producing a membrane on the bead surface, which resisted gastric pH and pepsin activity. The bifidobacteria immobilized in the mixed gel were more resistant to simulated gastrointestinal tract conditions (Guérin et al. 2003).

Except conventional polymers, polysaccharides (fructo-oligosaccharides, isomalto-oligosaccharides) and peptides may also be used as an outer coating layer (Chen et al. 2005). Introducing an additional enteric coating (made from methacrylic acid copolymer, which is not food-grade) together with the outer coating layer (mixture of sodium alginate and hydroxypropyl cellulose in the weight ratio 9:1) enabled 10^4- to 10^5-fold increase in cell survival in simulated gastrointestinal tract fluids. In addition, the use of the non-food grade toluene diisocyanate as a cross-linking agent provided membranes which were more resistance to breakage (Hyndman et al. 1993).

The release of encapsulated probiotic bacteria from calcium alginate and chitosan-coated-alginate–starch encapsulates (CCAS) under ex vivo and in vivo conditions have been reported (Iyer et al. 2004, 2005). In these studies, the release profiles of different bacteria, *L. casei* strain Shirota (LCS) and green fluorescent protein (GFP)-tagged *Escherichia coli* K12 (*E. coli* GFP$^+$ K-12), from encapsulates were investigated in porcine gastrointestinal contents by an ex vivo method. In another study by the same authors, calcium alginate and CCAS encapsulates were fed to mice and bacterial release at different sites in the gastrointestinal tract was monitored for up to 24 h. In the latter experiment, LCS was used as a model probiotic strain because of the specific selective media used that allowed differentiation of the inoculated bacteria from food and from the gastrointestinal tract microbiota. The results showed that there was no detectable release of encapsulated bacteria from the capsules in the acidic gastric contents. In contrast, there was a complete release of *E. coli* within 1 h of incubation in the small intestinal contents (pH 6.5–6.8) at 37° C, while it took nearly 8 h to completely release the *E. coli* in the colon contents (pH 6.9) under similar conditions. In the case of LCS, there was no significant release of LCS in gastric porcine contents (pH 2.5) even after 24 h of incubation. There was a complete release of LCS in the ileal contents (pH 6.8) after 8 h of incubation. As in the ileum, there was a complete release of LCS from capsules in colon contents, but it took approximately 12 h. The results reported indicate that while there was a complete release of *E. coli* GFP^+ from calcium alginate encapsulates within 1 h in porcine ileal contents ex vivo, it took approximately 8 h to completely release LCS from CCAS capsules. The difference between the release of *E. coli* and LCS was reported to be due to the chitosan coating of the capsules. *E. coli* GFP^+ was encapsulated with alginates while alginate capsules containing LCS were coated with chitosan polymer. It can be said that microencapsulation in alginate gel beads with or without coating effectively minimizes the bactericidal effects of the gastric pH and maximize the number of encapsulated bacterial cells reaching the ileum and subsequently to the colon.

10.3.6.3 Encapsulation of Probiotics in κ-Carrageenan

Carrageenan is a natural polysaccharides isolated from marine macroalgae, commonly used as food additives (see Sect. 3.2.1.4). Carrageenan dissolves at high temperatures (60–80° C) in concentrations of 2–5% (Klein and Vorlop 1985). Dispersion of the carrageenan gel into small droplets has to be carried out at elevated temperatures (40–45° C) and gelation occurs during cooling procedure down to room temperatures. After the beads are formed, K ions in the form of KCl are used to stabilize the gel, prevent swelling or to induce gelation (Krasaekoopt et al. 2003). Audet et al. (1988) reported inhibitory effect of KCl on some bacteria such as *Streptococcus thermophilus* and *L. bulgaricus*. The presence of monovalent ions such as Rb^+, Cs^+, and NH_4^+ makes stronger gels (Tosa et al. 1979). *Lactobacillus acidophilus* survived freezing, freeze-drying and storage in a freeze-dried form much better in 3 mm 4% (w/v) κ-carrageenan gel beads made in 0.3 M KCl than free cells (Tsen et al. 2002). Locus bean gum in ratio to carrageenan of 1:2 significantly increases the strength of the gel through specific interaction of its galactomannan chains with carrageenan. Carrageenan/locus bean gum mixture has been frequently tested for microbial encapsulation (Audet et al. 1990, 1991; Ouellette et al. 1994; Doleyres et al. 2002a, b, 2004). Encapsulated cells proliferate in high biomass concentration in dairy products and exhibited increased tolerance to stresses, such as freeze-drying, hydrogen peroxide and simulated gastrointestinal conditions.

10.3.6.4 Encapsulation of Probiotics in Chitosan

Chitosan is a positively charged, linear polysaccharide formed by deaceylation of chitin (see Sect. 3.2.1.5). It is water soluble below pH 6 and forms a gel by ionotropic gelation. The terms chitin and chitosan refer not to specific compounds, but to two types of copolymers containing the two monomer residues anhydro-*N*-acetyl-D-glucosamine and amino-D-glucosamine, respectively. Chitin is a polymer of β-(1,4)-2-acetamido-2-deoxy-D-glucopyranose and one of the most abundant organic materials. Chitosan, a polycation with amine groups, can be cross-linked by anions and polyanions, such as polyphosphates (Anal and Stevens 2005), $[Fe(CN)_6]^{4-}$ and $[Fe(CN_6)]^{3-}$ (Anal and Singh 2007), polyaldehydrocarbonic acid (Klein and Vorlop 1985), and sodium alginate (Anal et al. 2003). It is an important biomaterial in food and pharmaceutical applications due to its favorable properties, such as good biocompatibility, biodegradability and non-toxicity. However, chitosan's food-grade status is not clear in many countries and does not taste well in a free form. Furthermore, it exhibits inhibitory effects on different types of lactic acid bacteria (Groboillot et al. 1993). Thus, chitosan is mainly used as coating for conventional alginate gel beads (Krasaekoopt et al. 2003, 2004; Lee et al. 2004; Zhou et al. 1998; see also Sect. 10.3.6.2). Various chitosans (different molecular weights) in combination with alginate can be used to achieve high cell loadings (up to 10^{10} cfu g^{-1}, Zhou et al. 1998). Nevertheless, the viability of the encapsulated microorganisms depends on the way by which chitosan cross-links with alginate (whether they interact and form matrix together, i.e., chitosan is the inner polymer or chitosan creates an outer layer around alginate sphere, i.e., chitosan is the outer polymer). Calcium alginate–chitosan microcapsules can be made by one- or two-step processes, based on the presence or absence of Ca^{2+} in the receiving chitosan solution (Lacík 2004). The beads can be prepared in a way to differ in a level of homogeneity of the alginate concentration gradient through the cross-section of the bead by addition of sodium chloride to the calcium chloride solution. Capsules' mechanical strength and permeability strongly depend on the process of capsule preparation (Gaserod et al. 1998, 1999). In the one-step process (in the absence of Ca^{2+} in chitosan solution), chitosan is located only at the interface, as a thin-alginate–chitosan membrane with a weak mechanical resistance. The capsules were much stronger when the two-step protocol was used. This difference between two protocols of capsule formation is due to the ability of chitosan to penetrate through the membrane (Lacík 2004). The kinetics of membrane formation and the capsule parameters (like thickness, permeability and mechanical strength) depend on the concentration of components, molar masses of both, alginate and chitosan, reaction time, pH and ionic strength. Sprayed particles coated with chitosan are recommended as impressively effective vehicles in delivering viable bacterial cells to the colon and stable shells during refrigerated storage.

10.3.7 Submerged Co-extrusion

Seamless capsules containing probiotics are available from Morishita Jintan Co. Ltd in Japan. These capsules are composed of three layers: a core of freeze-dried probiotics in solid fat (m.p of 35°C), with an intermediate hard fat layer (m.p. of

40° C) and a gelatin–pectin outer layer (Asada et al. 2003, and http://www.jintan-world.com). They are made with a concentric, multi-nozzle via a submerged co-extrusion technique (see also Sect. 2.3.9 of this book). The size of the capsules is quite large (1.8–6.5 mm) and the technique is relatively expensive, which may be a barrier for use in many food products. Capsules with different bifidobacteria and lactobacillus strains are available, and these reach the intestine alive without being too much affected by stomach acid or oxygen. Other actives, such as fish oil, vitamin C and iron sulfate might be encapsulated as well in these kind of capsules.

10.3.8 Twin Screw Extrusion

Some publications have also shown that probiotics can be processed in a twin screw extruder at moderate pressures and low temperatures. Example, Van Lengerich (1999) disclosed that pellets with *Lactobacillus acidophilus* can be prepared by twin screw extrusion. First, cookies were ground and this flour was fed into the extruder at 4 kg/h, followed by mixing with water and citrus juice (7/1 w/w) at 0.8 kg/h, and feeding in the next barrel of the extruder a preblend of 0.118 kg of probiotics, 0.375 kg of vegetable fat and 0.188 kg of vegetable oil at 0.75 kg/h. The extruder was operating at 150 rpm, 45 bar and 20° C, and equipped with a 20×1 mm die. The product temperature reached 31° C, and the pellets were dried afterwards in a convection batch dryer for 1 h at 30° C to 5.9% moisture. Optionally, pellets can be coated with a 25% shellac solution in alcohol to give 5–10% shellac coating. The patent claims that the starch should have been preprocessed (i.e., mixed and heat treated) to avoid gelatinization of the starch and provide a pleasantly taste and texture. Van Lengerich (2000) also entrapped *Lactobacillus acidophilus* in 0.5–1 mm pellets by feeding into a twin screw extruder semolina/wheat gluten 70/25 (w/w) at 2.5 kg/h, vegetable oil at 0.29 kg/h, water at 0.06 kg/h and 20% (w/w) *Lactobacillus acidophilus* at 0.82 kg/h. All barrels of the extruder were kept at 21° C, and a screw speed of 67 rpm and a die with 40 circular openings of 0.5 mm each were used. The temperature of the product remained in this way below 40° C. After extrusion, the pellets were dried for about 30 min under vacuum or carbon dioxide to prevent access of oxygen. Jongboom-Yilmaz (2002) disclosed in her patent that probiotics can be extruded in destructurized potato starch and/or sugar at 100 rpm, 13–17% torque, 8–17 bars and 33–38° C (die temperature). Ten percent of glycerol might be added as a plasticizer, which improved the survival of probiotics during the process and afterwards during storage.

10.3.9 Compression Coating

Recently, compressing coating has been developed as a promising technique which permits the stabilization of lyophilized cells during storage (Chan and Zhang 2002, 2004; Ubbink et al. 2003). This technique involves compressing dried cell powder

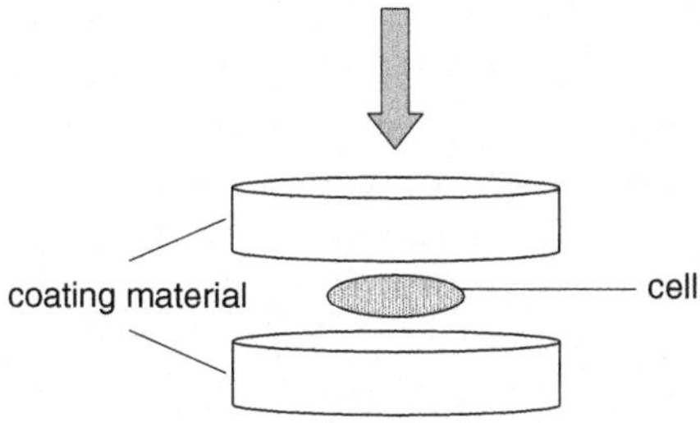

Fig. 10.3 Compression coating of cells (Chan and Zhang 2004)

into a core tablet or pellet with a 10 mm die, and then compressing coating material around the core to form the final compact. The compressed cell pellet should be positioned on the center of the die before the rest of the coating material is poured on the top of it and the punch applied, as shown in Fig. 10.3. In this way, two coatings may be formed: one enteric and the other an outer coating layer. Investigations of the bacteria immobilized by this procedure showed very good results with regard to cell protection against gastrointestinal tract in in vitro studies. Sodium alginate or pectin can be used as coating material with the addition of a binder compound (such as hydroxypropyl cellulose) to make a more rigid compact (Chan and Zhang 2002). Alternatively, 50% flour, 25% maltodextrin and 25% semi-humid pet food, or direct compressible starch, or 50% lactose and 50% maltodextrin can be used (Ubbink et al. 2003). An additional outer coating can be applied by dipping the pellets in a barrier solution or by fluid bed coating [e.g., using dipping in a melt of fat or fluid bed coating using an aqueous solution of 15% Sepifilm LP010 (= hydroxypropyl methylcellulose + 10% stearic acid)] (Ubbink et al. 2003). Above 90 MPa, the viability of microbial culture after the compression to form a pellet, gradually decreased with the pressure applied during compression procedure (upper punch pressure) (Chan and Zhang 2002). Since the compression pressure could have harmful effects on the cells during compaction, careful selection of a pressure which will be employed is needed. Pellets with probiotics might be useful as pharmaceuticals, food supplements or feed.

10.4 Food Applications

10.4.1 Challenges for Probiotics in Food Products

A number of technological challenges exists to successfully incorporate probiotics into foods and to maintain their viability:

1. *Stability of probiotics during processing and storage*. Processing of probiotic foods may involve mild heat treatment (e.g., low temperature and long-term pasteurization), pumping, homogenizing and stirring (incorporating air), freezing (frozen

products), addition of ingredients that can be antimicrobial (e.g., salt in cheese manufacture), drying (osmotic dehydration, e.g., powdered foods), packaging (oxygen ingress through packaging during storage), unfavorable storage conditions (e.g., post-acidification in yoghurt or presence of oxygen), large ice crystals formations (e.g., thawing and freezing of stored ice-cream) and the possible development of anti-microbial compounds secreted by the starter cultures during fermentation. In the past, culture companies select probiotic strains to withstand these conditions; however, the recent trend has been for these companies to focus the selection of strains on the basis of health-enhancing and therapeutic effects. Therefore the latest probiotic strains may have lost their ability to withstand unfavorable processing and storage conditions. Hence the viability of probiotic bacteria is of paramount importance in the marketability of probiotic incorporated products. In the development of functional foods, microencapsulation is especially used for incorporation and protecting viable cells into the products.

2. *Protection in the gastrointestinal tract and controlled release of probiotics in the intestines.* Most of the probiotics cannot stand the acid in the stomach (ranging from pH 2.5 to 3.5), and also the bile salts in the small intestine might be harmful. Microencapsulation and also components of the food matrix (like fat) may provide protection of probiotics against these harsh conditions. In addition, it is important that bacterial cells end up in large numbers in areas of the gastrointestinal tract where they are beneficial. Controlled release of bacterial cells from micro, encapsulates at the target site is therefore critical. It is beneficial for encapsulated bacteria to be released in the small intestine, where Peyer's patches exist, to activate the immune system. Therefore the polymers used as shell material for microencapsulation should be able to protect the bacteria in the acidic stomach and release the bacteria under the alkaline conditions in the small intestine. Many reports show that microencapsulation in, for example, alginate or pectin based beads can be used for controlled release of bioactive substances (Champagne and Kailasapathy 2008). Other examples are fat coated ones (see below).
3. *Clinical proof of health effects of the food product containing the probiotics.* The food matrix may affect the health benefit(s) of probiotics and ideally clinical studies using the final food product application should be performed to demonstrate them.

Co-encapsulation of probiotic cultures with certain food ingredients may be beneficial in two ways. First, it enables introducing multiple bioactive compounds. In addition, with the right selection of compounds, probiotic beneficial activity can be enhanced, prolonged or complemented by interactions between cells and co-encapsulated ingredients. Co-encapsulation can be performed by adding the second bioactive ingredient to the polymer solution, polymerizing solution or coating solution. Co-encapsulation with prebiotics, antioxidants, peptides or immune-enhancing polymers is becoming especially attractive in future perspectives. It has been determined that at least 3 g of prebiotics in a sample is needed to cause detectible activity improvement of the probiotic culture in the gastrointestinal tract (Krasaekoopt et al. 2003). This high amount is hardly possibly to achieve in real-life microencapsulation systems. On the other hand, some other compounds are active in much lower concentrations.

A combination with antioxidants is especially beneficial, due to the extending of probiotic stability in the gut and during storage caused by the effect of an antioxidant. Bioactive peptides, like bacteriocins, could enhance or complement the antimicrobial activities of probiotic bacteria.

Not only co-encapsulation, but also the presence of (ingredients in) a food product may improve the viability of probiotics (Ross et al. 2005), for example, by feeding the probiotics during storage, by the presence of probiotics which can be consumed during storage or upon consumption, by neutralizing partly the low pH in the stomach or by 'hiding' of probiotics in the food matrix during passage through gastrointestinal tract.

There are two classical ways of bacterial culture distribution in supply chains. One way is the storage and delivery of fresh, concentrated, chilled or frozen probiotic cultures for direct use. This has the advantage of very limited loss of viability, but the limit of a short storage time, similar to milk products. For fresh products containing probiotics, storage time is usually limited to 4 up to 6 weeks under refrigerated conditions. In fresh dairy products, the probiotics may multiply upon storage, even at low temperatures (except if the products get frozen), and may compensate for some probiotic deaths. The use of semipermeable encapsulates might be then a good option; feed is able to penetrate slowly into the encapsulates and the shell is still able to protect the probiotics against some harsh (sub-lethal) conditions in the gastrointestinal tract.

Another way of bacterial culture distribution in supply chains is the storage and delivery of dried probiotics, optionally in combination with microencapsulation techniques, which give microorganisms more stability and flexibility. The demands for probiotic stability are quite large when they do not multiply in the food product upon storage, as is often the case in non-dairy products, or when longer storage times are required. Probiotic viability in a food product depends on, for example, pH, storage temperature, oxygen levels, presence of competing microorganisms, presence of inhibitors (Mattila-Sandholm et al. 2002), and these factors are even more important when probiotics do not multiply. Bringing probiotics in a dormant state, by drying in the presence of additives and optionally coating them with an impermeable barrier during storage, might be a way to meet the demands for probiotics stability. For the food and pharmaceutical industries, a period of 1 year is often a minimum requirement to supply a marketable dry probiotic product. Capsule fillings, sachets and tablets with dried probiotics are very popular among consumers and inexpensive to produce, thus manufactured in the pharmaceutical or food supplement area. Application of dormant probiotics in both dry and liquid food products is possible, as discussed below in the following subsections which exemplify the potential of using microencapsulates containing probiotics in food products.

10.4.2 Yoghurt

It has been reported that microencapsulation using calcium-induced alginate–starch polymers (Godward and Kailasapathy 2003; Sultana et al. 2000), potassium-induced κ-carrageenan polymers (Adhikari et al. 2000, 2003) and whey protein polymers

(Picot and Lacroix 2004) have increased the survival and viability of probiotic bacteria in set and stirred yoghurts during storage. Kailasapathy (2006) reported that incorporation of calcium-induced alginate–starch microencapsulates containing probiotic bacteria (*L. acidophilus* and *B. lactis*) did not substantially alter the overall sensory characteristics of yoghurts. Microencapsulation also appears to provide anoxic regions inside the encapsulates thus reducing oxygen trapped inside the encapsulates which prevented the viability losses of oxygen-sensitive strains (Talwalkar and Kailasapathy 2003, 2004) in addition to protecting the cells against the detrimental effects of the acid environment in the yoghurt. McMaster et al. (2005) also showed increased oxygen tolerance by bifidobacteria in gel beads. The efficiency of microencapsulation in protecting the probiotic bacteria, however, depends on the oxygen sensitivity of the bacteria and the dissolved oxygen levels in the product. The addition of starch as a filler material in the alginate capsule matrix (Sultana et al. 2000), co-encapsulation with prebiotic substances such as inulin (Iyer and Kailasapthy 2005), or coating the microbeads with chitosan (Krasaekoopt et al. 2006) appear to improve the viability of probiotic cultures. A filler material used in preparing microencapsulated probiotic cultures is, for example, Hi Maize™ starch. Because of its cross-linked structure it will swell and absorb water but it will not gelatinize fully during pasteurization of yoghurt mix. This swollen starch therefore will contribute to increased viscosity and firmness. The formation of exopolysaccharides by the yoghurt starter cultures and probiotic cultures may contribute to prevention of syneresis and an increase in viscosity, combined with a better mouthfeel. The exopolysaccharides produced during fermentation may themselves form natural encapsulant for the yoghurt and probiotic bacteria.

The encapsulates above are semipermeable, and protect the still active probiotics against harsh conditions (oxygen, low pH of around 4). Another approach has been disclosed by Tessier (2005), who used granules composed of dormant, dehydrated lactic acid bacteria and coated with a solid fat in fermented milk (e.g., yoghurt, but also other liquid foods were claimed). The granules were coated on a fluidized bed in a 50/50 (w/w) mixture of stearic acid and palmitic acid, and had an average particle size between 150 and 200 μm. The encapsulated probiotics had no effect on the fermentation by other, non-encapsulated bacteria. Larger granules (1–3 mm) can also be used (Shin et al. 2002), but then the granules must have a density very close to that of the yoghurt. Furthermore, one may need to place the granules first at the bottom of the container prior to the filling of it with the yoghurt, which is a considerable manufacturing constraint.

10.4.3 Cheese

Among the traditional dairy foods, cheddar cheese has a markedly higher pH (4.8–5.6) than fermented milks and yoghurt (pH 3.7–4.3) and thus help in providing a more stable medium to support the long-term survival of acid-sensitive probiotic bacteria (Stanton et al. 1998). The metabolism of various lactic acid bacteria in cheddar cheese results in anaerobic environment within a few weeks of ripening, favoring the survival of probiotic bacteria (Van den Tempel et al. 2002). Furthermore, the matrix of cheddar cheese and its relatively high fat content offers protection to probiotic bacteria during passage through gastrointestinal tract (Vinderola et al. 2002). Thus it appears that microencapsulation may be only marginally beneficial in protecting probiotic bacteria in cheddar cheese. However, compared to yoghurt, cheddar cheese has a longer ripening, storage and shelf life during which the pH decrease, making the cheese acidic in nature during ripening. The combination of long maturation periods and acidic conditions could make it difficult for probiotic bacteria to survive during the 6–12 month ripening period. Additionally, compared to yoghurts, cheddar cheese also contains starter and non-starter lactic acid bacteria which may affect the survival of probiotic bacteria.

Dinakar and Mistry (1994) reported improved survival of *B. bifidum* in cheddar cheese over a 6 month ripening period. Gardiner et al. (2002) reported improved and increased survival as well as an increased growth rate of *L. paracasei* in cheddar cheese after 3 months of ripening. Similar results have been reported by McBrearty et al. (2001), Godward and Kailasapathy (2003) and Darukaradhya (2005). Cheese containing encapsulated *Bifidobacterium* was shown to possess similar flavor, texture and appearance compared to the control (Dinakar and Mistry 1994; Desmond et al. 2002). Kailasapathy and Masondole (2005) have reported that production of feta cheese incorporating encapsulated probiotic bacteria (*L. acidophilus* and *B. lactis*) is technologically feasible; however, selection of probiotic strains that are acid and salt tolerant and produces exo-polysaccharides as well as using food polysaccharides as shell materials for encapsulation will allow the production of a better quality feta cheese with greater survival rate of probiotic bacteria and an improved texture.

10.4.4 Frozen Desserts

Several studies have reported that probiotics entrapped in alginate or carrageenan beads have greater viability following freezing in dairy desserts (Kebary et al. 1998; Sheu et al. 1993; Godward and Kailasapathy 2003; Shah and Ravula 2000). In the manufacturing of frozen ice milk, probiotics microencapsulated with 3% calcium alginate are blended with milk and the mix is frozen continually in a freezer. The incorporation of microencapsulated probiotics has no measurable effect on the overrun and the sensory characteristics of the products with 90% probiotic survival (Sheu et al. 1993). Addition of encapsulated cultures (*L. acidophilus* and *B. infantis*) did not show any effect on the amount of air incorporated into the ice-cream (Godward and Kailasapathy 2003). The high fat content of ice-cream and the neutral pH of dairy desserts may be the main factors responsible for the additional protection provided to probiotic bacteria. However, the addition of cryoprotectants such as glycerol (Sheu et al. 1993; Sultana et al. 2000) seems to improve the viability of probiotic bacteria during freezing of the dairy desserts. The milk fat in ice-cream formulations may also act as an encapsulant material for probiotic bacteria during the homogenization of the ice-cream mix. The high total solids in ice-cream mix, including the fat (emulsion), may provide protection for the bacteria (Kailasapathy and Sultana 2003). However, full-fat ice-cream offered no extra protection for probiotic bacterial cultures (*L. acidophilus* LAFTI™ L10, *B. lactis* BLC-1 and *L. casei* subsp *paracasei* LCS-1) over the low-fat product during storage, with the low-fat formulation showing improved survival of all three cultures during the freezing process (Haynes and Playne 2002).

10.4.5 Powdered Formulations

In powdered milk products, the challenge is to protect the probiotics from the excessive heat and osmotic degradation during spray-drying. Improved viability upon conjointly spray-drying of milk and probiotics might be achieved by the use of a second drying step in two fluidized bed compartments operating at 60–90°C and a last, third compartment to cool to about 30°C (Meister et al. 1999).The addition of a thermoprotectant such as trehalose (Conrad et al. 2000) may help to improve the viability during drying and storage. Some studies have examined the stability of encapsulated probiotics in dried milk. Incorporation of the soluble fiber gum acacia into a milk-based medium prior to spray-drying the probiotic *L. paracasei* enhanced its viability during storage, compared with milk powder alone (Desmond et al. 2002). However, not all soluble fibers enhanced the probiotic viability during spray-drying of milk or milk powder storage, for example, inulin and polydextrose did not influence the viability (Corcoran et al. 2005). Freeze-drying of probiotics in micro-encapsulated hydrogel beads seems to be more stable than non-encapsulated ones during yoghurt incubation at room temperature (Kailasapathy and Sureeta 2004; Capela et al. 2006). Spray-coating of a freeze-dried culture seems to be more effective for additional protection (Siuta-Cruce and Goulet 2001). When a lipid coating is used, it may form a barrier to moisture and oxygen entry into the microcapsules. The nature of the packaging materials (e.g., yoghurt packaging) including their oxygen scavenging capacity, together with addition of antioxidants, desiccants, etc., may need to be considered for effective protection of probiotic cells during storage (Hsiao et al. 2004).

10.4.6 Meat Products

While dairy products are the most commonly used food vehicles for delivery of probiotics, their use in meat is not reported widely (Incze 1998; Chap. 13). Meat emulsion for the manufacture of small goods such as dry fermented sausages with their low water activity, pH, curing salts and competing starter culture organisms presents a challenging environment for the survival of introduced probiotics during processing. When *Lactobacillus plantarum* and *Pediococcus pentosaeccus* were immobilized in alginate micro capsules, the fermentation rate was much rapid with the encapsulated cells (Kearney et al. 1990). The rapid fermentation performance of the immobilized cells was caused by available nutrients (i.e., skim milk) and more protective re-hydration environment within the alginate capsules. Similar results can be obtained when microencapsulated probiotics are incorporated into a meat fermenting product mix. Muthukumarasamy and Holley (2006) showed that microencapsulated *Lactobacillus reuteri* can be used in dry fermented sausages to ensure that a desirable level of probiotic organisms is maintained in the final product at consumption without altering the sensory quality of these traditional small goods. In this study, alginate microcapsules prepared by either emulsion or extrusion were added to the salami batter (meat ingredients, starter cultures, cure mix, spice mix and salt) at 1% (w/w). The batter was stuffed in casings, transferred to a smoke house and allowed to ferment at 26°C and 88% relative humidity for 24 h, to reach pH less than 5.3. Fermentation was followed by drying at 13°C and 75% relative humidity for 25 days. It has been shown that *L. casei* cells when microencapsulated in alginate beads were more resistant to heat processing at 55–65°C (Mandal et al. 2006). This was also demonstrated when microencapsulated alginate beads containing cultures were heat treated to 55°C for 15 min; the encapsulated cells showed more stability than free cells in MRS broth acidified to pH 5.0 (Lemay et al. 2002). These data suggest that probiotic cells microencapsulated in alginate gel beads could be used in meat processing which require moderate heat treatments. For meat small goods where a meat emulsion is initially prepared (e.g., salami, sausages) the high fat in the system may also envelop the alginate gel particles containing the bacterial cells to provide additional protection to heat during processing.

10.4.7 Fermented Plant-Based (Vegetarian) Probiotic Products

With regard to plant-fermented products, probiotics are most frequently incorporated into soy products (Wang et al. 2002), although interest is increasing in the use of probiotics in fermented cereals (Charalampopoulos et al. 2003; Laine et al. 2003) and vegetable pickles (Savard et al. 2003). For stabilization of bifidobacteria during a traditional African-fermented corn product, the bacterial cells were encapsulated in mixed polymer (gellan/xanthan) beads (McMaster et al. 2005). Microencapsulation improved the survival of *L. rhamnosus* subjected to freezing in a cranberry juice concentrate and during storage of the frozen product (Reid et al. 2006). Microencapsulation can be of benefit to the stability of probiotic cultures; however, the way the bacteria are grown, harvested and dried for subsequent industrial use can be as important in promoting the viability of the cultures in food systems as the microencapsulation itself. Although the probiotic bacteria show good stability in products having a low water activity such as peanut butter ($a_w = 0.24$), spray-coating of *L. rhamnosus* using hard fat and incorporating into peanut butter formulations (incubated at 21°C), showed decreased cell viability (Belvis et al. 2006). In bakery applications, stabilizing viability of probiotics is a challenge, due to the high temperature treatment during processing. Microencapsulation by spray-coating in hard fat did not improve the survival and stability of added lactobacilli during bread-making (Belvis et al. 2006). However, microencapsulation in a whey protein particle was reported to be effective at enhancing the survival of probiotic lactobacilli during the heat treatment applied during biscuit manufacture (Reid et al. 2006).

10.4.8 Mayonnaise

Bifidobacterium bifidum and *Bifidobacterium infantis* survived only for 2 weeks in mayonnaise at pH 4.4 and 5° C (Khalil and Mansour 1998). However, within calcium alginate beads they survived for 12 and 8 weeks, respectively. This also resulted in lower total bacterial, yeast and mold counts. The mayonnaise containing encapsulated bifidobacteria also had a higher titratable acidity (due to acid production of the surviving bifidobacteria) and lower thiobarbituric acid (TBA, a measure for oxidation) values. These lower TBA values might be due to lower lypolytic activity as a result of lower bacterial, yeast and mold growth in the presence of the encapsulated bifidobacteria. Finally, the sensory properties were improved by the use of encapsulated bifidobacteria.

10.5 Future Perspectives

Despite the lack of industry standardization, and potential safety issues, there is obviously considerably potential for the benefits of probiotics. Ongoing basic research will continue to identify and characterize existing strains of probiotics, identify strain-specific properties, determine optimal doses needed for the aspired results, and assess their stability through processing and digestion. Parallel with the basic research, gene and industry-centered research are essential. Gene technology plays a role in developing new strains, with gene sequencing allowing an increased understanding of mechanisms and functionality of probiotics. The assessment of the industrial feasibility of a microencapsulation technology is mandatory for providing cost-effective, large-scale quantities of a probiotic product for specific clinical and/or commercial use. There are sequential steps, which from the identification of a possible probiotic strain, through laboratory tests, investigations in animal models and finally in humans, leads to its production and marketing.

The therapeutic effect of probiotic bacteria and their use in preventive medicine is increasingly being reported. As clinical evidence of the beneficial effects of probiotics accumulate, the food, nutraceutical and pharmaceutical industries will formulate new and innovative probiotic-based therapeutic products. New innovative ways of administering, delivering and controlled-releasing of probiotics will be developed in the near future. In addition to food and nutraceutical products, personal products, sports and health products, and cosmetics containing specific strains of probiotics are currently being either developed or planned and more innovative products will be developed in the future. Designer probiotic products delivering specific therapeutic strains will be the next phase of development. This will include food, pharmaceutical and nutraceutical products. These products may take the form of tablets, pills, re-constitutable single-serve sachet products, or convenient packs with instructions on how to prepare and administer them. Some food companies have already developed formulations to prepare probiotic yoghurts in the kitchen at home using a yoghurt maker.

An important issue in the development of functional foods is the stability and functionality of bioactive cultures. The viability of probiotic bacteria is important for their efficacy and a large number of reports have shown that many probiotic-based food products do not have the cell numbers in recommended number of viable cells (Iwana et al. 1993; Rybka and fleet 1997; Shah et al. 1995; Vinderola et al. 2000; Shah et al. 2000). Microencapsulation is an effective way of protecting and improving the viability of probiotic bacteria. It has been shown that non-protected cells consumed in a dried form have lower recovery levels in stools than those consumed in milk or cheese (Saxelin et al. 2003). The high viability losses that occur when free cells in a powder enter the stomach explains why microencapsulation is beneficial for the functionality of probiotics in nutraceuticals (Champagne and Fustier 2007). Microencapsulation or enteric-coated probiotic nutraceuticals may deliver the recommended number of viable cells. Microencapsulation offers the potential to reduce the adverse effects on probiotic viability of the food and gastrointestinal tract environment as well as during food or nutraceutical processing, storage and consumption. A number of efficient shell materials and controlled release trigger mechanisms have been developed in microencapsulation and this trend will continue, particularly with reference to food grade materials and the controlled and targeted release of probiotic bacteria in the gastrointestinal tract. For example, spray-drying of probiotics is not commercially used for probiotics yet (as far as we know), but this may change in the future and it might be combined with shell materials that do not only protect probiotics in a dry state but also in the gastrointestinal tract. Co-encapsulation with prebiotics, antioxidants, peptides, or immune-enhancing polymers might also be further explored. Furthermore, more research is needed of the stability and release of microencapsulated probiotics in food products.

The biological activity of probiotic bacteria is due in part to their ability to attach to enterocytes and thereby prevent binding of pathogens. The attachment of probiotic bacteria to receptors on the cell surface of intestinal epithelial cells can activate signaling processes leading to the synthesis of cytokines that affect the function of mucosal lymphocytes. Many of these receptors, such as, glycosphingolipids, mannosylated glycoproteins and TOLL, are already utilized by pathogens. This could be used to develop designer probiotic bacteria by coating with the selected receptor compound and targeting and directing the probiotic bacteria to areas in the gastrointestinal tract, such as the Payer's patches (small intestine) for maximum activation of the immune system. Further selection of suitable receptor polymers and microencapsulation can also help to direct the probiotic bacteria to access areas of medical interest such as tumors in the colon. More research is needed to study the adhesion properties of probiotic bacteria and the selection of polymers that can trigger successful adhesion to targeted intestinal cells and to design these polymers as capsular wall materials or coatings. This could achieve targeted delivery of probiotic bacteria to various sites within the gastrointestinal tract.

In addition to efficacious capsular wall materials or coatings, cell loading of the capsules is an important challenge. Capsules larger than 20–50 μm may influence the texture of the food products and hence the overall sensory characteristics. However, the microbial cells are already 1–5 μm in size and therefore could limit the cell loading within the capsules. Another challenge is to improve the heat resistance of these probiotic cells. There appears to be no commercial probiotic product available that is stable at high temperatures. Discovering or manipulating strains that are heat stable and developing new heat-insulating-encapsulating systems are two of the major challenges in this area of functional food development.

The sensory aspects of foods are critical in the acceptance of new products. Food scientists have generally tried to prevent sensory changes related to the addition of probiotics (Champagne et al. 2005), but in many instances there are no major changes in texture or organoleptic quality that significantly affect the sensorial properties of food (Kailasapathy 2006). An emerging marketing strategy is to develop food products that clearly show the microcapsules (possibly colored) distributed within the product. Then microencapsulation could also become a future marketing tool for the food and nutraceutical industry.

References

Adhikari K, Mustapha A, Grun IU, Fernando L (2000) Viability of microencapsulated bifidobacteria in set yoghurt during refrigerated storage. J Dairy Sci 83:1946–1951

Adhikari K, Mustapha A, Grun IU (2003) Survival and metabolic activity of microencapsulated *Bifidobacterium longum* in stirred yoghurt. J Food Sci 68:275–280

Anal AK, Singh H (2007) Recent advances in microencapsulation of probiotics for industrial applications and targeted delivery. Trends Food Sci Technol 18:240–251

Anal AK, Stevens WF (2005) Chitosan–alginate multilayer beads for controlled release of ampicillin. Int J Pharm 290:45–54

Anal AK, Bhopatkar D, Tokura S, Tamura H, Stevens WF (2003) Chitosan–alginate multilayer beads for gastric passage and controlled intestinal release of protein. Drug Dev Ind Pharm 29:713–724

Asada M, Hatano Y, Kamaguchi R, Sunohara H (2003) Capsules containing vital cells or tissues. WO03001927 A1, US2005283849 AA, US6982095 BB, & EP1407678 A4

Audet P, Paquin C, Lacroix C (1988) Immobilized growing lactic acid bacteria with κ-carrageenan-locus bean gum gel. Appl Microbiol Biotechnol 29:11–18

Audet P, Paquin C, Lacroix C (1990) Batch fermentations with a mixed culture of lactic acid bacteria with immobilized separately in κ-carrageenan-locus bean gum gel beads. Appl Microbiol Biotechnol 32:662–668

Audet P, Paquin C, Lacroix C (1991) Effect of medium and temperature of storage on viability of LAB immobilized on κ-carrageenan-locus bean gum gel bead. Biotechnol Tech 5:307–312

Belvis J, Tompkins TA, Wallace TA, Casavant L, Fortin C, Carson C (2006) Stability of probiotic bacteria in food stuffs. CIFIST Meeting, Montreal May30

Brian CS, Etzel MR (1997) Spray drying, freeze drying, or freezing of three different lactic acid bacteria species. J Food Sci 62(3):576–585

Capela P, Hay TKC, Shah NP (2006) Effect of cryoprotectants, prebiotics and microencapsulation on survival of probiotic organisms in yoghurt and freeze-dried yoghurt. Food Res Int 39:203–211

Carvalho AS, Silvia J, Ho P, Teixeira P, Malcata FX, Gibbs P (2002) Survival of freeze-dried *Lactobacillus plantarum* and *Lactobacillus rhamnosus* during storage in the presence of protectants. Biotechnol Lett 24:1587–1591

Carvalho AS, Silvia J, Ho P, Teixeira P, Malcata FX, Gibbs P (2003) Protective effect of sorbitol and monosodium glutamate during storage of freeze-dried LAB. Lait 83:203–210

Carvalho AS, Silvia J, Ho P, Teixeira P, Malcata FX, Gibbs P (2004a) Effect of various sugars added to growth and drying media upon thermotolerance and survival throughout storage of freeze-dried *Lactobacillus delbrueckii* spp. bulgaricus. Biotechnol Prog 20:248–254

Carvalho AS, Silvia J, Ho P, Teixeira P, Malcata FX, Gibbs P (2004b) Relevant factors for the preparation of freeze-dried LAB. Int Dairy J 14:835–847

Champagne CP, Fustier P (2007) Microencapsulation for the improved delivery of bioactive compounds into foods. Curr Opin Biotechnol 18:184–190

Champagne CP, Kailasapathy K (2008) Encapsulation of probiotics. In: Garti N (ed) Delivery and controlled release of bioactives in foods and nutraceuticals. Woodhead Publishing Limited, Cambridge, pp 344–369

Champagne CP, Gardner N, Brochu E, Beaulieu Y (1991) The freeze-drying of lactic acid bacteria. A review. Can Inst Food Sci Technol J 24(3/4):118–128

Champagne CP, Gardiner N, Roy D (2005) Challenges in the addition of probiotic cultures to foods. Crit Rev Food Sci Nutr 45:61–84

Chan ES, Zhang Z (2002) Encapsulation of probiotic bacteria *Lactobacillus acidophilus* by direct compression. Food Bioprod Process 80:78–82

Chan ES, Zhang Z (2004) Direct compression – novel method for encapsulation of probiotic cells. In: Nedović V, Willaert RG (eds) Focus on biotechnology, vol 8a, Fundamentals of cell immobilisation biotechnology. Kluwer, Dordrecht, pp 205–227

Chandramouli V, Kailasapathy K, Peiris P, Jones M (2004) An improved method of microencapsulation and its evaluation to protect Lactobacillus spp. in simulated gastric conditions. J Microbiol Methods 56:27–35

Charalampopoulos D, Pandiella SS, Webb C (2003) Evaluation of the effect of malt, wheat, and barley extracts on the viability of potentially probiotic lactic acid bacteria under acidic conditions. Int J Food Microb 82:133–141

Charteris WP, Kelly PM, Morelli M, Collins JK (2000) Effect of conjugated bile salts on antibiotic susceptibility of bile salt-tolerant *Lactobacillus* and *Bifidobacterium* isolate. J Food Protect 63:1369–1373

Chavez BE, Ledeboer AM (2007) Drying of probiotics: optimization of formulation and process to enhance storage survival. Drying Technol 25:1193–1201

Chen KN, Chen MJ, Lui JR, Lin CW, Chiu HY (2005) Optimization of incorporated prebiotics as coating materials for probiotic microencapsulation. J Food Sci 70:260–266

Conrad PB, Miller DP, Cielenski PR, de Pablo JJ (2000) Stabilization and preservation of *Lactobacillus acidophilus* in saccharide matrices. Cryobiology 41:17–24

Corcoran BM, Ross RP, Fitzgerald GF, Stanton C (2004) Comparative survival of probiotic lactobacilli spray-dried in the presence of prebiotic substances. J Appl Microbiology 96:1024–1039

Corcoran BM, Stanton C, Fitzgerald GF, Ross RP (2005) Survival of probiotic lactobacilli in acidic environment as enhanced in the presence of metabolic sugars. Appl Environ Microbiol 71:3060–3067

Crittenden R, Laitila A, Forssell P, Matto J, Saarela M, Mattila-Sandholm T, Myllarinen P (2001) Adhesion of bifidobacteria to granular starch and implication in probiotic technologies. Appl Environ Microbiol 67:3469–3475

Crittenden R, Weerakkody R, Sanguansri L, Augustin MA (2006) Synbiotic microcapsules that enhance microbial viability during non-refrigerated storage and gastro-intestinal transit. Appl Environ Microbiol 72:2280–2282

Darukaradhya J (2005) Enumeration and survival studies of free and encapsulated *Lactobacillus acidophilus* and *Bifidobacterium lactis* in Cheddar cheese. Master Hons. Thesis, University of Western Sydney, NSW, Australia, 2005

Del Piano M, Morelli L, Strozzi GP, Allesina S, Barba M, Deidda F, Lorenzini P, Ballare M, Montino F, Orsello M, Sartori M, Garello E, Carmagnola S, Pagliarulo M, Capurso L (2006) Probiotics: from research to consumer. Dig Liver Dis 38(2 Suppl):S248–S255

Desmond C, Stanton C, Fitzgerald GF, Collins K, Ross RP (2001) Environmental adaptation of probiotic lactobacilli towards improvement of performance during spray drying. Int Dairy J 11:801–808

Desmond C, Ross RP, O'Callaghan E, Fitzgerald G, Stanton C (2002) Improved survival of *Lactobacillus paracasei* NFBC 338 in spray-dried powders containing gum acacia. J Microbiol Methods 93:1003–10011

Diguet, S. (2000). Method for drying bacteria. WO01/36590.

Dinakar P, Mistry VV (1994) Growth and viability of *Bifidobacterium bifidum* in Cheddar cheese. J Dairy Sci 77:2854–2864

Doleyres Y, Lacroix C (2005) Technologies with free and immobilised cells for probiotic bifidobacteria production and protection. Int Dairy J 15(10):973–988

Doleyres Y, Fliss I, Lacroix C (2002a) Quantitative determination of the spatial distribution of pure- and mixed-stain immobilized cells in gel beads by immunofluorescence. Appl Microbiol Biotechnol 59:297–302

Doleyres Y, Fliss I, Lacroix C (2002b) *Bifidobacterium longum* ATCC 15707 cell production during free- immobilized cell cultures in MRS-whey permeate medium. Appl Microbiol Biotechnol 60:168–173

Doleyres Y, Fliss I, Lacroix C (2004) Continuous production of mixed lactic starters containing probiotics using immobilized cell technology. Biotechnol Prog 20:145–150

Durand H, Panes J (2001) Particles containing coated living micro-organisms, and method for producing same. WO0168808, & US 7157258 B2

Finegold SM, Sutter VL, Sugihara PT, Elder HA, Lehmann SM, Phillips RL (1977) Fecal microbial flora in seventh day adventist populations and controlled subjects. Am J Clin Nutr 30: 1781–1792

Garcia De Castro A, Bredholt H, Strom AR, Tunnaclife A (2000) Anhydrobiotic engineering of gram-negative bacteria. Appl Environ Microbiol 66:4142–4144

Gardiner GE, Bouchier P, O'Sullivan E, Collins JK, Fitzgerald G, Ross RP, Stanton C (2002) A spray-dried culture for probiotic cheddar cheese manufacture. Int Dairy J 12:749–756

Gaserod O, Smidsrod O, Skjak-Braek G (1998) Microcapsules of alginate-chitosan I. A quantitative study of the interaction between alginate and chitosan. Biomaterials 19:1815–1825

Gaserod O, Sannes A, Skjak-Braek G (1999) Microcapsules of alginate-chitosan II. A study of capsule stability and permeability. Biomaterials 20:773–783

Goderska K, Czarnecki Z (2008) Influence of microencapsulation and spray drying on the viability of *Lactobacillus* and *Bifidobacterium* strains. Pol J Microbiol 57(2):135–140

Godward G, Kailasapathy K (2003) Viability and survival of free, encapsulated and co-encapsulated probiotic bacteria in yoghurt. Milchwissenschaft 58:396–399

Gombotz WR, Wee S (1998) Protein release from alginate matrices. Adv Drug Deliv Rev 31: 267–285

Groboillot AF, Champagne CP, Darling GD, Poncelet D (1993) Membrane formation by interfacial cross-linking of chitosan for microencapsulation of *Lactococcus lactis*. Biotechnol Bioeng 42(10):1157–1163

Guérin D, Vuillemard JC, Subirade M (2003) Protection of *Bifidobacteria* encapsulated in polysaccharide-protein gel beads against gastric juice and bile. J Food Protect 66:2076–2084

Haynes N, Playne MJ (2002) Survival of probiotic cultures in low-fat ice cream. Aust J Dairy Technol 57(1):10–14

Hsiao HC, Lian WC, Chou CC (2004) Efficiency of packaging conditions and temperatures on viability of microencapsulated bifidobacteria during storage. J Sci Food Agric 84:134–139

Huis in't Veld J, Havenaar R (1991) Probiotics in man and animal. J Chem Technol Biotechnol 51:562–567

Hyndman CL, Groboillot AF, Poncelet D, Champagne CP, Neufeld RJ (1993) Microencapsulation of *Lactobacillus lactis* within cross-linked gelatin membranes. J Chem Technol Biotechnol 56:259–263

Incze K (1998) Dry fermented sausages. Meat Sci 49:S169–S177

Iwana H, Masuda H, Fujisawa T, Suzuki H, Mitsuoka T (1993) Isolation and identification of Bifidobacterium spp. In commercial yoghurts sold in Europe. Bifidobacteria Microflora 12(1):39–45

Iyer C, Kailasapthy K (2005) Effect of co-encapsulation of probiotics with prebiotics on increasing the viability of encapsulated bacteria under in vitro acidic and bile salt conditions and in yoghurt. J Food Sci 70(1):M18–M23

Iyer C, Kailasapathy K, Peiris P (2004) Evaluation of survival and release of encapsulated bacteria in ex vivo porcine gastrointestinal contents using a green fluorescent protein gene labelled *E. coli*. Lebensmittelwissenschaft und -Technologie 37:639–642

Iyer C, Phillips MK, Kailasapathy K (2005) Release studies of *Lactobacillus casei* strain shirota from chitosan-coated alginate-starch microcapsules in ex vivo porcine gastro-intestinal contents. Lett Appl Microbiol 41:493–497

Jacquot M, Pernetti M (2004) Spray coating and drying processes. In: Nedović V, Willaert RG (eds) Focus on biotechnology, vol 8a, Fundamentals of cell immobilisation biotechnology. Kluwer, Dordrecht, pp 343–356

Jongboom-Yilmaz, G. (2002). A method for preserving cells by processing the same into dry products. EP1213347

Kailasapathy K (2006) Survival of free and encapsulated probiotic bacteria and their effects on the sensory properties of yoghurt. Lebensmittelwissenschaft und -Technologie 39:1221–1227

Kailasapathy K, Chin JC (2000) Survival and therapeutic potential of probiotic organisms with reference to *Lactobacillus acidophilus* and *Bifidobacterium* spp. Immunol Cell Biol 78:80–88

Kailasapathy K, Masondole L (2005) Survival of free and microencapsulated *Lactobacillus acidophilus* and *Bifidobacterium lactis* and their effect on texture of feta cheese. Aust J Dairy Technol 60:252–258

Kailasapathy K, Sultana K (2003) Survival and beta-D-Galactosidase activity of encapsulated and free *Lactobacillus acidophilus* and *Bifidobacterium lactis* in ice cream. Aust J Dairy Technol 58:223–227

Kailasapathy K, Sureeta BS (2004) Effect of storage on shelf life and viability of freeze-dried and microencapsulated *Lactobacillus acidophilus* and *Bifidobacterium infantis* cultures. Aust J Dairy Technol 59:204–208

Kearney L, Lipton M, Loughlin A (1990) Enhancing the viability of *Lactobacillus plantarum* inoculum by immobilising the cells in calcium-alginate beads. Appl Environ Microbiol 56: 3112–3116

Kebary KMK, Hussein SA, Badawi RM (1998) Improving viability of bifidobacteria and the effect of frozen ice milk. Egypt J Dairy Sci 26:319–337

Khalil AH, Mansour EH (1998) Alginate encapsulated bifidobacteria survival in mayonnaise. J Food Sci 63(4):702–705

Kheadr E, Bernoussi N, Lacroix C, Fliss I (2004) Comparison of the sensitivity of commercial strains and infant isolates of bifidobacteria to antibiotics and bacteriocins. Int Dairy J 14:1041–1053

King VAE, Su JT (1993) Dehydration of *Lactobacillus acidophilus*. Process Biochem 28:47–52

King VAE, Su JT (1995) Studies on the effect of protectants on *Lactobacillus acidophilus* strain dehydrated under controlled low-temperature vacuum dehydration and freeze-drying by using response surface methodology. J Sci Food Agric 68:191–196

King VAE, Lin HJ, Liu CF (1998) Accelerated storage testing of freeze-dried and controlled low-temperature vacuum dehydrated Lactobacillus acidophilus. J Gen Appl Microbiol 44:161–165

Klein J, Vorlop DK (1985) Immobilization techniques: cells. In: Cooney CL, Humphrey AE (eds) Comprehensive biotechnology. Pergamon Press, Oxford, pp 542–550

Kopp-Hoolihan L (2001) Prophylactic and therapeutic uses of probiotics: a review. J Am Diet Assoc 101(2)):229–241

Krasaekoopt W, Bhandari B, Deeth H (2003) Evaluation of encapsulation techniques of probiotics for yoghurt. Int Dairy J 13:3–13

Krasaekoopt W, Bhandari B, Deeth H (2004) The influence of coating materials on some properties of alginate beads and survivability of microencapsulated probiotic bacteria. Int Dairy J 14:737–743

Krasaekoopt W, Bhandari B, Deeth HC (2006) Survival of probiotics encapsulated in chitosan-coated alginate beads in yoghurt from UHT-and conventionally treated milk during storage. Lebensmittelwissenschaft und -Technologie 39:177–183

Lacík I (2004) Polyelectrolyte complexes for microcapsule formation. In: Nedović V, Willaert RG (eds) Focus on biotechnology, vol 8a, Fundamentals of cell immobilisation biotechnology. Kluwer, Dordrecht, pp 103–120

Laine R, Salminen S, Benno Y, Ouwenhand AC (2003) Performance of bifidobacteria in oat-based media. Int J Food Microbiol 83:105–109

Lee K-Y, Heo T-R (2000) Survival of Bifidobacterium longum immobilised in calcium alginate beads in simulated gastric juices and bile salt solutions. Appl Environ Microbiol 66(2):869–873

Lee JS, Cha DS, Park HJ (2004) Survival of freeze-dried *Lactobacillus bulgaricus* KFRI 673 in chitosan-coated calcium alginate microparticles. J Agric Food Chem 52:7300–7305

Lemay MJ, Champagne CP, Garie'py C, Saucier L (2002) A comparison of the effect of meat formulations on the heat resistance of free or encapsulated cultures of *Lactobacillus* sakei. J Food Sci 67:3428–3434

Levy MC, Edwards-Levy F (1996) Coating alginate beads with cross-linked biopolymers: a novel method based on a transacylation reaction. J Microencapsul 13(2):169–183

Lourens-Hattingh A, Viljoen BC (2001) Yogurt as probiotic carrier food. Int Dairy J 11:1–7

Makhal S, Kanawjia SK (2003) "Probiocap": a novel scientific approach. Indian Food Ind 22(5): 44–50

Mandal S, Puniya AK, Singh K (2006) Effect of alginate concentrations on survival of microencapsulated *Lactobacillus casei* NCDC-298. Int Dairy J 16:1190–1195

Martinsen A, Skjak-Braek G, Smidsrod O (1989) Alginate as an immobilization material: I. Correlation between chemical and physical properties of alginate gel beads. Biotechnol Bioeng 33:79–89

Mattila-Sandholm T, Myllarinen P, Crittenden R, Mogensen G, Fonden R, Saarela M (2002) Technological challenges for future probiotic foods. Int Dairy J 12:173–182

McBrearty S, Ross RP, Fitzgerald GE, Collins JK, Wallace JM, Stanton C (2001) Influence of two commercially available bifidobacteria cultures on Cheddar cheese quality. Int Dairy J 11:599–610

McKnight CA, Ku A, Goosen MF (1988) Synthesis of chitosan-alginate microencapsule membrane. J Bioact Compat Polym 3(8):334–354

McMaster LD, Kokott SA, Reid SJ, Abratt VR (2005) Use of traditional African fermented beverages as delivery vehicles for *Bifidobacterium lactis* DSM 10140. Int J Food Microbiol 102:231–237

Meister N, Sutter A, Vikas M (1999) Dehydrated food containing lactic acid bacteria. EP0924993B1

Moubareck C, Gavini F, Vaugien L, Butel MJ, Doucer-Popularie F (2005) Antimicrobial susceptibility of *Bifidobacteria*. J Antimicrob Chem 55:38–44

Muthukumarasamy P, Holley RA (2006) Microbiological and sensory quality of dry fermented sausages containing alginate-microencapsulated *Lactobacillus reuteri*. Int J Food Microbiol 111:164–169

Myllarinen P, Forssell P, von Wright A, Alander M, Mattila-Sandholm T, Poutanen K (2000). Starch capsules containing microorganisms and/or polypeptides or proteins and a process for producing them. F11104405 (WO9952511A1)

O'Riordan K, Andrews D, Bruckle K, Konway P (2001) Evaluation of microencapsulation of *Bifidobacterium* strain with starch as an approach to prolonging the viability during storage. J Appl Microbiol 91:1059–1066

Ouellette V, Chevalier P, Lacroix C (1994) Continuous fermentation of a supplemented milk with immobilized *Bifidobacterium infantis*. Biotechnol Tech 8:45–50

Picot A, Lacroix C (2003) Effect of micronization on viability and thermotolerance of probiotic freeze-dried cultures. Int Dairy J 13:455–462

Picot A, Lacroix C (2004) Encapsulation of bifidobacteria in whey protein-based microcapsules and survival in simulated gastrointestinal conditions and in yoghurt. Int Dairy J 14:505–515

Poncelet D, Poncelet SB, Beaulieu C, Neufeld RJ (1993) Scale-up of gel bead and microcapsule production in cell immobilization. In: Goosen MFA (ed) Fundamentals of animal cell encapsulation and immobilization. CRC Press, Boca Raton, pp 532–547

Rao AV, Shiwnarain N, Maharaj I (1989) Survival of microencapsulated *Bifidobacterium pseudolongum* in simulated gastric and intestinal juices. Can Inst Food Sci Technol J 22:345–349

Reid AA, Champagne CP, Gardner N, Fustier P, Vuillemand JC (2006) Survival I food systems of *Lactobacillus rhamnosus* R 011 microentrapped in whey protein gel particles. J Food Sci 72:M031–M037

Rescigno M, Rotta G, Valzasina B, Riccardi-Castagnoli P (2001) Dendritic cells shuttle microbes across gut epithelial monolayers. Immunobiology 204:572–581

Ross RP, Desmond C, Fitzgerald GF, Stanton C (2005) Overcoming the technological hurdles in the development of probiotics foods. J Appl Microbiol 98:1410–1417

Rutherford WM, Allen JE, Schlameus HW, Mangold DJ, Harlow WW Jr., Lebeda JR (1993) Dried, rotary disc microspheres of microorganisms. WO93/17094

Rybka S, Fleet GH (1997) Populations of *Lactobacillus delbrueckii* ssp. *Bulgaricus*, *Streptococcus thermophilus*, *Lactobacillus acidophilus* and *Bifidobacterium* species in Australian yoghurts. Food Australia 49(10):471–475

Savard T, Gardner N, Champagne CP (2003) Croissance de cultures lactobacillus et de Bifidobacterium dans un jus de legumes et viabilite au cours de l'entreposage dans le jus de legumes fermente. Sciences des Alimentes 23:273–283

Saxelin M, Korpela R, Mayra-Makinen A (2003) Introduction: classifying functional dairy products. In: Mattila-Sandholm T, Saarela M (eds) Functional dairy products. CRC Press/Woodhead Pub. Ltd, Boca Raton, pp 1–15

Selmer-Olsen E, Sørhaug T, Birkeland SE, Pehrson R (1999) Survival of *Lactobacillus helveticus* entrapped in Ca–alginate in relation to water content, storage and rehydration. J Ind Microbiol Biotechnol 23:79–85

Shah NP, Ravula RR (2000) Microencapsulation of probiotic bacteria and their survival in frozen fermented desserts. Aust J Dairy Technol 55:139–144

Shah NP, Lankaputhra WEV, Britz ML, Kyle WSA (1995) Survival of *Lactobacillus acidophilus* and *Bifidobacterium bifidum* in commercial yoghurts during refrigerated storage. Int Dairy J 5(5):515–521

Shah NP, Ali JF, Ravula RR (2000) Populations of *Lactobacillus acidophilus, Bifidobacterium* spp and *Lactobacillus casei* in commercial fermented milk products. Biosci Microflora 19(1):35–39

Sheu TY, Marshall RT, Heymann H (1993) Improving survival of culture bacteria in frozen by microentrapment. J Dairy Sci 76:1902–1907

Shima M, Morita Y, Yamashita M, Adachi S (2006) Protection of *Lactobacillus acidophilus* from the low pH of a model gastric juice by incorporation in a w/o/w emulsion. Food Hydrocolloids 20:1164–1169

Shin Y-S, Baik B-H, Lee J-K (2002) Liquid yogurt with encapsulated lactic acid bacteria and method for producing the same. US6447823

Siuta-Cruce P, Goulet J (2001) Improving probiotic survival rates: microencapsulation preserves the potency of probiotic micro-organisms in food systems. Food Technol 55:36–42

Skjak-Braek G, Larsen B, Smidsrod O (1986) Tailoring of alginates by enzymatic modifications in vitro. Int J Biol Macromol 8:330–336

Smidsrod O, Skjak-Braek G (1990) Alginates as immobilisation matrix for cells. Trends Biotechnol 8:71–78

Stanton C, Gardiner G, Lynch PB, Collins JK, Fitzgerald GF, Ross RP (1998) Probiotic cheese. Int Dairy J 8:491–496

Sultana K, Godward G, Reynolds N, Arumugaswamy R, Peiris P, Kailasapathy K (2000) Encapsulation of probiotic bacteria with alginate-starch and evaluation of survival in simulated gastro-intestinal conditions and in yoghurt. Int J Food Microbiol 62:47–55

Talwalkar A, Kailasapathy K (2003) Effect of microencapsulation and oxygen toxicity in probiotic bacteria. Aust J Dairy Technol 58:36–39

Talwalkar A, Kailasapathy K (2004) A review of oxygen toxicity in probiotic yoghurts: influence on the survival of probiotic bacteria and protective techniques. Comp Rev Food Sci Food Safety 3:117–124

Tessier P (2005) Food product containing lactic bacteria granules. WO2005/070221, & US2007/0098847

Tosa T, Sato T, Mori T, Yamamoto K, Takata I, Nishida Y, Chibata I (1979) Immobilization of enzymes and microbial cells using carrageenan as matrix. Biotechnol Bioeng 21(2):1697–1700

Truelstrup Hansen LT, Allan-Wojtas PM, Jin YL, Paulson AT (2002) Survival of Ca–alginate microencapsulated *Bifidobacterium spp.* in milk and simulated gastrointestinal condition. Food Microbiol 19:35–45

Tsen J-H, Chen H-H, King VA-E (2002) Survival of freeze-dried *Lactobacillus acidophilus* immobilized in κ-carrageenan gel. J Gen Appl Microbiol 48(4):237–241

Ubbink JB, Schaer-Zammeretti P, Cavadini C (2003) Probiotic delivery system. WO03/075676A1

Van Lengerich (2000). Encapsulation of sensitive liquid components into a matrix to obtain discrete shelf-stable particles. WO00/21504

Van den Tempel T, Gundersen JK, Nielsen MS (2002) The micro-distribution of oxygen in Danablu cheese measure by a microsensor during ripening. Int J Food Microbiol 75:157–161

Van Lengerich H (1999) Encapsulation of components into edible products. WO99/48372

Viernstein H, Raffalt J, Polheim D (2005) Stabilisation of probiotic microorganisms. In: Nedović V, Willaert RG (eds) Focus on biotechnology, vol 8a, Fundamentals of cell immobilisation biotechnology. Kluwer, Dordrecht, pp 439–453

Vinderola CG, Bailo N, Reinheimer JA (2000) Survival of probiotic microflora in Argentinian yoghurts during refrigerated storage. Food Res Int 33(2):97–102

Vinderola CG, Costa GA, Regenhardt S, Reinheimer JA (2002) Influence of compounds associated with fermented dairy products on the growth of lactic acid starter and probiotic bacteria. Int Dairy J 12:579–589

Wang YC, Yu RC, Chou CC (2002) Growth and survival of bifidobacteria and lactic acid bacteria during the fermentation and storage of cultured soymilk drinks. Food Microbiol 19:501–508

Wu W-H, Roe WS, Gimino VG, Seriburi V, Martin D, Knapp SE (2000) Low melt encapsulation. WO0074499 C2, EP1408774 A4, & US 6153236 A

Würster DE (1950) Means for applying coatings to tablets or like. J Am Pharm Assoc 48(8):451–460

Zhou Y, Martins E, Groboillot A, Champagne CP, Neufeld RJ (1998) Spectrophotometric quantification of lactic bacteria in alginate and control of cell release with chitosan coating. J Appl Microbiol 84(3):342–348

Zubillaga M, Weill R, Postaire E, Goldman C, Caro R, Boccio J (2001) Effect of probiotics and functional foods and their use in different diseases. Nutr Res 21:569–579

Chapter 11
Bioprocess Intensification of Beer Fermentation Using Immobilised Cells

Pieter J. Verbelen, Viktor A. Nedović, Verica Manojlović, Freddy R. Delvaux, Ida Laskošek-Čukalović, Branko Bugarski, and Ronnie Willaert

11.1 Introduction

Beer production with immobilised yeast has been the subject of research for approximately 30 years but has so far found limited application in the brewing industry, due to engineering problems, unrealised cost advantages, microbial contaminations and an unbalanced beer flavor (Linko et al. 1998; Brányik et al. 2005; Willaert and Nedović 2006). The ultimate aim of this research is the production of beer of desired quality within 1–3 days. Traditional beer fermentation systems use freely suspended yeast cells to ferment wort in an unstirred batch reactor. The primary fermentation takes approximately 7 days with a subsequent secondary fermentation (maturation) of several weeks. A batch culture system employing immobilization could benefit from an increased rate of fermentation. However, it appears that in terms of increasing productivity, a continuous fermentation system with immobilization would be the best method (Verbelen et al. 2006). An important issue of the research area is whether beer can be produced by immobilised yeast in continuous culture with the same characteristic as the traditional method.

P.J. Verbelen and F.R. Delvaux
Centre for Malting and Brewing Science, Faculty of Bioscience Engineering KU Leuven, Kasteelpark Arenberg 22, 3001 Heverlee, Belgium

V.A. Nedović (✉) and I. Laskošek-Čukalović
Department of Food Technology and Biochemistry, University of Belgrade, Nemanjina 6, PO Box 127, 11081 Belgrade-Zemun, Serbia and Montenegro
e-mail: vnedovic@agrif.bg.ac.rs

V. Manojlović and B. Bugarski
Department of Chemical Engineering,University of Belgrade, Karnegijeva 4, 11000 Belgrade, Republic of Serbia

R. Willaert (✉)
Structural Biology Brussels, Vrije Universiteit Brussel, Flanders Institute for Biotechnology, Pleinlaan 2, B-1050 Brussel, Belgium
e-mail: ronnie.willaert@vib-vub.be

N.J. Zuidam and V.A. Nedović (eds.), *Encapsulation Technologies for Active Food Ingredients and Food Processing*,
DOI 10.1007/978-1-4419-1008-0_11,

In beer production, as opposed to a process such as bio-ethanol production, the goal is to achieve a particular balance of different secondary metabolites rather than the attainment of high yields of one product. Any alterations of the fermentation procedure can thus have serious implications on the flavor profile. At present, only beer maturation and alcohol-free beer production are obtained by means of commercial-scale immobilised yeast reactors, because in these processes no real yeast growth is required. Immobilised cell physiology control and fine-tuning of the flavor compounds formation during long-term fermentation processes remain the major challenges for successful application of immobilised cell technology on an industrial scale. The key factors for the implementation of this technology on an industrial level are carrier materials, immobilization technology and bioreactor design.

The purpose of this chapter is to summarise and discuss the main cell immobilization methods, process requirements, available carrier materials and bioreactor designs aimed for better yeast physiology control and fine-tuning of the flavor formation during beer fermentation process. Further, it will provide an overview on the latest important breakthroughs, accomplished in understanding of the effects of immobilization on yeast physiology, metabolism and fermentation behaviour.

11.2 Carrier Materials and Design

11.2.1 Cell Immobilization Methods and Carrier Materials

Generally, immobilised cell systems can be classified into four categories based on the physical mechanism of cell localisation and the nature of the support mechanisms: "attachment to a surface," "entrapment within a porous matrix," "containment

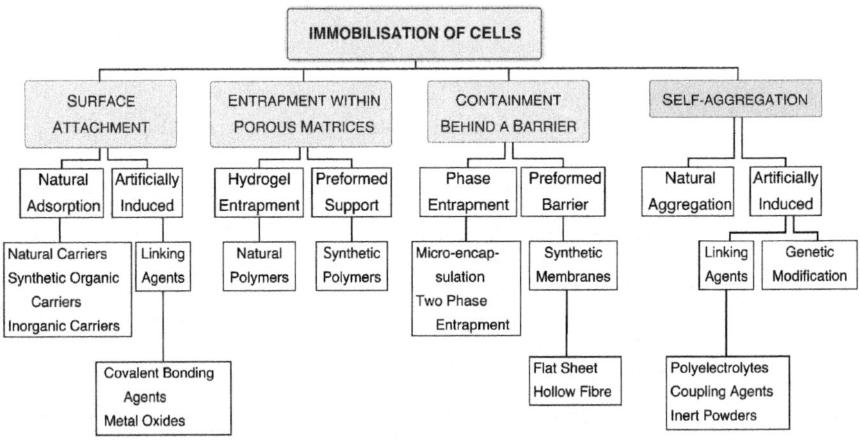

Fig. 11.1 Classification of immobilised cell systems according to the physical localisation and the nature of the micro-environment (Willaert and Baron 1996)

behind a barrier" and "self aggregation" (Karel et al. 1985; Willaert and Baron 1996) (Fig. 11.1). For beverage (and other food) applications, special attention must be paid to the selection of approved, food grade compounds or to the prevention of any leakages of undesired compounds into the beverages.

Cell immobilization by adsorption to a support material is a very popular method, because it is a simple, cheap and fast method. Micro-organisms adsorb spontaneously on a wide variety of organic and inorganic supports. Binding of cells occurs through interactions such as Van der Waals forces, ionic bonds, hydrogen bridges or covalent interactions. Microbial cells exhibit a dipolar character and behave as cations or anions, depending on the cell type and environmental conditions such as pH of the solution. Furthermore, cell physiology has a significant influence on the strength of the adhesion. Various rigid support materials are available mainly aimed at applications in packed-bed reactors. Diethylaminoethyl (DEAE)–cellulose supports have been successfully used on industrial scale for the production of alcohol-free beer and maturation of green beer (Lommi 1990). It is an inert, non-dissolving cellulose matrix, which has a non-uniform granular shape. Yeast cells are immobilised by ionic attraction. Compared to porous supports, the biomass loading capacity of DEAE–cellulose is considerably lower. Other selected materials are presented in Table 11.1.

Cell immobilization in porous matrices can be performed by two different basic methods. In the first method, indicated as gel entrapment, the porous matrix is synthesised in situ around the cells to be immobilised. In the second method, cells are allowed to move into a preformed porous matrix. Generally, both methods provide cell protection from the fluid shear and higher cell densities as compared to surface immobilization are reached. A drawback of these systems can be mass transfer limitations. However, understanding of mass transfer phenomena within entrapment matrices may allow one to simultaneously provide different conditions at the carrier surface and in the interior, which could be attractive for co-immobilization of different cell types performing consecutive processes (Willaert et al. 1999).

Over the last 30 years, most of the research concerning the immobilization of living microbial cells was focused on gel entrapment. Natural polysaccharides (e.g., alginate, chitosan, pectate and carrageenan), synthetic polymers (e.g., polyvinylalcohol, PVA) and proteins (e.g., gelatin, collagen) can be gelled into hydrophilic matrices under mild conditions, allowing cell entrapment with minimal loss of viability. As a result, very high biomass loadings can be achieved. Gels are mostly used in the form of spherical beads with diameters ranging from about 0.3 to 5 mm. However, a disadvantage of gels is their limited mechanical stability. It has been frequently observed that the gel structure is easily destroyed by the growth of the cells and carbon dioxide production. Moreover, calcium alginate gels are weakened in the presence of phosphates (which is present in wort). Long-term use of calcium alginate beads in continuous production (maturation) of beer resulted in loss of bead 3D-structure due to high phosphate contents in the wort/beer. However, several methods have been proposed for reinforcement of gel structures. For example, calcium alginate gel can be strengthened by reaction with polyethyleneimine, glutaraldehyde

Table 11.1 Carrier materials and reactor types for some selected beer fermentation processes using immobilised cells

Carrier material	Reactor type	Reference
Flavor maturation		
Calcium alginate beads	Fixed bed	Shindo et al. (1994)
DEAE-cellulose beads	Fixed bed	Pajunen and Grönqvist (1994)
Polyvinyl alcohol beads	Fixed bed	Smogrovicová et al. (2001)
Porous glass beads	Fixed bed	Linko et al. (1993), Aivasidis (1996)
Alcohol-free beer		
DEAE-cellulose beads	Fixed bed	Collin et al. (1991), Lommi (1990)
Porous glass beads	Fixed bed	Aivasidis et al. (1991)
Silicon carbide rods	Monolith reactor	Van De Winkel et al. (1991)
Acidified wort		
DEAE-cellulose beads	Fixed bed	Pittner et al. (1993)
Main fermentation		
Calcium alginate beads	Gas lift	White and Portno (1979), Onaka et al. (1985), Ryder and Masschelein (1985), Nedovic et al. (1993), Nedovic et al. (1997)
Calcium alginate beads	Fixed bed	Ryder and Masschelein (1985)
Calcium pectate beads	Gas lift	Smogrovicová et al. (1997), Smogrovicová and Dömény (1999)
κ-Carrageenan beads	Gas lift	Mensour et al. (1996), Mensour et al. (1997), Decamps et al. (2004)
Ceramic beads	Fixed bed	Inoue (1995)
Corncobs	Gas lift	Brányik et al. (2006a)
DEAE-cellulose beads	Fixed bed	Kronlöf et al. (1989), Andersen et al. (1999)
Gluten pellets	Fixed bed	Bardi et al. (1997)
Gluten pellets	Gas-lift (external-loop)	Manojlović et al. (2008)
Kieselguhr (diatomaceous earth)	Fixed bed	Narziss and Hellich (1971), Moll et al. (1973), Virkajärvi and Pohjala (2000)
Polyvinyl alcohol beads	Gas lift	Smogrovicová et al. (2001)
Polyvinyl alcohol Lentikats®	Gas lift	Smogrovicová et al. (2001), Bezbradica et al. (2007), Bugarski & Nedović (2007)
Polyvinyl chloride granules	Fixed bed	Moll et al. (1973)
Porous glass beads	Fixed bed	Virkajärvi and Krönlof (1998), Virkajärvi and Pohjala (2000)
Porous chitosan beads	Fluidised bed	Unemoto et al. (1998), Maeba et al. (2000)
Self-aggregation using super-flocculent yeast	Continuous stirred tank reactors	Coutts (1957), Linko et al. (1997)
Silicon carbide rods	Monolith reactor	Van De Winkel et al. (1993); Andries et al. (1996)
Spent grains	Gas lift	Brányik et al. (2002, 2004a)
Stainless steel fibre cloth	Gas lift	Verbelen et al. (2006)
Stainless steel wire spheres	Fluidised bed	Cross and Mavituna (1987)
Wood chips Aspen	Fixed bed	Pajunen et al. (2001)
Wood chips Beech	Fixed bed	Linko et al. (1997), Kronlöf and Virkajärvi (1999)

cross-linking, addition of silica, genepin and polyvinylalcohol, or by partial drying of the gel (Willaert and Baron 1996). For beer production, a food grade reinforcement method should be used. Gel or polymeric matrix beads (usually spherical in shape) could be further coated with an outer layer to create (micro)capsules. In those, the solid core may also be dissolved within the capsule to create the liquid media for cells. However, micro-encapsulation is generally too expensive to be used in beer production and only suited for non-growing cells (Raymond et al. 2004). Another approach is to contain cells within a compartment separated by a preformed membrane such as hollow fibre and flat membrane modules. Entrapment behind preformed membranes represents a gentle immobilization method since no chemical agents or harsh conditions are employed. Usually polymeric microfiltration or ultrafiltration membranes were used, although other types of membranes were also investigated, such as ceramic, silicone or ion exchange membranes. Mass transfer through the membrane is dependent on the pore size and structure as well as on the hydrophobicity/hydrophilicity and surface charge.

Unlike the in situ formed gels, preformed porous supports can be inoculated directly from the bulk medium. In these systems, cells are not completely separated from the effluent, similarly as in the adsorption method (Baron and Willaert 2004; Mavituna 2004). Cell immobilization occurs by attachment to the internal surfaces, self-aggregation and retention in dead-end pockets within the carrier material. Ideally, the colonised porous particles should retain some void spaces for flow so that mass transport of substrates and products could be achieved by both molecular diffusion and convection, avoiding mass transfer limitations. However, fluid flow within the support can be realised only if cell adhesion is not very strong so that excessive biomass could be washed out from the matrix. When high cell densities are obtained, convection is no longer possible and the particles behave as dense cell agglomerates with high diffusion limitations. Yet the cell densities represented per unit of support volume are lower than those achievable by gel entrapment since the preformed porous matrix material takes up significant volume fraction. As compared to the in situ gel particles, preformed porous carriers provide better mechanical properties and higher resistances to compression and disintegration.

Cell immobilization behind or within a porous barrier includes systems with a barrier formed around cells such as microcapsules, and systems with cells contained within a compartment separated by a preformed membrane such as hollow fibre and flat membrane modules. However, micro-encapsulation is generally too expensive to be used in beer production and only suited for non-growing cells (Raymond et al. 2004).

Cell immobilization by self-aggregation is based on the formation of cell clumps or floccules, which can be naturally occurring as in the case of flocculent yeast strains (or can be induced by addition of flocculating agents). It is the simplest and the least expensive immobilization method.

Yeast flocculation is a reversible, asexual and calcium dependent process in which cells adhere to form flocs consisting of thousands of cells (Stratford 1989; Bony et al. 1997; Jin and Speers 1999). Many fungi contain a family of cell wall glycoproteins (called "adhesines") that confer unique adhesion properties (Teunissen and Steensma 1995; Guo et al. 2000; Hoyer 2001; Sheppard et al. 2004). These molecules are required for the interactions of fungal cells with each other (flocculation and filamentation) (Teunissen and Steensma 1995; Lo and Dranginis 1998; Guo et al. 2000; Viyas et al. 2003), with inert surfaces such as agar and plastic (Gaur and Klotz 1997; Lo and Dranginis 1998; Reynolds and Fink 2001; Li and Palecek 2003) and with mammalian tissues/cells (Cormack et al. 1999; Staab et al. 1999; Fu et al. 2002; Li and Palecek 2003). They are also crucial for the formation of fungal biofilms (Baillie and Douglas 1999; Reynolds and Fink 2001; Green et al. 2004). The adhesin proteins in *S. cerevisiae* are encoded by genes including *FLO1*, *FLO5*, *FLO9* and *FLO10* (Verstrepen et al. 2004). These proteins are called flocculins (Caro et al. 1997), because these proteins promote cell–cell adhesion to form multicellular clumps that sediment out of solution. The *FLO1*, *FLO5*, *FLO9* and *FLO10* genes share considerable sequence homology. The member proteins of the adhesin family have a modular configuration that consists of three domains (A, B and C) and an N-terminal secretory sequence that must be removed as the protein moves through the secretory pathway to the plasma membrane (Hoyer et al. 1998). The N-terminal domain (A) is involved in sugar recognition (Kobayashi et al. 1998). The adhesins undergo several post-translational modifications. They move from the endoplasmic reticulum (ER), through the Golgi and pass through the plasma membrane and find their final destination in the cell wall where they are anchored by a glycosyl phosphatidylinositol (GPI) (Teunissen et al. 1993; Bidard et al. 1994; Hoyer et al. 1998; Bony et al. 1997). The GPI anchor is added to the C-terminus in the ER and mannose residues are added to the many serine and threonine residues in domain B in the Golgi (Udenfriend and Kodukula 1995; Bony et al. 1997; Frieman and Cormack 2003; De Groot et al. 2003). The *FLO1* gene product (Flo1p) has been localised at the cell surface by immunofluorescent microscopy (Bidard et al. 1995). The amount of Flo proteins in flocculent strains increased during batch yeast growth and the Flo1p availability at the cell surface determined the flocculation degree of the yeast. Flo proteins are polarly incorporated into the cell wall at the bud tip and the mother–daughter neck junction (Bony et al. 1997). The transcriptional activity of the flocculation genes is influenced by the nutritional status of the yeast cells as well as other stress factors (Verstrepen et al. 2003a). This implies that during beer fermentation, flocculation is affected by numerous parameters such as nutrient conditions, dissolved oxygen, pH, fermentation temperature, and yeast handling and storage conditions.

Continuous beer fermentation technology, using yeast flocculation and cell recycling, has been successfully exploited over almost 40 years by Dominion Breweries in New Zealand (Coutts 1957; Pilkington et al. 1998; Van de Winkel and De Vuyst 1997). This fermentation system consists of a hold-up vessel followed by two stirred tank fermentors for the primary fermentation. Subsequently, the flocculent yeast cells are separated from the green beer in a conical settler by gravity. Yeast is then recycled back into the hold-up vessel to increase the cell density and to achieve better control of the fermentation rates.

11.2.2 Carrier Selection

Various cell immobilization carrier materials have been tested and used for beer production. Selection criteria are summarised in Table 11.2. Depending on the particular application, reactor type and operational conditions, some selection criteria will be more appropriate. Examples of selected carrier materials for particular applications are tabulated in Table 11.1.

In one of the first ICT process for continuous beer fermentation, kieselguhr (diatomaceous earth) was selected as carrier material (Narziss and Hellich 1971; Moll et al. 1973). Later on, alginate hydrogel encapsulation became popular (e.g., White and Portno 1979; Hsu and Bernstein 1985; Onaka et al. 1985; Curin et al. 1987; Shindo et al. 1994; Nedovic et al. 1997; Nedovic et al. 2001; Nedovic et al. 2004). In addition, some new hydrogel materials were introduced, like κ-carrageenan, pectate gels and polyvinyl alcohol (PVA) (e.g., Mensour et al. 1996; Pilkington et al. 1999; Smogrovicová and Dömény 1999; Raymond et al. 2004; Nedovic et al. 2005a). The main advantage of the hydrogel entrapment method is the attainment of extremely high cell loadings and – consequently – providing high fermentation rates. However, in some cases cell proliferation and activity can be limited by low mass transfer rates within the matrices. The reduced cell growth in

Table 11.2 Selection criteria for yeast cell immobilization carrier materials for beer production (Nedovic et al. 2005b)

High cell mass loading capacity
Easy access to nutrient media
Simple and gentle immobilization procedure
Immobilization compounds approved for food applications
High surface area-to-volume ratio
Optimum mass transfer distance from flowing media to centre of support
Mechanical stability (compression, abrasion)
Chemical stability
Highly flexible: rapid start-up after shut-down
Sterilisable and reusable
Suitable for conventional reactor systems
Low shear experienced by cells
Easy separation of cells and carrier from media
Readily up-scalable
Economically feasible (low capital and operating costs)
Desired flavor profile and consistent product
Complete attenuation
Controlled oxygenation
Control of contamination
Controlled yeast growth
Wide choice of yeast

immobilised conditions can result in an insufficient free amino nitrogen consumption and as a consequence an unbalanced flavor profile of the final beer (Curin et al. 1987; Hayes et al. 1991). Recently, cheap carrier materials have been investigated since the use of these materials avoids the costly regeneration of used immobilization matrices. Wood chips showed good performance and the use of this material reduced the total investment cost by one third compared to more expensive carriers (Kronlöf et al. 2000).

An alternative preformed support material, based on spent grains, has been proposed recently for yeast cell immobilization for primary beer fermentation in an air-lift bioreactor (Brányik et al. 2001; Brányik et al. 2002; Brányik et al. 2004a; Brányik et al. 2004b; Brányik et al. 2006a). This is an interesting carrier for cell immobilization, since spent grains are a waste by-product from the brewing process. Recently, corncobs were also investigated as carrier material (Brányik et al. 2006a).

11.3 Bioreactor Design

In bioreactors for beverage production, immobilised cells are either mixed with suspended carriers or fixed on carrier particles or large surfaces. Beer production is based on the utilisation of yeast cells, while the acidification of wort employs lactic acid bacteria. Reactor configuration is related to the choice of cell carrier and various modifications and combinations of stirred tank, packed-bed, fluidised-bed, gas-lift and membrane reactors were proposed for different phases in beer production (Table 11.1). Critical issues related to the selection of reactor type and configuration are supply and removal of gases and solutes in the liquid phase and removal of excess biomass formed.

Most studies have been made in packed-bed bioreactor. Packed-bed bioreactors are characterised by a simple design, consisting of a column, which is packed with biocatalyst. The liquid flow is close to plug regime causing low shear rates. This enables usage of various, even fragile materials for cell immobilization (Obradovic et al. 2004). In packed-bed fermentors, high mass transfer restrictions, accumulation of carbon dioxide, non-uniform temperature profiles, flow channelling and stagnant zones were observed during primary fermentation. Therefore, different approaches for the adaptation of immobilised systems were investigated in order to correct the final beer quality. Packed-bed reactors have been selected for the production of alcohol-free or low-alcohol beers and for enhanced flavor maturation using immobilised cells. In these applications, conditions are anaerobic and yeast growth is limited. Preformed carrier materials are selected. Immobilization of cells can be by adsorption (e.g., DEAE-cellulose beads) or a combination of adsorption and entrapment (e.g., porous glass beads). These carrier materials need to be mechanically strong to withstand the high pressures in packed-bed reactors. However, the use of mechanically week materials, such as hydrogels, can be limited to lower bed heights and liquid flow rates due to possible compression of beads.

In the fluidised-bed bioreactors, particles with immobilised cells are fluidised in the liquid up-flow, while gas can be optionally supplied. As a consequence of particle fluidisation, moderate local mixing is established providing better mass and heat distribution with more uniform liquid flow throughout the reactor volume, as compared to packed-bed reactors. It is difficult to maintain low-density particles in fluidisation and prevent their washout. Particle movements and collisions in the fluidised state result in moderate shear stresses and abrasion, creating a need for relatively mechanically stable supports. The scaling up of fluidised-bed bioreactors meets problems due to the difficulties in controlling the bed expansion and may encounter hydrodynamic problems. In stirred tank reactors high aeration resulted in less balanced aroma profile of the final product. Beers produced in fluidised and stirred-tank fermentors had high concentrations of diacetyl, and low of higher alcohols and esters (Okabe et al. 1992; Mensour et al. 1997).

Gas-lift reactors are especially attractive since they apply pneumatic agitation with no mechanical devices. They are based on liquid circulation which can be effectively tuned to achieve an adequate flow regime, and optimal external mass transfer. Gas-lift reactors can be constructed as internal loop or external loop configuration. In general, the flow in the riser and down-comer section can be described as plug flow with axial dispersion (Obradovic et al. 2004). A gas-lift reactor that was introduced in beer fermentation studies by a Serbian group (Nedovic et al. 1993) retains the advantages of fluidised-beds, such as high loading of solids and good mass transfer properties and it is particularly suitable for applications with low-density carriers (Nedovic et al. 1997, 2004, 2005a, b). Efficient mixing and low shear rates make gas-lift reactors suitable for all types of immobilization materials.

As compared to conventional reactor types, the design of membrane reactors is relatively more complex and more expensive, mainly due to the high cost of the membrane material. Membrane reactors provide simultaneous bioconversion and product separation. A special design of a multichannel loop bioreactor has been used by the Belgian company Meura (Tournai) for production of lager and ale, and acidified wort (Masschelein et al. 1994). Yeast cells are immobilised in porous sintered silicon carbide rods perforated with 19 or 37 channels for fluid flow. This immobilization method can be regarded as containment behind a preformed barrier, and as entrapment in a porous preformed support.

11.4 The Impact of Immobilised Yeast Cell Systems on Beer Flavor

An overview of the flavor formation of various systems for the primary fermentation of beer demonstrates that flavor formation is dependent on the bioreactor system and carrier material (Willaert and Nedovic 2006). The mechanisms of the formation of these compounds in beer fermentation by freely suspended cell systems and changes that usually occur when immobilised systems are used, are explained in the following paragraphs.

11.4.1 Secondary Fermentation Using Immobilised Yeast Cells

One of the objectives of the maturation of green beer is the removal of diacetyl, an unwanted aroma compound. This vicinal diketone has a very low threshold (0.08–0.15 ppm) and imparts a buttery aroma to the beer (Wainwright 1973). During the primary fermentation, diacetyl is produced as a by-product in the synthesis pathway of isoleucine, leucine and valine (ILV pathway) (Fig. 11.2). Because the formation of α-acetolactate is related to the amino acid metabolism, more α-acetolactate will be produced with increasing yeast growth. Because brewer's yeast does not possess α-acetolactate decarboxylase activity, α-acetolactate is excreted from the cell and non-enzymatically converted to diacetyl by an oxidative decarboxylation. This step is the rate-limiting step and proceeds faster at high temperature and lower pH. Subsequently, diacetyl is re-assimilated in the yeast cell, which possesses the necessary enzymes (reductases) to reduce it to the flavor inactive acetoin and further to 2,3-butanediol (Bamforth and Kanauchi 2004). The reduction can occur fast when sufficient yeast is present.

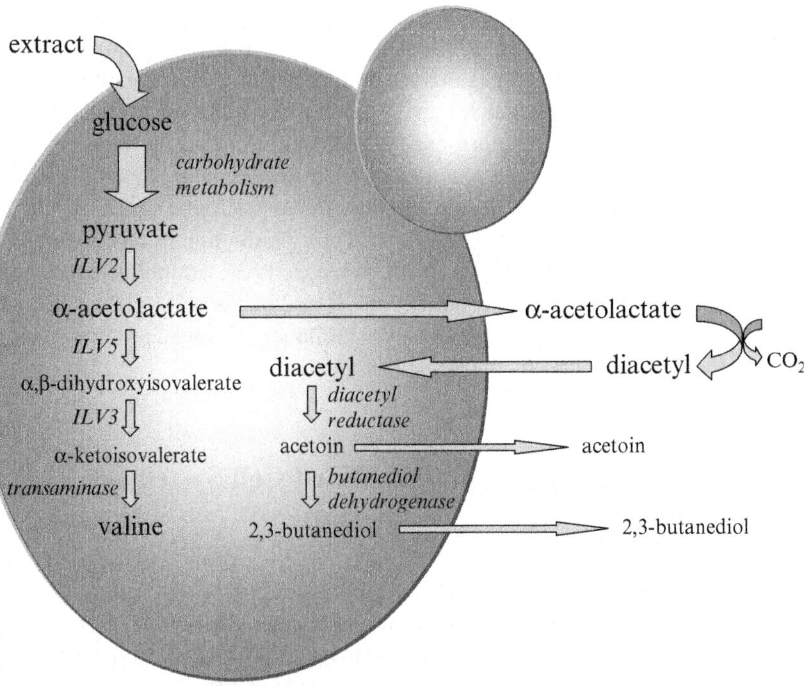

Fig. 11.2 Schematic presentation of the formation, re-assimilation and reduction of diacetyl. The grey oval shape represent a yeast cell, and the smaller circle its daughter cell

The traditional maturation process is performed at a low temperature and low yeast concentration, resulting in a very long maturation period of 3–4 weeks. Nowadays, several strategies have been developed to accelerate the diacetyl removal (Willaert 2007). However, using immobilised cell technology, this period could be further reduced to a few hours. Two continuous maturation systems have been implemented industrially so far: one at Sinebrychoff Brewery (Finland, capacity: 1 million hl per year) and another system, developed by Alfa Laval and Schott Engineering (Mensour et al. 1997). They are both composed of a separator (to prevent growing yeast cells in the next stages), an anaerobic heat treatment unit (to accelerate the chemical conversion of α-acetolactate to diacetyl, but also the partial directly conversion to acetoin), and a packed-bed reactor with yeast immobilised on DEAE-cellulose granules or porous glass beads (to reduce the remaining diacetyl), respectively (Yamauchi et al. 1995). Later on, the DEAE-cellulose carriers were replaced by cheaper wood chips (Virkajärvi 2002). Recently, the heat treatment has been replaced by an enzymatic transformation in a fixed-bed reactor, in which the α-acetolactate decarboxylase is immobilised in special multilayer capsules, followed by the reduction of diacetyl by yeast in a second packed-bed reactor (Nitzsche et al. 2001).

11.4.2 Alcohol-free or Low-alcohol Beer Using Immobilised Yeast Cells

The main objective during fermentation of alcohol-free beer is the reduction of wort carbonyl flavors by the activity of alcohol dehydrogenases of yeast, without the formation of alcohol (Van Iersel et al. 1999). Traditionally, alcohol-free beer is being produced by arrested batch fermentations. An alternative method is the removal of alcohol by using membrane, distillation or vacuum evaporation processes, although these have the disadvantage that the production cost is increased. Controlled ethanol production for low alcohol or alcohol-free beers has been successfully achieved by partial fermentation using immobilization reactors. The reduction of the wort aldehydes can be efficiently achieved by a short-contact time with the immobilised yeast cells. The process is performed at a low temperature to avoid undesirable cell growth and ethanol production and to maintain a good long-term yeast viability (Van Dieren 1995). A disadvantage of this process is the low production of desirable esters. Bavaria Brewery (the Netherlands) is using a packed-bed immobilised yeast bioreactor with a production capacity of 150,000 hl alcohol-free beer per annum (Mensour et al. 1997).

11.4.3 Primary Fermentation with Immobilised Yeast

During the main fermentation of beer, not only ethanol is being produced, but also a complex mixture of flavor-active secondary metabolites, of which the higher (or fusel) alcohols and esters are the most important. In addition, diacetyl and some sulphury compounds can cause off-flavors. Since this complex flavor profile is closely related to the amino acid metabolism and consequently to the growth of the yeast cells, differences in the growth metabolic state between freely suspended and immobilised yeast cell systems are most probably responsible for the majority of alterations in the beer flavor. For that reason, it is important that the physiological and metabolic state of the yeast in conventional batch systems is mimicked as much as possible during the continuous fermentation with immobilised yeast. In the continuous mode of operation, cells are not exposed to significant alterations of the environment, influencing the metabolism of the cells and consequently the flavor. Hence, the microbial population of continuous systems lacks the different growth phases of a batch culture. To imitate the batch process as much as possible, plug-flow reactors or a series of reactors can be used. As can be assumed, both the continuous mode of operation and the immobilization of yeast cells can influence the beer flavor.

11.4.3.1 Nutrient Uptake

An increase of glucose uptake and ethanol formation during glucose fermentation using immobilization systems is often stated (Navarro and Durand 1977; Doran and Bailey 1986; Galazzo and Bailey 1990). In conventional beer fermentations, a sequential uptake of the wort sugars is observed. First glucose, sucrose and fructose are taken up. After the uptake of glucose, the main wort sugar maltose (60–70%) is assimilated. Finally, maltotriose is slowly taken up (Hammond 1995). In immobilised yeast systems, alterations in the sugar uptake can occur, which can result in an altered beer taste (Van De Winkel et al. 1993; Willaert et al. 1999).

In a normal batch fermentation, the amino acids of the wort are categorised based on the sequential uptake pattern (Jones and Pierce 1964). Amino acids of Group A (arginine, asparagine, aspartic acid, glutamic acid, lysine, serine and threonine) are immediately taken up, while the uptake of amino acids of Group B (histidine, isoleucine, leucine, methionine and valine) is delayed. Amino acids of Group C (alanine, glycine, phenylalanine, tyrosine and tryptophan) are only taken up when Group A amino acids are depleted in the medium. Proline (Group D) is not utilised by the yeast. In immobilised yeast systems, the uptake pattern often differs from that of the traditional pattern (Ryder and Masschelein 1985; Shen et al. 2003a). Because amino acid metabolism is closely associated with flavor compounds, altered nitrogen uptake can have significant effects on beer flavor. Less uptake of amino acids were reported in a continuous packed-bed reactor (Ryder and Masschelein 1985). This was circumvented by using a fluidised bed reactor, suggesting that mass transfer limitations affected the nitrogen uptake (Cop et al. 1989). In a batch reactor with yeast immobilised on stainless steel fibre cloth, the final utilisation of Group A, B and C amino acids was more complete than in the standard free cell system (Shen et al. 2003a). In the system, described by Pajunen et al. (2001), little FAN ("*Free Amino Nitrogen*") was taken up by the cells, due to non-growth conditions, causing a high beer pH. Dunbar et al. (1988) reported that a continuous free cell system was characterised by a different uptake pattern than that proposed by Jones and Pierce (1964) for batch fermentations.

11.4.3.2 Higher Alcohols

The most important higher alcohols are isoamyl alcohol (3-methyl-1-butanol) and 2-phenyl ethanol which can be found around their threshold concentrations (60–70 and 25–125 ppm, respectively) in lager beer. Other aliphatic alcohols can also contribute to the alcoholic aroma of the beer (Meilgaard 1975). The aromatic alcohol 2-phenylethanol has a sweet rose-like aroma and is believed to mask the dimethyl sulphide (DMS) perception (Hegarty et al. 1995). Higher alcohols are synthesised by yeast during the fermentation via the catabolic (Ehrlich) or anabolic pathways (Genevois) of the amino acid metabolism (Chen 1978). In both pathways α-keto acids are formed, which are decarboxylated to aldehydes and further reduced (by an alcohol dehydrogenase) to higher alcohols. Higher alcohols are mainly produced during the active growth phase of the fermentation and are therefore influenced by each factor affecting the yeast growth, such as high levels of nutrients (amino acids, oxygen, lipids, zinc), increased temperature and agitation (Landaud et al. 2001).

The influence of immobilised cell systems on the production of higher alcohols differs from system to system. At the same degree of attenuation in batch and immobilised systems, the differences can be most probably ascribed to different levels of amino acid utilisation and yeast growth. In immobilised systems with enhanced or similar FAN uptake levels, the formation of higher alcohols was higher or equal to batch systems (Kronlöf et al. 1989; Shen et al. 2003a). Decreased higher alcohol production rates by immobilised as opposed to free-cell systems were attributed to limited cellular growth. The use of entrapped yeast was most of the time associated with FAN uptake limitations (Ryder and Masschelein 1985; Smogrovicová and Dömény 1999). However, new technologies introduced some new inclusion carriers with adjusted shape and size to overcome internal mass transfer restrictions, which gave similar higher alcohol concentrations compared to a conventional process (Nedovic et al. 2005a).

There is a trend of increased propanol yields, compared to isobutanol (Fig. 11.3). According to Pajunen et al. (2001), the overproduction was the result of a more active α-ketobutyrate pathway, which could also explain the increased concentrations of the vicinal diketone 2,3-pentanedione. Alternatively, continuous inflow of the amino acid threonine could be causing the stimulation of propanol production, as threonine is the precursor of propanol in the catabolic pathway (Chen 1978).

11.4.3.3 Esters

Esters are the most important flavor compounds in beer, since their low flavor thresholds can be crossed in lager beers. The major esters in beer are ethyl acetate (solvent-like, threshold: 33 ppm), isoamyl acetate (banana, threshold: 1.4 ppm), ethyl caproate (apple, threshold: 0.2 ppm), ethyl caprylate (apple, threshold: 0.9 ppm)

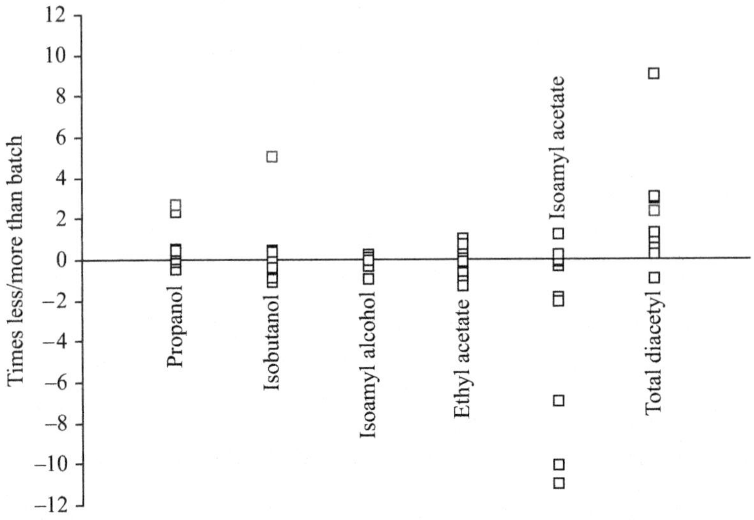

Fig. 11.3 Multiples of compound concentrations and their distribution in continuous beer fermentation systems using immobilised yeast systems compared to corresponding conventional, batch beer fermentation systems [data based on Brányik et al. (2006b), with permission]

and phenylethyl acetate (rosey, threshold: 3 ppm) (Meilgaard 1975). They are desirable components of beer when present in appropriate quantities. Esters are produced by yeast during the growth phase (60%) and the stationary phase (40%). They are formed by the condensation reaction between acetyl/acyl-CoA and higher alcohols catalysed by the alcohol acyltransferases of the yeast (Peddie 1990) (Fig. 11.4). Furthermore, it has been shown that the balance between ester-synthesising enzymes and esterases, which hydrolyse esters, might be important for the net rate of ester accumulation (Fukuda et al. 1998). Fundamentally, two factors determine the rate of ester formation: the availability of the two substrates (acetyl/acyl-CoA and fusel alcohols) and the activity of the enzymes. Ester formation is therefore influenced by temperature, hydrostatic pressure, growth rate, dissolved oxygen, fatty acids, carbon source and nitrogen concentration (Fig. 11.4, for a review, see Verstrepen et al. 2003b).

Immobilised yeast systems give variable amounts of esters, depending on the type of the system and operational conditions. However, the overall tendency in most continuous systems indicates a somewhat reduced ester formation (Fig. 11.4). It can be suggested that the most important factor affecting the ester formation is the oxygen availability during the fermentation, because immobilization itself stimulates the expression of *ATF1* (Shen et al. 2003b). Low oxygen concentrations, due to mass transfer limitations, trigger the formation of esters, by a direct effect upon the gene expression and by an indirect effect on substrate availability

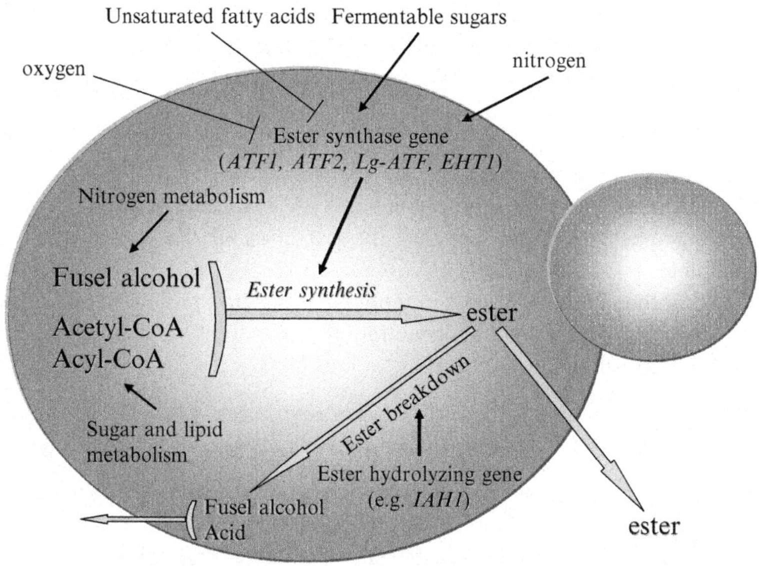

Fig. 11.4 Biochemical formation of esters (Verstrepen et al. 2003b)

(more acetyl-CoA is available for ester synthesis instead of fatty acid synthesis) (Masschelein et al. 1985; Van Iersel et al. 1999; Shen et al. 2003a; Shen et al. 2003b). High oxygen concentrations, due to continuous aeration of the wort in the reactor, result in poor ester formation (Virkajärvi et al. 1999; Wackerbauer et al. 2003).

Generally, the ester synthesis is a sensitive process, which is rather difficult to control, due to numerous influencing factors involved. The selection of an optimal yeast strain or the use of genetically modified strains could ameliorate the ester profile of beer, made by immobilised cell technology.

11.4.3.4 Diacetyl

In most cases, the amount of diacetyl formed by immobilised systems is much higher than in young beers of traditional free cell systems (Ryder and Masschelein 1985; Kronlöf et al. 1989; Van De Winkel et al. 1993; Brányik et al. 2006a). This has been explained by the following:

- Increased expression of Ilv2 enzyme (acetohydroxy acid synthetase) due to immobilization, resulting in increased α-acetolactate concentrations (Shindo et al. 1994).
- Alterations in the amino acid metabolism, caused by cell immobilization itself (Ryder and Masschelein 1985; Shindo et al. 1993; Shen et al. 2003a).

- Enhanced uptake of amino acids, due to rapid yeast growth, causing valine depletion and anabolic formation of the amino acid, resulting in increased α-acetolactate concentrations. The excess yeast growth is often a result of over-aeration (Brányik et al. 2006a).
- Continuous inlet of amino acids of Group A, which inhibit the uptake of valine (Dufour and Devreux 1986).
- Short residence times, meaning that the slow oxidative decarboxylation from α-acetolactate to diacetyl and thus the reduction of diacetyl by the yeast is incomplete (Okabe et al. 1994).
- Increasing the concentration of immobilised cells and prolonging the residence time leads to lower diacetyl concentrations (Shindo et al. 1994; Pajunen et al. 2001; Brányik et al. 2006a).

However, without lowering the fermentation rate, other strategies are possible to suppress these excessive diacetyl concentrations during continuous fermentation:

- Addition of the missing enzyme α-acetolactate decarboxylase, which is commercial available, to the wort (Hanneman 2002).
- Optimization of FAN-content, especially the valine concentration, of the wort (Pajunen et al. 2001; Petersen et al. 2004).
- The use of genetically modified yeast, encoding the α-acetolactate decarboxylase or overexpressing the *ILV5* gene, encoding the acetohydroxy acid reductoisomerase enzyme in the rate-limiting step of the ILV-pathway (Linko and Kronlöf 1991; Hammond 1995).

11.4.3.5 Control Strategies

The optimization of temperature, wort gravity, feed volume and wort composition seems to be an important tool for the control of the flavor-active compounds formation in immobilised beer fermentation systems (Table 11.3) [for a recent review of these effects, see Willaert and Nedovic (2006)]. Many researchers have concluded that the optimization of aeration during continuous fermentation is essential for the quality of the final beer (Virkajärvi et al. 1999). Oxygen is needed for the formation of unsaturated fatty acids and sterols, needed for growth (Depraetere et al. 2003). However, excess oxygen will lead to low ester production and to excessive diacetyl, acetaldehyde and fusel alcohol formation (Okabe et al. 1992; Wackerbauer et al. 2003; Brányik et al. 2004a). It is possible to adjust the flavor of the produced beer by ensuring the adequate amount of dissolved oxygen by sparging with a mixture of air, nitrogen or carbon dioxide (Kronlöf and Linko 1992; Brányik et al. 2004a). However, it remains difficult to predict the right amount of oxygen, because the oxygen availability to the immobilised yeast cells is dependent of external and internal mass transfer limitations.

Finally, the reactor design, the carrier and yeast strain can also have a dominant effect on flavor formation (Cop et al. 1989; Linko, et al. 1997; Smogrovicová and Dömény 1999; Tata et al. 1999; Virkajärvi and Pohjala 2000).

Table 11.3 Some operational parameters and their impact on flavor in immobilised beer fermentation systems

Parameter	Impact on flavor	Reference
Temperature	Increased diacetyl, higher alcohols and esters concentrations	Smogrovicová and Dömény (1999)
Wort gravity	Increased acetaldehyde; similar higher alcohols and esters	Virkajärvi et al. (2002)
Feed volume	Controlling attenuation, enhancing diacetyl removal	Pajunen et al. (2001)
Wort composition	Optimal FAN levels lower diacetyl concentration	Pajunen et al. (2001); Petersen et al. (2004)
Dissolved oxygen	Increased concentrations of diacetyl, acetaldehyde and higher alcohols; decreased esters levels	Kronlöf and Linko (1992)

11.5 Conclusions

The main benefit of immobilised cell technology (ICT) in the beverage industries is rendering high-productivity continuous fermentation feasible. Although research on immobilised cells is now approximately 30 years old, difficulties encountered in pilot-plant and full industrial-scale fermentation processes have not been solved yet. In fact, engineering problems linked to choice of the carrier and reactor design are complicated by the effects of immobilization on the flavor profile of the final product. Incomplete knowledge of the effects of immobilization on the physiology of brewers' yeast will lead to incomplete and partially empirical use of immobilised cell technology for processes based on alcoholic fermentation. These processes are very complex, linked to various side reactions important for flavor formation and final product quality.

Ongoing basic research is continuing to explore new materials as potential carriers for microbial cells and to identify and characterise changes in cell physiology and metabolism upon immobilization. Mass transfer limitations and process control are still issues that have to be resolved in order to obtain consistent quality of the beer. The assessment of the industrial feasibility of the immobilised fermentation technology is mandatory for providing cost-effective, large-scale applications. Since scale up is not always that easy and the beer quality should be as desired, piloting is required to fine tune the specifications. Manipulating yeast strains that have been genetically modified to develop an ability to produce excessive amounts of one and/or lower amounts of another flavor compound is another direction in the area of fermentation technology development.

Acknowledgements This work was supported by the Belgian Federal Science Policy Office, European Space Agency PRODEX program, the Institute for the Promotion of Innovation through Science and Technology in Flanders (IWT Vlaanderen), Serbian Ministry of Science and Technological Development (#371005 & #142075), Flanders Interuniversity Institute for Biotechnology (VIB) and the Research Council of the VUB.

References

Aivasidis A (1996) Another look at immobilized yeast systems. Cerevisia 21(1):27–32

Aivasidis A, Wandrey C, Eils H-G, Katzke M (1991) Continuous fermentation of alcohol-free beer with immobilized yeast cells in fluidized bed reactors. In: Proceedings European brewery convention congress, pp 569–576

Andersen K, Bergin J, Ranta B, Viljava T (1999) Main fermentation of beer. In: Proceedings of the 27th European brewery convention congress, pp 771–778

Andries M, Van Beveren PC, Goffin O, Masschelein CA (1996) Design and application of an immobilized loop bioreactor for continuous beer fermentation. In: Wijffels RH, Buitelaar RM, Bucke C, Tramper J (eds) Immobilized cells: Basics and applications. Elsevier, Amsterdam, pp 672–678

Baillie GS, Douglas LJ (1999) Role of dimorphism in the development of *Candida albicans* biofilms. J Med Microbiol 48:671–679

Bamforth CW, Kanauchi M (2004) Enzymology of vicinal diketone reduction in brewers' yeast. J Inst Brew 110(2):83–93

Bardi E, Koutinas AA, Kanellaki M (1997) Room and low temperature brewing with yeast immobilized on gluten pellets. Process Biochem 32:691–696

Baron GV, Willaert RG (2004) Cell immobilization in pre-formed porous matrices. In: Nedovic V, Willaert R (eds) Fundamentals of cell immobilization biotechnology. Kluwer Academic Plubishers, The Netherlands, pp 229–244

Bezbradica D, Obradovic B, Laskošek-Čukalović I, Bugarski B, Nedovic V (2007) Immobilization of yeast cells in PVA particles for beer fermentation. Process Biochem 42(9):1348–1351

Bidard F, Blondin B, Dequin S, Vezinhet F, Barre P (1994) Cloning and analysis of a *FLO5* flocculation gene from *Saccharomyces cerevisiae* yeast. Curr Gentetics 25:196–201

Bidard F, Bony M, Blondin B, Dequin S, Barre P (1995) The *Saccharomyces cerevisiae FLO1* flocculation gene encodes for a cell surface protein. Yeast 11:809–822

Bony M, Thines-Sempoux D, Barre P, Blondin B (1997) Localisation and cell surface anchoring of the *Saccharomyces cerevisiae* flocculation protein Flo1p. J Bacteriol 179:4929–4936

Brányik T, Vicente AA, Machado Cruz JM, Teixeira JA (2001) Spent grains – a new support for brewing yeast immobilization. Biotechnol Lett 23:1073–1078

Brányik T, Vicente A, Cruz JM, Teixeira J (2002) Continuous primary beer fermentation with brewing yeast immobilized on spent grains. J Inst Brew 108:410–415

Brányik T, Vicente AA, Cruz JMM, Texeira JA (2004a) Continuous primary fermentation of beer with yeast immobilized on spent grains – the effect of operational conditions. J Am Soc Brew Chem 62:29–34

Brányik T, Vicente AA, Kuncová G, Podrasky O, Dostálek P, Teixeira J (2004b) Growth model and metabolic activity of brewing yeast biofilm on the surface of spent grains: a biocatalyst for continuous beer fermentation. Biotechnol Prog 20:1733–1740

Brányik T, Vicente AA, Dostálek P, Teixeira JA (2005) Continuous beer fermentation using immobilized yeast cell bioreactor systems. Biotechnol Prog 21(3):653–663

Brányik T, Silva DP, Vicente AA, Lehnert R, Almeida e Silva JB, Dostálek P, Teixeira JA (2006a) Continuous immobilized yeast reactor system for complete beer fermentation using spent grains and corncobs as carrier materials. J Ind Microbiol Biotechnol 33:1010–1018

Brányik T, Vicente AA, Dostálek P, Teixeira J (2006b) Flavor formation in continuous fermentations. Proceedings Chair De Clerck, Leuven, p 15p

Caro LH, Tettelin H, Vossen JH, Ram AF, Van den Ende H, Klis FM (1997) In silicio identification of glycosyl-phosphatidylinositol-anchored plasma-membrane and cell wall proteins of *Saccharomyces cerevisiae*. Yeast 13:1477–1489

Chen EC-H (1978) The relative contribution of Ehrlich and biosynthetic pathways to the formation of fusel alcohols. J Am Soc Brew Chem 36(1):39–43

Collin S, Montesinos M, Meersman E, Swinkels W, Dufour JP (1991) Yeast dehydrogenase activities in relation to carbonyl compounds removal from wort and beer. In: Proceedings European brewery convention congress, pp 409–416

Cop J, Dyon D, Iserentant D, Masschelein CA (1989) Reactor design optimization with a view on the improvement of amino acid utilization and flavor development of calcium-alginate entrapped brewing yeast fermentations. In: Proceedings of the 22nd European brewery convention, Zurich, pp 315–322

Cormack BP, Ghori N, Falkow S (1999) An adhesin of the yeast pathogen Candida glabrata mediating adherence to human epithelial cells. Science 285(5427):578–582

Coutts MW (1957) A continuous process for the production of beer. UK Patent 872,391400

Cross PA, Mavituna F (1987) Yeast retention fermentors for beer production. In: Proceedings of the 4th European congress on biotechnology, Amsterdam, pp 199–200

Curin J, Pardonova M, Sedova H, Kahler M (1987) Beer production with immobilized yeast. Proceedings of the 21th EBC congress, pp 433–440

De Groot PWJ, Hellingwerf KJ, Klis FM (2003) Genome-wide identification of fungal GPI proteins. Yeast 20:781–796

Decamps C, Norton S, Poncelet D, Neufeld RJ (2004) Continuous pilot plant-scale immobilization of yeast in κ-carrageenan gel beads. AIChE J 50:1599–1605

Depraetere SA, Winderickx J, Delvaux FR (2003) Evaluation of the oxygen requirement of lager and ale yeast strains by preoxygenation. Master Brewers Assoc Americas Tech Quarterly 40(4): 283–289

Doran PM, Bailey JE (1986) Effects of immobilization on growth, fermentation properties, and molecular composition of *Saccharomyces cerevisiae* attached to gelatin. Biotechnol Bioeng 28:73–87

Dufour JP, Devreux A (1986) The use of amino acids analysis as a tool to control malting and brewing processes. EBC Symposium on Wort Production, Monograph XI. Nürnberg, pp. 227–248

Dunbar J, Campbell SI, Banks DJ, Warren DR (1988) Metabolic aspects of a commercial continuous fermentation system. In: Proceedings of the 20th convention of the institute of brewing (Asia and New Zealand Section), pp 151–158

Frieman MB, Cormack BP (2003) The omega-site sequence of glycosylphosphatidyl-inositol-anchored proteins in *Saccharomyces cerevisiae* can determine distribution between the membrane and the cell wall. Mol Microbiol 50:883–896

Fu Y, Ibrahim AS, Sheppard DC, Chen YC, French SW, Cutler JE, Filler SG, Edwards JE Jr (2002) *Candida albicans* Als1p: an adhesin that is a downstream effector of the EFG1 filamentation pathway. Mol Microbiol 44:61–72

Fukuda K, Yamamoto N, Kiyokawa Y, Yanagiuchi T, Wakai Y, Kitamoto K, Inoue Y, Kimura A (1998) Balance of activities of alcohol acetyltransferase and esterase in *Saccharomyces cerevisiae* is important for production of isoamyl acetate. Appl Environ Microbiol 64(10):4076–4078

Galazzo JL, Bailey JE (1990) Fermentation pathway kinetics and metabolic flux control in suspended and immobilized *Saccharomyces cerevisiae*. Enzyme Microb Technol 12:162–172

Gaur NK, Klotz SA (1997) Expression, cloning, and characterization of a *Candida albicans* gene, *ALA1*, that confers adherence properties upon *Saccharomyces cerevisiae* for extracellular matrix proteins. Infect Immun 65:5289–5294

Green CB, Cheng G, Chandra J, Mukherjee P, Ghannoum MA, Hoyer LL (2004) T-PCR detection of *Candida albicans ALS* gene expression in the reconstituted human epithelium (RHE) model of oral candidiasis and in model biofilms. Microbiology 150:267–275

Guo B, Styles CA, Feng Q, Fink GA (2000) A *Saccharomyces* gene family involved in invasive growth, cell–cell adhesion, and mating. Proc Natl Acad Sci USA 97:12158–12163

Hammond JRM (1995) Genetically-modified brewing yeast for the 21st century. Progress to date. Yeast 11:1613–1627

Hanneman W (2002) Reducing beer maturation time and retaining quality. Master Brewers Assoc Americas Tech Quarterly 39(3):149–155

Hayes AH, Power J, Ryder DS (1991) Immobilized cell technology for brewing: A progress report. Brewers Dig 66(11):28–33

Hegarty PK, Parsons R, Bamforth CW, Molzahn SW (1995) Phenyl ethanol – a factor determining lager character. In: Proceedings of the 25th European brewery convention, pp 515–522

Hoyer LL (2001) The *ALS* gene family of *Candida albicans*. Trends Microbiol 9:176–180

Hoyer LL, Payne TL, Hecht JE (1998) Identification of *Candida albicans ALS2* and *ALS4* and localization of ALS proteins to the fungal cell surface. J Bacteriol 180:5334–5343
Hsu W-P, Bernstein L (1985) A new type of bioreactor employing immobilized yeast. J Am Soc Brew Chem 43:101–103
Inoue T (1995) Development of a two-stage immobilized yeast fermentation system for continuous beer brewing. In: Proceedings European brewery convention congress, pp 25–36
Jin YL, Speers RA (1999) Flocculation of *Saccharomyces cerevisiae.* Food Res Int 31:421–440
Jones M, Pierce JS (1964) Absorption of amino acids from wort by yeasts. J Inst Brew 70:307–315
Karel SF, Libicki SB, Robertson CR (1985) The immobilization of whole cells: engineering principles. Chem Eng Sci 40:1321–1353
Kobayashi O, Hayashi N, Kuroki R, Sone H (1998) Region of Flo1 proteins responsible for sugar recognition. J Bacteriol 180:6503–6510
Kronlöf J, Linko M (1992) Production of beer using immobilized yeast encoding α-acetolactate decarboxylase. J Inst Brew 98:479–491
Kronlöf J, Virkajärvi I (1999) Primary fermentation with immobilized yeast. In: Proceedings European brewery convention congress, pp 761–770
Kronlöf J, Härkönen T, Hartwall P, Home S, Linko M (1989) Main fermentation with immobilized yeast. In: Proceedings of the 22nd European brewery convention, Zurich, pp 355–362
Kronlöf J, Virkajärvi I, Storgards EL, Londesborough J, Dymond G (2000) Combined primary and secondary fermentation with immobilized yeast. In: World brewing congress 2000 proceedings, Orlando, P-56
Landaud S, Latrille E, Corrieu G (2001) Top pressure and temperature control of the fusel alcohol/ester ratio through yeast growth in beer fermentation. J Inst Brew 10(2):107–117
Li F, Palecek SP (2003) *EAP1*, a *Candida albicans* gene involved in binding human epithelial cells. Eukaryotic Cell 2:1266–1273
Linko M, Kronlöf J (1991) Main fermentation with immobilised yeast. Proceedings of the 23rd european brewery convention, Lisbon, pp 353–360
Linko M, Suihko M-L, Kronlöf J, Home S (1993) Use of brewer's yeast expressing α-acetolactate decarboxylase in conventional and immobilized fermentations. Master Brewers Assoc Americas Tech Quarterly 30:93–97
Linko M, Virkajärvi I, Pohjala N, Lindborg K, Kronlöf J, Pajunen E (1997) Main fermentation with immobilized yeast – a breakthrough? Proceedings of the 26th European brewery convention, Maastricht, pp 385–394
Linko M, Haikara A, Ritala A, Penttilä M (1998) Recent advances in the malting and brewing industry. J Biotechnol 65:85–98
Lo WS, Dranginis AM (1998) The cell surface flocculin Flo11 is required for pseudohyphae formation and invasion by *Saccharomyces cerevisiae.* Mol Biol Cell 9:161–171
Lommi H (1990) Immobilized yeast for maturation and alcohol-free beer. Brewing Distilling Int 5:22–23
Maeba H, Unemoto S, Sato M, Shinotsuka K (2000) Primary fermenation with immobilized yeast in porous chitosan beads. Pilot scale trial. In: Proceedings 26th convention institute brewing Australia and New Zealand Section, Singapore, pp 82–86
Manojlović V, Agouridis N, Kopsahelis N, Kanellaki M, Bugarski B, Nedovic V (2008) Brewing by immobilized freeze dried cells in a novel gas flow bioreactor. Proceedings of the 2008 joint central European congress, 4th central European congress on food, 6th Croatian congress of food technologists, biotechnologists and nutritionists, Cavtat, Croatia, pp 327–334
Masschelein CA, Carlier A, Ramos-Jeunehomme C (1985) The effect of immobilization on yeast physiology and beer quality in continuous and discontinuous systems. Proceedings of the 20th European brewery convention, Helsinki, pp 339–346
Masschelein CA, Ryder DS, Simon J-P (1994) Immobilized cell technology in beer production. Crit Rev Biotechnol 14:155–177
Mavituna F (2004) Pre-formed carriers for cell immobilization. In: Nedovic V, Willaert R (eds) Fundamentals of cell immobilization biotechnology. Kluwer Academic Plubishers, The Netherlands, pp 121–139

Meilgaard MC (1975) Flavor chemistry of beer: Part II: flavor threshold of 239 aroma volatiles. MBAA Tech Quarterly 12(3):151–168

Mensour N, Margaritis A, Briens CL, Pilkington H, Russell I (1996) Applications of immobilized yeast cells in the brewing industry. In: Wijffels RH, Buitelaar RM, Bucke C, Tramper J (eds) Immobilized cells: Basics and applications. Elsevier, Amsterdam, pp 661–671

Mensour N, Margaritis A, Briens CL, Pilkington H, Russell I (1997) New developments in the brewing industry using immobilised yeast cell bioreactor systems. J Inst Brew 103:363–370

Moll M, Durand G, Blachere H (1973). Continuous production of fermented liquids. French Patent 73/23397; US Patent 4009286

Narziss L, Hellich P (1971) Ein Beitrag zur wesentlichen Beschleunigung der Gärung und Reifung des Bieres. Brauwelt 111:1491–1500

Navarro JM, Durand G (1977) Modification of yeast metabolism by immobilization onto porous glass. Eur J Appl Microbiol Biotechnol 4:243–254

Nedovic V, Obradovic B, Vunjak-Novakovic G, Laskošek-Čukalović I (1993) Kinetics of beer fermentation with immobilized yeast cells in an internal-loop air-lift bioreactor. Chem Ind 47:168–172

Nedovic VA, Pesic R, Laskošek-Čukalović I, Laketic D, Vunjak-Novakovic G (1997) Analysis of liquid axial dispersion in an internal loop gas-lift bioreactor for beer fermentation with immobilized yeast cells. In: Proceedings 2nd European conference on fluidization, Bilbao, pp 627–635

Nedovic V, Obradovic B, Laskošek-Čukalović I, Trifunovic O, Pesic R, Bugarski B (2001) Electrostatic generation of alginate microbeads loaded with brewing yeast. Process Biochem 37(1):17–22

Nedovic V, Bezbradica D, Obradovic B, Laskošek-Čukalović I, Bugarski B (2004) Primary beer fermentation by PVA-immobilized brewing yeast in a gas-lift bioreactor. In: World brewing congress 2004 CD Rom proceedings, San Diego, O-63

Nedovic VA, Cukalovic IL, Bezbradica D, Obradovic B, Bugarski B (2005a) New porous matrices and procedures for yeast cell immobilization for primary beer fermentation. In: Proceedings of the 30th European brewery convention, Prague, pp 401–413

Nedovic V, Willaert R, Laskošek-Čukalović I, Obradovic B, Bugarski B (2005b) Beer production using immobilized cells. In: Nedovic V, Willaert R (eds) Applications of cell immobilization biotechnology. Springer, Dordrecht, The Netherlands, pp 259–273

Nitzsche F, Höhn G, Meyer-Pittroff R, Berger S, Pommersheim R (2001) A new way for immobilized yeast systems: secondary fermentation without heat treatment. In: Proceedings of the 28th European brewery convention, Budapest, pp 486–494

Obradovic B, Nedovic V, Bugarski B, Willaert RG, Vunjak-Novakovic G (2004). In: Nedovic V, Willaert RG (eds) Focus on biotechnology, volume 8a: Fundamentals of cell immobilization biotechnology, Kluwer Academic Publishers, Dordrecht, pp 411–436

Okabe M, Katoh M, Furugoori F, Yoshida M, Mitsui S (1992) Growth and fermentation characteristics of bottom brewer's yeast under mechanical stirring. J Ferment Bioeng 73(2):148–152

Okabe M, Oda A, Park Y-S, Noguchi K, Okamoto T, Mitsui S (1994) Continuous beer fermentation by high cell-density culture of bottom brewer's yeast. J Ferment Bioeng 77(1):41–45

Onaka T, Nakanishi K, Inoue T, Kubo S (1985) Beer brewing with immobilized yeast. Nat Biotechnol 3:467–470

Pajunen E, Grönqvist A (1994) Immobilized yeast fermenters for continuous lager beer maturation. In: Proceedings 23rd convention institute of brew. Australia and New Zealand Section, Sydney, pp 101–103

Pajunen E, Tapani K, Berg H, Ranta B, Bergin J, Lommi H, Viljava T (2001) Controlled beer fermentation with continuous one-stage immobilized yeast reactor. In: Proceedings of the 28th European brewery convention, Budapest, pp 465–476

Peddie HAB (1990) Ester formation in brewery fermentations. J Inst Brew 96:327–331

Petersen EE, Margaritis A, Stewart RJ, Pilkington H, Mensour N (2004) The effects of wort valine concentration on the total diacetyl profile and levels late in batch fermentations with brewing yeast *Saccharomyces carlsbergensis*. J Am Soc Brew Chem 62(4):131–139

Pilkington H, Maragaritis A, Mensour N, Russell I (1998) Fundamentals of immobilised yeast cell for continuous beer fermentation: a review. J Inst Brew 104:19–31

Pilkington H, Maragaritis A, Mensour N, Sobczak J, Hancock I, Russell I (1999) Kappa-carrageenan gel immobilization of lager brewing yeast. J Inst Brew 105:398–404
Pittner H, Back W, Swinkels W, Meersman E, Van Dieren B, Lomni H (1993) Continuous production of acidified wort for alcohol-free-beer with immobilized lactic acid bacteria. In: Proceedings European brewery convention congress, pp 323–329
Raymond MC, Neufeld RJ, Poncelet D (2004) Encapsulation of brewing yeast in chitosan coated carrageenan microspheres by emulsification/termal gelation. Artif Cells Blood Substit Imobil Biotechnol 32:275–291
Reynolds TB, Fink GR (2001) Baker's yeast, a model for fungal biofilm formation. Science 291:878–881
Ryder DS, Masschelein CA (1985) The growth process of brewing yeast and the biotechnological challenge. J Am Soc Brew Chem 43(2):66–75
Shen H-Y, Moonjai N, Verstrepen KJ, Delvaux FR (2003a) Impact of attachment immobilization on yeast physiology and fermentation performance. J Am Soc Brew Chem 61(2):79–87
Shen H-Y, Moonjai N, Verstrepen KJ, Delvaux F, Delvaux FR (2003b) Immobilization of *Saccharomyces cerevisiae* induces changes in the gene expression levels of *HSP12*, *SSA3* and *ATF1* during beer fermentation. J Am Soc Brew Chem 61(4):175–181
Sheppard DC, Yeaman MR, Welch WH, Phan QT, Fu Y, Ibrahim AS, Filler SG, Zhang M, Waring AJ, Edwards JE Jr (2004) Functional and structural diversity in the Als protein family of *Candida albicans*. J Biol Chem 279:30480–30489
Shindo S, Sahara H, Koshino S (1993) Relationship of production of succinic acid and methyl citric acid pathway during alcohol fermentation with immobilized yeast. Biotechnol Lett 15(1):51–56
Shindo S, Sahara H, Koshino S (1994) Suppression of α-acetolactate formation in brewing with immobilized yeast. J Inst Brew 100:69–72
Smogrovicová D, Döméný Z (1999) Beer volatile by-product formation at different fermentation temperature using immobilized yeasts. Process Biochem 34:785–794
Smogrovicová D, Döméný Z, Gemeiner P, Malovíková A, Sturdík E (1997) Reactors for the continuous primary beer fermentation using immobilised yeast. Biotechnol Tech 11:261–264
Smogrovicová D, Döményi Navrátil M, Dvorák P (2001) Continuous beer fermentation using polyvinyl alcohol entrapped yeast. In: Proceedings European brewery convention congress, vol 50, pp 1–9
Staab JF, Bradway SD, Fidel PL, Sundstrom P (1999) Adhesive and mammalian transglutaminase substrate properties of *Candida albicans* Hwp1. Science 283:1535–1583
Stratford M (1989) Yeat flocculation: calcium specificity. Yeast 5:487–496
Tata M, Bower P, Bromberg S, Duncombe D, Fehring J, Lau V, Ryder D, Stassi P (1999) Immobilized yeast bioreactor systems for continuous beer fermentation. Biotechnol Progress 15:105–113
Teunissen AWRH, Steensma HY (1995) Review: the dominant flocculation genes of *Saccharomyces cerevisiae* constitute a new subtelomeric gene family. Yeast 11:1001–1013
Teunissen AW, Holub E, Van der Hucht J, Van den Berg JA, Steensma HY (1993) Sequence of the open reading frame of the *FLO1* gene from *Saccharomyces cerevisiae*. Yeast 9:423–427
Udenfriend S, Kodukula K (1995) How glycosylphosphatidylinositol-anchored membrane proteins are made. Ann Rev Biochem 64:563–591
Unemoto S, Mitani Y, Shinotsuka K (1998) Primary fermentation with immobilized yeast in a fluidized bed reactor. Master Brewers Assoc Americas Tech Quarterly 35:58–61
Van De Winkel L, De Vuyst L (1997) Immobilized yeast cell systems in today's breweries and tomorrow's. Cerevisia 22(1):27–31
Van De Winkel L, Van Beveren PC, Masschelein CA (1991) The application of an immobilized yeast loop reactor to the continuous production of alcohol-free beer. In: Proceedings European brewery convention congress, pp 307–314
Van De Winkel L, Van Beveren PC, Borremans E, Goossens E, Masschelein CA (1993) High performance immobilized yeast reactor design for continuous beer fermentation. In: Preceedings of the 24th European brewery convention congress, pp 307–314

Van Dieren B (1995) Yeast metabolism and the production of alcohol-free beer. In: EBC symposium immobilized yeast applications in the brewery industry, Monograph XXIV, Espoo, Finland, pp 66–76

Van Iersel MFM, Van Dieren B, Rombouts FM, Abee T (1999) Flavor formation and cell physiology during the production of alcohol-free beer with immobilized *Saccharomyces cerevisiae*. Enzyme Microb Technol 24(7):407–411

Verbelen PJ, De Schutter DP, Delvaux F, Verstrepen KJ, Delvaux FR (2006) Immobilized yeast cell systems for continuous fermentation applications. Biotechnol Lett 28:1515–1525

Verstrepen KJ, Derdelinckx G, Verachtert H, Delvaux FR (2003a) Yeast flocculation: what brewers should know. Appl Microbiol Biotechnol 61:197–205

Verstrepen KJ, Derdelinckx G, Dufour J-P, Winderickx J, Thevelein JM, Pretorius IS, Delvaux FR (2003b) Flavor-active esters: Adding fruitiness to beer. J Biosci Bioeng 96(2):110–118

Verstrepen KJ, Reynolds TB, Fink GR (2004) Origins of variation in the fungal cell surface. Nat Rev Microbiol 2:533–540

Virkajärvi I (2002) Some developments in immobilized fermentation of beer during the last 30 years. Brauwelt Int 20:100–105

Virkajärvi I, Krönlof J (1998) Long-term stability of immobilized yeast columns in primary fermentation. J Am Soc Brew Chem 56:70–75

Virkajärvi I, Pohjala N (2000) Primary fermentation with immobilized yeast: effects of carrier materials on the flavor of the beer. J Inst Brew 106:311–318

Virkajärvi I, Lindborg K, Kronlöf J, Pajunen E (1999) Effects of aeration on flavor compounds in immobilized primary fermentation. Monatschrift Brauwiss 52(9–12):25–28

Virkajärvi I, Vainikka M, Virtanen H, Home S (2002) Productivity of immobilized yeast reactors with very-high-gravity worts. J Am Soc Brew Chem 60(4):188–197

Viyas VK, Kuchin S, Berkely CD, Carlson M (2003) Snf1 kinases with different b-subunit isoforms play distinct roles in regulating haploid invasive growth. Mol Biol Cell 23:1341–1348

Wackerbauer K, Ludwig A, Möhle J, Legrand J (2003) Measures to improve long term stability of main fermentation with immobilized yeast. In: Proceedings of the 29th European brewery convention, pp 445–457

Wainwright T (1973) Diacetyl – a review. J Inst Brew 79:451–470

White FH, Portno AD (1979) The influence of wort composition on beer ester levels. In: Proceedings European brewery convention congress, pp 447–460

Willaert R (2007) The beer brewing process: wort production and beer fermentation. In: Huie YH (ed) Handbook of food products manufacturing. John Wiley & Sons, New Jersey, pp 443–506

Willaert RG, Baron GV (1996) Gel entrapment and micro-encapsulation: methods, applications and engineering principles. Rev Chem Eng 12:5–205

Willaert R, Nedovic VA (2006) Primary beer fermentation by immobilised yeast – a review on flavor formation and control strategies. J Chem Technol Biotechnol 81:1353–1367

Willaert R, Van de Winkel L, De Vuyst L (1999) Improvement of maltose and maltotriose brewery fermentation efficiency using immobilised cell technology. In: Proceedings of the 27th European brewery convention, Cannes, pp 663–671

Yamauchi Y, Okamoto T, Murayama H, Kajino K, Amikura T, Hiratsu H, Nagara A, Kamiya T, Inoue T (1995) Rapid maturation of beer using an immobilized yeast bioreactor. 1. Heat conversion of α-acetolactate. J Biotechnol 38:101–108

Chapter 12
Immobilization of Microbial Cells for Alcoholic and Malolactic Fermentation of Wine and Cider

Yiannis Kourkoutas, Verica Manojlović, and Viktor A. Nedović

12.1 Introduction

Wine- or cider-making is highly associated with biotechnology owing to the traditional nature of must fermentation.. Nowadays, there have been considerable developments in wine- or cider-making techniques affecting all phases of wine or cider production, but more importantly, the fermentation process. It is well-known that the transformation of grape must by microbial activity results in the production of wine, and the fermentation of apples (or sometimes pears) in the production of cider. In this process, a variety of compounds affecting the organoleptic profile of wine or cider are synthesized. It is also common sense that in wine- or cider-making, the main objective is to achieve an adequate quality of the product. The technological progress and the improved quality of the wines or ciders have been associated with the control of technical parameters. Herein, cell immobilization offers numerous advantages, such as enhanced fermentation productivity, ability for cell recycling, application of continuous configurations, enhanced cell stability and viability, and improvement of quality (Margaritis and Merchant 1984; Stewart and Russel 1986; Kourkoutas et al. 2004a).

The objective of the present chapter is to analyze and assess data on the impact of immobilization technologies of viable microbial cells on the alcoholic and malolactic fermentation (MLF) of wine and cider. The immobilized biocatalysts are evaluated for their scale-up ability and their potential future impact in industrial application is highlighted and assessed. Handicaps associated with maintenance of cell viability and

Y. Kourkoutas (✉)
Department of Molecular Biology and Genetics, Democritus University of Thrace, Alexandroupolis, GR, 68100, Greece
e-mail: ikourkou@mbg.duth.gr

V. Manojlović
Department of Chemical Engineering, Faculty of Technology and Metallurgy, University of Belgrade, Karnegijeva 4, 11000 Belgrade, Serbia

V.A. Nedović
Department of Food Technology and Biochemistry, Faculty of Agriculture, University of Belgrade, Nemanjina 6, PO Box 127, 11081 Belgrade-Zemun, Serbia

N.J. Zuidam and V.A. Nedović (eds.), *Encapsulation Technologies for Active Food Ingredients and Food Processing*,
DOI 10.1007/978-1-4419-1008-0_12,

fermentation efficiency during preservation and storage, constraining the industrial use of immobilized cell systems are discussed.

12.2 Cell Immobilization Methods

12.2.1 Prerequisites for Cell Immobilization

Various materials are available for cell immobilization. However, not all carriers are considered suitable for food production. The main prerequisites that should be fulfilled by cell immobilization supports include:

1. A large surface of the immobilization support, with functional groups for cells to adhere to,
2. Easy-handling and regeneration of the immobilization support,
3. Cost-effectiveness of the support and immobilization process,
4. Acceptance of immobilization support by the consumers and avoidance of negative effects on the final food product (e.g., off-flavor formations),
5. Retention of immobilized cell viability,
6. Avoidance of negative effects of cell immobilization on biological and metabolic activity of immobilized cells, and
7. Food-grade purity of the immobilization support.

12.2.2 Effects of Cell Immobilization

Cell immobilization might cause desirable or undesirable alterations in the metabolism of cells (Melzoch et al. 1994; Norton and D'Amore 1994; Walsh and Malone 1995). It might affect cell growth and physiology (Melzoch et al. 1994; Jamai et al. 2001), metabolic activity (Navarro and Durand 1977; Buzas et al. 1989; Hilge-Rotmann and Rehm 1990; Jamai et al. 2001), stress tolerance (Nolan et al. 1994; Norton and D'Amore 1994; Lodato et al. 1999), cells survival and viability (Argiriou et al. 1996; Kourkoutas et al. 2003a) and flavor formation (Bakoyianis et al. 1993; Bardi et al. 1997). It has been difficult to predict the type and magnitude of these possible metabolic changes due to immobilization as a number of parameters are involved, such as mass transfer limitations by diffusion (Webb et al. 1986), changes in the cell morphology and oxygen uptake (Shirai et al. 1988), altered membrane permeability (Brodelius and Nilsson 1983), media components deficiency (Chen et al. 1990), surface pressure and osmotic pressure changes (Vijayalakshmi et al. 1979), contacts between cells in limited space (Shuler 1985), etc. For example, immobilization of *Saccharomyces cerevisiae* resulted in increased ethanol production and glucose consumption, higher ploidy and RNA content (Doran and Bailey 1986), lower internal pH value (Galazzo et al. 1987) and altered glucose catabolic pathways compared to free cells in suspensions, and in increased enzyme activity and therefore productivity (Galazzo and Bailey 1990).

12.2.3 Advantages of Immobilized Cell Systems

The advantageous effects of immobilization applications could be summarized in the following:

1. Prolonged activity and stability of the immobilized cells, since the immobilization support may act as a protective agent against physicochemical changes (pH, temperature, heavy metals, solvents, etc.),
2. Higher cell densities than usually achieved, resulting in higher productivities and increased substrate uptake and yield,
3. Increased tolerance to high substrate concentration and final product inhibition,
4. Reduced risk of microbial contamination attributed to high cell densities and enhanced fermentation activity,
5. Ability for low-temperature fermentation and/or maturation for certain food products,
6. Ability for regeneration and re-use,
7. Versatility in the selection of bioreactors,
8. Ability for continuous operation bioreactor systems, and
9. Reduction of maturation times in certain circumstances.

12.3 Alcoholic Fermentation by Immobilized Cells

Cell immobilization in alcoholic fermentation is a rapidly expanding research area because of its attractive technical and economic advantages compared to the conventional free cell system (Sakurai et al. 2000; Sree et al. 2000; Shindo et al. 2001; Lu et al. 2002). However, for industrial wine and cider production it is important to identify a suitable support for cell immobilization that meets the prerequisites named above and results in overall improvement of the sensory characteristics of the final product. Another criterion that is sought after in an immobilized system is the suitability for use in low temperature fermentations, since low temperature fermentation leads to an improved quality product.

12.3.1 Alcoholic Fermentation of Wine Using Immobilized Cells

Immobilization supports have been grouped into four categories (Kourkoutas et al. 2004a):

1. Inorganic supports,
2. Organic supports,
3. Membrane systems,
4. Natural supports and cell aggregates.

12.3.1.1 Inorganic Supports

Kissiris (a cheap, porous volcanic mineral found in Greece, similar to granite, containing 70% SiO_2, 13% Al_2O_3, and other inorganic oxides) was successfully used as immobilization support of *S. cerevisiae* for wine production in both batch (Kana et al. 1989) and continuous bioreactor systems (Bakoyianis et al. 1992; Bakoyianis et al. 1993). Kissiris-supported biocatalyst showed increased ethanol productivity and biocatalytic stability for about 2.5 years during successive preservations at 0°C (Argiriou et al. 1996). Similar studies involved γ-alumina in the form of porous cylindrical pellets as immobilization support for wine making (Kana et al. 1989; Loukatos et al. 2000).

Although inorganic supports offered many advantages such as abundance, enhanced fermentation productivity and cost effectiveness, they were considered inappropriate for wine-making because they are characterized as unsuitable materials for human nutrition. Nevertheless, they seem to have a potential in production of alcohol (Bakoyianis and Koutinas 1996; Koutinas et al. 1997) or distillates (Loukatos et al. 2003), as mineral residues do not distill.

12.3.1.2 Organic Supports

Organic immobilization supports mainly include those composed of polysaccharides or other polymers. The most popular polysaccharides proposed in wine fermentation are alginates (Suzzi et al. 1996; Silva et al. 2002). Calcium alginate was used for immobilizing *Candida stellata* alone or in combination with *S. cerevisiae* to enhance glycerol formation in wine (Ciani and Ferraro 1996). The same immobilized biocatalyst along with an inoculum of *S. cerevisiae* was used at pilot scale and under non-sterile conditions, in order to control wild microflora (Ferraro et al. 2000). Most efforts have focused on application of alginates for the secondary fermentation of sparkling wines (Fumi et al. 1988; Busova et al. 1994; Colagrande et al. 1994). In fact, application of immobilized cell technology in the production of sparkling wines is the only case to have been commercially available, imitating the traditional Champagne method.

However, gel-like supports discourage scale-up and industrial applications, since they are mechanically unstable in high-capacity bioreactors. In addition, it is difficult to attain consumer acceptance, in case they are not of food-grade purity (e.g., polyvinyl alcohol).

12.3.1.3 Membrane Systems

References concerning the use of membrane systems in wine and cider production are scarce in literature. In a previous study, a single-vessel membrane bioreactor was found unsuitable for continuous dry wine production, as high levels of unfermented sugars were reported (Takaya et al. 2002). However, a double-vessel continuous membrane configuration resulted in sugar content <4 g/L, which was considered satisfactory for dry wine-making. Additionally, wine productivity in the latter was 28 times higher compared to batch systems.

12.3.1.4 Natural Supports and Cell Aggregates

Food-grade natural immobilization supports, such as delignified cellulosic materials (Bardi and Koutinas 1994), gluten pellets (Bardi et al. 1996), and brewer's spent grains (Mallouchos et al. 2007), were successfully used for ambient and low-temperature wine-making, resulting in improved profiles of volatile by-products (Bardi et al. 1997; Mallouchos et al. 2002; Mallouchos et al. 2003). The commercial potential of immobilized yeast on gluten pellets and delignified cellulosic material was evaluated by the production of freeze-dried biocatalysts without using a protecting medium during freezing and freeze-drying (Iconomopoulou et al. 2002; Iconomopoulou et al. 2003). The freeze-dried immobilized biocatalysts retained high viability and showed long operational stability during low-temperature wine-making. The possibility of storage of freeze-dried immobilized biocatalysts for long time intervals without any loss of cell viability and fermentation activity is of crucial importance for industrial application.

The use of fruit pieces in developing an immobilized biocatalyst that would meet all the required prerequisites was an obvious alternative. Fruit pieces are of food-grade purity, cheap, abundant and could be easily accepted by consumers. Apple (Kourkoutas et al. 2001), quince (Kourkoutas et al. 2003b) and pear pieces (Mallios et al. 2004) were proposed as immobilization supports due to ease in the immobilization technique and to the distinctive aromatic potential and improved sensory characteristics of the produced wines. The immobilized yeasts resulted in rapid fermentations, as wine production was effective in 12 h at 30°C and in 4 days at 10°C (Kourkoutas et al. 2003b). Apple-supported biocatalyst was able to reactivate and ferment after successively increased periods of storage (up to 120 days) at 30°C (Kourkoutas et al. 2003a). The ability of storage at ambient temperature for such long time periods is considered very attractive for the industrialization of the process.

Apple pieces were also used as support for immobilizing *Kluyveromyces marxianus* IMB3 for high-temperature wine-making at 45°C (Kourkoutas et al. 2004b). The fermented grape must contain 3–4% alcohol and special types of semi-sweet wines were produced by the addition of potable alcohol.

Additionally, fruit-based biocatalysts were found suitable for continuous wine-making (Kourkoutas et al. 2002a, b; Mallios et al. 2004), as the immobilized biocatalyst showed high operational stability up to 95 days (Kourkoutas et al. 2002a).

Taking into account that grape skins are the principal undesired solid wastes of the wine-making process and raisins are directly related to the main raw material of wine, their use as immobilization supports was an interesting approach (Mallouchos et al. 2002; Tsakiris et al. 2004a, b). They were both found suitable for low-temperature fermentation and a positive influence on wine aroma was evident. The main advantage, however, is that the grape skins and the extracted residues of raisins from wineries and potable producing plants, which are difficult to dispose of, could be alternatively beneficially exploited.

Natural immobilization of single microorganisms, such as flocculating yeasts and cell aggregates forming pellets or solid microspheres, offer practical potentiality as the metabolic state of the immobilized cells may remain unaltered, since artificial immobilization techniques may induce cell damage. This is the simplest and the least expensive immobilization method. However, interactions among cells are not easily controlled and cell aggregates are very sensitive to conditions usually dominating in fermentors. In *Saccharomyces cereivisiae*, a number of proteins called flocculins (Caro et al. 1997) are responsible for cell–cell adhesion and formation of aggregates.

A novel technique of co-immobilization a filamentous fungus and a flor yeast avoiding chemical crosslinkers or external supports was recently reported (Peinado et al. 2005; Peinado et al. 2006). The technique was based on reproducing conditions adequate for a forced symbiosis of *Penicillium* and *Saccharomyces cereivisiae*. As a result, formation of hollow biocapsules with walls composed of mycelium and trapped yeast cells was obtained after killing of *Penicillium* by the ethanol produced. The mycelium walls enclosed an inner space partially occupied by yeast cells. The yeast biocapsules were successfully used in must fermentation producing wine with increased acetaldehyde and isobutanol, but reduced ethyl acetate and acetoin contents compared to free cells fermentation.

12.4 Cell Immobilization and Malolactic Fermentation

MLF is a difficult and time-consuming process that does not always proceed favorably under the natural conditions of wine. Traditional MLF is used worldwide to produce high-quality wines, although delay or failure is not an unusual outcome. During MLF L-malic acid is converted to L-lactic acid and carbon dioxide by malolactic bacteria, principally of the genera *Oenococcus*, *Pediococcus* and *Lactobacillus*. Lactic acid is "less acidic" than malic acid, as it is a mono-acid compared to the two-acid malic acid. Consequently, total acidity is decreased during MLF (deacidification).

MLF usually results in microbial stability (Lonvaud-Funel 1995) and in improvement of the organoleptic properties. Besides deacidification, it induces a dramatic change in the organoleptic quality of wines, since the characteristic taste of malic acid disappears. MLF is usually encouraged by wine-producers in cold climates where ripe grapes contain a high malic acid content resulting in too acidic wines, in instances where it is required enhanced microbial stability (i.e., prior bottling), and in situations where the flavorful by-products (such as diacetyl) are essential in order to provide the recognizable sensory notes desired in certain wine types. MLF is crucial for wines originating from Burgundy (especially for Chardonnay variety), as well as for heavy red wines, as such originating from Bordeaux. However, MLF is not always advantageous, as in wines with very low total acidity; it may cause a further reduction affecting both flavor and biological stability (Lonvaud-Funel 1995; Versari et al. 1999). Hence, it should be discouraged and may even be considered destructive in grapes from warm regions having small amounts of malic acid, in wines of poor microbial stability, and in wines where by-products of MLF may provide an unpleasant character.

During recent years several attempts have been proposed to induce biological deacidification of wines by involving the use of immobilized cells on various matrices. The use of immobilized lactic acid bacteria (LAB) for controlling MLF is desirable, as it offers the following advantages:

1. As spontaneous MLF is time consuming and growth of malolactic bacteria microflora depends on physicochemical properties and composition of wine (e.g., fatty acids, ethanol and SO_2 may inhibit growth of malolactic bacteria), cell immobilization techniques aim to increase tolerance of malolactic bacteria.
2. The use of immobilized malolactic bacteria results in higher cell densities and hence, acceleration of the process.
3. The use of selected cultures results in development of characteristic organoleptic properties (Maicas et al. 1999).
4. The immobilized cultures allow reuse of cells and favor application of continuous process (Lonvaud-Funel 1995; Kosseva et al. 1998; Maicas et al. 2001).

The first efforts involved immobilization of *Lactobacillus casei* cells in polyacrylamide gels (Diviès and Siess 1976; Totsuka and Hara 1981; Rossi and Clementi 1984). Nevertheless, the proposed technology was practically restricted to laboratory use, since industrial application of the immobilized biocatalysts was never attempted due to safety regulations. The use of alginates (Spettoli et al. 1982; Naouri et al. 1991) hardly offered an alternative, because they proved mechanically unstable in long term industrial applications (e.g., when exposed to high pressures).

κ-Carrageenan, a naturally occurring polysaccharide isolated from sea wood, was used as support for immobilizing several LAB species, including *Lactobacillus* and *O. oeni* (formerly *Leuconostoc oenos*) cells (McCord and Ryu 1985; Crapisi et al. 1987a, b). *Lactobacillus* sp. cells were immobilized in 2% κ-carrageenan gel and subsequently used in a continuous-flow bioreactor to promote and control MLF. The immobilized cells showed enhanced operational stability in malic acid metabolism with the addition of 5% purified bentonite to the matrix and conversion rates up to 53.9% were reported (Crapisi et al. 1987a). The half-life of the bioreactor was extended to 46 days, while the minimum retention time for MLF accomplishment was decreased to about 2 h. Kosseva et al. (1998) reported an approximate 30% degradation of malic acid using immobilized *L. casei* cells on calcium pectate gels and chitopearls with an operational stability up to 6 months. The degradation of malic acid obtained using the immobilized cells was twice as high as compared to that obtained with the free cells. In another study, the possibility of using immobilised *O. oeni* cells on positively charged cellulose sponge [diethyl (DE) or diethylaminoethyl (DEAE) cellulose] for MLF of wine was investigated in a semi-continuous bioreactor (Maicas et al. 2001). The first 24 h, malic acid was almost completely metabolized, while in subsequent fermentations a drastic reduction of malolactic activity was evident. Recently, *Lactobacillus casei* ATCC 393 cells immobilized on delignified cellulosic material (DCM) were used for MLF of wine (Agouridis et al. 2005). As the repeated MLF batches proceeded, the activity of the immobilized biocatalyst was reduced from 80 to 2%. Acetic acid content was slightly reduced or remained stable and concentrations of the higher alcohols 1-propanol, isobutyl alcohol and amyl alcohol were decreased by 84, 23 and 11%, respectively, resulting in improved quality.

12.5 Cell Immobilization in Cider-Making

Cider-making is a complex process requiring two successive fermentations: (a) the alcoholic fermentation, and (b) the MLF. During cider production, the reduction of beverage acidity by bacteria inducing MLF is generally recognized as an important phase for flavor development. Therefore, attempts have been focused on simultaneous alcoholic and MLF.

A sponge-like material was proposed to immobilize both *S. cerevisiae* and *L. plantarum* for carrying out fermentation and partial maturation of cider (Scott and O'Reilly 1996). The sponge's open porous network promoted extensive and rapid surface attachment of the microorganisms. Fermentations carried out with immobilized yeast and sequential addition of LAB enhanced fermentation rate and had a positive effect on flavor development. In another study, a bi-reactor composed of an alcoholic fixed immobilized *Saccharomyces bayanus* bed bioreactor coupled to a second fixed immobilized *Leuconostoc oenos* bed bioreactor was used for dry as well as sweet ciders by controlling dilution rate of the influent (Simon et al. 1996). Malic acid metabolism and acetaldehyde content in cider produced by *Saccharomyces cerevisiae* and immobilised *L. oenos* was temperature dependent, while concentrations of ethyl acetate and methanol were influenced by the type of *L. oenos* inoculation (Cabranes et al. 1998). Co-immobilization of *Saccharomyces bayanus* and *L. oenos* in Ca-alginate matrix was proposed for simultaneous alcoholic and MLF of apple juice in a continuous packed bed bioreactor (Nedovic et al. 2000). The continuous process resulted in faster fermentation compared to the traditional batch process, whereas a reduction of higher alcohols and an increase in diacetyl, attributed to the altered metabolism of immobilized cells was evident. Production of soft or dry cider was possible by controlling the feeding flow rates. Cells of *Oenococcus oeni* immobilized in alginate beads resulted in reduced ethyl acetate and acetic acid content when used as starter culture to conduct MLF of cider compared to the free cells system (Herrero et al. 2001).

The situation of conducting alcoholic and MLFs simultaneously by the same microorganism has been sought after by wine scientists. The first attempts involved transferring DNA encoding the malolactic enzyme from *Lactobacillus delbrueckii* to *Saccharomyces serevisiae* (Williams et al. 1984). Later on, a *S. cerevisiae* strain was generated containing the metabolic machinery for malate degradation, which was able to grow at pH values <2.9 (Volschenk et al. 1997). However, the use of a novel yeast replacing all malolactic procedures is doubtful, but in no case efforts aiming at the accomplishment of such a splendid achievement should be distracted. Although there are some arguments concerning concurrent conduction of the two fermentations, it is not uncommon, especially among wine manufacturers in California (Kunkee 1997), who usually inoculate wine must with both malolactic bacteria and yeast.

12.6 Bioreactors in Wine and Cider-Making

Although there have been tremendous research activities, the main drawbacks constraining industrial application of immobilized bacteria for MLF apparently are: (a) inactivity of malolactic enzyme at wine pH (Colagrande et al. 1994), (b) instability of the required cofactor NAD which is particularly unstable in wine (Colagrande et al. 1994; Diviès et al. 1994), and (c) inhibition of LAB and MLF due to the high ethanol content (Rossi and Clementi 1984; Crapisi et al. 1987a). Nevertheless, they have already been widely studied at pilot plant production sizes. A summary of the main immobilization systems proposed for alcoholic and MLF of wine and cider is presented in Table 12.1. Although the high number of proposed immobilization supports and the advantages associated with the use of immobilized cells, industrial application is still limited. There are only a few attempts aimed at the application of immobilized cell technologies in wine and cider-making on a large scale (Colagrande et al. 1994; Diviès et al. 1994). Most of these efforts concern production of sparkling wine, and secondary MLF of wine. A wide range of fermentors varying shape and technical design are nowadays available to provide preferred flow/mixing and external mass transfer characteristics during processing. However, in the majority of wine and cider-making studies, packed bed configuration has been used, operating in batch, and in fewer cases, continuous mode on lab-to-pilot scale. The reason is a simple design, consisting of a column which is packed with the immobilized biocatalyst. However, for possible industrial applications, the use of mechanically unstable materials, such as hydrogels, should be limited to lower bed heights. Furthermore, packed-bed bioreactors are linked to several engineering problems associated with mass transfer limitations, accumulation of evolved gas (which can be in high concentrations mortal to living cells and reduces the useful volume), formation of preferential paths or channeling (causing concentration and temperature gradients), excessive pressure drop, as well as short circuiting and clogging.

In a recent study, a modification of packed-bed fermentor, a multi-stage fixed-bed tower (MFBT) bioreactor has been proposed for batch and continuous wine-making (Sipsas et al. 2009). It consists of a vertical cylindrical tank with five packed sections containing freeze-dried immobilized cells on gluten pellets. The MFBT bioreactor resulted in higher alcohol productivity compared to fermentations carried out in a packed-bed bioreactor and showed high operational stability. A relatively small (5,000–10,000 L) MFBT bioreactor (Fig. 12.1) (Loukatos et al. 2000; Koutinas et al. 1997) is proposed for the industrialization of immobilized cells in wine-making,

Table 12.1 Summary of the main immobilization supports proposed for alcoholic and malolactic fermentation of wine and cider

Immobilization support	Type of fermentation	Processing mode	Product	Reference
Inorganic supports				
Kissiris	AF	Continuous	Wine	Bakoyianis et al. (1992)
γ-alumina	AF	Batch and continuous	Wine	Bakoyianis et al. (1997); Loukatos et al. (2000)
Organic supports				
Alginates	AF	–	Sparkling wine	Fumi et al. (1987, 1988)
Alginates	AF	Batch	Wine	Ferraro et al. (2000)
Alginates	AF and MLF	Continuous	Cider	Simon et al. (1996)
Alginates	MLF	Batch	Cider	Cabranes et al. (1998)
Alginates	AF and MLF	Continuous	Cider	Nedovic et al. (2000)
Alginates	MLF	Batch	Cider	Herrero et al. (2001)
Calcium pectate gel	MF	Batch	Wine	Kosseva et al. (1998)
Chemically modified chitosan	MLF	Batch	Wine	Kosseva et al. (1998)
DEAE-cellulose	MLF	Batch/Semi-continuous	Wine	Maicas et al. (2001)
Cryogels of polyvinyl alcohol	AF	–	Champagne	Martynenko et al. (2004)
Pectate gels	MLF	Batch	Wine	Kosseva and Kennedy (2004)
Membrane systems				
Membrane bioreactor		Continuous	Wine	Takaya et al. (2002)
Membrane bioreactor	MLF	Continuous	Cider	Lovitt et al. (2006)
Natural supports				
Sponge-like material	AF and MLF	Batch	Cider	Scott and O'Reilly (1996)
Delignified cellulosic material	AF	Batch and continuous	Wine	Bardi and Koutinas (1994); Iconomou et al. (1996); Iconomopoulou et al. (2003)
Delignified cellulosic material	MLF	Batch	Wine	Agouridis et al. (2005)

Spent-grains	AF	Batch	Wine	Mallouchos et al. (2007)
Gluten pellets	AF	Batch	Wine	Bardi et al. (1996); Iconomopoulou et al. (2002)
Apple pieces	AF	Batch and continuous	Wine	Kourkoutas et al. 2001, 2002a); Kourkoutas et al. (2004b)
Quince pieces	AF	Batch and continuous	Wine	Kourkoutas et al. (2002b); Kourkoutas et al. (2003b)
Pear pieces	AF	Batch and continuous	Wine	Mallios et al. (2004)
Raisin berries	AF	Batch	Wine	Tsakiris et al. (2004a, b)
Cell aggregates				
Yeast biocapsules	AF	Batch	Wine	Peinado et al. (2006), Peinado et al. (2005)

AF Alcoholic fermentation; *MLF* Malolactic fermentation

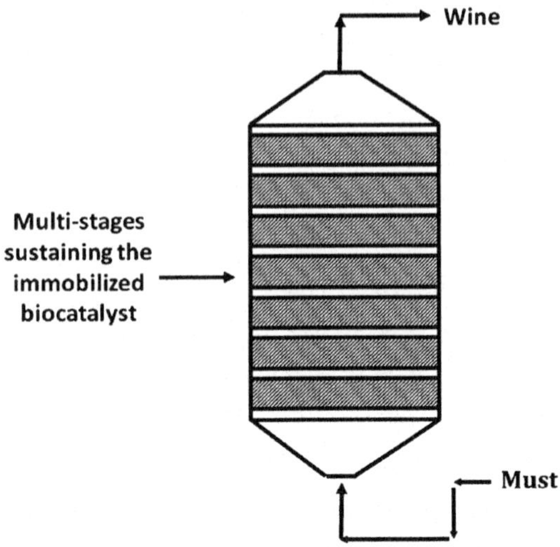

Fig. 12.1 Schematic diagram of MFBT bioreactor

as handling of the support at this scale could be performed without any problems, while cell immobilization could be carried out in the bioreactor. Application of the MFBT bioreactor at industrial scale eliminates insufficient mass transfer and enables support division, especially when mechanically unstable supports are used in order to minimize high pressures, which may result in support destruction and reduction of fermentation activity. Experiments concerning long term storage of the immobilized biocatalysts (Kourkoutas et al. 2003a) are very promising, since the preparation of new biocatalyst, emptying and filling of the bioreactor could be avoided when industrial production is halted. Taking into consideration the above discussion of technical problems, the scale-up of the proposed technology seems feasible.

Future research efforts should be directed towards application of other types of reactors (such as fluidized or gas-lift) already approved in beer production. A recent study was focused on the use of a membrane bioreactor for MLF of ethanol containing media (Lovitt et al. 2006).

12.7 Conclusions

Microbial cell immobilization can improve the efficiency of fermentation of alcoholic and MLF and the quality of the fermentation products. The selection of the suitable carrier and bioreactor system is a challenge and many issues should be taken into account, such as product quality, safety and stability, processing, investment and operating costs, as well as matters of legality (food grade purity, the use of cross linking agents, etc.). It is obvious that in alcoholic fermentation process a high number of technologies (cell adsorption, entrapment, and flocculation) and immobilization materials (both organic and inorganic) may have a positive impact on cell metabolism. The ability of storage for long time periods of freeze-dried immobilized biocatalysts produced by food-grade natural materials is considered very attractive for the industrialization of the process. Packed-bed reactors are still the first choice in the industrial sector, mainly due to the simplicity of the design and in combination with natural supports (such as fruit-based or cellulosic materials) seem to have potential in scale-up. The use of mechanically unstable materials, such as hydrogels, depends on innovations on the design of packed-bed columns (e.g., multiple sections are usually applied for biocatalysts division). Therefore, problems related to high pressures and insufficient mass transfers are minimized. Investment cost and difficulties in scale up are still the main issues constraining application of other bioreactor types in the fermentation processes.

In wine and cider production, the development of a fine taste is an undisputable goal and thus control of MLF is an important step. Immobilization in either gels or cellulosic materials aims to increase tolerance of malolactic bacteria and to accelerate the process. In cider-making, attempts have been focused on simultaneous alcoholic and MLF by co-immobilization of two different species or by the same microorganism, usually after genetic modification. Production of soft or dry cider is possible by controlling the feeding flow rates.

The main drawback inhibiting industrial use of immobilized biocatalysts for fermentation processes is related to maintenance of cell viability during production and storage, especially when production is halted, as wine and cider products are season dependent. Further application will depend on research results upon preservation of immobilized cells, as well as development of processes that can be readily scaled-up. Thus, research should be oriented towards the evolvement of reliable preservation and storage techniques that could be easily adopted by the industrial sector.

References

Agouridis N, Bekatorou A, Nigam P, Kanellaki M (2005) Malolactic fermentation in wine with *Lactobacillus casei* cells immobilized on delignified cellulosic material. J Agric Food Chem 53(7):2546–2551

Argiriou T, Kanellaki M, Voliotis S, Koutinas AA (1996) Kissiris-supported yeast cells: high biocatalytic stability and productivity improvement by successive preservations at 0°C. J Agric Food Chem 44:4028–4031

Bakoyianis V, Koutinas AA (1996) A catalytic multistage fixed-bed tower bioreactor in an industrial-scale pilot plant for alcohol production. Biotechnol Bioeng 49:197–203

Bakoyianis V, Kanellaki M, Kaliafas A, Koutinas AA (1992) Low temperature wine making by immobilized cells on mineral kissiris. J Agric Food Chem 40:1293–1296

Bakoyianis V, Kana K, Kalliafas A, Koutinas AA (1993) Low-temperature continuous wine making by kissiris-supported biocatalyst: Volatile byproducts. J Agric Food Chem 41:465–468

Bakoyianis V, Koutinas AA, Agelopoulos K, Kanellaki M (1997) Comparative study of kissiris, γ-alumina, and calcium alginate as supports of cells for batch and continuous wine-making at low temperatures. J Agric Food Chem 45:4884–4888

Bardi EP, Koutinas AA (1994) Immobilization of yeast on delignified cellulosic material for room temperature and low-temperature wine making. J Agric Food Chem 42:221–226

Bardi EP, Bakoyianis V, Koutinas AA, Kanellaki M (1996) Room temperature and low temperature wine making using yeast immobilized on gluten pellets. Process Biochem 31:425–430

Bardi E, Koutinas AA, Psarianos C, Kanellaki M (1997) Volatile by-products formed in low-temperature wine-making using immobilized yeast cells. Process Biochem 32:579–584

Brodelius P, Nilsson K (1983) Permealization of immobilized plant cells, resulting in release of intracellularly stored products with preserved cell viability. Eur J Appl Microbiol Biotechnol 17:275–280

Busova K, Magyar I, Janky F (1994) Effect of immobilized yeasts on the quality of bottle-fermented sparkling wine. Acta Alimentaria 23:9–23

Buzas Z, Dallmann K, Szajani B (1989) Influence of pH on the growth and ethanol production of free and immobilized *Saccharomyces cerevisiae* cells. Biotechnol Bioeng 34:882–884

Cabranes C, Moreno J, Mangas JJ (1998) Cider production with immobilized *Leuconostoc oenos*. J Inst Brew 104:127–130

Caro LH, Tettelin H, Vossen JH, Ram AF, Van den Ende H, Klis FM (1997) In silicio identification of glycosyl-phosphatidylinositol-anchored plasma-membrane and cell wall proteins of *Saccharomyces cerevisae*. Yeast 13:1477–1489

Chen C, Dale MC, Okos MR (1990) Minimal nutritional requirements for immobilized yeast. Biotechnol Bioeng 36:993–1001

Ciani M, Ferraro L (1996) Enhanced glycerol content in wines made with immobilized Candida stellata cells. Appl Environ Microbiol 62:128–132

Colagrande O, Silva A, Fumi MD (1994) Recent applications of biotechnology in wine production. Biotechnol Prog 10:2–18

Crapisi A, Nuti MP, Zamorani A, Spettoli P (1987a) Improved stability of immobilized *Lactobacillus* sp. cells for the control of malolactic fermentation in wine. Am J Enol Vitic 38:310–312

Crapisi A, Spettoli P, Nuti MP, Zamorani A (1987b) Comparative traits of *Lactobacillus brevis*, *Lactobacillus fructivorans* and *Leuconostoc oenos* immobilized cells for the control of malolactic fermentation in wine. J Appl Bacteriol 63:513–521

Diviès C, Siess MH (1976) Étude du catabolisme de l'acide L-malique para *Lactobacillus casei* emprisonée dans un gel de polyacrylamide. Ann Microbiol (Paris) 127:525–539

Diviès C, Cachon R, Cavin J-F, Prévost H (1994) Theme 4: Immobilized cell technology in wine production. Crit Rev Biotechnol 14:135–153

Doran PM, Bailey JE (1986) Effects of immobilization on growth, fermentation properties, and macromolecular composition of *Saccharomyces cerevisae* attached to gelatin. Biotechnol Bioeng 28:73–87

Ferraro L, Fatichenti F, Ciani M (2000) Pilot scale vinification process using immobilized *Candida stellata* cells and *Saccharomyces cerevisiae*. Process Biochem 35:1125–1129

Fumi MD, Trioli G, Colagrande O (1987) Immobilization of *Saccharomyces cerevisiae* in calcium alginate for sparkling wine processes. Biotechnol Lett 9:339–342

Fumi MD, Trioli G, Colombi MG, Colagrande O (1988) Immobilization of *Saccharomyces cerevisiae* in calcium alginate gel and its application to bottle-fermented sparkling wine production. Am J Enol Vitic 39:267–272

Galazzo J, Bailey JE (1990) Growing *Saccharomyces cerevisae* in calcium alginate beads induces cell alteration, which accelerate glucose conversion to ethanol. Biotechnol Bioeng 36:417–426

Galazzo JL, Shanks JV, Bailey JE (1987) Comparision of suspended and immobilized yeast metabolism using ^{31}P Nuclear Magnetic Resonance spectroscopy. Biotechnol Tech 1:1–6

Herrero M, Laca A, Garcia LA, Díaz M (2001) Controlled malolactic fermentation in cider using *Oenococcus oeni* immobilized in alginate beads and comparison with free cell fermentation. Enzyme Microb Technol 28:35–41

Hilge-Rotmann B, Rehm HJ (1990) Comparison of fermentation properties and specific enzyme activities of free and calcium-alginate-entrapped *Saccharomyces cerevisiae*. Appl Microbiol Biotechnol 33:54–58

Iconomopoulou M, Psarianos K, Kanellaki M, Koutinas AA (2002) Low temperature and ambient temperature wine making using freeze-dried immobilized cells on gluten pellets. Process Biochem 37:707–717

Iconomopoulou M, Kanellaki M, Soupioni M, Koutinas AA (2003) Effect of freeze-dried cells on delignified cellulosic material in low-temperature wine making. Appl Biochem Biotechnol 104:23–36

Iconomou L, Kanellaki M, Voliotis S, Agelopoulos K, Koutinas AA (1996) Continuous wine making by delignified cellulosic materials supported biocatalyst. An attractive process for industrial applications. Appl Biochem Biotechnol 60:303–313

Jamai L, Sendide K, Ettayebi K, Errachidi F, Hamdouni-Alami O, Tahri-Jouti MA, McDermott T, Ettayebi M (2001) Physiological difference during ethanol fermentation between calcium alginate-immobilized *Candida tropicalis* and *Saccharomyces cerevisiae*. FEMS Microbiol Lett 204:375–379

Kana K, Kanellaki M, Papadimitriou A, Psarianos C, Koutinas AA (1989) Immobilization of *Saccharomyces cerevisiae* on γ-alumina pellets and its ethanol production in glucose and raisin extract fermentation. J Ferment Bioeng 68:213–215

Kosseva MR, Kennedy JF (2004) Encapsulated lactic acid bacteria for control of malolactic fermentation in wine. Artif Cells Blood Substit Immobil Biotechnol 32:55–65

Kosseva M, Beschkov V, Kennedy JF, Lloyd LL (1998) Malolactic fermentation in Chardonnay wine by immobilized *Lactobacillus casei* cells. Process Biochem 33:793–797

Kourkoutas Y, Komaitis M, Koutinas AA, Kanellaki M (2001) Wine production using yeast immobilized on apple pieces at low and room temperatures. J Agric Food Chem 49:1417–1425

Kourkoutas Y, Koutinas AA, Kanellaki M, Banat IM, Marchant R (2002a) Continuous wine fermentation using a psychrophilic yeast immobilized on apple cuts at different temperatures. Food Microbiol 19:127–134

Kourkoutas Y, Douma M, Koutinas AA, Kanellaki M, Banat IM, Marchant R (2002b) Room and low temperature continuous wine making using yeast immobilized on quince pieces. Process Biochem 39:143–148

Kourkoutas Y, Kanellaki M, Koutinas AA, Banat IM, Marchant R (2003a) Storage of immobilized yeast cells for use in wine-making at ambient temperature. J Agric Food Chem 51:654–658

Kourkoutas Y, Komaitis M, Koutinas AA, Kaliafas A, Kanellaki M, Marchant R, Banat IM (2003b) Wine production using yeast immobilized on quince biocatalyst at temperatures between 30 and 0°C. Food Chem 82:353–360

Kourkoutas Y, Bekatorou A, Banat IM, Marchant R, Koutinas AA (2004a) Immobilization technologies and support materials suitable in alcohol beverages production: a review. Food Microbiol 21:377–397

Kourkoutas Y, McErlean C, Kanellaki M, Hack CJ, Marchant R, Banat IM, Koutinas AA (2004b) High-temperature wine making using the thermotolerant yeast strain *Kluyveromyces cerevisiae* IMB3. Appl Biochem Biotechnol 112:25–35

Koutinas AA, Bakoyianis V, Argiriou T, Kanellaki M, Voliotis S (1997) A qualitative outline to industrialize alcohol production by catalytic multistage fixed bed tower (MFBT) bioreactor. Appl Biochem Biotechnol 66:121–131

Kunkee RE (1997) A heady concoction of alcoholic and malolactic fermentations. Nat Biotechnol 15:224–225

Lodato P, Segovia De Huergo M, Buera MP (1999) Viability and thermal stability of a strain of *Saccharomyces cerevisiae* freeze-dried in different sugar and polymer matrices. Appl Microbiol Biotechnol 52:215–220

Lonvaud-Funel A (1995) Microbiology of the malolactic fermentation: molecular aspects. FEMS Microbiol Lett 126:209–214

Loukatos P, Kiaris M, Ligas I, Bourgos G, Kanellaki M, Komaitis M, Koutinas AA (2000) Continuous wine making by γ-alumina-supported biocatalyst. Quality of the wine and distillates. Appl Biochem Biotechnol 89:1–13

Loukatos P, Kanellaki M, Komaitis M, Athanasiadis I, Koutinas AA (2003) A new technological approach for distillate production using immobilized cells. J Biosci Bioeng 95:35–39

Lovitt R, Jung I, Jones M (2006) The performance of a membrane bioreactor for the malolactic fermentation of media containing ethanol. Desalination 199:435–437

Lu ZX, Lu FX, Bie XM, Fujimura T (2002) Immobilization of yeast cells with polymeric carrier cross-linked using radiation techniques. J Agric Food Chem 50:2798–2801

Maicas S, Gil J-V, Pardo I, Ferrer S (1999) Improvement of volatile composition of wines by controlled addition of malolactic bacteria. Food Res Intern 32:491–496

Maicas S, Pardo I, Ferrer S (2001) The potential of positively charged cellulose sponge for malolactic fermentation of wine, using *Oenococcus oeni*. Enzyme Microb Technol 28:415–419

Mallios P, Kourkoutas Y, Iconomopoulou M, Koutinas AA, Psarianos C, Marchant R, Banat IM (2004) Low temperature wine-making using yeast immobilized on pear pieces. J Sci Food Agric 84:1615–1623

Mallouchos A, Reppa P, Aggelis G, Koutinas AA, Kanellaki M, Komaitis M (2002) Grape skins as a natural support for yeast immobilization. Biotechnol Lett 24:1331–1335

Mallouchos A, Komaitis M, Koutinas AA, Kanellaki M (2003) Wine fermentation by immobilized and free cells at different temperatures. Effect of immobilization and temperature on volatile by-products. Food Chem 80:109–113

Mallouchos A, Loukatos P, Bekatorou A, Koutinas A, Komaitis M (2007) Ambient and low temperature wine-making by immobilized cells on brewer's spent grains: Effect on volatile composition. Food Chem 104(3):918–927

Margaritis A, Merchant FJA (1984) Advances in ethanol production using immobilized cell systems. Crit Rev Biotechnol 2:339–393

Martynenko NN, Gracheva IM, Sarishvili NG, Zubov AL, El'Registan GI, Lozinsky VI (2004) Immobilization of champagne yeasts by inclusion into cryogels of polyvinyl alcohol: Means of preventing cell release from the carrier matrix. Appl Biochem Microbiol 40:158–164

McCord JD, Ryu DDY (1985) Development of malolactic fermentation process using immobilized whole cells and enzymes. Am J Enol Vitic 36:214–218

Melzoch K, Rychtera M, Habova V (1994) Effect of immobilization upon the properties and behavior of *Saccharomyces cerevisiae* cells. J Biotechnol 32:59–65

Naouri P, Bernet N, Chagnaud P, Arnaud A, Galzy P (1991) Bioconversion of L-malic acid into L-lactic acid using a high compacting multiphase reactor (HCMR). J Chem Technol Biotechnol 51:81–95

Navarro JM, Durand G (1977) Modification of yeast metabolism by immobilization onto porous glass. Eur J Appl Microbiol 4:243–254

Nedovic VA, Durieux A, Van Nederveide L, Rosseels P, Vandegans J, Plaisant AM, Simon J-P (2000) Continuous cider fermentation with co-immobilized yeast and *Leuconostoc oenos* cells. Enzyme Microb Technol 26:834–839

Nolan AM, Barron N, Brady D, McAree T, McHale L, McHale AP (1994) Ethanol production at 45°C by an alginate-immobilized thermotolerant strain of *Kluyveromyces marxianus* following growth on glucose-containing media. Biotechnol Lett 16:849–852

Norton S, D'Amore T (1994) Physiological effects of yeast cell immobilization applications for brewing. Enzyme Microb Technol 16:365–375

Peinado RA, Moreno JJ, Maestre O, Mauricio JC (2005) Use of a novel immobilization yeast system for winemaking. Biotechnol Lett 27:1421–1424

Peinado RA, Moreno JJ, Villalba JM, González-Reyes JA, Ortega JM, Mauricio JC (2006) Yeast biocapsules: A new immobilization method and their applications. Enzyme Microb Technol 40:79–84

Rossi J, Clementi F (1984) L-malic acid catabolism by polyacrylamide gel entrapped *Leuconostoc oenos*. Am J Enol Vitic 36:100–102

Sakurai A, Nishida Y, Saito H, Sakakibara M (2000) Ethanol production by repeated batch culture using yeast cells immobilized within porous cellulose carriers. J Biosci Bioeng 90:526–529

Scott JA, O'Reilly AM (1996) Co-immobilization of selected yeast and bacteria for controlled flavour development in an alcoholic cider beverage. Process Biochem 31:111–117

Shindo S, Takata S, Taguchi H, Yoshimura N (2001) Development of a novel carrier using natural zeolite and continuous ethanol fermentation with immobilized *Saccharomyces cerevisiae* in a bioreactor. Biotechnol Lett 23:2001–2004

Shirai Y, Hashimoto K, Yamaji H, Kawahara H (1988) Oxygen uptake rate of immobilized growing hybridoma cells. Appl Microbiol Biotechnol 29:113–118

Shuler ML (1985) Immobilized whole cell bioreactors: potential tools for directing cellular metabolism. World Biotechnol Rep 2:231–239

Silva S, Ramon-Portugal P, Silva P, Texeira MF, Strehaiano P (2002) Use of encapsulated yeast for the treatment of stuck and sluggish fermentations. J Intern des Sci de la vigne et du vin 36:161–168

Simon JP, Durieux A, Pinnel V, Garré V, Vandegans J, Rosseels P, Godan N, Plaisant AM, Defroyennes J-P, Foroni G (1996) Organoleptic profiles of different ciders after continuous fermentation (encapsulated living cells) versus batch fermentation (free cells). In: Wijffels RH, Buitelaar RH, Bucke C, Tramper J (eds) Immobilized cells: Basics and applications. Elsevier BV, Amsterdam, pp 615–621

Sipsas V, Kolokythas G, Kourkoutas Y, Plessas S, Nedovic VA, Kanellaki M (2009) Comparative study of batch and continuous multi-stage fixed-bed tower (MFBT) bioreactor during wine-making using freeze-dried immobized cells. J Food Eng 90:495–503

Spettoli P, Bottacin A, Nuti MP, Zamorani A (1982) Immobilization of *Leuconostoc oenos* ML 34 in calcium alginate gels and its application to wine technology. Am J Enol Vitic 33:1–5

Sree NK, Sridhar M, Suresh K, Banat IM, Rao IV (2000) High alcohol production by repeated batch fermentation using an immobilized osmotolerant *Saccharomyces cerevisiae*. J Ind Microbiol Biotechnol 24:222–226

Stewart GG, Russel I (1986) One hundred years of yeast research and development in the brewing industry. J Inst Brew 92:537–558

Suzzi G, Romano P, Vannini I, Turbanti L, Domizio P (1996) Cell-recycle batch fermentation using immobilized cells of flocculent *Saccharomyces cerevisiae* wine strains. World J Microbiol Biotechnol 12:25–27

Takaya M, Matsumoto N, Yanase H (2002) Characterization of membrane bioreactor for dry wine production. J Biosci Bioeng 93:240–244

Totsuka A, Hara S (1981) Decomposition of malic acid in red wine by immobilized microbial cells. Hako Kogaku Zasshi 59:231–237

Tsakiris A, Bekatorou A, Psarianos C, Koutinas AA, Marchant R, Banat IM (2004a) Immobilization of yeast on dried raisin berries for use in dry white wine-makin. Food Chem 87:11–15

Tsakiris A, Sipsas V, Bekatorou A, Mallouchos A, Koutinas AA (2004b) Red wine making by immobilized cells and influence on volatile composition. J Agric Food Chem 53:1357–1363

Versari A, Parpinello GP, Cattaneo M (1999) *Leuconostoc oenos* and malolactic fermentation in wine: a review. J Ind Microbiol Biotechnol 23:447–455

Vijayalakshmi M, Marcipar A, Segard E, Broun GB (1979) Matrix bound transition metal for continuous fermentation tower packing. Ann NY Acad Sci 326:249–254

Volschenk H, Viljoen M, Grobler J, Petzold B, Bauer F, Subden RE, Young RA, Lonvaud A, Denayrolles M, Van Vuuren HJJ (1997) Engineering pathways for malate degradation in *Saccharomyces cerevisiae*. Nat Biotechnol 15:253–257

Walsh PK, Malone DM (1995) Cell growth in immobilization matrices. Biotechnol Adv 13:13–43

Webb C, Fukuda H, Alkinson B (1986) The production of cellulose in a spouted bed fermentor using cells immobilized in biomass support particles. Biotechnol Bioeng 28:41–50

Williams SA, Hodges RA, Strike TL, Snow R, Kunkee RE (1984) Cloning the gene for malolactic fermentation of wine from *Lactobacillus delbrueckii* in *Escherichia coli* and yeasts. Appl Environ Microbiol 47:288–293

Chapter 13
Immobilization of Cells and Enzymes for Fermented Dairy or Meat Products

Claude P. Champagne, Byong H. Lee, and Linda Saucier

13.1 Introduction

Historically, we can find fermented products in almost all cultural backgrounds around the world. Notably, there are many different milk or meat-based foods and this chapter will focus on them (Kosikowski 1982; Wood 1998). Cheese, yoghurt, sour cream, kefir, or cultured butter are probably the most common fermented dairy products, but many regional varieties exist (Farnworth 2004). Fermented meats are typically found as dry sausages (Lüke 1998). Yeasts are mostly involved in the manufacture of bread and alcoholic beverages, which are basically cereal- or fruit-based products. In fermented meat and milk, the main microorganisms used are the lactic acid bacteria (LAB). Yeast and molds are rather involved in ripening. Therefore, the LAB will constitute the main focus of this chapter.

In addressing the potential of immobilized cell technology (ICT) or immobilized enzyme technology (IET) in dairy and meat fermentations or enzymatic processes, we will consider a wide scope of "immobilization" procedures. Therefore, for the purpose of this manuscript, ICT and IET will include adsorption of enzymes or bacteria on surfaces as well as microentrapment or encapsulation in particles of various natures.

In the dairy and meat sectors, lactic starters are now inoculated, which plays a major role in standardizing characteristics of a product. As a rule, starter cultures are added to milk or meat as free cells. However, there are two instances where encapsulated cultures are used commercially (production of LAB and yoghurt), and many potential applications have been proposed. One aim of this chapter is, therefore, to present the current practices with respect to immobilized LAB as well as opportunities for expanded use.

C.P. Champagne (✉) and B.H. Lee
Food R&D Center, Agriculture and Agri-Food Canada, 3600 Casavant, Saint-Hyacinthe, QC, Canada, J2S 8E3
e-mail: claude.champagne@agr.gc.ca

L. Saucier
Department of Animal Science, Paul Comtois Building, 2425 Agriculture Street, Université Laval, Quebec City, QC, G1V 0A6 Canada

N.J. Zuidam and V.A. Nedović (eds.), *Encapsulation Technologies for Active Food Ingredients and Food Processing*,
DOI 10.1007/978-1-4419-1008-0_13, © Springer Science+Business Media, LLC 2010

In the last 15 years, probiotic cultures have also been added to foods for their health benefits. Probiotics can be described as "live microorganisms which, administered in adequate amounts, confer a beneficial physiological effect on the host" (Araya et al. 2002). Adding probiotic bacteria to foods presents a variety of challenges (Champagne et al. 2005). Similarly, protective cultures may be used to improve the safety of a product and its shelf life by competitive exclusion (Hammes and Hertel 1998; Eie et al. 2007). How immobilization (or encapsulation) can be helpful in improving the stability of probiotics and protective cultures in fermented foods will be examined. Please see Chap. 10 also for more information on the microencapsulation of probiotics.

Although the LAB constitute the cornerstone of dairy and meat fermentations, some processes also include enzymes. Rennet and lipases in cheese manufacture represent the best examples. As for the LAB, the enzymes are typically added in free form. But there are circumstances where encapsulation prior to use in the manufacturing process can be very useful and this will also be addressed.

It must be mentioned that many fermentation processes of meat or dairy by-products have been developed, notably, whey fermentation for bioproducts (for example, ethanol or organic acids). ICT has extensively been used for the production of such bioproducts (Norton and Vuillemard 1994). However, these fermentations are designed to generate specialty products or ingredients, and the milk or dairy components are not typically reintroduced into the fermented product. Since there are no fermented foods involved in such applications, this chapter will not address ICT/IET of this type of specialty ingredients in great detail.

High fructose corn syrup and amino acids are the best known examples where IET is used commercially in food processes (Cheetham 1988; see also Chap. 14 of this book). In the dairy sector, enzymes with potential applications include catalase, glucose oxidase, alkaline phosphatase, lysozyme, glucose isomerase, lactoperoxidase, superoxide dismutase, and sulfhydryl oxidase (Fox 1993; Lee 1996), but the principal immobilization of lactases, lipases, and proteinases will be discussed.

Several free enzymes are utilized for meat applications (Table 13.1), but the use of encapsulated enzymes in fermented meat products is largely under-studied. On the other hand, (micro) encapsulation techniques are used for other purposes (Table 13.2). The use of bromelain and papain for meat tenderization is well documented but to our knowledge, none are used encapsulated (Toren 2007). Several enzymes, notably proteinase and lipase have also been tested for the acceleration of dry fermented sausage production but again as free molecules (Naes et al. 1994; Blom et al. 1996). It is nonetheless interesting to mention here that even before muscle is converted to meat, enzymes are added to animal feeds to improve their nutritional and metabolic efficacy. Enzymes used for that purpose represent 6% of the whole enzyme market (Anonymous 2003). The use of phytase to reduce phosphorous emission in the environment from animal feces has been commercially used notably in pig and fish production (Haefner et al. 2005). Although phytase must be released in the gut to be effective, some are encapsulated to provide a protective effect during the processing steps (Benchabane et al. 2004).

Immobilization requires a specific processing step which adds to the costs. Therefore, there must be a clear advantage to carry out immobilization of the bioactive components.

Table 13.1 Examples of enzymes used in various meat related applications

Application	Enzymes	References
Tenderization	Bromelain Papain Ficin Actinidin	TenderIn™ World Technology Ingredients Toren (2007)
Restructured meat and improvement of homogenate gel microstructure	Fibrinogen Trombin Transglutaminase Tyrorinase	Fibrimex®, ex Sonac Loenen Tseng et al. (2006) Lantto et al. (2007)
Fish protein solubilization	Bacterial whole cells	Venugopal (1994)
Enzymatic additives for better feed conversion	Phytase Amylase Fibrozyme® Vergpro®	Anonymous (2003)

Table 13.2 Examples of microencapsulation technologies used for various meat processing applications

Materials	Technology	Matrix	Purpose	References
Meat pigment	Spray-drying	Carbohydrate	Color retention	Shahidi and Pegg (1991)
Volatile meat flavor	Entrapment	Starches	Flavor retention	You-Jin et al. (2003)
Acidulant, antioxidant, salts	Spray-drying or fluidized bed spray-coating	Lipid based coating; cores of variable composition	Direct acidification	Lemay et al. (2002b), Meatshure®, ex Balchem Corporation
Antimicrobials		Various packaging film	Safety and shelf life improvement	Kerry et al. (2006)
Bacteriophage		Under patent application	Safety by controlling pathogens in animal through feed	Murthy (2007) GangaGen Life Sciences Inc.

It can be an economic advantage or it can be that the ICT/IET solves a problem which is encountered when using free cells or enzymes. Having this in mind, the current chapter will not be subdivided into various classes of dairy products or enzymes, but rather on the basis of problems which are solved by ICT/IET or by advantages which are provided by the technology.

13.2 For the Production/Inoculation of Lactic Cultures

The application of ICT for the production of concentrated cultures is indirectly linked to milk or meat fermentations because it initially targets the supplier of cultures used at the dairy or meat processing plants. Indeed, fermented meat and dairy industries tend to rely on specialized suppliers for their cultures. At these suppliers' production plants, many technological steps in the production process of the lactic cultures are detrimental to the cell viability – oxygen during fermentation, centrifugation or filtration pressures, freezing, or drying. Furthermore, some processing steps require expensive equipment (centrifuges, filtration units, and freeze-drying units) that ought to be optimized. One of the two known current industrial applications of ICT is the production of lactic/probiotic concentrated cultures in order to address these problems. Basically, cells are micro entrapped into gel particles and added to the growth medium (Fig. 13.1). Two techniques for the immobilization of lactic cultures in gels are available, i.e., extrusion and emulsion techniques, and a description of the two is presented in Table 13.3. If one wished to (1) carry out the entrapment procedure directly into the fermentation medium, (2) remain on a small scale, or (3) carry out a coating step with chitosan, then the extrusion method is best. Otherwise, the emulsion technique seems more appropriate.

The cells grow within the gel particles and the concentrated biomass is obtained simply by recovering the particles. Populations close to 10^{11} CFU/mL of gel are obtained with mesophilic organisms (Champagne et al. 1992), particularly if fermentations conducted under continuous culture technology are practiced (Prevost and Divies 1987), but are slightly lower with thermophilic starters (Table 13.4). Once the ICT cultures are recovered, they can be sold to food processors and used in the microencapsulated form in the fermented meat or milk products.

However, there are also disadvantages to this biomass production technology for the culture suppliers. A bead-production step must be added in the concentrated starter/probiotic production process. Secondly, overall yields can be lower (Table 13.4). With *Lactococcus lactis*, the total bead-entrapped population in the fermentor

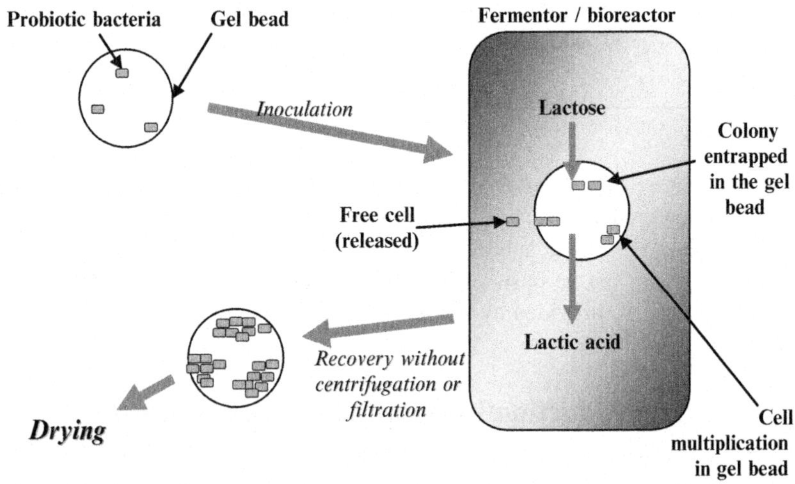

Fig. 13.1 Production of concentrated suspensions of lactic or probiotic cultures in gel beads (Champagne 2006)

Table 13.3 Comparative features of extrusion and emulsion techniques for gel entrapment for biomass production of lactic cultures (adapted from Krasaekoopt et al. 2003)

Parameter	Extrusion	Emulsion
Polymers	Alginate, carrageenan, whey protein	Alginate, carrageenan, agar, gelatin
Technical feasibility	Difficult to scale up	Easy to scale up
Simplicity	High	Low
Done in the fermentation medium itself	Yes	No
Add a coat to the bead	Yes	May be difficult
Survival of microorganisms	5–95%, depending on pH of $CaCl_2$ solution	80–95%
Size of bead	Gravity only: 1–5 mm. Electromagnetic, or vibrating systems, rotating disk: 100 μm to 1 mm	25 μm to 2 mm

Table 13.4 Production of *Lactobacillus rhamnosus* by ICT in alginate beads

Treatment	Content in fermentor	Population (CFU) per mL or g	Total population in fermentor (CFU/L)
Free cell-control	1,000 mL	3.0×10^{9}	3.0×10^{12}
Immobilized cell system			
In the beads (immobilized)	50 g	3.0×10^{10}	1.5×10^{12}
In the broth (free cells)	950 mL	1.1×10^{9}	1.0×10^{12}

Data taken from Champagne et al. (2007)

represented only 25% of that obtained in free-cell fermentation produced in similar conditions and, in the ICT system, free cells represented about half of the population in the fermentor (Morin et al. 1992). This was also observed with thermophilic cultures (Champagne et al. 1993). Third, some lactobacilli do not grow well in alginate gels (Lamboley et al. 2003).

A feature specific to cultures grown in gel beads is that most of the biomass is located on the surface, principally because of mass-transfer limitations of substrates and fermentation products. Cells are therefore released from the beads into the surrounding medium (Champagne et al. 1993). Although this feature is undesirable when the ICT process is used for biomass production, as described above for the industrial manufacturers of starters/probiotics, it becomes desirable when, for example, a dairy plant wishes to use the system for continuous inoculation of milk.

This property led to the second ICT starter process – continuous inoculation of milk. In this inoculation approach, the dairy-based substrate is continuously injected in a bioreactor containing gel particles harboring the bacterial cultures

(Lamboley et al. 2001). As cells are released from the particles, the medium exiting the bioreactor becomes strongly inoculated. After an adaptation period of a few days, the continuous inoculation by multiple-strain mesophilic cultures is achieved, and strain ratios have been shown to be stable up to 53 days (Lamboley et al. 1997). The same phenomenon was demonstrated with yoghurt cultures (Prevost and Divies 1985). Thus, an important feature of this ICT process is standardization of strain ratios in milk inoculation. This obviously is very helpful in ensuring reproducible fermentation rates and constant quality of the finished product.

The strain ratios of a mesophilic cheese mixed starter cultures can be adjusted by changing the pH of the medium, the dilution rate of milk, the bead content in the bioreactor or the incubation temperature. In response to pH or temperature parameters, the strain ratios in the ICT system followed patterns similar to free-cell cultures. Thus, lowering the pH of the medium below 6.0, and increasing the incubation temperature above 32°C, tends to be more detrimental to *Lactococcus lactis* ssp. *cremoris* than to *Lactococcus lactis* ssp. *lactis* (Lamboley et al. 1997). The great advantage of this technique is that bulk starter preparation can be eliminated. Thus, the ICT technique is a much more economical method for milk inoculation than the traditional starter technology preparation processes, and even more economical compared to the direct inoculation procedure with frozen or freeze-dried cultures. Furthermore, the technique can potentially be used to inoculate milk with probiotics such as bifidobacteria (Doleyres and Lacroix 2005).

However, an important disadvantage of the continuous inoculation ICT technique is the potential attacks of the released cells by bacteriophages. Indeed, bead-entrapped cells in the bioreactor are protected against phages (Steenson et al. 1987), but not the free cells released into the milk. Milk for cheese making is not sterilized and bacteriophages from raw milk can survive pasteurization. Therefore phages in raw milk can contaminate the bioreactor and implant themselves (Macedo et al. 1999). Hence, in considering the application of ICT for milk inoculation, a certain number of parameters must be examined. First, the technology appears to be better adapted to yoghurt manufacture, rather than cheese production, because (1) the yoghurt milk is heated at temperatures which destroy bacteriophages prior to fermentation, and because (2) the thermophilic cultures used in yoghurt processing are simply less prone to bacteriophage attack than mesophilic LAB used in cheese making. Another aspect to consider is the problem related to the storage of the beads when the plant is not manufacturing fermented milks. Data show that the beads can be stored for 48 h at 4°C in peptone water, but a reactivation/stabilization period of 4 h is recommended. Finally, many studies examining the ICT continuous inoculation process were carried out in supplemented whey media, and it is to be expected that the use of different medium might influence the behavior of the cultures.

ICT-based continuous inoculation of milk can actually reduce the overall fermentation times. In yoghurt and cheese making batch processes, milk is typically inoculated between 1 and 5×10^7 CFU/mL. In ICT bioreactors designed for continuous milk inoculation, the dilution rate (D) can be adjusted to generate much higher free cell populations in the effluents than the populations initially found after inoculation in batch processes. Thus, for a yoghurt culture, an effluent population of 1×10^8 CFU/mL can be achieved at $D=10\ h^{-1}$ (Prevost and Divies 1988b) and even higher populations are reached at lower D values (Prevost and Divies 1988a). Obviously, higher inoculation rates results in shorter fermentation times (Prevost and Divies 1985). The same strategy was applied to a mesophilic starter culture, resulting in a 50% shorter fermentation times for a fresh cheese (Philadelphia, "fromages frais") or sour cream type fermentations (Prevost and Divies 1987).

It is noteworthy that the ICT continuous inoculation technique not only serves to stabilize strain inoculation, but the cells released from the beads have different properties and physiology compared to free cells. It is known that cells growing at high cell density in biofilm have increased resistance to adverse environmental conditions (Bower and Daeschel 1999). ICT cultures are exposed to comparable high cell density environments. Hence, it can be hypothesized that cells of LAB released from the beads should be less sensitive to a variety of residual antibiotics or sanitizers in milk than are the traditional free cells (Doleyres and Lacroix 2005). This property has not yet been exploited and deserves further examination.

13.3 For Shorter Fermentation Times

13.3.1 Immobilized Cells

As mentioned previously, ICT inoculation can reduce the fermentation time in fermented milks. Contrary to the dairy sector, there is little research on encapsulation of starter, probiotic or protective cultures in meat fermentation (Työppönen et al. 2003; Saucier and Champagne 2005). As a rule, the specific acidifying rate of free cells is higher than that of ICT cultures in alginate gel beads (Champagne et al. 1992) because of mass transfer limitations. However, there is one instance where this was not observed. Kearney et al. (1990) carried out meat fermentation with free and immobilized cultures of *Lactobacillus plantarum* and *Pediococcus pentosaceus*. The fermentation time required to obtain a pH of 5.0 was reduced from 45 to 28 h with the alginate-encapsulated cells. The authors suggest that the greater fermentation performance of the ICT culture was linked to the microenvironment enabling greater availability of nutrients (i.e., skim milk) and more protective hydration conditions within the beads. It is well known that the composition of the hydrating medium will influence the viability of the cultures (De Valdez et al. 1985). Since dry sausage formulations contain salts and other antimicrobial agents which are detrimental to the viability of the cells, it is not surprising that the microenvironment in the gel particles during hydration might be more favorable, for greater viability, than in the batter itself as free cells. Furthermore, a drying process, to various water activity levels, is often carried out after the fermentation step to improve shelf life.

13.3.2 Immobilized Lactase or Proteases

Lactase (β-D-galactosidase) converts lactose into glucose and galactose. The major applications of lactase have been in fluid and powdered milk to alleviate lactose maldigestion and to improve milk sweetness (Lee 1996). In Asia where lactose intolerance is predominant, lactase treatment is compulsory to manufacture yogurt or cheese. Lactose is also a hygroscopic sugar that has a strong tendency to absorb flavors and odors and causes many defects in refrigerated foods such as crystallization in dairy foods, development of sandy or gritty mouth feel (Panesar et al. 2006).

With respect to fermented milks, in addition to its health benefit for the populations which do not digest lactose well, lactose hydrolysis would have the following benefits: (1) faster acidification rates by LAB (Gaudreau et al. 2005), (2) higher viable counts (Shah et al. 1997) and (3) synthesis of oligosaccharides as prebiotics (Hung et al. 2001).

Various systems have been designed for immobilized lactase: permeabilized cells in alginate beads (Genari et al. 2003), immobilization on cellulose (Roy and Gupta 2003), boronate- or chelate-epoxy beads (Pessela et al. 2003) or chitin beads (Illanes et al. 2000), as well as corn grits (Siso et al. 1994).

Although numerous hydrolysis systems have been investigated, only a few of them have been scaled up with success. Centrale del Latte of Milan, Italy, utilized the SNAM Progetti technology process which makes use of a neutral lactase from *Saccharomyces* (*Kluyveromyces*) *lactis* entrapped in cellulose triacetate fibers. Sumitomo Chemical, Japan, has developed an immobilized β-D-galactosidase preparation of fungal origin on the rugged surface of an amphoteric ion-exchange resin of phenol formaldehyde polymer and this technology was used by Drouin Cooperative Butter Factory for producing pasteurized milk and hydrolysed whey (Honda et al. 1993). A rotary column reactor has been developed by Snow Brand's factory that could be used as a stirred tank reactor and a packed bed reactor, since this apparatus may be used in both types of processing. Pasteurization of immobilized β-D-galactosidase using glycerol or propylene glycol was effective without inactivation of β-D-galactosidase. Major problems associated with the immobilized enzyme system still are microbial contamination, protein adherence and channeling. Therefore, for long-term operations, using immobilized enzymes, periodic washing and medium pasteurization are indispensable during operation.

In summary, in deciding if IET should be adopted for a process of lactose hydrolysis of milk used for lactic fermentations, one must ascertain if at least one of the following benefits can be obtained: less lactose maldigestion problems, higher acidification rates, higher populations and increased oligosaccharides levels. However, since scale up presents many challenges, the IET process has only been commercially applied to unfermented milk(s).

Proteinase treatment of milk has also been shown to be beneficial to the subsequent acidification rate in yoghurt cultures (Hemantha-Kumar et al. 2001). However, no proteolytic IET are reported in the literature for specific improvements in fermented milk(s).

Protease immobilization could also be used for a continuous coagulation process. The major proteinase used in dairy is rennet. Immobilization techniques have been investigated for recycling rennet, other milk clotting enzymes and microbial chymosin. Cheryl et al. (1975) used a continuous coagulation system for milk using pepsin and rennet in porous and alkylamine glass incorporated into a fluidized-bed reactor with the overall superior performance. Other continuous cheese-making process using an immobilized rennet enzyme reactor through a spiral flow path has been patented (Goldberg and Chen 1989). In this process, spiral flow path is formed by a spirally wound microporous sheet containing the immobilized rennet. Anprung et al. (1989) found that the most active immobilized rennin resulted from using 50-mesh river-bed sand as a carrier by covalent bonding that retained 82.6% of its original activity after 4 months.

13.4 For Accelerated Flavor Development

Two immobilized systems are available: attenuated cells and encapsulated enzymes. Both are available on the market and potential users must contact suppliers for benefits and limits.

Since LAB contain the main enzymes for cheese ripening and lactose hydrolysis, they are a logical source for "encapsulated" enzymes. However, adding more starter cells affects the technological process (acidification rate, final pH) which can be undesirable. One option is to reduce the acidification property (attenuation) of the cultures while maintaining the enzymatic components active. Contrary to proteinases which are located at the cell surface, peptidases are intracellular constituents. The literature shows that the extent of LAB autolysis affects the levels of various peptides in cheese extracts (Azarnia et al. 2006). Therefore, attenuated cultures can be seen as adsorbed or encapsulated enzymes, a sophisticated form of IET and its success requires two components: (1) a process to control the acidifying ability of the cultures, and (2) the autolysis ability to release the enzyme into the cheese matrix during ripening. Experience shows that attenuating the cultures in itself is difficult (Law 1999), and autolysis is variable as well (O'Reilly et al. 2002). Nevertheless, some of these problems have been overcome and there are commercial products in the market (Law 1999).

Since many reactions are required to generate cheese flavor, it is unlikely that a single enzyme can possibly generate all the flavor compounds required for a typical cheese flavor profile. Consequently, the development of an adequate flavor might require microencapsulation in multi-enzymes systems (Azarnia et al. 2006). Peptidases seem to be critical components in the effectiveness of these accelerated ripening systems, but proteinases can also be present (Kheadr et al. 2000). The addition of unencapsulated proteases for ripening directly into the milk has two major drawbacks: (1) proteolysis begins too early which can affect gel structure or yields, and (2) it is estimated that 95% of the added enzyme is actually lost in

Table 13.5 Some enzyme encapsulation techniques for accelerated cheese ripening

Encapsulation matrix/ technology	Effectiveness/benefit	References
Liposomes – phospholipids	– Pure lipids may be too expensive for economic viability – Prevented bitter off-flavors	Law (1999) Kheadr et al. (2000), (2003)
Gum beads composed of alginate, carrageenan, pectin, etc.	– 50% encapsulation efficiency, 90% recovery in cheese – Increased proteolysis but no effect on sensory	Kailasapathy and Lam (2005) Anjani et al. (2007)
High melting point milk fat particles prepared by spray coating	– 40% encapsulation efficiency, 74% recovery in cheese – Slower enzyme release than gums	Kailasapathy and Lam (2005)
Water-in-oil emulsions	– 80–90% encapsulation efficiency – Eightfold increase in diacetyl concentration during ripening	Magee and Olson (1981)

the whey (Law 1999). This has led to a number of studies which have examined encapsulation for accelerated cheese ripening (Table 13.5). Microencapsulated peptidases seem particularly promising (Kailasapathy et al. 2006). Although, some microencapsulation methods of cheese ripening have been effective, their industrial use remains limited because other approaches are available to cheese manufacturers, such as inoculation with specialty cultures (Azarnia et al. 2006).

Lipases are added into many type of cheeses (e.g., Romano, Provolone, Parmesan, Feta) to modify flavors. They have traditionally been added as free enzymes. However, as for ripening enzymes, losses occur in the whey. As mentioned previously, many reactions are involved in the development of Cheddar cheese flavor; one research team in particular has added microencapsulated lipase in combination with various proteases in order to generate multiple flavor elements (Kheadr et al. 2003). In general, cheeses made with a mixture of microencapsulated lipase and bacterial proteases were preferred to control cheeses. Even microencapsulated lipase alone benefits Cheddar cheese ripening (Kheadr et al. 2002).

13.5 For Protection During Food Processing

13.5.1 For Protection of Cells During Fermentation and Drying in Specialty Sausages

Probiotic bacteria are mostly added to dairy products, but there are novel applications for these cultures, and addition to dry sausages has been proposed (Työppönen et al. 2003). In one study, free or alginate-microencapsulated cells of *Lactobacillus reuteri* were added to the meat batter and viability followed during processing. Following the fermentation and drying periods, free cell counts dropped by 2.6 log units, whereas alginate-microencapsulated *L. reuteri* were reduced by approximately 0.5 log (Muthukumarasamy and Holley 2006). Furthermore, no significant difference in sensory quality was found between the control and sausages containing either unencapsulated or microencapsulated *L. reuteri*. This is another example where one of the concerns of the use of ICT cultures, i.e., negative impacts on sensory properties, did not occur. Unfortunately, while micro-encapsulation increased survival of *L. reuteri* and *Bifidobacterium longum*, it reduced their inhibitory action against *Escherichia coli* O157:H7 (Muthukumarasamy and Holley 2007).

13.5.2 For Enhanced Survival of Cells to Heating

In dairy processes, starters are always added to milk after heat treatment if one is applied. In meat fermentation, the raw batter is not heat treated prior to fermentation. So, the meat must be of excellent microbial quality in order to have an optimal control of the fermentation process. In fresh cheeses, a typical pasteurization treatment is applied (73°C for 15 s), but in yoghurt, the heat treatment is usually above 85°C. Free cells of LAB cannot survive these processes, so attempts were made to improve heat resistance using encapsulation techniques. Mandal et al. (2006) showed that *Lactobacillus casei* cells microencapsulated in alginate particles were more resistant to heat processes at 55–65°C. It was also found, in a MRS broth acidified to pH 5.0, that LAB encapsulated in alginate beads were more resistant to a heat treatment of 55°C for 15 min than were free-cells (Lemay et al. 2002a). Limited data suggest that microencapsulation of LAB in spray-coated particles could even be better than alginate beads; Goulet and Wozniak (2002) report enhanced survival of spray-coated lactobacilli to heat treatment at 50°C for up to 7 h in a simulated dry food process.

All these data point to enhanced survival of ICT or encapsulated cultures to moderate heating treatments. In practice, however, this enhanced survival is still not sufficient to enable inoculation prior to the higher heat treatment (73–95°C) used in dairy processes. With respect to meats, unexpectedly, when alginate-microencapsulated *Lactobacillus casei* were added in an acidified chicken meat model, the improved resistance conferred in the broth model did not extend to the solid matrix (Lemay et al. 2002a). In a more recent study, the efficacy of alginate-microencapsulated LAB to survive various heat treatments applied in sausage processing was confirmed by Muthukumarasamy et al. (2006). These data demonstrate the potential of microencapsulation in gel particles for enhanced survival of lactic culture to a thermal processing step, but that the food matrix itself can affect the results.

13.5.3 For Enhanced Survival of Cells to Freezing

Fermented milks can be frozen and are termed "frozen yoghurt" or "frozen desserts." In some instances, probiotic bacteria are included in the dairy mix. Freezing LAB or probiotics in unfermented media generates losses in viability ranging from

10 to 90% as a function of strain and conditions (Champagne et al. 2005). Thus, in deciding if ICT is useful one must examine the stability of the selected probiotic strain. If viability losses are greater than 90% (1 log) then ICT cultures become economically advantageous. The high total solids level in ice cream mix including fat (emulsion) may provide some protection to the cells (Kailasapathy and Sultana 2003). However, viability losses are greater with fermented products (Laroia and Martin 1991). Thus, a 1 log reduction in *Bifidobacterium longum* viable counts was observed following freezing in a low acid yogurt (pH 5.85) but a 2 log reduction was recorded when the pH was 4.47 (Modler and Villa-Garcia 1993). Hence, the low pH of fermented milks enhances the detrimental effect of freezing.

Many studies in various dairy desserts show that probiotics microencapsulated in alginate or carrageenan beads have lower viability losses following freezing (Sheu and Marshall 1993; Shah and Ravula 2000). This is particularly evident when cryoprotectants, such as glycerol, are added in the gel particles (Sheu et al. 1993). Microencapsulation enables the use of such protective compounds in the beads, without having to add them to the ice milk mix, where they would most likely affect sensory properties. Therefore, the microenvironment provided by beads seems to be a critical aspect for a successful application and this is particularly the case with low pH fermented products.

Unfortunately, this approach also has its drawbacks. With large beads, an effect on the texture of the fermented foods is to be expected which, in a number of applications, is undesirable. In frozen desserts, it was found that gel particles had to be 30 μm in diameter or less in order to avoid a detectable effect on texture (Sheu et al. 1993). Grinding the ICT beads after a freeze-drying step could generate such particle sizes (Lemay et al. 2002a) but the effect on sensory properties requires further investigation.

13.5.4 For Protection Against Bacteriophage Attack

Bacteriophages are viruses which attack specifically prokaryotic cells including lactic cultures. Extensive bacteriophage development during a lactic fermentation significantly affects acidification rates and can even provoke a complete stop in acid production (Fig. 13.2). Therefore, they are a constant concern in cheese making plants. At the dairy plant level, means to protect starter cultures against bacteriophage infection typically include plant design and production set up (preventing cheese whey from contaminating milk), starter tank design and operation, culture rotations, use of direct vat set (DVS) cultures and sanitation. It was shown that LAB in alginate beads are protected from bacteriophages (Steenson et al. 1987) and that acidification is maintained in phage-contaminated milk (Fig. 13.2). Interestingly, phage counts increased during the fermentation, presumably on released cells or on those growing at the surface of the beads. Although, this ICT property would theoretically be of great value to the industry, commercial application is still lacking. This might be due to the lower specific activity of ICT cultures (Champagne et al.

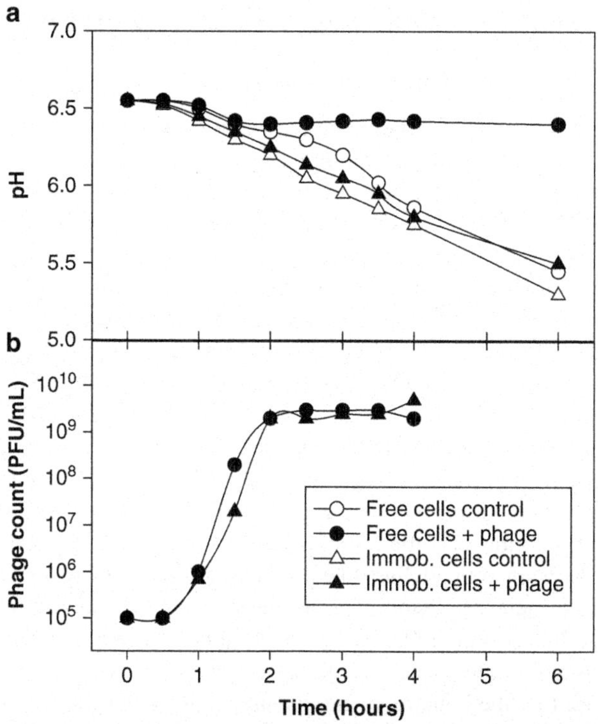

Fig. 13.2 Effect of microentrapment of lactococci into alginate beads and of bacteriophage contamination of the bioreactor on (**a**) milk acidification, and (**b**) bacteriophage growth in the system (Champagne et al. 1992)

1992) or to the potentially detrimental effect on texture resulting from the presence of the beads. This being said, the data showing phage protection were obtained on rather large beads (1–2 mm in diameter) and it remains to be determined if the smaller particles could also enable satisfactory acidification in the presence of phages. Reducing particle size diminishes the benefits of microencapsulation on survival to freezing (Sheu et al. 1993). So, it can be questioned whether the same would occur with phage protection.

All data on phage protection have been obtained with alginate as the matrix. In some countries there might be legislations that prohibit the use of non-dairy ingredients. In this instance, extrusion particles made from whey protein instead of alginate (Reid et al. 2005) may be considered.

In fermented meats, bacteriophages are not so much of a concern mainly because the phage particles do not spread easily in a solid matrix and therefore limits the viral infection. Hence, the benefit of ICT towards phage attacks would not be of great value in meat products.

13.6 For Improved Stability of Cells During Storage

Although attractive, the trend of adding probiotic cultures to fermented milk is facing several challenges including maintaining the viability of the cells in the product and in the gastrointestinal tract (Muthukumarasamy et al. 2006; see also Chap. 10). The same statement applies to protective cultures used to improve food safety by means of competitive exclusion and production of inhibitory compounds (ex. bacteriocin). Since viability of probiotics is considered critical in their functionality and because there are numerous reports of their viability losses in fermented dairy products (Champagne et al. 2005), many studies have examined the benefits of microencapsulated probiotic cultures during the storage of fermented milk products and passage into the gastrointestinal tract (Muthukumarasamy et al. 2006). With respect to enhanced stability, the ICT benefits are not convincing in frozen products (Sheu et al. 1993) or cheese (Gobbetti et al. 1998; Kailasapathy and Masondole 2005). As for survival to the freezing process itself, in deciding if ICT is useful one must examine the stability of the given probiotic strain. If viability losses are greater than 90% (1 log) then ICT cultures become economically advantageous. However, a very different picture is obtained in yoghurt-like products. Although there are discrepancies, most teams report increased viability of probiotics when they are microencapsulated in gel beads (Adhikari et al. 2000; Anjani et al. 2004). It was assumed that microencapsulated cells were better protected against acidity, but Talwalkar and Kailasapathy (2003) observed that the benefits of microentrapment only occurred when there is oxygen present in the medium. Microencapsulation thus appears to provide a microenvironment having reduced oxygen levels, which prevent viability losses to oxygen-sensitive strains (Talwalkar and Kailasapathy 2004). Data from McMaster et al. (2005) also show an enhanced degree of oxygen tolerance by bifidobacteria in gel beads. Thus, ICT is clearly beneficial in fermented milk products to improve storage stability of probiotic and protective cultures.

In one study, the addition of probiotic cultures either in the free or encapsulated state did not significantly affect appearance and color, acidity, flavor and after taste of the yogurts during storage. There were, however, significant differences in texture (smoothness) of the yoghurts (Kailasapathy 2006). In one current commercial ICT application, encapsulated bifidobacteria are added as beads into yoghurt. They can be easily detected both visually and in the mouth feel. In this application the use of capsules is clearly stated on the label, and ICT almost appears as a marketing benefit for product differentiation.

13.7 For the Inhibition of Undesirable Flora

Catalase decomposes hydrogen peroxide into oxygen and water. In some processes, hydrogen peroxide is added to raw milk to activate the lactoperoxidase-thiocyanate preservation system (Boots and Floris 2006). Catalase is then added to remove the residual peroxide. It would therefore seem logical to have an immobilized enzyme system to remove peroxide from milk in a continuous fashion.

Lysozyme is marketed to prevent clostridia overgrowth and gas production in a certain number of ripened cheeses, particularly the "swiss-type" varieties. It also has potential for the preservation of other foods including meats (Bower et al. 1998).

Although microencapsulated antimicrobials are used in bakery, to our knowledge, no immobilized catalase or microencapsulated lysozyme has been used for these purposes so far. Should the lysozyme negatively affect the lactic cultures, then spray-coating with a fat which would only melt at a cooking stage of the cheese making process could be considered. Spray-coating is typically carried out by spraying a coating material (e.g., fat-based) in a fluidized bed system on a core powder containing the cell or enzyme components (Fig. 13.3). It is not cell immobilization process as such, but rather an encapsulation technology. In some applications, it is more appropriate to use coated bioactives than to use microentrapped or immobilized ones.

13.8 Conclusions

It can be seen that immobilization of cells and enzymes has the potential to address many problems in food processing. However, particular conditions are often required before one can consider ICT or IET and a summary of those which were presented throughout the chapter is presented in (Table 13.6). Some processes are already used

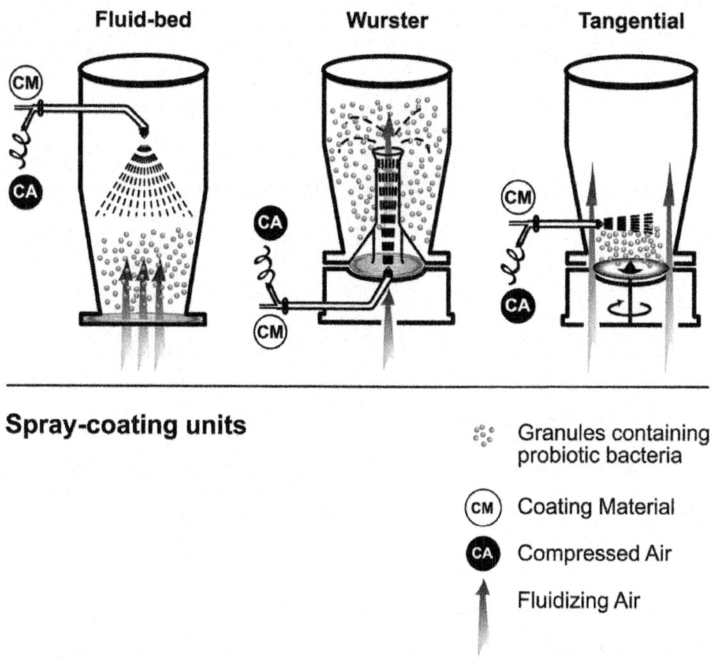

Fig. 13.3 Various systems which enable the spray-coating of cultures in a fluidized bed system

Table 13.6 Conditions to choose ICT or IET bioactives over free-cell of free enzymes in fermented dairy or meat products

Application	Condition
All applications	Encapsulation or immobilization adds to the cost. There must be one of the following benefits: – Protecting the stability ingredient – Enhanced functionality of the ingredient – Enhanced retention of the ingredient in the end-product – Product differentiation
Biomass production of starter cultures	– A strain is difficult to produce with traditional processing methods – ICT confers a functionality (protection against phages or GI conditions, stability during storage)
Continuous inoculation of milk with starter or probiotic cultures	– Milk is heated to destroy phages (ex. yoghurt) – Cultures are less prone to phage attack (processes which do not generate whey, thermophilic cultures) – Specialty cultures are to be added (probiotics) – Few production interruptions occur (weekends) – Need for cultures having specific functionalities generated by immobilization
Meat fermentations	– With alginate particles
Attenuated cells for flavor enhancement	– Cultures must autolyse in the given cheese matrix (process, pH, salt level etc.) to release enzymes
Survival to heating	– For foods that are heterogeneous in appearance or texture – Encapsulation by spray-coating is more efficient than immobilization into alginate bead for this application – Encapsulation by spray-coating with a fats having high melting point temperatures is preferable if one whishes release of the contents at a given temperature
Protection against bacteriophage attack	– With whey-based particles – In ripened cheeses because beads can dissolve
Survival to freezing Stability of probiotics during storage	– If a strain shows unacceptable viability losses (>1 log) to the freezing process or and/or over the storage period
Inhibition of undesirable flora	– Spray-coated with fats in fluidized bed systems with compounds which enable a triggered delivery of the compounds at a late stage of processing

commercially, particularly with flavor acceleration, stability of probiotics and protective cultures in foods. In the future, it can be expected that technological applications of ICT or IET will increase, as novel benefits will also appear. Thus, immobilization (encapsulation) of probiotics may not only have technological benefits, but may also improve the functionality of the cultures by enabling improved survival of the gastric environment as well as controlled release of the cells in the gastro-intestinal tract (Muthukumarasamy et al. 2006). Similarly, protective culture survival to antimicrobial systems used in food requires further investigation. Valuable enzymes which could have beneficial health effects, such as α- or β-D-galactosidases (for soy or milk carbohydrates, respectively), could also benefit from encapsulation in the same fashion. Thus, continued research in ICT and IET is warranted not only to improve technological processes and the safety of foods, but also to enhance the functionality of valuable bioactive ingredients in functional foods.

References

Adhikari K, Mustapha A, Grün IU, Fernando L (2000) Viability of microencapsulated bifidobacteria in set yogurt during refrigerated storage. J Dairy Sci 83:1946–1951

Anjani K, Iyer C, Kailasapathy K (2004) Survival of co-encapsulated complementary probiotics and prebiotics in yoghurt. Milchwissenschaft 59:396–399

Anjani K, Kailasapathy K, Phillips M (2007) Microencapsulation of enzymes for potential application in acceleration of cheese ripening. Int Dairy J 17:79–86

Anonymous (2003) Les enzymes déclenchent de nouveaux processus biologiques. Feeding Times 8:1–37

Anprung P, Chengaengsatityaporn S, Thunpithayakui C (1989) Immobilized rennin for cheese making. Asian Food J 4:107–110

Araya M, Morelli L, Reid G, Sanders ME, Stanton C, Pineiro M, Ben Embarek P (2002) Guidelines for the evaluation of probiotics in food. Joint FAO/WHO Working Group Report on Drafting Guidelines for the Evaluation of Probiotics in Food, London (ON, Canada)

Azarnia S, Robert N, Lee B (2006) Biotechnological methods to accelerate cheddar cheese ripening. Crit Rev Biotechnol 26:121–143

Benchabane S, Vandenberg GW, Subirade M (2004) Optimization of the production of phytase microcapsules by spray drying. Aquac Assoc Canada 9:84–87

Blom H, Hagen BF, Pedersen BO, Holck AL, Axelsson L, Naes H (1996) Accelerated production of dry fermented sausage. Meat Sci 43:S229–S242

Boots JW, Floris R (2006) Lactoperoxidase: From catalytic mechanism to practical applications. Int Dairy J 16:1272–1276

Bower CK, Daeschel MA (1999) Resistance responses of microorganisms in food environments. Int J Food Microbiol 50:33–44

Bower CK, Daeschel MA, McGuire J (1998) Protein antimicrobial barriers to bacterial adhesion. J Dairy Sci 81(10):2771–2778

Champagne CP (2006) Starter cultures biotechnology: The production of concentrated lactic cultures in alginate beads and their applications in the nutraceutical and food industries. Chem Ind Chem Eng Quarterly 12(1):11–17

Champagne CP, Morin N, Couture R, Gagnon C, Jelen P, Lacroix C (1992) The potential of immobilized cell technology to produce freeze-dried, phage-protected cultures of *Lactococcus lactis*. Food Res Int 25(6):419–427

Champagne CP, Girard F, Rodrigue N (1993) Production of concentrated suspensions of thermophilic lactic acid bacteria in calcium alginate beads. Int Dairy J 3(3):257–275

Champagne CP, Roy D, Gardner N (2005) Challenges in the addition of probiotic cultures to foods. Crit Rev Food Sci Nutr 45(1):61–84

Champagne CP, Gardner NJ, Lacroix C (2007) Fermentation technologies for the production of exopolysaccharide-synthesizing *Lactobacillus rhamnosus* concentrated cultures. Electron J Biotechnol 10(2) April 10. Found at: http://www. ejbiotechnology.info/content/vol10/issue2/full/10/

Cheetham PSJ (1988) Recent developments in enzyme technology as applied to foods and food processing. In: King RD, Cheetham PSJ (eds) Food biotechnology. Elsevier Applied Science, London, pp 117–171

Cheryl M, Van Wyk PJ, Olson NF, Richardson T (1975) Continuous coagulation of milk using immobilized enzymes in a fluidized-bed reactor. Biotechnol Bioeng 17:585–598

De Valdez GF, De Giori GS, De Ruiz Holgado AP, Oliver G (1985) Effect of the rehydration medium on the recovery of freeze-dried lactic acid bacteria. Appl Environ Microbiol 50:1339–1341

Doleyres Y, Lacroix C (2005) Technologies with free and immobilized cells for probiotic bifidobacteria production and protection. Int Dairy J 15:973–988

Eie T, Larsen H, Sørheim O, Pettersen MK, Hansen AÅ, Wold JP, Naterstad K, Mielnik M (2007) New technologies for extending shelf life. Ital J Food Sci 19:127–152

Farnworth ER (2004) The beneficial health effects of fermented foods – Potential probiotics around the world. J Nutraceuticals Funct Med Foods 4:93–117

Fox PF (1993) Exogenous enzymes in dairy technology – a review. J Food Biochem 17: 173–199

Gaudreau H, Champagne CP, Jelen P (2005) The use of crude cellular extracts of Lactobacillus delbrueckii ssp. bulgaricus 11842 to stimulate growth of a probiotic *Lactobacillus rhamnosus* culture in milk. Enzyme Microb Technol 36(1):83–90

Genari AN, Passos FV, Passos FML (2003) Configuration of a bioreactor for milk lactose hydrolysis. J Dairy Sci 86(9):2783–2789

Gobbetti M, Corsetti A, Smacchi E, Zocchetti A, De Angelis M (1998) Production of Crescenza cheese by incorporation of Bifidobacteria. J Dairy Sci 81:37–47

Goldberg BS, Chen RY (1989) Continuous cheese-making process utilizing an immobilized rennet enzyme reactor. US Patent 4(801):463

Goulet J, Wozniak J (2002) Probiotic stability: A multifaceted reality. Innov Food Technol, February, 14–16

Haefner S, Knietsch A, Scholten E, Braun J, Lohscheidt M, Zelder O (2005) Biotechnological production and applications of phytases. Appl Microbiol Biotechnol 68:588–597

Hammes WP, Hertel C (1998) New developments to meat starter cultures. Meat Sci 49:S125–S138

Hemantha-Kumar HR, Monteiro PV, Bhat GS, Ramachandra-Rao HG (2001) Effects of enzymatic modification of milk proteins on flavour and textural qualities of set yoghurt. J Sci Food Agric 81(1):42–45

Honda Y, Kako M, Abiko K, Sogo Y (1993) Hydrolysis of lactose in milk. In: Tanaka A, Tosa T, Kobayashi T (eds) Industrial application of immobilized biocatalysis. Marcel Dekker, New York, pp 209–234

Hung MN, Xia Z, Lee BH (2001) Molecular and biochemcial analysis of two beta-galactosidases from Bifidobacterium infantis HL96. Appl Environ Microbiol 67:4256–4263

Illanes A, Wilson L, Tomasello G (2000) Temperature optimization for reactor operation with chitin-immobilized lactase under modulated inactivation. Enzyme Microb Technol 27(3–5): 270–278

Kailasapathy K (2006) Survival of free and encapsulated probiotic bacteria and their effect on the sensory properties of yoghurt. LWT-Food Sci Technol 39(10):1221–1227

Kailasapathy K, Lam SH (2005) Application of encapsulated enzymes to accelerate cheese ripening. Int Dairy J 15:929–939

Kailasapathy K, Masondole L (2005) Survival of free and microencapsulated *Lactobacillus acidophilus* and *Bifidobacterium* and their effect on texture of feta cheese. Aust J Dairy Technol 60:252–258

Kailasapathy K, Sultana K (2003) Survival and beta-D-galactosidase activity of encapsulated and free *Lactobacillus acidophilus* and *Bifidobacterium lactis* in ice cream. Aust J Dairy Technol 58:223–227

Kailasapathy K, Anjani K, Seneweera S (2006) Recent trends in accelerated cheese ripening using microencapsulated enzymes. Aust J Dairy Technol 61(2):78–80

Kearney L, Upton M, McLoughlin A (1990) Meat fermentations with immobilized lactic acid bacteria. Appl Microbiol Biotechnol 33:648–651

Kerry JP, O'Grady MN, Hogan SA (2006) Past, current and potential utilisation of active and intelligent packaging systems for meat and muscle-based products: A review. Meat Sci 74:113–130

Kheadr EE, Vuillemard JC, El Deeb SA (2000) Accelerated Cheddar cheese ripening with encapsulated proteinases. Int J Food Sci Technol 35:483–495

Kheadr EE, Vuillemard JC, El Deeb SA (2002) Acceleration of Cheddar cheese lipolysis by using liposome-entrapped lipases. J Food Sci 67(2):485–492

Kheadr EE, Vuillemard JC, El-Deeb SA (2003) Impact of liposome-encapsulated enzyme cocktails on Cheddar cheese ripening. Food Res Int 36:241–252

Kosikowski FV (1982) Cheese and fermented milk foods, 3rd edn. Brooktondale, New York, p 711

Krasaekoopt W, Bhandari B, Deeth H (2003) Evaluation of encapsulation techniques of probiotics for yoghurt. Int Dairy J 13:3–13

Lamboley L, Lacroix C, Champagne CP, Vuillemard JC (1997) Continuous mixed strain mesophilic lactic starter production in supplemented whey permeate medium using immobilized cell technology. Biotechnol Bioeng 56:502–516

Lamboley L, Lacroix C, Sodini I, Lemay MJ, Champagne CP (2001) Influence of inoculum composition and low KCl supplementation on the biological and rheological stability of an immobilized cell system for mixed mesophilic lactic starter production. Biotechnol Prog 17:1071–1078

Lamboley L, St Gelais D, Champagne CP, Lamoureux M (2003) Growth and morphology of thermophilic dairy starters in alginate beads. J Gen Appl Microbiol 49:205–214

Lantto R, Puolanne E, Kruus K, Buchert J, Autio K (2007) Tyrosinase-aided protein cross-linking: Effects on gel formation of chicken breast myofibrils and texture an water-holding of chicken breast meat homogenate gels. J Agric Food Chem 55:1248–1255

Laroia S, Martin JH (1991) Effect of pH on survival of *Bifidobacterium bifidum* and *Lactobacillus acidophilus* in frozen fermented dairy desserts. Cult Dairy Prod J 26(4):13–24

Law BA (1999) Technology of cheesemaking. Sheffield Academic Press – CRC Press, Sheffield, p 177

Lee BH (1996) Fundamentals of food biotechnology. VCH Publishers, New York

Lemay MJ, Champagne CP, Gariépy C, Saucier L (2002a) A comparison of the effect of meat formulation on the heat resistance of free or encapsulated cultures of *Lactobacillus sakei*. J Food Sci 67:3428–3434

Lemay MJ, Choquette J, Delaquis PJ, Gariépy C, Rodrigue N, Saucier L (2002b) Antimicrobial effect of natural preservatives in a cooked and acidified chicken meat model. Int J Food Microbiol 78:217–226

Lüke FK (1998) Fermented sausages. In: Wood BJB (ed) Microbiology of fermented foods, 2nd edn. Blackie Academic & Professional, London, pp 441–483

Macedo MG, Champagne CP, Vuillemard JC, Lacroix C (1999) Establishment of bacteriophages in an immobilized cells system used for continuous inoculation of lactococci. Int Dairy J 9:437–445

Magee EL Jr, Olson NF (1981) Microencapsulation of cheese ripening systems: Production of diacetyl and acetoin in cheese by encapsulated bacterial cell-free extract. J Dairy Sci 64:616–621

Mandal S, Puniya AK, Singh K (2006) Effect of alginate concentration on survival of micro-encapsulated *Lactobacillus casei* NCDC-298. Int Dairy J 16:1190–1195

McMaster LD, Kokott SA, Slatter P (2005) Micro-encapsulation of Bifidobacterium lactis for incorporation into soft foods. World J Microbiol Biotechnol 21:723–728

Modler HW, Villa-Garcia L (1993) The growth of *Bifidobacterium longum* in a whey-based medium and viability of this organism in frozen yogurt with low and high levels of developed acidity. Cult Dairy Prod J 28(1):4–8

Morin N, Bernier-Cardou M, Champagne CP (1992) Production of *Lactococcus lactis* biomass by immobilized cell technology. J Ind Microbiol 9:131–135

Murthy K (2007) Bacteriophage treatment – a natural "on-farm" approach to improving food safety. Symposium on advances in antimicrobial interventions for quality control of meat and poultry products, Canadian Meat Council, Toronto, Canada, September 13–14

Muthukumarasamy P, Holley RA (2006) Microbiological and sensory quality of dry fermented sausages containing alginate-microencapsulated *Lactobacillus reuteri*. Int J Food Microbiol 111(2):164–169

Muthukumarasamy P, Holley RA (2007) Survival of *Escherichia coli* O157:H7 in dry fermented sausages containing micro-encapsulated probiotic lactic acid bacteria. Food Microbiol 24(1):82–88

Muthukumarasamy P, Allan-Wojtas P, Holley RA (2006) Stability of *Lactobacilus reuteri* in different types of microcapsules. J Food Sci 71:M20–M24

Naes H, Holck AL, Axelsson L, Andersen HJ, Blom H (1994) Accelerated ripening of dry fermented sausage by addition of a *Lactobacillus* proteinase. Int J Food Sci Technol 29: 651–659

Norton S, Vuillemard S (1994) Food bioconversions and metabolite production using immobilized cell technology. Crit Rev Biotechnol 14(2):193–224

O'Reilly CE, O'Connor PM, Murphy PM, Kelly AL, Beresford TP (2002) Effects of high-pressure treatment on viability and autolysis of starter bacteria and proteolysis in Cheddar cheese. Int Dairy J 12:915–922

Panesar R, Panesar PS, Singh RS, Kennedy JF, Bera MB (2006) Production of lactase-hydrolyzing milk using ethanol permeabilized yeast cells. Food Chem 101:786–790

Pessela BCC, Mateo C, Fuentes M, Vian A, Garcia JL, Carrascosa AV, Guisan JM, Fernandez-Lafuente R (2003) The immobilization of a thermophilic beta-galactosidase on Sepabeads supports decreases product inhibition. Complete hydrolysis of lactose in dairy products. Enzyme Microb Technol 33(2–3):199–205

Prevost H, Divies C (1985) Continuous yoghurt production with *Lactobacillus bulgaricus* and *Streptococcus thermophillus* entrapped in Ca-alginate. Biotechnol Lett 7:247–252

Prevost H, Divies C (1987) Fresh fermented cheese production with continuous pre-fermented milk by a mixed culture of mesophioic lactic streptococci entrapped in Ca-alginate. Biotechnol Lett 9:789–794

Prevost H, Divies C (1988a) Continuous pre-fermentation of milk by entrapped yoghurt bacteria. I. Development of the process. Milchwissenschaft 43:621–625

Prevost H, Divies C (1988b) Continuous pre-fermentation of milk by entrapped yoghurt bacteria. II. Data for optimization of the process. Milchwissenschaft 43:716–719

Reid AA, Vuillemard JC, Britten M, Arcand Y, Farnworth E, Champagne CP (2005) Microentrapment of probiotic bacteria in a Ca2+-induced whey protein gel and effects on their viability in a dynamic gastro-intestinal model. J Microencapsul 22:603–619

Roy I, Gupta MN (2003) Lactose hydrolysis by LactozymTM immobilized on cellulose beads in batch and fluidized bed modes. Process Biochem 39(3):325–332

Saucier L, Champagne CP (2005) Cell immobilization technology and meat processing. In: Nedovic V, Willaert R (eds) Applications of cell immobilisation biotechnology, series "Focus on Biotechnology", vol 8B. Springer–Kluwer, Dordrecht, pp 337–350

Shah NP, Ravula RR (2000) Microencapsulation of probiotic bacteria and their survival in frozen fermented dairy desserts. Aust J Dairy Technol 55:139–144

Shah NP, Warnakulsuriya EV, Lankaputhra WEV (1997) Improving viability of *Lactobacillus acidophilus* and *Bifidobacterium* spp. in yogurt. Int Dairy J 7:349–356

Shahidi F, Pegg PB (1991) Encapsulation of the pre-formed cooked cured-meat pigment. J Food Sci 56:2500–1504

Sheu TY, Marshall RT (1993) Microentrapment of lactobacilli in calcium alginate gels. J Food Sci 54:557–561

Sheu TY, Marshall RT, Heymann H (1993) Improving survival of culture bacteria in frozen desserts by microentrapment. J Dairy Sci 76:1902–1907

Siso MIG, Freire MA, Ramil E, Belmonte ER (1994) Covalent immobilization of beta-galactosidase on corn grits. Process Biochem 29:7–12

Steenson LR, Klaenhammer TR, Swaisgood HE (1987) Calcium alginate-immobilized cultures of lactic streptococci are protected from bacteriophages. J Dairy Sci 70:1121–1127

Talwalkar A, Kailasapathy K (2003) Effect of microencapsulation on oxygen toxicity in probiotic bacteria. Aust J Dairy Technol 58:36–39

Talwalkar A, Kailasapathy K (2004) A review of oxygen toxicity in probiotic yogurts: Influence on the survival of probiotic bacteria and protective techniques. Compr Rev Food Sci Food Saf 3:117–124

Toren JF (2007) Composition and method for tenderizing meat. US Patent 7250184

Tseng T-F, Tsai C-M, Yang J-H, Chen M-T (2006) Porcine blood plasma tansglutaminase combined with thrombin and fibrinogen as a binder in restructured meat. Asian-Australas J Anim Sci 19:1054–1058

Työppönen S, Petäjä E, Mattila-Sandholm T (2003) Bioprotectives and probiotics for dry sausages. Int J Food Microbiol 83:233–244

Venugopal V (1994) Production of fish protein hydrolyzates by microorganisms. In: Martin AM (ed) Fisheries processing: Biotechnological applications. Chapman & Hall, London, pp 223–243

Wood BJB (1998) Microbiology of fermented foods, 2nd edn. Blackie Academic and Professional, London

You-Jin J, Vasanthan T, Temelli F, Song G-K (2003) The suitability of barley and corn starches in their native and chemically modified forms for volatile meat flavour encapsulation. Food Res Int 36:349–355

Chapter 14
Encapsulates for Food Bioconversions and Metabolite Production

Véronique Breguet, Vojislav Vojinovic, and Ian W. Marison

14.1 Introduction

The control of production costs in the food industry must be very strict as a result of the relatively low added value of food products. Since a wide variety of enzymes and/or cells are employed in the food industry for starch processing, cheese making, food preservation, lipid hydrolysis and other applications, immobilization of the cells and/or enzymes has been recognized as an attractive approach to improving food processes while minimizing costs. This is due to the fact that biocatalyst immobilization allows for easier separation/purification of the product and reutilization of the biocatalyst. The advantages of the use of immobilized systems are many, and they have a special relevance in the area of food technology, especially because industrial processes using immobilized biosystems are usually characterized by lower capital/energy costs and better logistics. The main applications of immobilization, related to the major processes of food bioconversions and metabolite production, will be described and discussed in this chapter.

Immobilization of cells permits higher cell densities to be attained in bioreactors, and thus results in higher enzyme productivity (Krisch and Szajani 1996). The immobilization of either cells or enzyme to achieve the enzymatic process makes re-utilization and continuous operation possible, and also precludes the need to separate the active biocatalyst from the surrounding medium following the process. Encapsulation of whole cells instead of pure enzyme often leads to better enzyme stability as well as cofactor regeneration. These advantages can drastically reduce

V. Breguet
Baxter BioScience Manufacturing Sarl, Neuchatel, Switzerland

V. Vojinovic
Nestec SA (Nestle Research Centre), Lausanne, Switzerland

I.W. Marison (✉)
School of Biotechnology, Dublin City University,
Glasnevin, Dublin 9, Ireland
e-mail: ian.marison@dcu.ie

N.J. Zuidam and V.A. Nedović (eds.), *Encapsulation Technologies for Active Food Ingredients and Food Processing*,
DOI 10.1007/978-1-4419-1008-0_14, © Springer Science+Business Media, LLC 2010

production costs, thereby enabling large-scale application of cells. Enzymes, and cells producing the enzymes, have been immobilized by various methods, such as gel entrapment, physical absorption and covalent attachment on various supports (Katchalski-Katzir 1993).

The selection of the immobilization method and support depends on the biocatalyst (enzyme or microorganism) and on the bioprocess itself (aerobic or anaerobic, production of small soluble metabolites or macromolecules). Hydrogels, such as calcium-alginate, κ-carrageenan, calcium-pectate, sepharose, dextrans and others are frequently used for cell immobilization by entrapment (Santoyo et al. 1996). Their interaction with the biomass and possible effects on the cell morphology and metabolism are relatively well known and diffusion coefficients for small organic molecules in hydrogels are relatively high. Immobilization of whole cells can protect them, and the associated enzyme activities, from shear forces resulting in higher viability and bioconversion rates, particularly for continuous and long-term applications. Although biomass is efficiently retained with the entrapment technique, the immobilization of proteins (enzymes) in hydrogels is often difficult due to leaching (exodiffusion) resulting from the relatively small spatial molecular volume when compared with the microparticle porosity. Hydrogels, and the production of beads under aseptic conditions, are costly, and may limit their use on an industrial scale. A major problem in using hydrogel beads is their chemical and mechanical stability in culture media over prolonged operational times. Cell growth can clog the pores of hydrogels, or cause rupture of the material. Stability of the beads may be increased by cross-linking, for example, with glutaraldehyde or complexation with poly-L-lysine (Leng et al. 2006). This, however, introduces additional mass transfer resistance, and may also introduce cytotoxicity issues. Entrapped cells are normally grown in packed-bed, fluidized-bed, gas-lift, or stirred tank reactors (Bodalo-Santoyo et al. 1999; Hernandez et al. 2001; Talabardon et al. 2000).

Surface immobilization by adsorption has the advantage of being very simple to carry out. In order to increase cell loads and make them comparable to entrapment techniques, highly porous materials, such as pretreated cellulose, are frequently used. (Tisnadjaja et al. 1996). A disadvantage of this method is the relatively low interaction between the support and the biocatalyst, resulting in biomaterial release to the surroundings. However, when using adsorbed cells, there is no additional mass transfer barrier as in the case of the entrapped or encapsulated cells. A serious problem in using cells immobilized by adsorption in submerged cultures is the mechanical damage to the cells caused by the abrasion between the carrier particles. This may be avoided by the use of solid state fermentations in which the cells are immobilized on a solid carrier.

Covalent binding, between a functional group of the biocatalyst and the support, is a technique used for immobilization of enzymes (Bodalo-Santoyo et al. 1999). Many different types of support are used, including cotton cloth, chitosan, alginate, silica hydrogel and agarose to name but a few. The release of enzyme to the surroundings is thus drastically reduced compared to physical adsorption techniques, which extends the duration of the process and also reduces the cost since the loss of biocatalyst is reduced.

14.2 Reactor Design for Immobilized Biocatalysts

Immobilized biocatalysts are generally quite stable, they are indeed frequently more stable than non-immobilized forms. This observation has enabled the design and construction of reactors that can be adapted to immobilized cells/enzyme (Fig. 14.1).

Current reactors for submerged immobilized biocatalysts include stirred-tank, packed-bed, fluidized-bed, gas-lift and stirred-tank reactors with a recycling device

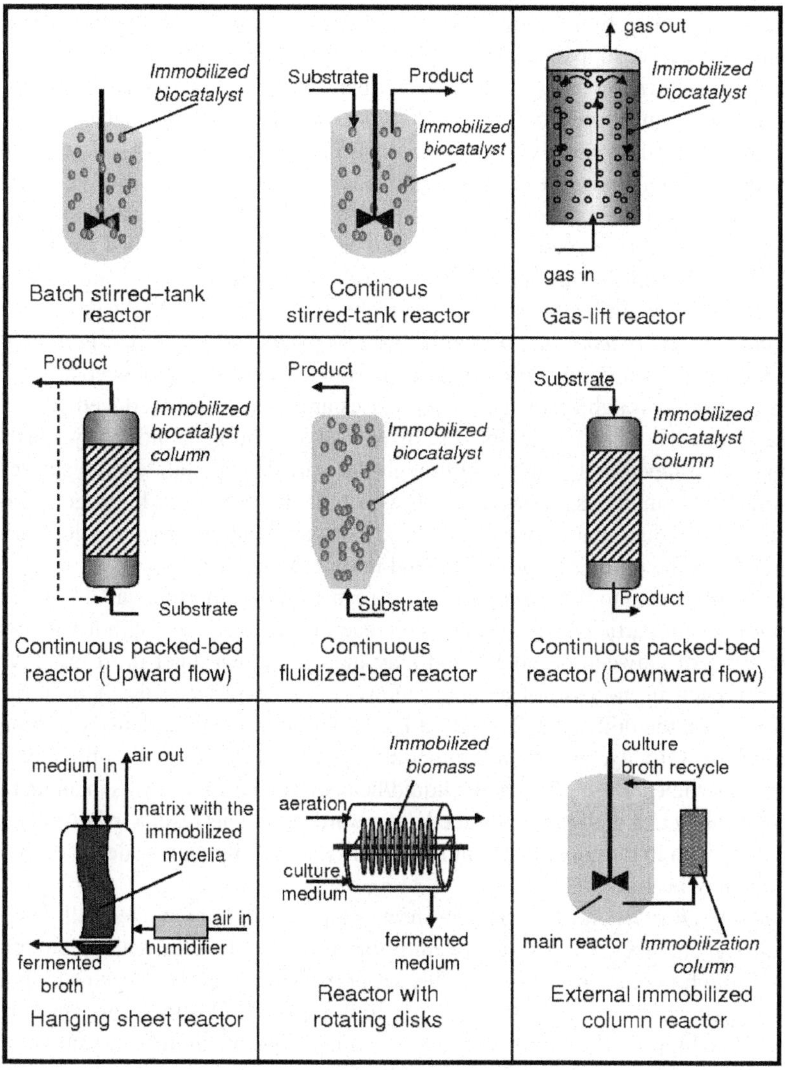

Fig. 14.1 Bioreactor design used with immobilized biocatalyst

operating in batch, fed-batch continuous and perfusion (continuous with cell recycle such as by the incorporation of a microfiltration/ultrafiltration membrane) modes (Katchalski-Katzir 1993). Common reactors for solid state fermentation include packed-bed reactors, hanging sheet reactors, rotating disk and external recirculation column reactors.

Stirred tank reactors are commonly used for processes with cells entrapped in hydrogels or adsorbed on the surface of solid porous beads for both aerobic and anaerobic processes. They provide good mixing conditions but are limited by the rheological properties of the immobilization support suspension in the culture medium, as well as by relatively high shear stress and abrasion that may damage the immobilization support and the cells growing on their surface.

Packed-bed reactors consist of columns filled with the fixed layer of biocatalyst immobilized by entrapment or by adsorption. These reactors provide high cell densities and high volumetric productivities. The shear stresses are low, which makes the culture of filamentous fungi possible. However, a major problem with packed-bed columns is due to mass and heat transfer limitations, which may reduce the performance of the reactor significantly. Oxygen supply may also be difficult in packed-bed reactors. The use of feed that contains insoluble matter, as well as cell growth, may lead to clogging of the reactor. Reactor maintenance between operation cycles is complicated.

Fluidized-bed reactors are a modification of packed-bed reactors, in which the energy needed for mixing is provided by the flow of medium introduced at the base of the reactor. Through this way, shear stresses are significantly reduced compared to mechanically stirred reactors. Fluidized-bed operation is normally used for hydrogel-entrapped cell cultures, although the abrasion of cells on the surface of the solid beads may pose a problem. However, as the density of hydrogels is only slightly higher than that of water, the maximum applicable velocity is low, which in turn limits the efficiency of mixing and mass transfer.

An improvement to fluidized-bed reactors can be achieved using gas-lift operation to achieve microparticle fluidization. Such reactors are normally cylindrical with an aspect ratio of between 6 and 12, and may incorporate an inner draft tube. A gas (air or oxygen in the case of aerobic cultures) is introduced at the bottom of the reactor within the draft tube. The resulting reduction in medium density, combined with the gas velocity, results in the medium rising, and entrains the movement of the microparticles. De-gassing at the liquid/headspace interface results in an increase in fluid density and gravitation-induced downward flow. This results in better mixing conditions than in the equivalent fluidized-bed systems, although scale-up of gas-lift reactors may be complex.

Hanging-sheet reactors consist of a sheet of fabric with immobilized cells hanging in a vertical chamber. The substrate is introduced at the top of the reactor and moves down through the action of gravity. This set-up provides a good oxygen supply and low shear stress (little mechanical damage to the cells), and is therefore very useful for aerobic bioprocesses with filamentous fungi. The medium viscosity is very much improved compared to submerged fungal cultures since the cells are mostly retained in the solid phase. High product concentrations can be achieved. However, this reactor design is susceptible to high concentration and temperature gradients.

Local temperature gradients caused by the heat transfer limitations and the heat released by cell metabolism may cause the drying of mycelium, even when water-saturated air is supplied. Together, these problems may result in a significant reduction in volumetric productivity.

Rotating disk reactors are an improvement of the hanging sheet systems. The reactor is normally a horizontal cylinder with disks (or drums) attached to a central rotating axis. The reactor is partially filled with nutritive medium, so that the mycelium immobilized on the rotating element is periodically submerged in the medium and exposed to the air (or oxygen).

External immobilized column reactors with re-circulation consist of two connecting vessels of different dimensions filled with culture medium which is continuously re-circulated through the system. The smaller vessel is filled with the immobilization support, which is generally a cellulose fiber material. The operation begins with a batch culture of cells in the larger vessel, until the desired biomass concentration is reached. At this point, the recirculation pump is turned on and the broth is forced through the immobilization column, where cells are retained by adsorption. This design allows for long-term operation, since the immobilized biomass is constantly renewed by the free cells growing in the larger vessel. Immobilization is straightforward and the reactor can be operated in batch or continuous mode. As a result of continuous flow of culture broth through the immobilization column, clogging may be a problem for long-term operation.

These relatively new types of reactor have the following advantages over other methods (Katchalski-Katzir 1993): (1) they allow high biocatalyst loading; (2) they prolong enzyme activity or cell life; (3) they have the ability to be used in recycle mode; (4) they can be used with high flow rates; (5) they operate with reduced capital and operating costs; (6) they have reduced energy requirements; (7) they produce less waste products; (8) they generally have high yields; (9) they are readily scalable.

Furthermore, automatization of the process leads to considerable reduction of operating costs, mainly through reduced enzyme levels and labor costs resulting from the higher production yields achieved.

The choice of reactor is generally defined by the process under consideration. Thus, batch processes might involve direct enzyme addition to the substrate (e.g., whole milk), since they will not be recycled, whereas continuous processes (e.g., glucose production) will require microencapsulates which are physically robust, from which the enzyme does not leach over the duration of use, in which the enzyme is stable under the conditions of pH, temperature and ionic strength employed and which can be operated at the required scale. Pre-requisites are that the enzyme be adsorbed or immobilized throughout the microencapsulate matrix, that the loading density is as high as possible, that the porosity of the microencapsulates allows both substrates and products to diffuse readily through them to avoid undue mass transfer limitations and product inhibition. Thus, the choice and scale of reactor will depend on (1) how much enzyme is required to carry out the process in the desired time (process kinetics); (2) the mechanical stability of the microencapsulates, since there are significant shear forces associated with stirred tank reactors and certain types of fluidized-beds. For packed-bed applications the flow rate, and thus kinetics, will be dictated by the pressure drop across the packed-bed, consequently elastic hydrogel matrices are less suited to these types of reactors; (3) the density of the microencapsulates, since hydrogels have a specific density only slightly higher than that of the reaction medium in most cases, therefore would be unsuitable for use in fluidized-beds unless ceramic or inert inorganic particles are co-immobilized to achieve a density which avoids wash-out; (4) whether the process is batch, fed-batch or continuous, with or without enzyme recycle; (5) whether purified enzymes or whole cells are used. In general, purified enzymes would be used for continuous processes to avoid excessive cost associated with enzyme consumption; (6) the productivity required. Generally continuous processes provide much higher rates of substrate conversion than batch or fed-batch processes; however, this must be counter-balanced by the requirement for sufficient enzyme concentration and residence time in the reactor to allow complete conversion, while avoiding product accumulation to inhibitory levels.

14.3 Sugar Conversions

In everyday terminology, sugar refers to sucrose, also referred to as “table sugar”, “saccharose” or “dextrose”, and is mainly used to alter the flavor (sweetening) and properties (preservation, texture) of food and beverages. Commercially produced table sugar is derived from sugarcane or sugar beet. More specifically, the term sugar refers to carbohydrates which may be mono-, di- or poly-saccharides.

Monosaccharides are the simplest carbohydrates, with the general chemical structure $(CHO)_n$. They consist of a single sugar moiety, are generally colorless, water soluble and crystalline solids, some of them having a sweet taste. Monosaccharides are classified by the number of carbon atoms they contain, the major families being pentoses and hexoses. In the food industry, the most common monosaccharides are glucose, fructose and galactose, which are hexoses. Monosaccharides are the building blocks of disaccharides such as sucrose (common sugar) or lactose (milk sugar) and polysaccharides (such as cellulose and starch).

14.3.1 Glucose Production

Glucose, in crystalline form or in syrups, is largely required in the food industry to sweeten beverages (Wang et al. 2007) or to adjust the physico-chemical properties of food products, especially for confectionary. Commercially, glucose is generally produced through enzymatic hydrolysis of starch. Many crops are used as a source of starch, such as maize, rice, wheat and potato. The first step for starch hydrolysis is a heat treatment (1–2 h at 100°C), in the presence of amylase to allow starch solubilization and to lead to the formation of oligosaccharides containing 5–10 glucose units. After heating, a second enzyme is added to hydrolyze the oligosaccharides into glucose units. This second step is termed "saccharification" (Rebroš et al. 2006), and the most commonly used enzyme is glucoamylase.

Another possibility to produce glucose is through the enzymatic hydrolysis of sucrose to glucose and fructose through the action of invertase, a relatively expensive enzyme compared to amylases and glucoamylases (Altinok et al. 2006). Immobilization of amylase, glucoamylase and invertase enzymes has been shown to be essential for the economic production of glucose, particularly glucose syrups, since (1) the enzyme may be re-used a number of times or used continuously; (2) it is relatively easy to separate the product from the reaction medium; (3) it is relatively easy to recover the enzyme; and (4) there is generally an improvement in enzyme stability (Rebroš et al. 2006; Wang et al. 2007).

Industrial processes using such immobilized biosystems are generally characterized by lower capital and energy costs (Rebroš et al. 2006). Due to the low added value of glucose for use in the food industry, the minimization of costs is essential (Karaa et al. 2005). In the case of glucose production, the enzymes may be immobilized to increase process efficiency. Moreover, immobilization has been reported to lead to changes in the pH and temperature profiles of the enzymes and to enhance pH and thermo-stability (Brijak 2003). The methods generally used for immobilization of the enzymes, including adsorption, entrapment, chemical cross-linking and covalent attachment, are non-exhaustively listed in the Table 14.1.

Table 14.1 Enzyme immobilization techniques for the production of glucose

Technique	Principle	References
Adsorption	Montmorillonite clay support	Sanjay and Sugunan (2007)
	Zirconia support	Reshmi et al. (2007)
	Alumina support	Reshmi et al. (2006)
Entrapment	Calcium-alginate beads	Ertan et al. (2007) Konsoula and Liakopoulou-Kyriakides (2006) Dey et al. (2003)
	Chitosan-alginate	Sankalia et al. (2007)
	Acrylated epoxidized soybean resin	Kahraman et al. (2006)
	Polyacrylamaide gels	Raviyan et al. (2003)
	Agar gels	Stefanova et al. (1998)
Cross-linking	Cross-linking with glutaraldehyde	Hamerska-Dudra et al. (2007)
	Chemically activated poly (2-hydroxyethyl methacrylate) beads	Altinok et al. (2006)
	Bonding between enzyme ester groups and *N*-(acryloxy)-succinimide	Chen et al. (1998)
	Amide bonds between amino groups of protein and phtaloyl chloride functionalized glass beads	Kahraman et al. (2007)
	Condensation of the epoxy groups of glycidyl methacrylate with amino groups of enzyme	Danisman et al. (2004)

Adsorption is a relatively easy and fast process; however, it often results in significant enzyme leaching and loss. This may be overcome by covalently linking the enzyme and support. However, care must be taken to achieve high enzyme loading without significant loss of activity.

The main problem with enzyme entrapment in a matrix is due to diffusional limitations, mainly diffusion of substrate to the enzyme, or of the product from the matrix to the surrounding. This results from the hydrolytic enzymes being subject to feedback (competitive) inhibition by the products (glucose) of the reaction. Thus, while entrapment allows for protection of the enzymes, allowing constant activity over long periods of time, it is essential to use small microparticle sizes with excellent fluid dynamics within the reactor to achieve high conversion rates.

Thus, the nature of the polymer matrix is generally less important than the requirement for small, monodisperse microencapsulates which are mechanically robust, such that they can be employed in a wide range of reactors such as stirred-tank, packed- and fluidized-beds at high liquid velocities. The matrix must provide the chemically defined support necessary to adsorb or immobilize the enzyme to achieve the desired loading density (specific activity, units/gram or volume of microencapsulate), while providing a high porosity and low mass transfer resistance (high rates of diffusion of substrates and products). Adsorption onto inorganic support matrices (Table 14.1) generally provides lower enzyme loadings than other techniques, due to the limited surface available, but has the advantage of having a high specific density which allows them to be used in a wide range of reactors due to their physical stability (Sanjay and Sugunan 2007; Reshmi et al. 2007). Enzyme desorption can be a problem when used for continuous processes, and the presence of enzyme on the surface of the particles makes them susceptible to shear forces. Entrapment can have the advantage of physically retaining the enzyme within a defined porosity matrix; however, it can be difficult to produce microencapsulates with high enzyme loadings due to the exodiffusion of the enzyme from encapsulates during the production process (Ertan et al. 2007). For example, when using alginate, the enzyme would be mixed with the sodium alginate solution and extruded into a calcium chloride bath. The rate of calcium complexation, and therefore gelation, of the alginate is rapid but even so >50% of the enzyme added will exodiffuse and not be retained within the calcium alginate microencapsulates. Another difficulty is related to the mechanical properties since hydrogels deform and change properties and porosity as a function of pH and ionic strength, resulting in abrasion and increased porosity, and leading to enzyme loss.

Chemical cross-linking to microencapsulates is best suited for long-term mechanical strength, such as required for continuous processes, with high enzyme loading (Hamerska-Dudra et al. 2007; Chen et al. 1998). These have the lowest loss of enzyme through exodiffusion, providing that a suitable immobilization chemistry can be found which minimizes activity loss, through enzyme denaturation.

14.3.2 Fructose Production

Fructose, an isomer of glucose, has a sweet taste similar to cane sugar, but with a "fruity aroma". It is naturally found in many foods, such as fruits, honey, some vegetables (beets, sweet potatoes, onions, etc.). Fructose is the sweetest naturally occurring sugar, and is estimated to be twice as sweet as glucose, and 50% more than sucrose. This sugar has the property to enhance flavor, color and product stability, and is thus widely used instead of sucrose for the sweetening of foods and beverages. Moreover, fructose has a Generally Recognized As Safe (GRAS) status, and is thus a safe alternative sweetener, which is of great advantage to the food industry (Singh et al. 2007), particularly due to its beneficial role for diabetics and its low cariogenic nature. As a result the demand is ever increasing, and efficient production requires development.

There are two principle processes for fructose production. The first involves the enzymatic hydrolysis of sucrose to glucose and fructose through the action of invertase, followed by the isomerization of glucose into fructose using glucose isomerase (Singh et al. 2007). The latter is the final step in the production of High Fructose Syrups, which is traditionally performed in packed-bed reactors using immobilized glucose isomerase (Brijak 2003; Salehi et al. 2005). Once again the main advantages of using immobilized glucose amylase are that they allow reuse of the enzyme and continuous operation and while avoiding enzyme contamination of the processed product. Moreover, the pH and temperature dependence of immobilized glucose isomerase is lowered, leading to more stable and controlled processes.

The second common way to produce fructose is through the degradation of inulin, a reserve carbohydrate found in the roots and tubers of many plants. Inulin is a polyfructan consisting of linear chains of β-2,1-linked D-fructofuranose terminating at the reducing end with a glucose residue. The enzymatic degradation of inulin molecules by inulinase allows the production of syrups containing more than 75% D-fructose (Nakamura et al. 1995). To allow the continuous production of fructose from inulin, the use of inulinase in an immobilized form in packed-bed column reactors is used. As with processes using amylases and glucoamylases, the immobilization methods are numerous and based on similar techniques, such as adsorption on a macroporous support (Gupta et al. 1992), cross-linking or covalent bond formation between the support and the enzyme (Brijak 2003; Nakamura et al. 1995; Singh et al. 2007), or entrapment in a polymer matrix (Gill et al. 2006).

14.3.3 Lactose Production

The main carbohydrate present in milk is lactose, a disaccharide composed of one galactose and one glucose unit. Lactose is poorly soluble, and not very sweet, and is also not easily digested by a significant fraction of the population. Lactose intolerance is caused by low levels of β-D-galactosidase (commonly named lactase) in the digestive system, and the consequent inability to digest this disaccharide.

Lactase is used to hydrolyze the lactose present in milk, in order to obtain glucose and galactose, which may find many applications in the food industry. Lactose is known to readily crystallize, particularly at concentrations exceeding 20–25%, which limits certain processes in the dairy industry. As a consequence, the treatment of milk (or milk products) with this enzyme reduces the lactose content, thereby avoiding the solubility and/or crystallization problems, while simultaneously increasing the sweetness. Moreover, it is known that cheese ripening is accelerated using hydrolyzed milk (Panesar et al. 2006). Finally, treatment of milk with lactase enables consumption of milk products by lactose-intolerant people. Lactose in milk and cheese whey may also be hydrolyzed using the same enzyme, thus converting whey into a sweet syrup which can then be used in the confectionary, dairy, baking and soft drink industries, in a similar way to glucose and fructose syrups obtained from sucrose or starch. Hydrolysis of whey lactose has the added advantage of converting a low-grade waste product from the dairy industry, having a BOD in the range 30,000–50,000 mg/L to a higher added value product. More recently, the enzymatic hydrolysis of whey lactose has also been proposed for the production of bioethanol.

β-D-galactosidase (lactase) can be obtained from a wide variety of sources, such as microorganisms, plants and animals. Generally, microorganisms are preferred, since they are easier to handle, and offer a higher production yield. Commercially available lactases generally originate from yeasts, fungi or bacteria (Mahoney 1997), with the optimal conditions of enzyme activity highly dependant on the type of microorganism. The difficulty and expense related to the release of active enzyme in high yield from cells and the cost of purification are major constraints affecting further industrial applications.

To overcome this problem, immobilization and encapsulation offer several ways to improve the process economics: (1) by immobilization of cells producing lactase in a bioreactor to achieve higher enzyme yields; (2) by immobilization of lactase for the hydrolysis of lactose to high yields and at high rates; (3) by immobilization of whole cells (containing lactase) to directly achieve lactose hydrolysis at reduced costs, compared to the use of purified enzyme.

For example, in the simplest process, β-D-galactosidase is added directly to whole milk, and when the hydrolysis of lactose has attained the desired level, the enzyme is deactivated by heat treatment. Since the enzyme is denatured and cannot be re-used, the resulting operating costs are very high. One solution is the hydrolysis of skimmed milk with immobilized lactase followed by the addition of cream to adjust the fat content (Hirohara et al. 1981).

Lactase and cells producing this enzyme have been immobilized by various methods, such as gel entrapment, physical absorption and covalent attachment on various supports. Successful applications based on physical adsorption of lactase (or cells producing lactase) have been designed, using different kinds of carrier, such as cotton cloth (Sharma and Yamazaki 1984), chitosan (Champluvier et al. 1988), polyvinyl chloride (Bakken and Hill 1990) and silica (Bakken and Hill 1992).

The advantage of immobilizing whole cells instead of pure lactase for the direct hydrolysis of lactose is that enzyme activity is conserved, and beads are mechanically stable and protect cells in the reactor where the reaction takes place. This is generally achieved through entrapment of microorganisms in a gel matrix.

β-D-galactosidases can be used in many different applications in order to hydrolyze lactose in milk, whey and whey permeate. The choice of the technology, immobilization method, support, enzyme or cell strain depends on the conditions (pH, temperature, continuous vs. batch) under which the process is undertaken and is similar to that described for glucose production in Sect. 14.3.1.

14.4 Use of Immobilized Cells for the Production of Organic Acids

14.4.1 Lactic Acid

Lactic acid is extensively used in the food industry as a preservative, antioxidant or acidifier. In recent years, it became interesting as a primary material for the production of biodegradable plastics. Although it is possible to produce by chemical synthesis, over 95% of the world lactic acid production is through biotechnological processes. The use of microorganisms for the production of lactic acid has the advantage that stereochemically pure D- or L-lactic acid may be produced. Lactic acid is readily produced using lactic acid bacteria, which readily ferment lactose present in significant amounts in whey, a by-product of the cheese industry. The use of immobilized *Lactobacillus casei*, *Lactobacillus delbrueckii* and *Rhizopus oryzae* for lactic fermentation has been reported.

Lactobacillus casei, immobilized in calcium-pectate beads (Panesar et al., 2007), or PEI coated porous glass (Poraver) beads (Senthuran et al. 1999), was used for fermentation of whey or enriched whey permeates. Due to the good stability at low pH, calcium-pectate gels can be used as a carrier for lactic acid production over prolonged periods of time. In a laboratory, batch reactor up to 33 g/L of lactate was produced with 95% yield over 30 h per batch. As a result of the high stability of the beads, up to 16 batch cultures could be performed, with virtually no loss of conversion yields or rates (Panesar et al. 2007).

Lactic acid production by *Lactobacillus casei* immobilized by adsorption on Poraver beads has been reported using an external column reactor. Both freely suspended cells and adsorbed cells in the external column are active and contribute to the bioconversion of lactose. The high volumetric productivity was shown to be the direct result of the action of the immobilized cells, rather than an eventual increase in suspended cell density compared to submerged cultures. Lactate concentrations of up to 90 g/L were attained using this system, during a fermentation time of about 30 h (Senthuran et al. 1999).

Lactobacillus delbrueckii can be used for lactic acid production from glucose. Cells entrapped in calcium-alginate beads have been used for the conversion of corn cob residue, a cheap and abundant residue from xylose production. This residue is a porous cellulosic material that can be hydrolyzed enzymatically to yield a glucose solution. Addition of wheat bran hydrolyzate improves culture performance. After 48 h of batch culture up to 49 g/L of lactate with 95% conversion could be achieved, and 12 successive batches were possible. In a continuous culture with immobilized cells an increase in dilution rate increased the productivity, but reduced the yield and the residual concentration. It was suggested that a compromise dilution rate of 0.13 h^{-1} could be used to give a 92.4% yield and 44.2 g/L lactate concentration. A simultaneous saccharification and fermentation process can be used to avoid the feedback inhibition of enzymes used for cellulose hydrolysis. Enzymatic breakdown of cellulose is compatible with the lactic acid fermentation in terms of pH and temperature, and authors propose two coupled reactors with re-circulation. The hydrolysis product is pumped from the enzymatic reactor through a filter to the immobilized cell reactor for the lactic acid fermentation. In this way 55.6 g/L lactic acid concentration is achieved in 60 h, with a yield from cellulose of 91%. This could be improved by performing a fed-batch fermentation involving multiple additions of cellulose substrate and hydrolytic enzymes. In this way, 108 g/L lactate could be produced in 90 h with 88% yield from cellulose.

The use of another cheap cellulosic material, called loofa sponge or vegetable sponge, as a support for the immobilization of filamentous fungi for lactic acid production has been described. The material is cut into pieces, washed and autoclaved, then used for the immobilization of *Rhizopus oryzae*. This is achieved by inoculating the reactor with spores and cultivating the mycelium during 24 h. The fermentation was conducted in shake flasks. Growth on enriched rice starch hydrolysate medium yielded 81 g/L of product in 10 h. The immobilized mycelium could be reused several times, with the reactor performance falling from cycle to cycle. After ten cycles, volumetric productivity fell by half, and the final lactic acid concentration to about 20 g/L (Ganguly et al. 2007).

14.4.2 Citric Acid

Citric acid is a common food additive especially in soft drinks, mainly as an acidifier, but also as a preservative and an antioxidant. It is often used as part of a formula for effervescent tablets. Nowadays it is produced in biotechnological processes, most frequently using *Aspergillus niger* strains, although *Yarrowia lypolitica* has been identified as a potential producer of citric acid in high concentrations from glycerol, a by-product from biodiesel production.

Citric acid production using *Aspergillus niger* entrapped in calcium-alginate beads (Bayraktar and Mehmetoglu 2000; Demirel et al. 2005), polyacrylamide gel (Horitsu et al. 1985; Wang and Liu 1996), adsorbed on cellulose materials (Sakurai et al. 1997; Sankpal et al. 2001; Tisnadjaja et al. 1996), or adsorbed on polyurethane (Lee et al. 1989; Ricciardi et al. 1997; Wang 2000) has been reported. Alginate entrapment of fungi for citric acid production is normally done by suspending conidiaspores in Na-alginate and extruding this suspension into a $CaCl_2$ solution

(Bayraktar and Mehmetoglu 2000; Demirel et al. 2005; Gupta and Sharma 1994; Hamamci et al. 1988; Tsay and To 1987). Fermentation using entrapped filamentous fungi proved to be difficult because of filamentous growth and clogging of the pores of the gel, but also because of the physical damage to the beads caused by the growth (Demirel et al. 2005). In addition citric acid is often used as a dissolving agent for alginate gels as a result of its ability to form stable chelates with Ca^{2+} ions, which results in reduction of bead stability during the fermentation. Addition of silicone oil to the fermentation enhances O_2 supply thus increasing productivity, but the re-usability of alginate immobilized mycelia is reduced even further (Ates et al. 2002). Generally speaking, very low productivity and citric acid concentrations could be achieved using alginate beads with 2–3 g/L formed after 4 days, with practically no re-usability of the beads (Bayraktar and Mehmetoglu, 2000; Demirel et al. 2005).

Gupta and Sharma (1994) described a double horizontal reactor with combined submerged Ca-alginate immobilized and surface stabilized *A. niger* cells. The set-up allowed very good productivities and yields on sugarcane molasses but, due to complex operation, was not implemented industrially.

The use of a rotating disk reactor with *A. niger* immobilized on polyurethane foam disks allowed eight batches with no loss of activity (Wang 2000), and production of up to 85 g/L in 96 h with about 72% product yield.

Sankpal et al. (2001) investigated different set-ups and modes of operation of bioreactors containing *A. niger* mycelium immobilized on a woven cotton or in submerged culture. Immobilized cell culture was performed in a hanging sheet reactor with a spirally wound woven cotton placed in a cylindrical chamber with a constant supply of humidified air, and the nutritive medium dripped on the top of the mycelium support. The nutritive medium used in the experiments was either rich medium containing sucrose as the main carbon source or sugar cane juice. Immobilization increased the volumetric productivity and yield as compared to submerged cultures. Continuous mode of operation increased productivity, even further while maintaining the high citric acid yield. Four repeated batches with no reduction of performance yielded 60 g/L citric acid in each 9 h cycle. However, as biomass increases, the specific productivity is reduced. When operated in continuous mode, a residence time of 20 h resulted in nearly 100% conversion of the 50 g/L sucrose, to yield 65 g/L citrate in the exiting medium (the product is somewhat concentrated due to evaporative concentration). Sugar cane medium can be used, but some technical difficulties related to irregular and thick mycelium growth causing mass transfer limitations were noted (Sankpal et al. 2001).

Due to its ability to convert both galactose and glucose to citric acid, the yeast *Candida guilliermondii* has been used for the production of citric acid from whey hydrolyzates. Tisnadjaja et al. (1996) demonstrated the use of sawdust as a cheap material for adsorption of yeast cells and operation in a bubble column reactor. They used rich medium with glucose as carbon source, and attained much higher volumetric productivities, but lower yields and specific productivity, compared to suspension cultures. This may have been the result of a low culture pH. The final citrate concentration was lower than 10 g/L.

Adsorption on synthetic porous cellulose beads, and cultivation on rich medium containing glucose in submerged culture in shake flasks resulted in 60 g/L citric acid, but only 50% yield. In addition, re-usability was poor, with less than half the citric acid concentration achieved after the second of two consecutive 8-day batch runs (Sakurai et al. 1997).

14.4.3 Gluconic Acid

Gluconic acid is a product of the oxidation of glucose, which is used extensively as a pH regulator in food industry. *Aspergillus niger*, grown on low nitrogen-containing medium, has been reported to convert glucose to gluconic acid. Sankpal et al. (1999) described a system for continuous gluconic acid production using hanging sheet reactors with the mycelium immobilized on a woven cotton cloth hung in a chamber with controlled humidity, temperature and air flow. The defined culture medium, containing glucose as carbon source, was dripped onto the top of the cloth. A tenfold increase in productivity, compared to a submerged culture, with a final product concentration of up to 125–140 g/L, was reported. Falling conversion rates and yield could be partially overcome by increasing the residence time progressively. It was noted that when aerated, mycelium growth was uneven resulting in an uneven nutrient and air supply and severe reduction of the overall volumetric productivity. If a pure oxygen stream was used instead of air, mycelium grows evenly and continuous operation of up to 60 days was possible. The inherent problem of pH, concentration and temperature gradients, which might reduce reactor performance, could be solved by careful process design and optimization.

14.4.4 Acetic Acid and Other Carboxylic Acids

Acetic acid is one of the most important organic acids with broad use in the food industry, as acidifying additive and/or preservative. Although most of the market demand for acetic acid is satisfied by chemical synthesis, all of the acetic acid used in the food industry must be of biological origin and is produced using acetogenic bacteria. Of the other major carboxylic acids, propionic acid is used as a food preservative in the form of sodium salt, while butyric acid is used as a flavoring additive, either in free or in ester forms.

The laboratory-scale production of acetic acid using entrapped *Acetobacter aceti* cells in calcium-alginate beads and cells adsorbed on cellulose beads has been reported (Krisch and Szajani 1996). After 70 h of culture on rich medium with glucose and ethanol as substrates, up to 30–34 g/L could be attained using both immobilization systems, which is comparable to free cell fermentations. It was also shown that immobilization improved culture tolerance to suboptimal temperatures and pH, although no data were reported for long-term operation.

The anaerobic homofermentative bacterium *Clostridium formicoaceticum* is capable of converting fructose from corn syrup and other fructose containing substrates to acetic acid in high yields and with good specific productivity. External fibrous-bed columns containing spirally wound terry cloth were used in an experimental set-up for adsorption of *Clostridium formicoaceticum*. A comparison of batch, fed-batch and continuous modes of operation revealed that cell immobilization gave higher yields, volumetric productivities and acetate concentrations. Up to 78 g/L of acetate could be achieved with fed-batch fermentations using such immobilized cells (Huang et al. 1998).

An external column reactor set-up was also used for a mixed-culture fermentation of whey permeate using *Clostridium thermolacticum* and *Clostridium thermoautotrophicum* co-immobilized by adsorption on cotton fabric (Talabardon et al. 2000). Being a heterofermentative bacterium, *Clostridium thermolacticum* is capable of converting lactose almost entirely to lactic acid under growth-limiting conditions. *Clostridium thermoautothropicum* is a homoacetogen that can assimilate lactate and produce acetic acid. The yields and productivity of the co-immobilized co-culture were relatively low with only 25 g/L acetate produced. The authors suggest that process optimization might improve performance, with the use of cheap substrates rendering the process economically viable even with lower yields. Both of the microorganisms used are sporogenous bacteria, which means that, theoretically, the process could be stopped and re-continued depending on the availability of whey supply (Talabardon et al. 2000). A similarly conceived co-culture of homolactic *Lactobacillus lactis* and homoacetogenic *Clostridium formicoaceticum* has been used for the conversion of whey lactose or corn meal hydrolyzate to acetic acid with almost 100% yield and 76 g/L of product (Huang et al. 2002).

Huang et al. (2002) used a fibrous-bed reactor for the fermentation of hydrolyzed corn meal for the production of acetic, propionic and butyric acids. *P. acidipropionici* was used for propionic acid production, and *Clostridium tyrobutylicum* for butyric acid production. Propionic acid could be produced on enriched whey, glucose medium and corn syrup hydrolyzate, with immobilization increasing volumetric productivity, while maintaining the yield resulting in up to 40 g/L propionate. Butyric acid could also be produced on glucose medium and pentose-containing corn meal hydrolyzate with immobilized *Clostridium tyrobutylicum* giving higher yields compared to free cell fermentations, with significantly increased volumetric productivity, reactor performance and with a final concentration of up to 45 g/L (Huang et al. 2002; Zhu et al. 2002). All three processes could be operated continuously for several months, even with corn meal hydrolyzate, which has a high content of insoluble material. No column fouling was observed, probably as a result of the way that the cotton fabric was packed, with sufficient space between the winding layers for the insoluble particles to pass through.

14.5 Amino Acids

Amino acids are the building blocks of proteins, being one of the three largest groups of nutrients. The daily nutritional requirement for amino acids is met by the intake of proteins which are hydrolyzed during digestion. However, some amino acids have important applications in the food industry, such as the artificial sweetener aspartame (aspartyl-phenylalanine-1-methyl ester) which is produced from aspartic acid and phenylalanine. Sodium glutamate, or mixtures of amino acids, is often used as flavor enhancers. Depending on the use and on the particular amino acid, production of amino acids is normally done by (1) enzymatic conversion of precursors to the desired product, for example, L-Asp production from fumaric acid using the aspartase from *Escherichia coli*; (2) hydrolysis of readily available and cheap proteins followed by extraction of the desired amino acid; (3) fermentation methods; or (4) chemical methods (DL-methionine is produced by chemical synthesis from raw materials such as acrolein, hydrocyanic acid and methyl mercaptan).

14.5.1 Production of Amino Acids Using Immobilized Whole Cells

Chao et al. (1999) immobilized recombinant *Escherichia coli,* which overproduces aspartase, aspartate aminotransferase or aromatic amino acid aminotransferase, in calcium-alginate beads and used these immobilized cells for the simultaneous production of phenylalanine and aspartic acid from phenylpyruvic acid, fumaric acid and ammonia. The beads were treated with glutaraldehyde and hexamethylenediamine in order to enhance their mechanical resistance. Fermentation in shake flasks resulted in product yields that were comparable to those of free cell cultures though a reduced specific productivity was observed, indicating mass transfer limitations or loss of viability. Immobilized preparations could be re-used for up to six cycles with virtually no activity loss for either of the products (Chao et al. 1999). Longer duration cultures led to rupture of the beads. However, long time operation using recombinant organisms was also limited by the genetic instability of the recombinant strain.

Recombinant *Escherichia coli* expressing transaminase, for the production of L-Phe from phenylpyruvic acid, was immobilized in κ-carrageenan beads as a whole cell catalyst for the production of L-phenylalanine from L-aspartic acid and phenylpyruvate (Leng et al. 2006). The treatment of cells with 0.05% glutaraldehyde prior to immobilization enhanced enzyme activity and operational stability. The gel cubes, containing immobilized cells, were used in packed-bed reactors operating in both batch and continuous modes. A conversion yield of about 90% was obtained in batch mode, falling to 88% after ten repeated batches with the same immobilized cells. When operated in continuous mode with a residence time of 10 h, 20.6 g/L of product could be obtained. After 40 days of continuous operation the conversion was still higher than 88%. A mutant Pseudomonas strain, selected for low L-cystein desulfhydrase activity, was entrapped in calcium-alginate beads and used to produce L-cystein by the enzymatic conversion of the chemically synthesized precursor, DL-2-amino thiazoline 4-carboxylic acid. The presence of sorbitol in the feed increased the stability of the enzymes. The developed process comprised two packed-bead columns in series, followed by an ion-exchange column for separation of L-cystein. The residual L-cystein degrading enzyme was responsible for 23% product loss. In the presence of 40% sorbitol in the feed at 37°C, 60% of the conversion was maintained after 1,000 h of continuous cultivation (Ryu et al. 1997).

Calcium-alginate bead entrapment of a *Pseudomonas* sp. which expresses aminoacylase was used for the production of L-alanine by the enzymatic, stereospecific deacetylation of a chemically produced an *N*-acyle derivative. The production process must be performed under non-growing conditions otherwise cells used the *N*-acyle alanine for biomass formation, thus reducing the product yield. Cells were immobilized in κ-carrageenan beads (Santoyo et al. 1998), or calcium-alginate beads cross-linked with glutaraldehyde (Hernandez et al. 2001; Santoyo et al. 1996). Total conversion was attained after approximately 5 h with up to 4 g/L of L-alanine produced in batch mode using κ-carrageenan beads. However, less than 6 h of continuous operation was possible with κ-carrageenan beads because of physical damage to the beads. Experiments with alginate showed lower conversion, probably because the process was not optimized. Continuous cultures with alginate beads were possible; however, after 48 h only 82% of the activity was conserved due to loss of enzyme activity, rather than cell leakage or damage to the beads. Pretreatment of the alginate beads with up to 20 mM glutaraldehyde solution increased the half-life of the enzymatic activity, and even resulted in increased activity, while higher glutaraldehyde concentrations were found to inhibit the enzyme.

The possibility to use cyanobacteria (a mutant of *Anabaena* Sp-287) immobilized in calcium alginate beads for the overproduction of D-Ala and L-His has also been investigated (Thomas and Shanmugasundaram 1992). The use of microorganisms capable of fixing atmospheric nitrogen and releasing either NH_3 or amino acids would be expected to make the process more cost effective. However, technical difficulties, such as correct illumination, still hinder the use of photosynthetic microorganisms in industrial biotechnology.

Tryptophan production from serine and indole using whole recombinant tryptophanase-overproducing *Escherichia coli* cells immobilized in κ-carrageenan beads has been described (Mateus et al. 2004). The reaction was conducted in a simple stirred vessel and up to 10 mM L-tryptophan could be obtained in about 2 h, with almost 100% conversion.

14.5.2 Amino Acid Production Using Immobilized Enzymes

N-acetyl-amino acid racemase and L- or D-aminoacylase can be used for the separation of amino acid stereoisomers. Chemically synthesized *N*-acetyl derivatives of an amino acid can be stereoselectively deacetylated and the produced L- or D-amino acid readily separated from the precursor using standard chromatographic methods. The remaining *N*-acetyl derivative is racemized using the racemase enzyme, and the process is repeated. *N*-acetyl-amino acid racemase and aminoacylase have been immobilized on a number of supports (Tokuyama and Hatano 1996; Tsai et al. 1992). Very good yields and operational stabilities were obtained with these immobilized preparations, with half-lives of 10–30 days and the initial yield of close to 100%.

Bodalo-Santoyo et al. (1999) described a reactor with fluidized- and packed-beds, containing enzymes immobilized by covalent linkage on controlled pore glyceryl glass for separation of D,L-valine from chemically produced *N*-acyl valine, using enantio-specific hydrolase, the unhydrolyzed D-isomer chemically racemized. The system could be operated in packed-bed and fluidized-bed modes, although conversion dropped after 10 h.

Oikawa et al. (1999) described a system with two packed-bed columns containing immobilized enzymes (glutamate racemase and L-glutamate oxidase) for the production of D-glutamate from L-glutamate. The enzymes were immobilized on Chitopearl 2505 and 2506 by adsorption followed by glutaraldehyde cross-linking. The process was divided into two phases (and two columns) such that both enzymes could operate under optimal conditions. As a result there was negligible loss of activity after 10 days of continuous operation (Oikawa et al. 1999).

Co-immobilization of D-amino acid oxidase and catalase on glutaraldehyde-agarose beads for the production of α-ketoglutaric acid (AKG) from D,L-alanine, or for separation of alanine enantiomers have been described (Buto et al. 1994; Fernandez-Lafuente et al. 1998). Catalase was used to remove the H_2O_2 and prevent oxidative damage to the system. A maximum AKG yield of 86% was obtained, due to part of the AKG formed being converted to the oxidation product phenyl acetic acid. Immobilization clearly increased the stability of the enzymes, with virtually no loss in activity after re-using the beads 20 times. Proteases may be immobilized in agarose gels (Pedroche et al. 2006, 2004) for protein digestion and production of hydrolyzates for food supplements and for medical purposes.

14.6 Hydrolysis of Triglycerides

14.6.1 Lipases in Food and Flavor Making

As described previously, the use of enzymes to improve traditional chemical processes for food manufacture has been extensively studied and developed in the past few years, due to the technological and economical advantage they offer. Most lipase-mediated food processes embody biocatalysis of lipase, especially fats and oil (Jachmanian et al. 1997). During storage, important changes arise from the hydrolysis of triglycerides catalyzed by lipases, leading to the formation of fatty acids. Esters of short chain fatty acids are extremely important aroma compounds, and are hence important compounds for the characteristic flavor of products. Moreover, it is often desirable to alter the fatty acid composition of naturally occurring triglycerides in fat with the required melting characteristics.

The modification of fats and other lipids may be achieved through (1) hydrolysis, (2) esterification, (3) inter or trans-esterification. All of these reactions are catalyzed by lipases (Wisdom et al. 1984). Inter-esterification processes consist in promoting an uncontrolled migration of fatty acid groups between the triglyceride molecules (Mukherjee and Kiewitt 1996). This process allows the formation of fats suitable for use in emulsions, margarine, artificial creams and ice-creams (Benjamin and Pandey 1998). A known application of this technology is the production of a cocoa-butter substitute from palm oil, where palmitoyl moieties are replaced by stearoyl groups (Katchalski-Katzir 1993). The esterification process allows the formation of short and medium chain esters, which are known to be flavor molecules, which are of great use for food flavoring.

Fermentation is induced by microorganisms, which also have lipolytic activity. As a consequence fermented food from fruits, vegetable, cereals, oilseed and meat is also affected by this process. The addition of lipase to these preparations can also be an alternative to alter the taste of these products, especially in terms of acidity (Katchalski-Katzir 1993).

The cost of microbial production of lipase is often prohibitive. Hence, processes that do not require the presence of lipase in the final product are more economic if the lipase is immobilized, and therefore re-cycled a number of times. The choice of the support for lipase immobilization is of central importance since it determines the enzymatic activity. Since lipase acts on organic products, the types of support are different than for reactions taking place in aqueous conditions. Several applications use diatomaceous earth (diatomite) as a support for lipase immobilization, while others use porous silica-based supports (Wisdom et al. 1984). The method used to immobilize the enzyme usually involves precipitation of the lipase from solution onto the support surface. Precipitation is induced by addition of chilled acetone as precipitant. Thus the enzyme remains associated with the support and is unable to dissolve in either the organic solvent, water or aqueous solution. The immobilized enzyme can then be re-used, allowing the process to run continuously, which significantly decreases the production costs.

14.7 Conclusions

Immobilization of both cells and enzymes is an important tool in the food industry for carrying out specific bioconversions and for the formation of metabolites. Indeed as a result of the increasing need for so-called – white and green biotechnologies – increasing numbers of bioconversions and/or molecular synthesis in the food industry are carried out using natural biocatalysts. In the case of products such as yoghurt, cheese and the like, whole microbial cells may be added to the food material during processing, since they are ultimately found in the product. On the other hand, many situations arise where purified enzymes are used to carry out specific reactions within the process. Since enzymes are relatively expensive and by their nature subject to loss of activity and denaturation under many of the environmental conditions found in food processing, the ability to encapsulate the enzymes has resulted in a significant stimulation to their use. This is due to encapsulated enzymes enabling (1) the design of continuous processes with increased productivity; (2) re-use of the enzymes over many weeks or indeed months, resulting in lower enzyme consumption; (3) increased stability and activity of enzymes for bioconversions, which allows extended use; (4) higher enzyme and cell densities which result in the maintenance of viability/activity for metabolite production; (5) physical protection of the enzymes and cells from mechanical shear forces found in reactor conditions.

In order to achieve this, enzymes and cells may be adsorbed onto the surface of particles, where the particles have a higher specific density than hydrogels, and thereby may be used in fluidized-bed systems, and where the loading density is limited by the available surface area. In cases where the enzymes have a relatively large molecular weight, or where whole cells are used, and require protection from the denaturing forces found under the process conditions used, then entrapment in a range of polymers is most suitable. However, these polymers have relatively high porosity and low mechanical stability and cannot be used in packed-bed operation at high flow rates or where enzyme leaching is a problem. Where significant stability is required it is essential to covalently cross-link the enzymes to the polymers in such a way as to retain activity over extended periods. However, the range of polymers, reactor types and immobilization methods and chemistries means that it is now possible to tailor the encapsulation technique to almost any process in order to get the optimum activity, stability and cost.

References

Altinok H, Aksoy S, Tumturk H, Hasirci N (2006) Covalent immobilization of invertase on chemically activated poly(2-hydroxyethyl methacrylate) microbeads. Russ Chem Bull 66:1860–1864

Ates S, Dingil N, Bayraktar E, Mehmetoglu U (2002) Enhancement of citric acid production by immobilized and freely suspended Aspergillus niger using silicone oil. Process Biochem 38:433–436

Bakken A, Hill CJ (1990) Use of a novel immobilized beta galactosidase reactor to hydrolyze the lactose constituent of skim milk. Biotechnol Bioeng 36:293–309

Bakken A, Hill CJ (1992) Hydrolysis of lactose in skim milk by immobilized beta galactosidase. Biotechnol Bioeng 39:408–417

Bayraktar E, Mehmetoglu U (2000) Production of citric acid using immobilized conidia of Aspergillus niger. Appl Biochem Biotechnol 87:117–125

Benjamin S, Pandey A (1998) Candida rugosa lipases: Molecular biology and versatility in biotechnology. Yeast 14:1069–1087

Bodalo-Santoyo A, Gomez-Carrasco JL, Gomez-Gomez E, Bastida-Rodriquez J, Maximo-Martin MF, Hidalgo-Montesinos AM (1999) Production of optically pure L-valine in fluidized and packed bed reactors with immobilized L-aminoacylase. J Chem Technol Biotechnol 74:403–408

Brijak J (2003) Alpha-amylase and beta-amylase immobilisation on acrylic carriers. Biochem Eng J 16:347–355

Buto S, Pollegioni L, Dangiuro L, Pilone MS (1994) Evaluation of D-amino-acid oxidase from Rhodotorula gracilis for the production of alpha-keto acids – a reactor system. Biotechnol Bioeng 44:1288–1294

Champluvier B, Kamp B, Rouxhet P (1988) Immobilization of beta galactosidase retained in yeast: Adhesion of the cells on a support. Appl Microbiol Biotechnol 27:464–469

Chao YP, Fu HY, Lo TE, Chen PT, Wang JJ (1999) One-step production of D-*p*-hydroxyphenylglycine by recombinant Escherichia coli strains. Biotechnol Prog 15:1039–1045

Chen JP, Sun YM, Chu DH (1998) Immobilization of alpha-amylase to a composite temperature-sensitive membrane for starch hydrolysis. Biotechnol Prog 14:473–478

Danisman T, Tan S, Kacar Y, Ergene A (2004) Covalent immobilization of invertase on microporous pHEMA-GMA membrane. Food Chem 85:431–466

Demirel G, Yaykasli KO, Yasar A (2005) The production of citric acid by using immobilized Aspergillus niger A-9 and investigation of its various effects. Food Chem 89:393–396

Dey G, Singh B, Banerjee R (2003) Immobilization of alpha-amylase produced by Bacillus circulans GRS 313. Braz Arch Biol Technol 46:167–176

Ertan F, Yagar H, Balkan B (2007) Optimization of alpha-amylase immobilization in calcium alginate beads. Prep biochem biotechnol 37:195–204

Fernandez-Lafuente R, Rodriguez V, Guisan JM (1998) The coimmobilization of D-amino acid oxidase and catalase enables the quantitative transformation of D-amino acids (D-phenylalanine) into alpha-keto acids (phenylpyruvic acid). Enzyme Microb Technol 23:28–33

Ganguly R, Dwivedi P, Singh RP (2007) Production of lactic acid with loofa sponge immobilized Rhizopus oryzae RBU2-10. Bioresour Technol 98:1246–1251

Gill P, Manhas R, Singh P (2006) Hydrolysis of inulin by immobilized thermostable extracellular exoinulinase from Aspergillus fumigatus. J Food Eng 76:369–375

Gupta S, Kaur M, Kaur N, Singh RP (1992) A comparison of properties of inulases of Fusarium-oxysporum immobilized on various supports. J Chem Technol Biotechnol 53:293–296

Gupta S, Sharma CB (1994) Continuous production of citric-acid from sugarcane molasses using a combination of submerged immobilized and surface-stabilized cultures of Aspergillus-niger-kcu-520. Biotechnol Lett 16:599–604

Hamamci H, Gursoy G, Hang YD (1988) Production of citric-acid using immobilized dried reactivated Aspergillus-niger. Abstracts of Papers of the American Chemical Society 196, 92-Mbtd.

Hamerska-Dudra A, Bryjaka J, Trochimczuka A (2007) Immobilization of glucoamylase and trypsin on crosslinked thermosensitive carriers. Enzyme Microb Technol 41:197–204

Hernandez H, Asanza ML, Alcaraz I, Hidalgo AM, Maximo MF, Bastida J (2001) Production of optically pure L-alanine in batch and continuous stirred tank reactors using immobilized Pseudomonas sp. Biotechnol Lett 23:887–891

Hirohara H, Yamamoto H, Kawano E, Nagase T (1981) Continuous hydrolysis of lactose in skim milk and acid whey by immobilized lactase of Aspergillus oryzae. Enzyme Eng 6:295

Horitsu H, Adachi S, Takahashi Y, Kawai K, Kawano Y (1985) Production of citric-acid by Aspergillus-niger immobilized in polyacrylamide gels. Appl Microbiol Biotechnol 22:8–12

Huang YL, Mann K, Novak JM, Yang ST (1998) Acetic acid production from fructose by Clostridium formicoaceticum immobilized in a fibrous-bed bioreactor. Biotechnol Prog 14:800–806

Huang YL, Wu ZT, Zhang LK, Cheung CM, Yang ST (2002) Production of carboxylic acids from hydrolyzed corn meal by immobilized cell fermentation in a fibrous-bed bioreactor. Bioresour Technol 82:51–59

Jachmanian I, Schulte E, Mukherjee K (1997) Substrate selectivity in esterification of less common fatty acids catalysed by lipases from different sources. Appl Microbiol Biotechnol 44:563–567

Kahraman M, Bayramoglu G, Kayaman-Apohan N, Gungor A (2007) alpha-Amylase immobilization on functionalized glass beads by covalent attachment. Food Chem 104:1385–1392

Kahraman M, Kayaman-Apohan N, Ogan A, Gungor A (2006) Optimization of alpha-amylase immobilization in calcium alginate beads. J appl polym sci 100:4757–4761

Karaa A, Osmana B, Yavuzb H, Beşirlia N, Denizlib A (2005) Immobilization of alpha-amylase on Cu2+ chelated poly(ethylene glycol dimethacrylate-*n*-vinyl imidazole) matrix via adsorption. Reactive funct polym 62:61–68

Katchalski-Katzir E (1993) Immobilized enzyme-learning from past successes and failures. Trends Biotechnol 11:471–478

Konsoula Z, Liakopoulou-Kyriakides M (2006) Starch hydrolysis by the action of an entrapped in alginate capsules alpha-amylase from Bacillus subtilis. Process Biochem 41:343–349

Krisch J, Szajani B (1996) Effects of immobilization on biomass production and acetic acid fermentation of Acetobacter aceti as a function of temperature and pH. Biotechnol Lett 18:393–396

Lee YH, Lee CW, Chang HN (1989) Citric-acid production by Aspergillus-niger immobilized on polyurethane foam. Appl Microbiol Biotechnol 30:141–143

Leng Y, Zheng P, Sun ZH (2006) Continuous production of L-phenylalanine from phenylpyruvic acid and L-aspartic acid by immobilized recombinant Escherichia coli SW0209-52. Process Biochem 41:1669–1672

Mahoney R (1997) Lactose: Enzymatic modification. In: Fox F (ed) Adv Dairy chem, vol 3. Chapman & Hall, London, pp 77–125

Mateus DMR, Alves SS, Da Fonseca MMR (2004) Kinetics of L-tryptophan production from indole and L-serine catalyzed by whole cells with tryptophanase activity. J Biosci Bioeng 97:289–293

Mukherjee K, Kiewitt I (1996) Enrichment of very long chain mono-unsaturated fatty acids by lipase-catalysed hydrolysis and transesterification. Appl Microbiol Biotechnol 44:557–562

Nakamura T, Ogata A, Shitara A, Nakamura A, Ohta K (1995) Continuous production of fructose syrups from inulin by immobilized inulinase from Aspergillus-Niger Mutant-817. J Ferment eng 80:164–169

Oikawa T, Watanabe M, Makiura H, Kusakabe H, Yamade K, Soda K (1999) Production of D-glutamate from L-glutamate with glutamate racemase and L-glutamate oxidase. Biosci Biotechnol Biochem 63:2168–2173

Panesar P, Panesa RR, Singh R, Kennedy J, Kumar H (2006) Microbial production, immobilization and applications of β-D-galactosidase. J Chem Technol Biotechnol 81:530–543

Panesar PS, Kennedy JF, Knill CJ, Kosseva MR (2007) Applicability of pectate-entrapped Lactobacillus casei cells for L(+) lactic acid production from whey. Appl Microbiol Biotechnol 74:35–42

Pedroche J, Yust MD, Lqari H, Megias C, Giron-Calle J, Alaiz M, Vioque J, Millan F (2006) Production of Brassica carinata protein hydrolyzates with a high Fischer's ratio using immobilized proteases. J Agric Food Chem 54:7621–7627

Pedroche J, Yust MM, Lqari H, Giron-Calle J, Vioque J, Alaiz M, Millan F (2004) Production and characterization of casein hydrolysates with a high amino acid Fischer's ratio using immobilized proteases. Int Dairy J 14:527–533

Raviyan P, Tang J, Rasco B (2003) Thermal stability of alpha-amylase from Aspergillus oryzae entrapped in polyacrylamide gel. J Agric Food Chem 51:5462–5466

Rebroš M, Rosenberga M, Mlichováa Z, Krištofíkováa L, Palucha M (2006) A simple entrapment of glucoamylase into LentiKats® as an efficient catalyst for maltodextrin hydrolysis. Enzyme Microb Technol 39:800–804

Reshmi R, Sanjay G, Sugunan S (2006) Enhanced activity and stability of alpha-amylase immobilized on alumina. Catal Commun 7:460–465

Reshmi R, Sanjay G, Sugunan S (2007) Immobilization of alpha-amylase on zirconia: A heterogeneous biocatalyst for starch hydrolysis. Catal Commun 8:393–399

Ricciardi A, Parente E, Volpe F, Clementi F (1997) Citric acid production front glucose by Aspergillus niger immobilized in polyurethane foam. Ann Microbiol Enzimol 47:63–76

Ryu OH, Ju JY, Shin CS (1997) Continuous L-cysteine production using immobilized cell reactors and product extractors. Process Biochem 32:201–209

Sakurai A, Itoh M, Sakakibara M, Saito H, Fujita M (1997) Citric acid production by Aspergillus niger immobilised on porous cellulose beads. J Chem Technol Biotechnol 70:157–162

Salehi Z, Sohrabi M, Kaghazchi T, Bonakdarpour B (2005) Application of down flow jet loop bioreactors in implementation and kinetic determination of solid–liquid enzyme reactions. Process Biochem 40:2455–2460

Sanjay G, Sugunan S (2007) Enzymes immobilized on montmorillonite K 10: Effect of adsorption and grafting on the surface properties and the enzyme activity. Appl Clay Sci 35:62–75

Sankalia M, Mashru R, Sankalia J, Sutariya V (2007) Reversed chitosan-alginate polyelectrolyte complex for stability improvement of alpha-amylase: Optimization and physicochemical characterization. Eur J Pharm Biopharm 65:215–232

Sankpal NV, Joshi AP, Kulkarni BD (2001) Citric acid production by Aspergillus niger immobilized on cellulose microfibrils: Influence of morphology and fermenter conditions on productivity. Process Biochem 36:1129–1139

Sankpal NV, Joshi AP, Sutar II, Kulkarni BD (1999) Continuous production of gluconic acid by Aspergillus niger immobilized on a cellulosic support: Study of low pH fermentative behaviour of Aspergillus niger. Process Biochem 35:317–325

Santoyo AB, Rodriguez JB, Carrasco JLG, Gomez EG, Rojo IA, Teruel LMA (1996) Immobilization of Pseudomonas sp BA2 by entrapment in calcium alginate and its application for the production of L-alanine. Enzyme Microb Technol 19:176–180

Santoyo AB, Rodriguez JB, Carrasco JLG, Gomez EG, Rojo IA, Teruel MLA (1998) Production of optically pure L-alanine by immobilized Pseudomonas sp. BA2 cells. J Chem Technol Biotechnol 73:197–202

Senthuran A, Senthuran V, Hatti-Kaul R, Mattiasson B (1999) Lactic acid production by immobilized Lactobacillus casei in recycle batch reactor: A step towards optimization. J Biotechnol 73:61–70

Sharma S, Yamazaki H (1984) Preparation of hydrophobic cotton cloth. Biotechnol Lett 6:301–306

Singh R, Dhaliwal R, Puri M (2007) Production of high fructose syrup from Asparagus inulin using immobilized exoinulinase from Kluyveromyces marxianus YS-1. J Ind Microbiol Biotechnol 34:649–655

Stefanova M, Tonkova A, Dobreva E, Spasova D (1998) Agar gel immobilization of Bacillus brevis cells for production of thermostable alpha-amylase. Folia Microbiol 43:42–46

Talabardon M, Schwitzguebel JP, Peringer P, Yang ST (2000) Acetic acid production from lactose by an anaerobic thermophilic coculture immobilized in a fibrous-bed bioreactor. Biotechnol Prog 16:1008–1017

Thomas SP, Shanmugasundaram S (1992) Amino-Acid Overproduction by Analog Resistant Mutants of the Nitrogen-Fixing Cyanobacterium Anabaena Sp-287. Appl Biochem Biotechnol 37:111–121

Tisnadjaja D, Gutierrez NA, Maddox IS (1996) Citric acid production in a bubble-column reactor using cells of the yeast Candida guilliermondii immobilized by adsorption onto sawdust. Enzyme Microb Technol 19:343–347

Tokuyama S, Hatano K (1996) Overexpression of the gene for *N*-acylamino acid racemase from Amycolatopsis sp TS-1–60 in Escherichia coli and continuous production of optically active methionine by a bioreactor. Appl Microbiol Biotechnol 44:774–777

Tsai YC, Lin CS, Tseng TH, Lee H, Wang YJ (1992) Production and immobilization of d-aminoacylase of Alcaligenes-faecalis da1 for optical resolution of *N*-acyl-DL-amino acids. Enzyme Microb Technol 14:384–389

Tsay SS, To KY (1987) Citric-acid production using immobilized conidia of Aspergillus-niger tmb-2022. Biotechnol Bioeng 29:297–304

Wang F, Guoa C, Liua H, Liu C (2007) Reversible immobilization of glucoamylase by metal affinity adsorption on magnetic chelator particles. J Mol Catal B Enzym 48:1–7

Wang JL (2000) Production of citric acid by immobilized Aspergillus niger using a rotating biological contactor (RBC). Bioresour Technol 75:245–247

Wang JL, Liu P (1996) Comparison of citric acid production by Aspergillus niger immobilized in gels and cryogels of polyacrylamide. J Ind Microbiol 16:351–353

Wisdom R, Funnill P, Lilly M (1984) Enzymic interesterification of fats: Factors influencing the choice of support for immobilized lipase. Enzyme Microb Technol 6:443–446

Zhu Y, Wu ZT, Yang ST (2002) Butyric acid production from acid hydrolysate of corn fibre by Clostridium tyrobutyricum in a fibrous-bed bioreactor. Process Biochem 38:657–666

Index

D

E

N

O

P

CPSIA information can be obtained
at www.ICGtesting.com
Printed in the USA
LVOW04*1133090316
478429LV00011B/178/P

9 781441 910073